EMERGING CANCER THERAPY

Wiley Series in
BIOTECHNOLOGY AND BIOENGINEERING

Significant advancements in the fields of biology, chemistry, and related disciplines have led to a barrage of major accomplishments in the field of biotechnology. The Wiley Series in Biotechnology and Bioengineering will focus on showcasing these advances in the form of timely, cutting-edge textbooks and reference books that provide a thorough treatment of each respective topic.

Topics of interest to this series include, but are not limited to, protein expression and processing; nanotechnology; molecular engineering and computational biology; environmental sciences; food biotechnology, genomics, proteomics, and metabolomics; large-scale manufacturing and commercialization of human therapeutics; biomaterials and biosensors; and regenerative medicine. We expect these publications to be of significant interest to practitioners both in academia and industry. Authors and editors were carefully selected for their recognized expertise and their contributions to the various and far-reaching fields of biotechnology.

The upcoming volumes will attest to the importance and quality of books in this series. I would like to acknowledge the fellow co-editors and authors of these books for their agreement to participate in this endeavor. Lastly. I would like to thank Ms. Anita Lekhwani, Senior Acquisitions Editor at John Wiley & Sons, Inc. for approaching me to develop such a series. Together, we are confident that these books will be useful additions to the literature that will not only serve the biotechnology community with sound scientific knowledge, but also with inspiration as they further chart the course in this exciting field.

Anurag S. Rathore
Amgen Inc.
Thousand Oaks, CA, USA

Titles in series

Quality by Design for Biopharmaceuticals: Principles and Case Studies / Edited by Anurag S. Rathore and Rohin Mhatre

Emerging Cancer Therapy: Microbial Approaches and Biotechnological Tools / Edited by Arsenio Fialho and Ananda Chakrabarty

EMERGING CANCER THERAPY

Microbial Approaches and Biotechnological Tools

Edited by

ARSÉNIO M. FIALHO

ANANDA M. CHAKRABARTY

WILEY

A JOHN WILEY & SONS, INC., PUBLICATION

Published by John Wiley & Sons, Inc., Hoboken, New Jersey
Published simultaneously in Canada

Library of Congress Cataloging-in-Publication Data:

Emerging cancer therapy : microbial approaches and biotechnological tools / edited by Arsenio Fialho and Ananda Chakrabarty.
p. ; cm.
Includes bibliographical references and index.
ISBN 978-0-470-44467-2 (cloth)
1. Cancer–Treatment. 2. Viruses–Therapeutic use. 3. Bacteria–Therapeutic use. 4. Microbial biotechnology. I. Fialho, Arsenio. II. Chakrabarty, Ananda M., 1938–
[DNLM: 1. Neoplasms–therapy. 2. Antineoplastic Agents. 3. Bacteria–immunology. 4. Biotechnology. QZ 266 E53 2010]
RC271.V567E64 2010
616.99′406–dc22

2009051007

Printed in Singapore

10 9 8 7 6 5 4 3 2 1

CONTENTS

PREFACE

The major approaches to cancer therapy today are surgery, radiation therapy, immunotherapy, and chemotherapy including the use of antimetabolites, rationally designed or randomly screened small molecule drugs, monoclonal antibodies, and combination thereof. While these therapies have contributed significantly to the savings of millions of lives and reduced pain and suffering, cancers still take a major toll in our lives as a deadly, mostly incurable, disease.

It has been known for more than a hundred years that microorganisms/viruses, particularly pathogenic bacteria that cause infections in human bodies, allow cancer regression, sometimes with astounding results. Not only live bacteria, but also bacteria-free sterile growth media of such bacteria, allow tumor shrinkage and regression, implying the role of soluble, secreted bacterial products having anticancer activity. Much efforts have, therefore, been expended to develop either live microorganisms/viruses, with or without additional cloned genes that encode toxins targeted specifically to cancer cells, or bacterial products with cancer-killing ability. This book addresses the recent advances in the use of microorganisms/viruses and their products, with or without genetic intervention, in cancer therapy, including human clinical trials. Seventeen different chapters address various facets of the use of live microorganisms, high/low molecular weight products derived from microorganisms, and microbial products fused to cancer-targeting molecules such as monoclonal antibodies or fragments of antibodies.

All therapies, including anticancer drugs, must reach the bedside and the global market to be useful. This requires that the therapies, whether diagnostic agents or drugs, be protected from copying. Such protection is usually afforded through creation of intellectual property rights, often through patenting. Patenting a drug could, however, be tricky and expensive as the criteria for patenting are still debatable, and patent infringement cases abound. Indeed, the U.S. Supreme Court, on June 1, 2009, accepted to consider a case, known as re Bilski, which purports to patent a business method for hedging the risks of commodity trading, for example, the price of natural gas or electricity, for both the suppliers and the consumers. The U.S. Court of Appeals for the Federal Circuit (CAFC), in an en banc decision, rejected the Bilski claims as nonpatentable subject matter involving no machines or material transformations that are believed to be required under U.S. patent laws for process patents. The Supreme Court's decision to take up the case is thus believed to be important to determine patent eligibility of business methods, including medical diagnostic processes. Thus, the last chapter is devoted to address the fundamental concepts in patenting of drugs, including patent infringement or eligibility cases. Hopefully, this will give the readers a sense of what it takes to bring a drug, including anticancer drugs, to the market. The editors hope that the readers will find all the chapters useful, timely, relevant, and informative.

A.M. Fialho
A.M. Chakrabarty

CONTRIBUTORS

Jozef Anné, Laboratory of Bacteriology, Rega Institute for Medical Research, Katholieke Universiteit Leuven, Leuven, Belgium

Donald J. Buchsbaum, Department of Radiation Oncology, University of Alabama at Birmingham, Birmingham, AL

Jason S. Buhrman, Massachusetts General Hospital, Harvard Medical School, MA and University of Illinois at Chicago School of Medicine, IL

Ananda M. Chakrabarty, Department of Microbiology and Immunology, University of Illinois College of Medicine, Chicago, IL

Suresh Chawrai, Department of Chemistry, University of Cambridge, Cambridge, UK

Tooba A. Cheema, Massachusetts General Hospital, Harvard Medical School, MA

Lin-Chi Chen, Department of Medical Oncology, Nevada Cancer Institute, Las Vegas, NV

Nam H. Dang, Division of Hematology/Oncology, University of Florida Cancer Center Clinical Trials Office, University of Florida, Gainesville, FL

Ursula Elsässer-Beile, Department of Urology, Experimental Urology, University Hospital Freiburg, Freiburg, Germany

Lynn Feun, Hematology/Oncology, Sylvester Comprehensive Cancer Center, University of Miami School of Medicine, Miami, FL

Arsénio M. Fialho, Institute for Biotechnology and Bioengineering (IBB), Center for Biological and Chemical Engineering, Instituto Superior Tecnico, Lisbon, Portugal

Geng-Feng Fu, Jiangsu Provincial Center for Disease Prevention and Control, Nanjing, China

Giulia Fulci, Massachusetts General Hospital, Harvard Medical School, MA

Thomas S. Griffith, Department of Urology, University of Iowa, Iowa City, IA

Ashley L. Hilchie, Department of Microbiology and Immunology, Faculty of Medicine, Dalhousie University, Halifax, Nova Scotia, Canada

Robert M. Hoffman, AntiCancer, Inc. and Department of Surgery, University of California, San Diego, CA

David W. Hoskin, Departments of Microbiology and Immunology, and Pathology, Faculty of Medicine, Dalhousie University, Halifax, Nova Scotia, Canada

Bi Hu, Center for Public Health Research, Medical School, Nanjing University, Nanjing, China

Sergey A. Kaliberov, Department of Radiation Oncology, University of Alabama at Birmingham, Birmingham, AL

Timothy P. Kresowik, Department of Urology, University of Iowa, Iowa City, IA

M. Tien Kuo, Molecular Pathology, M.D. Anderson Cancer Center, Houston, TX

Finian J. Leeper, Department of Chemistry, University of Cambridge, Cambridge, UK

Paulo Cesar Maciag, Advaxis Inc., North Brunswick, NJ

Jingxuan Pan, Department of Pathophysiology, Zhongshan School of Medicine, Sun Yat-Sen University, Guangzhou, People's Republic of China

Yvonne Paterson, University of Pennsylvania Department of Microbiology, Philadelphia, PA

Sandra Rivera, Advaxis Inc., North Brunswick, NJ

John Rothman, Advaxis Inc., North Brunswick, NJ

George P.C. Salmond, Department of Biochemistry, University of Cambridge, Cambridge, UK

Padma Sampath, Departments of Surgery and Immunology, University of Pittsburgh, Pittsburgh, PA

Niramol Savaraj, Hematology/Oncology, Sylvester Comprehensive Cancer Center, University of Miami School of Medicine, Miami, FL

Vafa Shahabi, Advaxis Inc., North Brunswick, NJ

Steve H. Thorne, Departments of Surgery and Immunology, University of Pittsburgh, Pittsburgh, PA

Lieve Van Mellaert, Laboratory of Bacteriology, Rega Institute for Medical Research, Katholieke Universiteit Leuven, Leuven, Belgium

Jörg Vollmer, Coley Pharmaceutical GmbH—A Pfizer Company, Düsseldorf, Germany

Anna Maria Elisabeth Walenkamp, Department of Medical Oncology, University Medical Center Groningen, University of Groningen, Groningen, The Netherlands

Anu Wallecha, Advaxis Inc., North Brunswick, NJ

Medhi Wangpaichitr, Hematology/Oncology, Sylvester Comprehensive Cancer Center, University of Miami School of Medicine, Miami, FL

Ming Q. Wei, Division of Molecular and Gene Therapies, Griffith Institute for Health and Medical Research, School of Medical Science, Griffith University, Queensland, Australia

Neil R. Williamson, Department of Biochemistry, University of Cambridge, Cambridge, UK

Philipp Wolf, Department of Urology, Experimental Urology, University Hospital Freiburg, Freiburg, Germany

Chung Jing Wu, Hematology/Oncology, Sylvester Comprehensive Cancer Center, University of Miami School of Medicine, Miami, FL

Gen-Xing Xu, Center for Public Health Research, Medical School, Nanjing University, Nanjing and Jiangsu Research Center for Gene Pharmaceutical Engineering and Technology, Suzhou, China

Sai-Ching Jim Yeung, Departments of General Internal Medicine, Ambulatory Treatment and Emergency Care, and Endocrine Neoplasia and Hormonal Disorders, The University of Texas, M.D. Anderson Cancer Center, Houston, TX

Yan Yin, Center for Public Health Research, Medical School, Nanjing University, Nanjing, China

Ming You, Hematology/Oncology, Sylvester Comprehensive Cancer Center, University of Miami School of Medicine, Miami, FL

PART I

LIVE/ATTENUATED BACTERIA AND VIRUSES AS ANTICANCER AGENTS

1

SALMONELLA TYPHIMURIUM MUTANTS SELECTED TO GROW ONLY IN TUMORS TO ERADICATE THEM IN NUDE MOUSE MODELS

Robert M. Hoffman

AntiCancer, Inc. and Department of Surgery, University of California, San Diego, CA

INTRODUCTION

The use of bacteria in cancer therapy dates back to Coley in the 1890s, when he observed that streptococcal (erysipelas) infection was associated with regression of soft tissue sarcomas. Coley then reported on the efficacy of infecting erysipelas in a series of 10 sarcoma patients (1). Coley's pioneering work was the basis for modern studies using bacteria to treat cancer. In the middle part of the last century, Malmgren et al. (2) showed that anaerobic bacteria had the ability to survive and replicate in necrotic tumor tissue with low oxygen content. Several approaches aimed at utilizing bacteria for cancer therapy have subsequently been described (3–15).

Bifidobacterium longum, an obligate anaerobe, has been shown to selectively grow in hypoxic regions of tumors following i.v. administration. This effect was demonstrated in 7,12-dimethylbenzanthracene-induced rat mammary tumors by Yazawa et al. (14, 15) as well as the Lewis lung carcinoma. Vogelstein et al. (16) created a strain of *Clostridium novyi*, also an obligate anaerobe, which was depleted of its lethal toxin. This strain of *C. novyi* was termed *C. novyi* NT. Following i.v. administration, the *C. novyi* NT spores germinated in the

Emerging Cancer Therapy: Microbial Approaches and Biotechnological Tools, Edited by Arsénio M. Fialho and Ananda M. Chakrabarty

avascular regions of tumors in mice, causing damage to the surrounding viable tumor (16). Combined with conventional chemotherapy or radiotherapy, intravenous *C. novyi* NT spores caused extensive tumor damage within 24 h (16).

Following attenuation by purine and other auxotrophic mutations, the facultative anaerobe *Salmonella typhimurium* was used for cancer therapy (11, 17, 18). These genetically modified bacteria replicated in tumors to levels more than 1000-fold greater than in normal tissue (11). *S. typhimurium* was further modified genetically by disrupting the *msb*B gene to reduce the incidence of septic shock (11). The *msb*B mutant of *S. typhimurium* has been tested in a Phase I clinical trial to determine its efficacy on metastatic melanoma and metastatic renal cell carcinoma (19). To raise the therapeutic index, *S. typhimurium* was further attenuated by deletion of the *pur*I and *msb*B genes (19). The new strain of *S. typhimurium*, termed VNP20009, could then be safely administered to patients (19). More studies are needed to completely characterize the safety and efficacy of the bacteria and to improve its therapeutic index.

Mengesha et al. utilized *S. typhimurium* as a vector for gene delivery by developing a hypoxia-inducible promoter (HIP-1) to limit gene expression to hypoxic tumors. HIP-1 was able to drive gene expression in bacteria residing in human tumor xenografts implanted in mice (20). Genes linked to the HIP-1 promoter showed selective expression in tumors (20). Yu et al. used green fluorescent protein (GFP)-labeled bacteria to visualize tumor targeting abilities of three pathogens: *Vibrio cholerae*, *S. typhimurium*, and *Listeria monocytogenes* (21, 22).

Targeted Therapy with a *S. typhimurium* Leucine–Arginine-Dependent Strain

We initially developed a mutant of *S. typhimurium*, termed A1, which selectively targeted tumors in nude mouse models (23). In contrast, normal tissue was rapidly cleared of infecting bacteria, even in immunodeficient athymic mice. *S. typhimurium* A1 is auxotrophic (leu/arg dependent) but receives sufficient nutritional support from tumor tissue. After inoculation with wild-type *S. typhimurium*, the mice died within 2 days. The longest lived mice were those inoculated with auxotroph A1 which survived as long as control uninfected mice. The bacteria were labeled with GFP so that their tumor-targeting efficacy could be imaged *in vivo*.

To observe the intracellular replication and virulence of *S. typhimurium*-GFP in a human prostate cancer cell line *in vitro*, the PC-3 human prostate cancer cells were labeled with red fluorescent protein (RFP) in the cytoplasm with retroviral RFP and GFP in the nucleus by means of a vector with GFP fused to histone H2B. The dual-color cancer cells and GFP bacteria have enabled visualization of the interaction between bacteria and cancer cells by fluorescence imaging. The quantitative ability of *S. typhimurium* to kill prostate cancer cells was determined with the MTT method and observed to be dose dependent (23).

Efficacy of *S. typhimurium* A1 on Human Prostate Cancer Growing Subcutaneously in Nude Mice

To observe the interaction of prostate cancer cells with bacteria, we used PC-3 human prostate cancer cell line expressing RFP, so their response to the bacteria could be visualized *in vivo*. To evaluate efficacy of *S. typhimurium* A1, 10 NCR nude mice, 6–8 weeks, were implanted subcutaneously (s.c.) on the mid-right side with 2×10^6 RFP-labeled PC-3 human prostate cancer cells. Bacteria were grown and harvested at late log phase and then diluted in PBS and injected directly into the tail vein (5×10^7 cfu/100 μL PBS). Tumor size was determined from fluorescence imaging at each time point after infection. *S. typhimurium* A1 selectively colonized the PC-3 tumor and suppressed its growth (24).

Isolation of a More Tumor-Virulent Strain of *S. typhimurium* A1, Termed A1-R

To enhance tumor virulence, *S. typhimurium* A1 has been passaged by injection in nude mice transplanted with the HT-29 human colon tumor. Bacteria, expressing GFP, isolated from the infected tumor, were then cultured. The reisolated A1 was termed A1-R. The ability of A1-R to adhere to tumor cells was evaluated in comparison with the parental A1 strain *in vitro*. The number of A1-R bacteria attached to HT-29 human colon cancer cells was approximately six times higher than parental A1.

The virulence of GFP-labeled *S. typhimurium* A1 and A1-R in human prostate cancer cells was compared *in vitro* under fluorescence microscopy. Both strains infected dual-color PC-3 cancer cells. Whereas almost all cells were infected and dead after 2 h with A1-R, it took 24 h to get the same result with A1. Thus, the virulence of A1-R was greatly increased (25).

A1-R Has Enhanced Virulence against Prostate Cancer in Nude Mouse Models

GFP-labeled *S. typhimurium* A1 and A1-R (5×10^7 cfu/100 μL) were administered (i.v.) to PC-3-bearing nude mice. The biodistribution of the bacteria in tumor tissue was determined at day 4. A1-R had 100 times greater colony-forming units in PC-3 tumor tissue than A1. This result also suggested that A1-R has greater tumor virulence than A1 (25).

To compare bacterial infection in the tumor with infection in normal tissue, A1-R bacteria (5×10^7 cfu/100 μL) were administered intravenously in PC-3-bearing nude mice. On day 4 after injection, the tumor, liver, and spleen were removed. The tissues were homogenized and plated on LB agar plates. After overnight growth at 37°C, the colony-forming units were counted. The ratio of tumor to normal tissue was approximately 10^6, indicating a very high degree of tumor targeting by A1-R (25).

Efficacy of A1-R on Breast Tumor Growth in an Orthotopic Nude Mouse Model

Treatment with A1-R resulted in significant efficacy in nude mice with s.c. MARY-X human xenografts. Bacteria (5×10^7 cfu/100 μL) were inoculated intravenously in MARY-X-bearing nude mice. Tumor growth was monitored by caliper measurement in two dimensions. The infected tumors regressed by day 5 after infection, and complete regression occurred by day 25. In orthotopic models of MARY-X, A1-R was also efficacious following a single i.v. injection. The destruction of the tumor in treated mice was visualized by whole-body imaging. The difference in tumor volume between the treated group, which showed quantitative regression, and the control was statistically significant ($p < 0.05$) (24).

The survival of the A1-R-treated animals was prolonged with a 50% survival time of 13 weeks compared with 5 weeks of control animals. Forty percent of the mice survived as long as the control non-tumor-bearing mice. Tumors that were eradicated did not regrow. In contrast, the parental *S. typhimurium* A1 was less effective than A1-R. Tumor growth was only slowed after A1 i.v. injection and not eradicated (24).

Treatment of an Orthotopic Human Pancreatic Tumor in Nude Mice with *S. typhimurium* A1-R

A1-R GFP could invade and replicate intracellularly *in vitro* in XPA1 human pancreatic cancer cells expressing GFP in the nucleus and RFP in the cytoplasm. Intracellular bacterial infection led to cell fragmentation and cell death (26).

On day 0, XPA1 was transplanted on the pancreas of nude mice. On day 7, the tumor was exposed and observed by fluorescence imaging. Tumor size was measured by fluorescent area (mm^2). Three mice were treated with a low concentration of A1-R (10^7 cfu/mL); three were treated with a high concentration (10^8 cfu/mL); and three were used as untreated controls. Tumor volume (mm^3) was calculated with the formula $V = ½ \times (\text{length} \times \text{width}^2)$. The bacteria were injected into the tumor. On day 14, the tumor was exposed again, and the size was measured to determine the efficacy of treatment (26).

Before treatment, the average tumor size (fluorescent area) on day 7 was $3.2 \pm 1.9\,mm^2$ in the untreated group, 3.1 ± 1.4 in the high-bacteria-dose group, and 3.5 ± 0.75 in the low-bacteria-dose group. On day 14, after 7 days of treatment, the tumor fluorescence area was $19.9 \pm 4.3\,mm^2$ in the untreated group, 2.2 ± 0.89 in the high-bacteria-concentration treatment group, and 12.7 ± 6.5 in the low-bacteria-concentration treatment group (26).

We have also demonstrated the efficacy of locally as well as systemically administered A1-R on liver metastasis of pancreatic cancer. Mice treated with A1-R, given locally via intrasplenic injection or systemically via tail vein injection, had a much lower hepatic and splenic tumor burden as compared to control mice (27). Systemic treatment with intravenous A1-R also increased survival time. All results were statistically significant.

Experimental Lymph Node Metastasis Cured by Specific Targeting by *S. typhimurium* A1-R

A new experimental model of lymph node metastasis was developed. To obtain an experimental metastasis in the axillary lymph node, XPA1-RFP human pancreatic cancer cells were injected into the inguinal lymph node in nude mice. Just after injection, cancer cells were imaged trafficking in the efferent lymph duct to the axillary lymph node. Metastasis in the axillary lymph node was subsequently formed. A1-R bacteria were then injected into the inguinal lymph node to target the axillary lymph node metastasis. Just after bacterial injection, a large amount of bacteria were visualized around the axillary lymph node metastasis. By day 7, all lymph node metastases had been eradicated in contrast to growing metastases in the control group. There were very few bacteria in the lymph node by day 7, and no bacteria were detected after day 10. This route of administration was therefore able to deliver sufficient bacteria to eradicate the lymph node metastasis after which the bacteria became undetectable. The average tumor size (fluorescent area) in the axillary lymph nodes on day 0 was $0.4 \pm 0.19\,mm^2$ in the treatment group and 0.46 ± 0.08 in the untreated group, respectively. On day 7, it was $0\,mm^2$ in the treatment group and 0.98 ± 0.17 in the untreated group (28).

We then tested bacterial therapy strategy for spontaneous lymph node metastasis from a fibrosarcoma tumor growing in the footpad. At first, only A1-R bacteria were injected in the footpad in nude mice in order to determine any adverse effects. No infection, skin necrosis, body weight loss, or animal death was observed. Then, HT-1080-GFP–RFP human fibrosarcoma cells were injected into the footpad of additional nude mice. The presence of popliteal lymph node metastasis was determined by weekly imaging. Once the metastasis was detected, A1-R bacteria were injected s.c. in the footpad. Bacteria are small particles, and when injected s.c., the lymph system immediately collects them from the site of injection. The lymph system is well known as a drainage route for bacterial infection. We observed the injected bacteria trafficking in the lymphatic channel. The popliteal region was exposed just after bacteria injection, and a large amount of GFP bacteria targeting the popliteal lymph node metastasis was observed by fluorescence imaging. Dual-color labeling of the cancer cells distinguished them from the GFP bacteria. After treatment, the popliteal lymph node was observed every week by fluorescence imaging. One mouse was used to image the bacteria by exposing the popliteal lymph node on day 7. GFP bacteria invading the lymph node metastasis were observed. All lymph node metastases shrank, and five out of six were eradicated within 7–21 days after treatment in contrast to growing metastases in the control group (28).

S. typhimurium A1-R Therapy for Experimental Lung Metastasis

To obtain experimental lung metastasis, HT-1080 GFP-RFP cells were injected into the tail vein of experimental nude mice (day 0). On day 4 and day 11, A1-R bacteria were injected into the tail vein. On day 16, all animals were sacrificed, and the lungs were imaged to determine the efficacy of bacteria therapy on lung metastasis. To observe the lung metastasis at lower magnification, an RFP filter was used (excitation 545 nm, emission 570–625 nm). In the bacterial treatment group, only a few cancer cells were observed in contrast to multiple metastases in the control (untreated) group. The number of meta stases on the surface of the lung was significantly lower in the treatment group than in the control group ($p < 0.005$). There were no significance differences between the treated and untreated groups in body weight and primary tumor size (28).

Targeting of Primary Bone Tumor and Lung Metastasis of High-Grade Osteosarcoma in Nude Mice with *S. typhimurium* A1-R

Mice were transplanted with 143B-RFP human osteosarcoma cells in the tibia and developed primary bone tumor and lung metastasis. Seven days after tumor injection, RFP tumor was confirmed inside the tibia. After three weekly injections of bacteria, the bone tumor size and lung metastasis were examined on day 28. The bone tumor size (RFP fluorescence area) was $231.7 \pm 70\,mm^2$ in the untreated group and $94.6 \pm 23\,mm^2$ in the treated group ($p < 0.05$). The lung was excised, and the metastases on the surface were counted. The number of metastasis was 52 ± 30 in the untreated group and 2.3 ± 2.1 in the treated group ($p < 0.05$). *S. typhimurium* A1-R therapy was therefore effective for primary and metastatic osteosarcoma (29).

Targeted Therapy of Spinal Cord Glioma with *S. typhimurium* A1-R

Spinal cord tumors are highly malignant and often lead to paralysis and death mainly due to their infiltrative nature, high recurrence rate, and limited treatment options. In this study, we measured the antitumor efficacy of *S. typhimurium* A1-R, administered systemically or intrathecally, to spinal cord cancer in orthotopic nude mouse models. Tumor fragments of human U87-RFP glioma were implanted by surgical orthotopic implantation into the dorsal site of the spinal cord. Five and 10 days after transplantation, eight mice in each group were treated with A1-R (2×10^7 cfu/200 µL i.v. or 2×10^6 cfu/10 µL intrathecal injection). The untreated mice showed progressive paralysis beginning 6 days after tumor transplantation and developed complete paralysis between 18 and 25 days. The mice treated intravenously with A1-R had an onset of paralysis at approximately 11 days and at day 30, five mice developed complete paralysis, while three other mice had partial paralysis. Mice treated via intrathecal injection of A1-R had an onset of paralysis at approximately 18 days, and one mouse was still not paralyzed at day 30. Only one mouse developed complete paralysis at day 30 in the intrathecal treatment group. The intrathecally treated animals had a significant increase in survival over the i.v.-treated group as well as the control group (30).

Screening for *Salmonella* Promoters Differentially Activated in the PC-3 Prostate Tumor

We have used a high-throughput method to screen for *S. typhimurium* promoters that are selectively activated in tumors in the mouse. A random library of *S. typhimurium* with DNA cloned upstream of a promoterless GFP were injected intravenously in nude mice with s.c. human PC-3 prostate tumors as well as control nude mice. GFP-positive *S. typhimurium* clones from tumor, spleen, and liver, and *in vitro* growth in LB medium, were isolated by fluorescence-activated cell sorting (FACS). Active promoters in all environments were amplified by PCR and identified by DNA microarray hybridization. Among promoters identified as preferentially induced in tumors, and not induced in any of the other environments (spleen, liver, or *in vitro*), were those of at least five genes known to be controlled by the fumerate and nitrate reduction global regulator (FNR). At least five other genes with unknown regulation were also enriched in tumors. The natural tendency of *S. typhimurium* to target tumors preferentially over other tissues, combined with the use of promoters preferentially induced in the tumor environment versus other environments, may allow the exquisitely tumor-specific expression of fusion proteins on the surface or secreted by *S. typhimurium* for highly selective tumor therapy (31, 32).

CONCLUSION

Our goal is to develop tumor-targeting *S. typhimurium* strains that can kill primary and metastatic cancer without toxic effects to the host and without the need for combination with toxic chemotherapy. Toward this goal, we have developed a new strain of *S. typhimurium*, A1-R, that has greatly increased antitumor efficacy but maintains its original auxotrophy for leu–arg that prevents it from mounting a continuous infection in normal tissues. A1-R was able to effect cures in monotherapy on mouse models of metastatic human cancer. We have also identified candidate *S. typhimurium* tumor-specific promoters that may enhance the antitumor efficacy of A1-R by driving expression of toxins that could be excreted in the tumors. Future studies will be aimed to bring bacterial treatment of cancer to the clinic.

REFERENCES

1. Coley W.B. 1893. The treatment of malignant tumors by repeated inoculations of erysipelas. With a report of ten original cases. Clin. Orthop. Relat. Res. **1991**: 3–11.

2. Malmgren R.A., and Flanigan C.C. 1955. Localization of the vegetative form of *Clostridium tetani* in mouse tumors following intravenous spore administration. Cancer Res. **15**: 473–478.
3. Gericke D., and Engelbart K. 1964. Oncolysis by Clostridia. II. Experiments on a tumor spectrum with a variety of Clostridia in combination with heavy metal. Cancer Res. **24**: 217–221.
4. Moese J.R., and Moese G. 1964. Oncolysis by Clostridia. I. Activity of *Clostridium butyricum* (M-55) and other nonpathogenic Clostridia against the Ehrlich carcinoma. Cancer Res. **24**: 212–216.
5. Thiele E.H., Arison R.N., and Boxer, G.E. 1964. Oncolysis by Clostridia. III. Effects of Clostridia and chemotherapeutic agents on rodent tumors. Cancer Res. **24**: 222–233.
6. Kohwi Y., Imai K., Tamura Z., and Hashimoto Y. 1978. Antitumor effect of *Bifidobacterium infantis* in mice. Gann. **69**: 613–618.
7. Kimura N.T., Taniguchi S., Aoki K., and Baba T. 1980. Selective localization and growth of *Bifidobacterium bifidum* in mouse tumors following intravenous administration. Cancer Res. **40**: 2061–2068.
8. Fox M.E., Lemmon M.J., Mauchline M.L., Davis T.O., Giaccia A.J., Minton N.P., and Brown J.M. 1996. Anaerobic bacteria as a delivery system for cancer gene therapy: in vitro activation of 5-fluorocytosine by genetically engineered clostridia. Gene Ther. **3**: 173–178.
9. Lemmon M.J., van Zijl P., Fox M.E., Mauchline M.L., Giaccia A.J., Minton N.P., and Brown J.M. 1997. Anaerobic bacteria as a gene delivery system that is controlled by the tumor microenvironment. Gene Ther. **4**: 791–796.
10. Brown J.M., and Giaccia A.J. 1998. The unique physiology of solid tumors: opportunities (and problems) for cancer therapy. Cancer Res. **58**: 1408–1416.
11. Low K.B., Ittensohn M., Le T., Platt J., Sodi S., Amoss M., Ash O., Carmichael E., Chakraborty A., Fischer J., Lin S.L., Luo X., Miller S.I., Zheng L., King I., Pawelek J.M., and Bermudes D. 1999. Lipid A mutant *Salmonella* with suppressed virulence and TNFalpha induction retain tumor-targeting in vivo. Nat. Biotechnol. **17**: 37–41.
12. Clairmont C., Lee K.C., Pike J., Ittensohn M., Low K.B., Pawelek J., Bermudes D., Brecher S.M., Margitich D., Turnier J., Li Z., Luo X., King I., and Zheng L.M. 2000. Biodistribution and genetic stability of the novel antitumor agent VNP20009, a genetically modified strain of *Salmonella typhimurium*. J. Infect. Dis. **181**: 1996–2002.
13. Sznol M., Lin S.L., Bermudes D., Zheng L.M., and King I. 2000. Use of preferentially replicating bacteria for the treatment of cancer. J. Clin. Invest. **105**: 1027–1030.
14. Yazawa K., Fujimori M., Amano J., Kano Y., and Taniguchi S. 2000. *Bifidobacterium longum* as a delivery system for cancer gene therapy: selective localization and growth in hypoxic tumors. Cancer Gene Ther. **7**: 269–274.
15. Yazawa K., Fujimori M., Nakamura T., Sasaki T., Amano J., Kano Y., and Taniguchi S. 2001. *Bifidobacterium longum* as a delivery system for gene therapy of chemically induced rat mammary tumors. Breast Cancer Res. Treat. **66**: 165–170.

16. Dang L.H., Bettegowda C., Huso D.L., Kinzler K.W., and Vogelstein B. 2001. Combination bacteriolytic therapy for the treatment of experimental tumors. Proc. Natl. Acad. Sci. U S A. **98**: 15155–15160.
17. Hoiseth S.K., and Stocker B.A. 1981. Aromatic-dependent *Salmonella typhimurium* are non-virulent and effective as live vaccines. Nature. **291**: 238–239.
18. Pawelek J.M., Low K.B., and Bermudes D. 1997. Tumor-targeted *Salmonella* as a novel anticancer vector. Cancer Res. **57**: 4537–4544.
19. Toso J.F., Gill V.J., Hwu P., Marincola F.M., Restifo N.P., Schwartzentruber D.J., Sherry R.M., Topalian S.L., Yang J.C., Stock F., Freezer L.J., Morton K.E., Seipp C., Haworth L., Mavroukakis S., White D., MacDonald S., Mao J., Sznol M., and Rosenberg S.A. 2002. Phase I study of the intravenous administration of attenuated *Salmonella typhimurium* to patients with metastatic melanoma. J. Clin. Oncol. **20**: 142–152.
20. Mengesha A., Dubois L., Lambin P., Landuyt W., Chiu R.K., Wouters B.G., and Theys J. 2006. Development of a flexible and potent hypoxia-inducible promoter for tumor-targeted gene expression in attenuated *Salmonella*. Cancer Biol. Ther. **5**: 1120–1128.
21. Yu Y.A., Timiryasova T., Zhang Q., Beltz R., and Szalay A.A. 2003. Optical imaging: bacteria, viruses, and mammalian cells encoding light-emitting proteins reveal the locations of primary tumors and metastases in animals. Anal. Bioanal. Chem. **377**: 964–972.
22. Yu Y.A., Shabahang S., Timiryasova T.M., Zhang Q., Beltz R., Gentschev I., Goebel W., and Szalay A.A. 2004. Visualization of tumors and metastases in live animals with bacteria and vaccinia virus encoding light-emitting proteins. Nat. Biotechnol. **22**: 313–320.
23. Zhao M., Yang M., Li X.M., Jiang P., Baranov E., Li S., Xu M., Penman S., and Hoffman R.M. 2005. Tumor-targeting bacterial therapy with amino acid auxotrophs of GFP-expressing *Salmonella typhimurium*. Proc. Natl. Acad. Sci. U S A. **102**: 755–760.
24. Zhao M., Yang M., Ma H., Li X., Tan X., Li S., Yang Z., and Hoffman R.M. 2006. Targeted therapy with a *Salmonella typhimurium* leucine-arginine auxotroph cures orthotopic human breast tumors in nude mice. Cancer Res. **66**: 7647–7652.
25. Zhao M., Geller J., Ma H., Yang M., Penman S., and Hoffman R.M. 2007. Monotherapy with a tumor-targeting mutant of *Salmonella typhimurium* cures orthotopic metastatic mouse models of human prostate cancer. Proc. Natl. Acad. Sci. U S A. **104**: 10170–10174.
26. Nagakura C., Hayashi K., Zhao M., Yamauchi K., Yamamoto N., Tsuchiya H., Tomita K., Kishimoto H., Bouvet M., and Hoffman R.M. 2009. Efficacy of a genetically-modified *Salmonella typhimurium* against metastatic human pancreatic cancer in nude mice. Anticancer Res. **29**: 1873–1878.
27. Yam C., Zhao M., Hayashi K., Ma H., Kishimoto H., McElroy M., Bouvet M., and Hoffman R.M. 2009. Monotherapy with a tumor-targeting mutant of *S. typhimurium* inhibits liver metastasis in a mouse model of pancreatic cancer. J Surg. Res. DOI: 10.1016/j.jss.2009.02.023 (in press).
28. Hayashi K., Zhao M., Yamauchi K., Yamamoto N., Tsuchiya H., Tomita K., and Hoffman R.M. 2009. Cancer metastasis directly eradicated by targeted therapy with a modified *Salmonella typhimurium*. J. Cell. Biochem. **106**: 992–998.

29. Hayashi K., Zhao M., Yamauchi K., Yamamoto N., Tsuchiya H., Tomita K., Kishimoto H., Bouvet M., and Hoffman R.M. 2009. Systemic targeting of primary bone tumor and lung metastasis of high-grade osteosarcoma in nude mice with a tumor-selective strain of *Salmonella typhimurium*. Cell Cycle. **8**: 870–875.
30. Kimura H., Zhang L., Zhao M., Hayashi K., Tsuchiya H., Tomita K., Bouvet M., Wessels J., and Hoffman R.M. 2010. Targeted therapy of spinal cord glioma with a genetically-modified *Salmonella typhimurium*. Cell Prolif. **43**: 41–48.
31. Arrach N., Zhao M., Porwollik S., Hoffman R.M., and McClelland M. 2008. *Salmonella* promoters preferentially activated inside tumors. Cancer Res. **68**: 4827–4832.
32. Arrach N., Cheng P., Zhao M., Santiviago C.A., Hoffman R. M., and McClelland M. 2010. High-throughput screening for *Salmonella* avirulent mutants that retain targeting of solid tumors. Cancer Res. **70**: 2165–2170.

2

THE USE OF LIVING *LISTERIA MONOCYTOGENES* AS AN ACTIVE IMMUNOTHERAPY FOR THE TREATMENT OF CANCER

John Rothman,[1] Anu Wallecha,[1] Paulo Cesar Maciag,[1] Sandra Rivera,[1] Vafa Shahabi,[1] and Yvonne Paterson[2]

[1]*Advaxis Inc., North Brunswick, NJ*

[2]*University of Pennsylvania Department of Microbiology, Philadelphia, PA*

INTRODUCTION

Listeria monocytogenes is a gram-positive facultative intracellular bacterium responsible for causing listeriosis in humans and animals (1–3). *L. monocytogenes* is able to infect both phagocytic and nonphagocytic cells (4–6). Due to this intracellular growth behavior, *L. monocytogenes* triggers potent innate and adaptive immune responses in an infected host that are required for the clearance of the organism (7). This ability, to induce efficient immune responses using multiple simultaneous and integrated mechanisms of action, has encouraged efforts to develop this bacterium as a recombinant antigen delivery vector to induce protective cellular immunity against cancer or infection. *Listeria* infection also involves other systems which are not essentially a part of the immune system but which support immune function to affect a therapeutic outcome, such as myelopoesis and vascular endothelial cell function (8–16). This chapter discusses the multiple simultaneous mechanisms of action

Emerging Cancer Therapy: Microbial Approaches and Biotechnological Tools, Edited by Arsénio M. Fialho and Ananda M. Chakrabarty induced by bioengineered iatrogenic *Listeria* infection and how they may be optimized to induce therapeutic responses as active immunotherapy.

LISTERIAL DISEASE

Pathogenesis

To survive within the host, *L. monocytogenes* activates a set of virulence genes, which have been identified using biochemical and molecular genetic approaches. These genes include: *actA*, *hly*, *inlA*, *inlB*, *inlC*, *plcA*, and *plcB*, which are regulated by a pluripotential transcriptional activator, PrfA (17). *Listeria* lacking *prfA* are avirulent as they lack the ability to survive within the infected host (18, 19). Several other proteins such as Ami, Auto, ActA, and Vip are also essential for the full virulence of *L. monocytogenes* (20–22).

L. monocytogenes surface proteins termed as invasins interact with the receptors present on host cell plasma membranes to subvert signaling cascades, leading to bacterial internalization in nonphagocytic cells. Among these are 24 internalins present in the *L. monocytogenes* genome that are believed to contribute to host cell invasion; of these, internalins A (InlA) and B (InlB) are the most well characterized (23–31).

Listeria must escape the host cell phagolysosome to become virulent. Upon infection, less than 10% of the *L. monocytogenes* typically escapes into the host cell cytosol. This is mediated by listeriolysin O (LLO), a pore-forming hemolysin (32), and phospholipases (PlcA and PlcB). LLO is a member of a group of cholesterol-dependent cytolysins (CDC) and was the first major virulence factor of *L. monocytogenes* identified (33–35). In the cytoplasm, *L. monocytogenes* replicates and uses ActA, another major virulence factor, to polymerize host cell actin, support its motility, and spread from cell to cell (6, 36–38).

Epidemiology

Listeria is a ubiquitous environmental pathogen found in the soil, on leafy vegetables, in meat, and in dairy products; typical exposure to *L. monocytogenes* does not result in disease (39). *Listeria* is not laterally transmissible and is only pathogenic when ingested. Globally, listeriosis is a rare disease and its prevalence has declined wherever food control measures have been implemented. In the United States, the attack rate is estimated to be around one per million, leading to 2500 cases and around 700 deaths per year with infection more common in children, the elderly, and pregnant women, who tend to have less competent immune function (1, 2, 40–48).

Unlike most human foodborne infections, which are associated with a high incidence rate counterbalanced by low morbidity and mortality, the situation is opposite for clinically presented human listeriosis, which is a rare but potentially fatal infection associated with a 30% mortality, even when an antimicrobial treatment is administered (2, 48). This is probably because individuals with healthy immune systems eliminate infection before clinical signs are apparent.

Innate Immunity and *Listeria* Infection

Innate immunity plays an essential role in the clearance of *L. monocytogenes* and control of the infection at early stages. Severe combined immunodeficiency (SCID) mice have been observed to clear infection with attenuated *L. monocytogenes* vaccine strains (for instance, Lm-LLO-E7) through innate immune mechanisms (Y. Paterson, unpublished observations). Upon intraperitoneal (IP) or i.v. inoculation, *L. monocytogenes* are cleared from the blood primarily by splenic and hepatic macrophages (49).

Cytokine, Chemokine, Costimulatory Molecules, and Similar Responses

Hepatic Kupffer cells clear most of the circulating bacteria and are the major source of interleukin (IL)-6 as a consequence of LLO (50, 51). Neutrophils are rapidly recruited to the site of infection by the cytokine IL-6 and other chemoattractants where they secrete IL-8, colony-stimulating factor (CSF)-1, and monocyte chemotaxtic protein 1 (MCP-1), which then attract macrophages to the infection foci (52). Granulocytes are replaced by large mononuclear cells, and within 2 weeks, the lesions are completely resolved (51). Mice in which granulocytes are depleted are unable to survive to *L. monocytogenes* administration (53–56). *Listeria* replicates within hepatocytes that are then lysed by the granulocytes, which migrate to the site of infection, releasing the intracellular bacteria to be phagocytosed and killed by neutrophils (53). Mast cells are not infected but are activated by *L. monocytogenes* and rapidly secrete tumor necrosis factor (TNF)-α and induce neutrophils recruitment, and their depletion results in higher titers of *L. monocytogenes* in liver and spleen (57).

L. monocytogenes or sublytic doses of LLO in human epithelial Caco-2 cells induce the expression of IL-6 that reduces bacterial intracellular growth (58) and causes overexpression of inducible nitric oxide synthase (NOS) (59). NO appears to be an essential component of the innate immune response to *L. monocytogenes*, having an important role in listericidal activity of neutrophils and macrophages (60), with a deficiency of inducible NOS (iNOS) causing susceptibility to infection (61).

Impairing the recruitment of myelomonocytic cells by blockade of the type 3 complement receptor (62) or diminishing chemokine receptor 2 (CCR2) (63) results in an enhanced susceptibility to *L. monocytogenes* infection (62, 63) and significantly decreased levels of IL-6 (50). CD18-deficient mice are more resistant to listeriosis due to neutrophilia, with faster clearance of *L. monocytogenes* in the liver and spleen, milder inflammatory and necrotizing lesions, and higher levels of IL-1β and G-CSF, leukocytosis, and impaired transendothelial neutrophil migration (64, 65). Similarly, mice deficient in lymphocyte function-associated antigen 1 (LFA-1 or CD11a/CD18) have an increased resistance to *L. monocytogenes* infection and neutrophilia, and upon infection, they show a higher infiltration of neutrophils in the liver in a LFA-1-independent way (64).

CCR2-deficient mice have impaired macrophage recruitment, which would be induced by the CCR2 ligand MCP-1, and thus, they are very susceptible to listeriosis and rapid death by *L. monocytogenes* infection. They also lack a subset of dendritic cells (Tip-DCs) that are the predominant source of TNF and iNOS in the spleens of infected mice (13,63), adding to their inability to clear primary bacterial infection, although $CD8^+$ and $CD4^+$ T cell responses to *L. monocytogenes* antigens are preserved (13). In the T cell zone in the spleen, Tip-DC can result from monocyte differentiation in the presence of *L. monocytogenes* (14).

L. monocytogenes-infected macrophages produce TNF-β, IL-18, and IL-12, all of which are important in inducing the production of interferon (IFN)-γ, and subsequent killing and degradation of *L. monocytogenes* in the phagosome (66). IL-12 deficiency results in an increased susceptibility to listeriosis (67,68), which can be reversed through administration of IFN-γ (67). Resistance to *L. monocytogenes* is conferred, in part, through the release of TNF-α and IFN-γ (69, 70), and deficiency in either of these cytokines or their receptors increases susceptibility to infection (71). Natural killer (NK) cells are the major source of IFN-γ in early infection (72). Upon reinfection, memory $CD8^+$ T cells have the ability to produce IFN-γ in response to IL-12 and IL-18 in the absence of the cognate antigen (73). $CD8^+$ T cells colocalize with the macrophages and *L. monocytogenes* in the T cell area of the spleen where they produce IFN-γ independent of antigen (74). $CD8^+$ T cells are also associated with Lm lesions in the liver (74). IFN-γ production by $CD8^+$ T cells depends partially on the expression of LLO (75).

IFN-γ plays an important role in antitumor responses obtained by *L. monocytogenes*-based vaccines. Although produced initially by NK cells, IFN-γ levels are subsequently maintained by $CD4^+$ T helper cells for a longer period (76). Dominiecki et al. (77) used a tumor that is insensitive to IFN-γ to show that *L. monocytogenes* vaccines require IFN-γ for effective tumor regression and that IFN-γ is specifically required for tumor infiltration of lymphocytes but not for trafficking to the tumor. Paterson et al. (unpublished) have demonstrated that the difference between the IFN-γ-insensitive and sensitive TC-1 tumors is the induced upregulation of multiple chemokines and their receptors, which may explain why lymphocytes infiltrate these wild-type tumors more efficiently. IFN-γ also inhibits angiogenesis at the tumor site in the early effector phase following vaccination (76).

IL-18 is also critical to resistance to *L. monocytogenes*, even in the absence of IFN-γ, and is required for TNF-α and NO production by infected macrophages (78). A deficiency of caspase-1 impairs the ability of macrophages to clear *L. monocytogenes* and causes a significant reduction in IFN-γ production and listericidal activity that can be reversed by IL-18. Recombinant IFN-γ injection restores innate resistance to listeriosis in caspase-$1^{-/-}$ mice (79).

Caspase-1 activation precedes the cell death of macrophages infected with *L. monocytogenes*, and LLO-deficient mutants that cannot escape the phagolysosome have an impaired ability to activate caspase-1 (80).

LLO secreted by *L. monocytogenes* causes specific gene upregulation in macrophages, resulting in significant IFN-γ transcription and secretion (81). Cytosolic LLO activates a potent type I IFN response to invasive *L. monocytogenes* independent of Toll-like receptors (TLR) without detectable activation of nuclear factor (NF)-κB and mitogen-activated protein kinase (MAPK) (82). One of the IFN I-specific apoptotic genes, TNF-related apoptosis-inducing ligand (TRAIL), is upregulated during *L. monocytogenes* infection in the spleen (83). Mice lacking TRAIL are also more resistant to primary listeriosis coincident with lymphoid and myeloid cell death in the spleen.

Pathogen-Associated Molecular Patterns

Nucleotide-binding oligomerization domain (NOD) proteins recognize peptidoglycans present in the bacterial cell wall and are believed to recognize bacteria in the cytosol. Degraded *L. monocytogenes* in the phagolysosomes of macrophages, but not intact bacteria, induce a TLR-independent IFN-β transcriptional response similar to the response observed with cytosolic *L. monocytogenes*, which is dependent on NOD2 (84). NOD1, however, is crucial for IL-8 production and NF-κB activation initiated by *L. monocytogenes* in human endothelial cultures (85).

TLRs are also important components of innate immunity, recognizing conserved molecular structures on pathogens and signaling through adaptor molecules, such as MyD88, to induce NF-κB activation and transcription of several proinflammatory genes. They have a role in the recognition of *L. monocytogenes* at the cell surface. TLR2 recognizes bacterial peptidoglycan lipoteichoic acid and lipoproteins present in the cell wall of gram-positive bacteria, including *L. monocytogenes*. Thus, TLR2-deficient mice are slightly more susceptible to listeriosis (86). TLR5 recognizes bacterial flagellin and may be involved in *L. monocytogenes* recognition; however, flagellin expression is downregulated at 37°C in most isolates, and the role of TLR5 in listeriosis *in vivo* is uncertain, since TLR5 is not required for innate immune activation against this bacterial infection (87).

Although a single TLR has not been shown to be essential in innate immune responses to *L. monocytogenes*, the adaptor molecule MyD88, which is used by signal transduction pathways of all TLRs, besides IL-1 and IL-18, is critical to the defenses against *L. monocytogenes* since infection is lethal in MyD88-deficient mice. $MyD88^{-/-}$ mice have a severely impaired ability to produce IL-12, IFN-γ, TNF-α, and NO following infection, and while MyD88 is not required for MCP-1 production or monocyte recruitment following infection, it is essential for *L. monocytogenes*-induced IL-12 and TNF-α production and monocyte activation (13). Mice deficient in the NOD receptor interacting protein kinase 2 (RIP2) are impaired in their ability to defend against infection and have decreased IFN-γ production by NK and T cells, which is partially attributed to a defective IL-12 signaling (88). NF-κB activates several genes involved in innate immune responses, and mice lacking the p50 subunit of NF-κB are also highly susceptible to *L. monocytogenes* infections (89).

CpG motifs act as pathogen-associated molecular patterns (PAMPs) and immune stimulators (90–94). *In vivo* and *in vitro* studies have shown that treatment with CpG oligodeoxynucleotides (ODN) can improve the resistance of normal neonatal mice and pregnant mice to lethal *L. monocytogenes* infection (95–96).

ADAPTIVE IMMUNITY

MHC Class I Responses

L. monocytogenes secretes a limited number of proteins into the cytosol of the host cell, which are rapidly degraded by the proteosome, resulting in MHC Class Ia-restricted peptide antigens (97, 98). Certain secreted proteins, such as p60 and LLO, are rapidly degraded because their amino-termini contain destabilizing residues as defined by the N-end rule (99, 100). LLO may also be degraded in a proteosome-dependent fashion as it contains a PEST-like sequence (101). The rapid proteosome-mediated degradation of a potentially toxic protein such as LLO enhances host cell survival and generates peptide fragments that enter the MHC Class I antigen processing pathway.

After intravenous inoculation of *L. monocytogenes*, MHC Class Ia-restricted T cell responses to reach peak frequencies in approximately 8 days (102). In experiments in which mice were treated with antibiotics to curtail the duration of the infection, it was found that the magnitude of T cell responses is independent of the quantity or the duration of *in vivo* antigen presentation (103–108), since even with marked differences in the number of viable bacteria and inflammatory response, the expansion and contraction of $CD8^+$ T cells is similar in mice treated with antibiotics 24 h after infection and in mice that are untreated (108). This is consistent with *in vitro* studies of *L. monocytogenes*-specific $CD8^+$ T cell proliferation, in which brief antigen exposure is followed by prolonged proliferation that does not require further exposure to antigen (9). It has been speculated that antigen-independent T cell proliferation is driven by cytokines such as IL-2; however, Wong et al. (109) showed that endogenous IL-2 production by $CD8^+$ T cells is required for Ag-independent expansion following TCR stimulation *in vitro*, but not *in vivo*.

Cell-Mediated Immune Responses to Heat-Killed and Irradiated Lm

Unlike priming with live infection, heat-killed *L. monocytogenes* does not induce a protective immune response. It was believed for years that this was due to the inability of killed bacteria to enter the cytosol of phagocytic antigen presenting cell (APC) impairing the access of antigen to MHC Class I pathway.

We now know that immunization of mice with heat-killed bacteria does result in proliferation of antigen-specific $CD8^+$ T cells but does not result in the differentiation of primed T cells into effector cells (110). Recently, studies have shown that vaccination with irradiated *L. monocytogenes* efficiently activated DCs and induced protective T cell responses (111).

$CD4^+$ T Cells Responses

L. monocytogenes infection also results in the generation of robust MHC Class II-restricted $CD4^+$ T cell responses and shifts the phenotype of $CD4^+$ T cells to Th-1 (112–114). Expansion of these cells was found to be synchronous with the expansion of the $CD8^+$ T cell responses (115). $CD4^+$ T cells produce copious amounts of Th-1 cytokines that contribute to bacteria clearance. Immunization with *L. monocytogenes* has been shown to result in the generation of "high quality" effector $CD4^+$ T cells capable of secreting multiple cytokines such as IFN-γ and TNF-α, or three cytokines such as TNF-α, IFN-γ, and IL-2 (116), coincident with the generation of a memory $CD4^+$ T cell response. $CD4^+$ T cell-mediated protective immunity requires T cell production of IFN-γ, whereas $CD8^+$ T cells mediate protection independently of IFN-γ (71, 117). Production of IFN-γ from $CD4^+$ T cells likely activates macrophages to become more bactericidal. This is supported by *in vitro* studies showing that treatment of macrophages with IFN-γ prevents bacterial escape from the phagosome (118).

$CD4^+$ T cell help is required for the generation and maintenance of functional $CD8^+$ T cell memory against *L. monocytogenes* (119). The kinetics of bacterial clearance and magnitude of the primary antigen-specific $CD8^+$ T cell responses are similar in MHC Class II-deficient mice and wild-type mice, and the kinetics of contraction and numbers of memory $CD8^+$ T cell responses are also similar between the two strains. However, when examined up to 60 days postinfection, a reduction in the number of memory $CD8^+$ T cells, generated in the absence of $CD4^+$ T cell help, was evident at later time points. When rechallenged months after the primary infection, mice that lacked $CD4^+$ T cells were not able to eliminate bacteria due to their inability to mount a vigorous secondary response required for killing infected cells. Memory $CD8^+$ T cells generated in the absence of $CD4^+$ T cells were ineffective (119–122). In viral infection studies, it has been shown that memory cytotoxic T lymphocytes (CTLs) cannot be converted back to effectors without $CD4^+$ cells or exogenous cytokines (123, 124). It is not known whether this help is provided by direct $CD4^+$–$CD8^+$ T cell interaction or by the cytokine milieu generated by the activated $CD4^+$ T cells.

Induction of $T_{\gamma\delta}$ Cells

Skeen et al. (125) has reported that infection of mice intraperitoneally with *L. monocytogenes* caused a local induction of $CD4^+$ $T_{\gamma\delta}$ cells associated with

IL-17 secretion in the peritoneal cavity; however, no changes were observed in the splenic or lymph node T cell populations after these injections. When peritoneal T cells from *L. monocytogenes*-immunized mice were restimulated *in vitro*, the induced $T_{\gamma\delta}$ cells exhibited a greater expansion potential than the $T_{\alpha\beta}$ cells. Modifications that abrogate the virulence, such as heat-killed or *hly* mutations, eliminate the inductive effect for $T_{\gamma\delta}$ cells. Depletion of either $T_{\alpha\beta}$ or $T_{\gamma\delta}$ cells *in vivo* impairs the resistance to primary infection; however, the memory response is unaffected by the depletion of $T_{\gamma\delta}$ cells, consistent with the presumed effects of IL-17 to shift $CD4^+$ Treg phenotypic cells to Th-17 phenotypes and supporting the hypothesis that this T cell subset forms an important line of defense in innate immunity (125).

NONCLASSICAL IMMUNE FUNCTIONS

Effects of *Listeria* on Myeloid Cells

Listeria infection stimulates the expansion of myeloid cells and biases myeloid cell lineages to proliferate and mature into effective terminally differentiated immune cells. Accelerated maturation of hematopoietic progenitor cells along the myeloid lineage, as demonstrated by the upregulation of CD13, CD14, and costimulatory signals, occurs in response to *Listeria* infection. Cytokines such as GM-CSF, IL-6, IL-8, IL-10, IL-12, and TNF-α were found to be induced by *L. monocytogenes* infection, which indicates that infection of human stem cells (HSC) affects the differentiation of $CD34^+$ hematopoietic progenitors (11). Bone marrow composition changes dramatically during infection, leading to an increase of myeloid cells, which peak after 1 week of infection (9). In addition, a water-soluble monocytosis-producing activity (MPA) extracted from *L. monocytogenes* has been found to be able to stimulate proliferation of promonocytes *in vivo*. An acceleration of both the generation time of monocyte precursors and the half-time of blood monocytes has been observed in *L. monocytogenes*-treated mice when compared with control mice (126). Elevated levels of various CSFs in the serum were quantitated subsequent to *L. monocytogenes* infection, with the great bulk of serum colony-stimulating activity represented in M-CSF and G-CSF, and with measurable GM-CSF. The increase in serum CSFs occurred before the peak in bone marrow GM progenitors and before the reduction in bacterial numbers, which follows the onset of specific cell-mediated immunity (8). The rise in serum CSF concentration correlated with monocyte production during *L. monocytogenes* infection (16).

L. monocytogenes induces the release of IL-2, IL-6, IL-12, and TNF-α from DCs and the subsequent upregulation within these cells of CD40, B7-H1 program death-ligand 1 (PD-L1), CD86 (B7-2), and B7-DC (PD-L2) that results in the maturation and activation of high-affinity T cells (12). In addition, the production of IL-12, IL-6, and TNF-α is most efficiently triggered by cytosolic *L. monocytogenes*. Costimulatory molecules induced by cytosolic entry regulate T cell proliferation and the number of functional T cells generated. DC-produced cytokines (IL-12 and IL-10) are the major factors determining the proportion of IFN-γ-producing T cells. LLO is required for optimal T cell priming and cytokine production that result in functionally therapeutic CTL responses (127).

In humans, *Listeria* infection is known to cause maturation of DCs, which display high levels of CD83, CD25, MHC Class II, and CD86 (10). Kinetic studies showed that infection with *L. monocytogenes* results in a tissue-specific expansion of conventional DC (cDC) and plasmacytoid DC (pDC), followed by an upregulation of CD80 and CD86 on cDC in spleen and mesenteric lymph nodes. Expansion of pDC is more prolonged than cDC. pDC upregulates CD86 and MHC-II but do not affect CD80 by themselves. In the spleen and lymph nodes, cDC is an important source of IL-12 (15). Serbina et al. (14) has shown that *Listeria* infection stimulates monocyte recruitment through a MyD88-related mechanism and that these monocytes differentiate into Tip-DCs.

Direct Invasion of *L. monocytogenes* of Vascular Endothelial Cells

Infection of endothelial cells by *L. monocytogenes* is an essential step in the pathogenesis of listeriosis. *L. monocytogenes* can infect endothelial cells *in vivo* by direct bacterial invasion of the endothelial cells mediated by internalins (27) or cell-to-cell spread from infected mononuclear phagocytes that adhere to vascular endothelium (128–130). *In vitro* data show that *L. monocytogenes* can invade and replicate within cultured human umbilical vein endothelial cells (HUVEC) (129, 131). Infection of HUVEC cells with *L. monocytogenes* stimulates an inflammatory phenotype on these cells and induces the upregulation of surface adhesion molecules. This event appears to be mediated by LLO, as similar effects are observed in the nonpathogenic strain *Listeria innocua*, which does not make LLO, unless it is engineered to produce large amounts of LLO. It is believed that *L. monocytogenes*-induced upregulation of surface adhesion molecules on endothelial cells probably occurs by two different mechanisms: LLO dependent and LLO independent (132–134). The uptake of *L. monocytogenes* by endothelial cells provokes the signaling pathways to induce the synthesis of inflammatory chemokines and cytokines such as IL-6, IL-8, MCP-1, and GM-CSF (135). LLO also transduces other events in endothelial cells, such as upregulation of intracellular adhesion molecule 1 (ICAM-1), vascular cell adhesion molecule 1 (VCAM-1), and selectins (132), and activation of NF-κB (135).

Purified LLO has been shown to induce the expression of the adhesion molecules E-selectin and ICAM in human vascular endothelial cells in association with the secretion of the chemokine IL-8 and MCP-1 (135). Additional vascular effects of LLO are its ability to induce Ca^{2+} influx across endothelial cell membranes with the subsequent penetration of *L. monocytogenes* directly into the vascular endothelial cells and beyond (136, 137).

RECOMBINANT *L. MONOCYTOGENES* AS A VACCINE VECTOR

Construction of Recombinant *L. monocytogenes* Strains

A variety of live, attenuated *L. monocytogenes* strains that express viral and tumor antigens, including human papilloma virus (HPV)-16 E7 (138), Her-2/neu (139, 140), high molecular weight melanoma-associated antigen (HMW-MAA) (141), influenza nucleoprotein (NP) (142), and PSA (143, 144), have been created that express the antigen as a fusion protein. These recombinant strains generate antigen-specific $CD4^+$ and $CD8^+$ T cell responses in mice. These particular antigens were expressed in *L. monocytogenes* from an episomal origin, but expression can also be from the chromosome. Plasmid-based strategies have the advantage of multicopy expression, which appears to be more efficacious as a function of the greater amount of antigen–LLO fusion protein expressed, but rely on complementation for the maintenance of the plasmid *in vivo*. Most of the episomal expression systems are based on the fusion of a tumor-associated antigen (TAA) to a nonhemolytic fragment of *hly* (truncated LLO) that maintains the adjuvant properties of LLO. The retention of plasmid by *L. monocytogenes in vivo* is achieved by the complementation of the *prfA* gene from the plasmid in a *prfA*-negative mutant background (138, 140–142). Without *prfA* complementation, this mutant cannot escape the phagolysosome and is destroyed by macrophages and neutrophils. As a result, it cannot grow intracellularly or present antigens to the immune system. In early studies, it was found that including a copy of *prfA* in the plasmid ensures the *in vivo* retention of the plasmid in *L. monocytogenes* (138, 145, 146). Another approach is based on the *in vitro* and *in vivo* complementation of D-alanine racemase in both *Escherichia coli* and *L. monocytogenes* strains deficient in D-alanine racemase (*dal*) and D-alanine amino transferase (*dat*) (144, 147). In this way, a *L. monocytogenes* vaccine strain can be developed, which is completely devoid of antibiotic selection markers.

Chromosomal integration techniques can utilize either a phage-based system, with a site-specific integrase to integrate a gene into the genome (148), or allelic exchange into a known chromosomal locus (149). Recombinant strains based on chromosomal integration have been shown to be somewhat more virulent (149) than similar episomal recombinants and are thus less suitable to be used as human vaccines' backbones without further attenuation. This can be achieved by the deletion of virulence genes such as *actA* and *inlB* from the *L. monocytogenes* chromosome, which, in combination, limit *L. monocytogenes* growth in the liver, a principal target organ of infection by the wild-type (WT) organism (150). This *L. monocytogenes* Δ*actA* Δ*inlB* was further modified to create another platform referred to as "killed but metabolically active" (KBMA) by the deletion of both *uvrA* and *uvrB* genes, which encode the DNA repair enzymes of the nucleotide excision repair pathway. KBMA vaccines (*Lm* Δ*actA* Δ*inlB* Δ*uvrAB* vaccine strains) are sensitive to photochemical inactivation by combined treatment with the synthetic psoralen S-59 and long-wave UV light (151). However, despite these attempts to develop vaccine strains, there is a continued need for improvements that either enhance the potency or reduce the toxicity of *L. monocytogenes*-based vaccines in order to facilitate their clinical development.

Advaxis Inc. (http://www.advaxis.com/) has recently filed a patent on a *L. monocytogenes* dual delivery strategy, expressing one LLO-TAA fusion protein via a multicopy plasmid and a different LLO-TAA fusion protein from the genome. This ability to deliver two different antigens allows for the simultaneous attack against disparate mechanisms, which may result in a synergistic response.

LLO as an Adjuvant in *L. monocytogenes*-Based Immunotherapy

Studies by Paterson et al. have shown that genetic fusion of antigens to a nonhemolytic truncated form of LLO results in enhancing their immunogenicity and antitumor efficacy (138, 140). The immunogenic nature of LLO may result from the targeting of LLO for ubiquitin proteosome-mediated degradation by the presence of PEST sequences at the amino-terminus of the protein (152). Schnupf et al. have shown that LLO is a substrate of the ubiquitin-dependent N-end rule pathway, which recognizes LLO through its N-terminal Lys residue (99). Removal of the PEST sequence from LLO–antigen fusions partially abrogates the ability of vaccine to induce full tumor regression in mice (152). The fusion of antigens to LLO also seems to facilitate the secretion of the antigen (138, 142), increase antigen presentation with a profound influence on the $CD8^+$ T cell activation (152) (Table 2.1A), and reduce the percentage of Tregs infiltrating the tumor (Table 2.1B), and helps to stimulate the maturation of DCs (12).

TABLE 2.1A. *L. monocytogenes*-LLO-E7, a Live *Listeria* Construct That Delivers an LLO-HPV-16 E7 Fusion Protein, Generates Four Times More Activated Tumor Infiltrating Lymphocytes (TIL) Than Does an Identical Construct That Secretes Only the Antigen

Vaccine Group	E7/Db Tetramer-Positive Activated $CD8^+$ T Cells in the Spleen (%)	E7/Db Tetramer-Positive Activated $CD8^+$ T Cells in the Tumor (%)
Lm-E7	12.80	9.40
Lm-LLO-E7	8.20	36.80

TABLE 2.1B. *L. monocytogenes*-LLO-E7, a Live *Listeria* Construct That Delivers an LLO-HPV-16 E7 Fusion Protein, Reduces Intratumoral Tregs by 85.6% and the Identical Vaccine That Secretes Just the Antigen reduces Tregs by 37.3%

Vaccine Group	$CD25^+$ $CD4^+$ $FoxP3^+$ Tregs in the Spleen (%)	$CD25^+$ $CD4^+$ $FoxP3^+$ Tregs in the Tumor (%)
Lm-E7	6.70	11.80
Lm-LLO-E7	4.20	1.70

Fusion of LLO to tumor antigens delivered by other vaccine modalities, such as viral vectors (153) and DNA vaccines (154), also enhances their therapeutic efficacy. Recently, Neeson et al. (155) have shown that LLO has adjuvant properties when used in the form of a recombinant protein. The chemical conjugation of LLO to lymphoma immunoglobulin idiotope induced a potent humoral and cell-mediated immune response, and promoted epitope spreading (ES) after lymphoma challenge. These effects were greater than that observed with idiotope–keyhole limpet hemocyanin (KLH) conjugations.

LLO is a potent inducer of inflammatory cytokines, such as IL-6, IL-8, IL-12, IL-18, TNF-α, and IFN-γ; GM-CSF; as well as NO, chemokines, and costimulatory molecules that are important for innate and adaptive immune responses (12, 13, 113, 156, 157). The proinflammatory cytokine-inducing property of LLO is thought to be a consequence of the activation of the TLR4 signal pathway (158). One evidence of the high Th-1 cytokine-inducing activity of LLO is that protective immunity to *L. monocytogenes* can be induced with killed or avirulent *L. monocytogenes* when administered together with LLO, whereas the protection is not generated in the absence of LLO (159). Macrophages in the presence of LLO release IL-1α, TNF-α, IL-12, and IL-18 (156), which in turn activate NK cells to release IFN-γ (156), resulting in enhanced macrophage activation.

A newly observed, and as yet poorly described property of LLO, is its ability to induce epigenetic modifications affecting control of DNA expression. Hamon et al. have published (160) that extracellular LLO can induce a dephosphorylation of the histone protein H3 and a similar deacetylation of the histone H4 in early phases of *Listeria* infection. This epigenetic effect resulted in reduced transcription of certain genes involved in immune function, thus providing a mechanism by which LLO may regulate the expression of gene products required for immune responses. Another genomic effect of LLO is its ability to increase NF-κB translocation in association with the expression of ICAM and E-selectin, and the secretion of IL-8 and MCP-1 (135). Another signaling cascade affected by LLO is the MAPK pathway, resulting in increase of Ca^{2+} influx across the cell membrane, which appears to facilitate the entry of *Listeria* into endothelial cells and their subsequent infection (136, 137, 161).

Protective and Therapeutic Tumor Immunity

Recombinant *L. monocytogenes* expressing TAAs fused to LLO are capable of generating potent antigen-specific immune responses and have shown profound antitumor efficacy in preclinical settings. Antitumor effects have been demonstrated using murine transplantable tumors that can undergo regression after administration of *L. monocytogenes*-based vaccines that express full antigens or fragments of antigen fused to truncated LLO. The rapid uptake of *L. monocytogenes* into cells does not allow for humoral immunity and opsonization to develop, and this allows for repeated administration as a vaccine without loss of activity due to neutralizing humoral immune responses directed against the vector.

Preclinical studies using a recombinant *L. monocytogenes* strain expressing LLO-HPV-16 E7 have demonstrated both prophylactic and therapeutic efficacy against E7-expressing tumors (138). In addition, a *L. monocytogenes* vaccine strain expressing LLO fused to a chimeric Her-2/neu protein was able to induce anti-Her2/neu CTL responses in mice, show prolonged growth stasis or eradication of Her-2/neu expressing tumors, prevent onset of tumors in Her-2/neu transgenic animals, as well as prevent growth of Her-2/neu expressing lung and brain tumors (140, 162, 163). Recently, Advaxis described a *L. monocytogenes* expressing LLO-PSA that, in a murine tumor model for PSA, caused regression of more than 80% of tumors (143, 144). Maciag et al. have shown that *L. monocytogenes* targeting the activated pericytes present in tumor vasculature with an LLO-HMW-MAA-directed attack has potent anti-angiogenesis effect on the tumors (141). *L. monocytogenes* (Lm)-based vaccines have also been studied in melanoma models using TRP-2 as the target antigen, which induced long-lasting therapeutic tumor protection against both subcutaneous (s.c.) tumors and metastatic tumor nodules in the lungs (164).

The ability of *L. monocytogenes*-based vaccines to break tolerance has been examined using transgenic mouse models for HPV-16 E6/E7 (165, 166) and Her-2/neu (162, 163). *L. monocytogenes*-based constructs expressing E7 fusions, such as Lm-LLO-E7 and Lm-ActA-E7, can impact the growth of autochthonous tumors that arise in HPV 16 E6/E7 transgenic mice (165). In Her-2/neu transgenic mice, which were treated with *L. monocytogenes*-based vaccines expressing an LLO fusion with one of five overlapping fragments of the Her2 antigen, all the Lm-LLO-Her2 constructs were capable of slowing or halting the tumor growth or eliminating tumors, despite the fact that $CD8^+$ T cells from the immunized mice were of lower avidity than those arising from wild-type mice (162). *L. monocytogenes*-based Her2/neu vaccines also delayed or completely prevented the appearance of spontaneous autochthonous tumors in the transgenic Her2/neu mice (163). Thus, these vaccines are able to overcome tolerance to self-antigens and expand autoreactive T cells otherwise too low in number and avidity to drive antitumor responses.

The response shown in Figure 2.1, in which partial or complete tumor regression is seen in normal and transgenic animals treated with an HPV-16 E7-directed *L. monocytogenes* vaccine, has been seen with a large number of vaccines directed against other antigens, including Her2/neu, PSA, HMW-MAA, and others, in normal animals (Table 2.2). It is important to note that animals that experience tumor regression upon immunization show resistance to rechallenge with the same tumors, suggesting a strong memory response elicited by the vaccine (141, 145).

It is noteworthy that the utility of live Lm-LLO vaccines is not limited to cancer vaccines. Various infectious disease *L. monocytogenes* vaccines have been made and tested successfully in animals, for the induction of cell-mediated immunity against pathogens, including vaccines for the treatment of

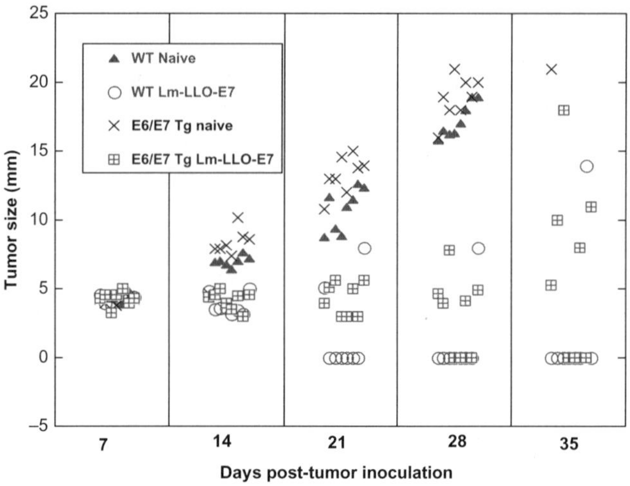

Figure 2.1 The antitumor effects of ADXS11-001 (Lm-LLOE7; formerly Lovaxin C) in normal and transgenic mice. Animals were inoculated s.c. with TC-1 (2×10^5) tumor cells. When tumors reached a size of 5 mm, mice were immunized twice with 1-week interval, with either ADXS11-001 or left untreated. Tumors were measured once a week with an electronic caliper (triangles = normal, X = transgenic; progress rapidly). The treated animals are either tumor free (circles = normal: 7 of 8, squares = transgenic 3 of 8) or have dramatically slower tumor growth.

bacterial and viral diseases, such as HIV, leishmaniasis, tularemia, and herpes simplex (Table 2.3).

CROSS-PRESENTATION AND EPITOPE SPREADING

In addition to specific antitumor CTL responses to the engineered antigen delivered by *Listeria*, immunization with attenuated *Listeria* can also impair the growth of tumors that do not contain epitopes present in the vaccine, presumably by means of ES. ES is an important mechanism by which therapeutic agents can increase their efficacy by engendering immune responses to different epitopes than the target epitope(s) and has been associated with the use of live *Listeria* vaccines more than once (155, 167, 172). The phenomenon of ES is believed to occur as a result of the release of antigens from the tumor cells killed by vaccine-induced T cells, which are then phagocytosed by APCs and presented to naive T cells in draining lymph nodes, which are then primed to respond to them. ES broadens the immune response to include unidentified tumor antigens in the context of therapeutic vaccines, thereby creating many more tumor targets than that for which the vaccine was engineered.

Perhaps the earliest evidence for this effect was seen when Liau et al. (167) used lymphocytic choriomeningitis virus NP (LCMV-NP) as a pseudotumor

TABLE 2.2. Various *L. monocytogenes* Vaccines and Responses in Tumor Models

Vaccine Name	Design	Strain Modification	Antigen	Impact on Tumor Growth	Reference
Lm-LLO-NP	Plasmid	*prfA*⁻	NP, influenza (pseudotumor antigen)	Regression	(145, 146)
Lm-NP	Chromosomal integration	WT	NP, LCMV	Protective immunity	(167)
Lm-LLO-E7	Plasmid	*prfA*⁻	E7, HPV-16	Regression	(138)
Lm-PEST-E7					(152)
Lm-ActA-E7					(165)
Lm-dd-TV3	Plasmid	*dal*⁻ *dat*⁻	E7, HPV-16	Regression	(147)
Lm-v1/Lm-v2 DNA vaccines	Plasmid	*dal*⁻ *dat*⁻	E7, HPV-16	Slow tumor growth	(168)
Lm-LLO-EC1 EC2, EC3, IC1, IC2	Plasmid	*prfA*⁻	Rat Her-2/neu, breast cancer	Regression/stasis	(140)
Lm-hHer2/neu chimera	Plasmid	*prfA*⁻	Human Her-2/neu breast cancer	Regression/stasis	(139)
E1-rLm	Chromosomal integration	WT	E1, CRPV	Protective immunity	(169)
Lm-Trp2	Chromosomal integration	WT	TRP-2, melanoma	Protective immunity	(164)
Lm-LLO-PSA	Plasmid	*prfA*⁻	PSA	Regression	(143)
*daldat*Δ*actA*142		*dal*⁻ *dat*⁻ *actA*⁻			(144)
actA⁻/*inlB*⁻AH1-A5	Inserted at phage integration site	*actA*⁻*InlB*⁻	Gp70 epitope	Reduction of metastatic disease	(150)
Dmpl2GK20	Plasmid	*mpl2*⁻	*E. coli* beta-galactosidase epitope	Protective immunity	(170)
Lm-LLO-Mage-b	Plasmid	*prfA*⁻	Mage-b	Reduction of metastatic disease	(171)
Lm-LLO-HMW-MAA-C	Plasmid	*prfA*–	HMW-MAA	Regression/slow tumor growth	(141)
Lm-LLO-flk-E1 Lm-LLO-flk-I1	Plasmid	*prfA*⁻	VEGF receptor 2	Regression	(172)

TABLE 2.3. Various *L. monocytogenes* Vaccines and Responses in Infectious Disease Models

Vaccine Name	Design	Strain Modification	Antigen	T Cell Responses	Reference
Lm-gag	Chromosomal integration	WT	HIV-gag	CD4, CD8	(149)
Lm-*daldat*-gag	Chromosomal integration	*dal*$^-$ *dat*$^-$	HIV-gag	CD8	(173)
Lm-LLO-L1	Inserted at phage integration site	WT	HPV-16, L1	CD4, CD8	(174)
Δ*actA*-LACK-Lm	Chromosomal integration	*actA*$^-$	*Leishmania* LACK	CD4	(175)
Lm/IgC	Inserted at phage integration site	*actA*$^-$	*Francisella tularensis* IgC	CD4, CD8	(176)
Lm Δ *actA* pHSVgB	Plasmid	*actA*$^-$	HSV-1 peptide $gB_{498–505}$	CD8	(177)

antigen to investigate recombinant *L. monocytogenes* as a tumor vaccine against s.c. and intracerebral challenges with a NP-expressing glioma cell line. These authors showed that vaccination with recombinant *L. monocytogenes*-NP stimulated protection against s.c., but not intracerebral, tumor challenge in an antigen-specific, CD8$^+$ T cell-dependent manner. After rejection of s.c. tumors, enhanced antitumor immunity was achieved via ES that then permitted complete resistance against lethal intracerebral challenge with tumor cells and with the untransfected parental tumor. Unlike the CD8-dependent immune responses against s.c. tumors, this expanded intracerebral immunity against endogenous TAAs is dependent on both CD4$^+$ and CD8$^+$ T cells.

Recently, Seavey et al. (172) used *Listeria* vaccines expressing fragments of the murine vascular endothelial growth factor receptor 2 (VEGFR2) gene (also known as fetal liver kinase [FLK]-1) to target tumor vasculature endothelial cells in a murine breast tumor line that overexpresses Her2/neu. Interestingly, immunization of mice in which these tumors had been established resulted in impaired tumor vasculature after immunization, and slowing or eradication of the tumors, accompanied by ES to the Her2/neu antigen. Comparing responses in wild-type mice with transgenic mice that are profoundly tolerant to Her2/neu, these investigators showed that immune responses to her2/neu were required for antitumor efficacy of the *L. monocytogenes*-FLK-1 vaccines and that Her2/neu-specific CTL infiltrated the tumors in wild-type mice in high numbers (172).

Attenuation of *L. monocytogenes* to Increase Safety

When considering the use of live *Listeria* vaccines for the treatment of cancer, safety must be the first and overriding concern, and strategies to design *L. monocytogenes* vaccines have focused on how to construct the least virulent, yet most effective vaccine strains. Many of the virulence deletion mutants are rapidly cleared from the host compared to wild type. As discussed above, in the context of antibiotic termination, there is substantial evidence to suggest that prolonged *in vivo* survival of *L. monocytogenes* is not necessary to induce effective immune responses, which is consistent with attenuated strains retaining their potent immunogenicity.

Irreversible deletion of multiple virulence genes is an appealing way to attenuate *L. monocytogenes*. The deletion of *actA* and *inlB* genes significantly reduces the toxicity of the vector *in vivo*, yet the ability to induce innate and adaptive immune responses is retained (150). Mutants attenuated by deletion of the *actA* and *plcB* genes have been tested in humans and found to be reasonably safe (178). The use of auxotrophic mutants that require exogenous factors for *in vivo* and *in vitro* growth is another strategy to attenuate bacteria. A *dal*/*dat* mutant that is unable to synthesize D-alanine, an essential component of the bacterial cell wall, has also been tested as a vaccine vector (179). Complementing this mutant with a plasmid carrying a tightly regulated D-alanine racemase gene results in an attenuated bacterium that is cleared very rapidly *in vivo*. Although avirulent in mice, these mutant strains remain immunogenic as evidenced by potent $CD8^+$ T cells responses (147, 180). The ability of KBMA *L. monocytogenes* that are unable to replicate but still have sufficient metabolic activity to be able to deliver antigens to the immune system were described by Brockstedt et al. (151). This attenuated *L. monocytogenes* retains the ability to enter the cell, escape the phagolysosome, and express antigens that induce functional $CD4^+$ and $CD8^+$ T cell responses (150). Live irradiated *L. monocytogenes*, unlike heat killed, efficiently activates DCs via TLRs and induced protective T cell responses in mice (111). Cross-presentation of irradiated listerial antigens to $CD8^+$ T cells involved transporter associated with antigen processing (TAP) and proteosomes-dependent cytosolic antigen processing (111).

LISTERIA LLO-AG FUSIONS AND THE TUMOR MICROENVIRONMENT

LLO-Ag *Listeria* Vaccines Create a Favorable Intratumoral Milieu

Infiltration of tumors by antigen-specific cytotoxic $CD8^+$ T cells is critical for tumor regression (181); however, the tumor microenvironment is under the influence of many other factors, which might limit the effector activity of these cells. Lm-LLO-Ag-based vaccines are excellent for tumor immunotherapy because they not only induce strong T cell responses but also switch the intratumoral milieu from a suppressed to a less tolerant and immune active state.

The reasons for the ability of *L. monocytogenes* to induce such a potent $CD8^+$ T cell response are not entirely characterized. However, the fusion of the Ag to LLO seems to play a major role (138) as deletion of the PEST domain of LLO can cause the numbers of antigen-specific $CD8^+$ TILs to decrease, compromising the efficacy of the vaccine (152). Depletion of $CD8^+$ T cells during the effector phase inhibits the therapeutic effect of *L. monocytogenes* vaccination against tumors (138, 140, 145), and Hussain et al. (182) have speculated that the ability of the vaccine to induce a specific chemokine profile in the $CD8^+$ T cells may have a role in this response. Souders et al. has shown that Lm-LLO-Ag vaccines can also induce the production of T cells in transgenic mice, which, although of low avidity, were capable of infiltrating the tumors and preventing tumor growth (166). Thus, Lm-LLO-Ag vaccines use multiple mechanisms that allow antigen-specific $CD8^+$ T cell tumor infiltration.

Tumor-specific $CD4^+$ T helper cells are produced and migrate to the tumor similarly to CTLs following Lm-LLO-Ag vaccination (77, 145, 182). It is not of great consequence that $CD4^+$ T cells can lyse antigen/MHC-II expressing tumor cells (155, 182–185) as most tumors only express MHC Class I molecules. However, the paracrine functions of $CD4^+$ T helper cells to promote rejection of MHC-II-negative tumors is likely to be important (145, 182, 186). The $CD4^+$ T cell response to *L. monocytogenes* infection has been shown to be primarily of the Th-1 type, and the production of the antitumoral cytokines IFN-γ, TNF-α, and IL-2 is consistent with this action.

Listeria LLO-Ag Vaccines and Endogenous Immune Inhibition

The accumulation of regulatory T cells (Tregs) and their apparent ability to protect tumors from immune attack and promote their growth represents a formidable challenge to traditional cancer immunotherapeutics. It is currently felt that vaccination strategies that are therapeutically ineffective, despite generating large numbers of antigen-specific activated $CD4^+$ and $CD8^+$ T cells, fail in part as a result of the inhibitory effects of intratumoral Tregs and similar innate sources of immune inhibition (187). Lm-LLO-Ag vaccines, however, seem to function by decreasing the population of Tregs (and possibly other inhibitory cells) and, thus, reducing innate immune inhibition within the tumors while simultaneously stimulating a strong attack against the tumor (Figure 2.2).

L. monocytogenes-based vaccines that secrete LLO-Ag fusions, but not vaccines that secrete only an antigen, have been shown to uniquely prevent large numbers of Tregs within tumors (143, 189). This is also supported by studies showing that immunization with Lm-LLO-E7 fusion protein resulted in fewer Tregs in the tumors when compared to Lm-E7 that secretes only the antigen (182, 188). These studies also indicated that inhibitory cytokines associated with Tregs, such as IL-10 and TGF-β, were reduced by LLO-Ag secreting *L. monocytogenes* vaccines but not by those that secrete the antigen alone.

(a)

% intratumoral Tregs by vector type

Lm-E7		Lm-LLO-E7	
Spleen	Tumor	Spleen	Tumor
6.42	12.18	4.45	2.33
6.99	12.20	3.88	2.08
6.89	14.94	4.47	1.10
6.52	8.92	3.90	1.36

(b)

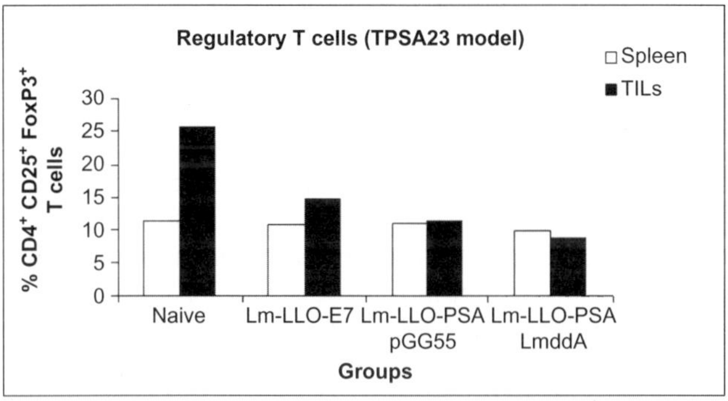

Figure 2.2 (A) Percent of intratumoral Tregs in a TC-1 model following treatment with Lm-LLO-E7. Decreases are seen with the secretion of LLO-E7 but not with E7 alone. (B) Tregs in TPSA-23 tumors (a PSA^+ TRAMP-C tumor model) treated with Lm-LLO-PSA. Two different *Listeria* backbones referred as LmΔdal dat actA (144) and Lm-LLO-PSA (143) expressing LLO-PSA were tested against naive and Lm-LLO-E7 controls. A nonspecific effect of LLO–antigen secreting constructs is typical and can be seen in the Lm-LLO-E7 group, with a greater effect seen when the antigen is specific to the tumor type.

Interestingly, immunization with a *L. monocytogenes* expressing an irrelevant antigen fused to LLO resulted in a lesser reduction of Tregs within tumors when compared to an antigen-specific *L. monocytogenes* vaccine, suggesting that this phenomenon is at least partially antigen dependent (143, 189). Lm-LLO-based vaccines thus seem superior to other vaccine strategies due, at least in part, to their ability to inhibit Treg accumulation within tumors. Thus, fusion to LLO is a key component in *L. monocytogenes*-based vaccines and may have important implications as an adjuvant for antitumor vaccination strategies.

Recently, a novel subset of $CD4^+$ T cells called Th-17 cells that may be induced by *L. monocytogenes* vaccination strategies (190) has been defined, which are phenotypically opposite from Tregs in that they promote autoimmune responses and thus may be antitumoral. However, regulation of Treg and Th-17 levels appears to be linked, such that Tregs are induced by the presence of TGF-β, while Th-17 cells are induced by IL-6 and TGF-β, and importantly, Tregs may be induced to become Th-17 cells if a source of IL-6 is provided (191). Lm-LLO-Ag vaccines may provide such a source of IL-6 from infected Kupffer cells, as discussed above, early in infection (50). *L. monocytogenes*-infected macrophages, which are known to be present within the tumor, are also a source of IL-6 (Y. Paterson, unpublished data). Whether the presence of IL-6 early in infection induces a Th-17 phenotype over the Treg phenotype within tumors has yet to be determined. However, TGF-β levels are significantly lower in tumors from Lm-LLO-E7-vaccinated mice compared to controls (188), and depletion of TGF-β does affect the efficacy of the vaccine (138).

L. monocytogenes Within the Tumor

L. monocytogenes has been found to accumulate within tumors (192, 193) and to persist there for at least 10 days while being cleared from the spleen and the liver after just 3 days (194). It appears that *Listeria* home to tumors as a function of their size, in part due to the impaired vasculature bed within the tumor and the privileged immunosuppressive microenvironment that impairs bacterial clearance, allowing for replication, as well as carriage in infected tumor-infiltrating monocytes. This persistence within tumors suggests that immune responses to the infection itself at the site of the tumor, independent of antigen-specific effects, may play a role in the potent antitumoral effect of vaccines. For example, macrophages activated by *L. monocytogenes* may home to the tumor and secrete a variety of tumoricidal cytokines, including IL-6, IL-12, IL-1, and TNF-α. In addition, infected macrophages would serve as a source of listerial antigens such as LLO, especially those infected with a bioengineered strain that secretes an LLO–antigen fusion protein. LLO, in turn, induces a Th-1-type cytokine profile with secretion of the proinflammatory cytokines IL-12, IL-18, IFN-γ, as well as IL-1, IL-6, and TNF-α.

Clinical Experience

The first clinical trial of *Listeria* was performed by Dr. Elizabeth Hohmann's group (178) in 2002, using an *actA/plcB*-deleted strain that did not carry an antigen. This agent was attenuated by a factor of approximately 10^3, and it was given to 20 healthy volunteers in single escalating oral doses from 10^6 to 10^9 cfu as inpatients. No serious adverse effects were observed, and the volunteers displayed no fever or positive blood cultures, although they shed in their stool for up to 4 days, without diarrhea (178). Minor liver perturbations were seen in 3 out of 20 patients. In summary, this study showed that it was safe to orally administer a live, attenuated *L. monocytogenes* strain in humans, without serious adverse effects or long-term sequelae (178).

As *L. monocytogenes* has been extensively studied as a cancer immunotherapeutic, two other clinical trials have been conducted in patients with cancer. Anza Therapeutics, formerly Cerus, has completed a trial of a modified *L. monocytogenes* devoid of antigen as a potential therapy for liver metastases associated with colon cancer (NCT00327652). The results from this trial have not been published as yet.

Advaxis Inc. has also concluded their first human clinical trial using a live, attenuated *L. monocytogenes* vaccine (195). The vector ADXS11-001 (Lm-LLO-E7) secretes an LLO-HPV-16 E7 fusion protein and targets HPV-associated diseases (138). In this clinical trial, a total of 15 patients with histologically confirmed progressive, recurrent, metastatic squamous cell carcinoma of the cervix, who had failed prior chemotherapy, radiotherapy, and/or surgery, were enrolled in the study (195). These patients were divided in three cohorts of five patients, and each cohort received two i.v. infusions of ADXS11-001 21 days apart at doses of 1×10^9, 3.3×10^9, or 1×10^{10} cfu, each dose followed after 5 days by ampicillin. A fourth cohort scheduled to receive a dose of 3.3×10^{10} cfu of ADXS11-001 was not enrolled as dose-limiting toxicity was observed in the third dosage group (1×10^{10} cfu) (195). Doses were extrapolated from animal data, and patients were maintained as inpatients for 5 days following each dose. Toxicity associated with ADXS11-001 administration appeared to be confined predominantly to a flu-like syndrome comprised of fever, chills, nausea with and without vomiting, myalgia, and headache. Diastolic hypotension was observed in the group receiving the highest dose of 1×10^{10} cfu. Importantly, in almost all patients, the side effects of ADXS11-001, including dose-limiting hypotension, promptly responded to symptomatic treatment without the need for early antibiotics (195). This pattern of adverse events closely resembles what has been observed following the administration of IL-2 and may be related to the strong innate response associated with a high level of infectious particles.

Although the Advaxis trial was not designed to evaluate efficacy, tumor assessment using the response evaluation criteria in solid tumors (RECIST) and survival data from 13 eligible patients were analyzed. Stable disease was reported in seven (53.8%) and progressive disease in five (38.5%) patients, while an unconfirmed partial tumor response was observed in one patient (7.7%). A reduction in total tumor size was observed in four patients (30.8%), with no dose–response effect observed. Overall, the median survival was 347 days for all the patients and over 2 years for the three patients currently alive (195). These are encouraging results, taking into consideration that, historically, the survival of patients with previously treated metastatic, refractory, or recurrent ICC is very poor, with a median survival time of 6–7 months (196). In conclusion, these trials show the feasibility of using live, attenuated *L. monocytogenes* for cancer immunotherapy in humans, making it an actual alternative treatment in the future.

CONCLUSIONS

Many elements of the *Listeria* life cycle and our response to it make *L. monocytogenes* a very attractive vaccine candidate. It appears to be more effective *in vivo* than other vectors, such as vaccinia, peptides, or DNA-based approaches, and lacks the drawbacks associated with other vectors, such as *Salmonella*, *Shigella*, *Legionella*, *Lactococcus*, and *Mycobacterium* (Bacillus Calmette–Guerin). Production is straightforward, and the bacterium can be grown under standard BSL2 laboratory conditions. Genetic manipulation of this organism is well established, allowing construction of new recombinant vaccine strains. A single recombinant *L. monocytogenes* strain can be manipulated to express multiple gene products using either plasmid or chromosomal expression systems. There is extensive knowledge about the life cycle, genetics, and immunological characteristics of the organism. Taken together, this provides the ability to design and build potent, specific, and safe vaccine platforms.

The broad diversity of the various mechanisms of action makes *Listeria* a very attractive vaccine vector. It is not merely that *Listeria* activates $CD4^+$ and $CD8^+$ T cells, and APC, as well as the strong stimulation of cytokines, chemokines, and costimulatory molecules that make *L. monocytogenes* useful. This efficacy is augmented by the ability to deliver the strong stimulation of an antigen or an antigen–LLO fusion to both the endogenous and exogenous pathways of antigen presentation from within an APC directly. In addition, *L. monocytogenes*' ability to alter the development of myeloid cell lines from the bone marrow to the peripheral effector cells may also play a role in its efficacy. The ability of *L. monocytogenes* to infect vascular endothelial cells and trigger changes that support tumor infiltration of activated cells are also very useful for a therapeutic antitumor response. Finally, the ability of *Listeria* vaccines that secrete LLO–antigen fusions to reduce intratumoral Tregs and to alter the tumor microenvironment in salutary ways unlike any other vaccine vector makes *L. monocytogenes* a potentially unique therapeutic. Taken together, the effects of live *Listeria* vectors, especially those that secrete LLO–antigen fusions, comprise a complex and compelling interrelated group of simultaneous and integrated mechanisms of action.

REFERENCES

1. Lecuit M. 2007. Human listeriosis and animal models. Microbes Infect. **9**: 1216–1225.
2. Lorber B. 1997. Listeriosis. Clin. Infect. Dis. **24**: 1–9.
3. Vázquez-Boland J.A., Kuhn M., Berche P., Chakraborty T., Domínguez-Bernal G., Goebel W., González-Zorn B., Wehland J., and Kreft J. 2001. *Listeria* pathogenesis and molecular virulence determinants. Clin. Microbiol. Rev. **14**: 584–640.
4. Camilli A., Tilney L.G., and Portnoy D.A. 1993. Dual roles of plcA in *Listeria monocytogenes* pathogenesis. Mol. Microbiol. **8**: 143–157.
5. Gaillard J.L., Berche P., Mounier J., Richard S., and Sansonetti P. 1987. In vitro model of penetration and intracellular growth of *Listeria monocytogenes* in the human enterocyte-like cell line Caco-2. Infect. Immun. **55**: 2822–2829.
6. Tilney L.G., and Portnoy D.A. 1989. Actin filaments and the growth, movement, and spread of the intracellular bacterial parasite, *Listeria monocytogenes*. J. Cell Biol. **109**: 1597–1608.

7. Paterson Y., and Maciag P.C. 2005. *Listeria*-based vaccines for cancer treatment. Curr. Opin. Mol. Ther. **7**: 454–460.
8. Cheers C., Haigh A.M., Kelso A., Metcalf D., Stanley E.R., and Young A.M. 1988. Production of colony-stimulating factors (CSFs) during infection: separate determinations of macrophage-, granulocyte-, granulocyte-macrophage-, and multi-CSFs. Infect. Immun. **56**: 247–251.
9. de Bruijn M.F.V.V.W., Ploemacher R.E, Bakker-Woudenberg I.A., Campbell P.A, van Ewijk W., and Leenen P.J. 1998. Bone marrow cellular composition in *Listeria monocytogenes* infected mice detected using ER-MP12 and ER-MP20 antibodies: a flow cytometric alternative to differential counting. J. Immunol. **217**: 27–39.
10. Kolb-Mäurer A., Gentschev I., Fries H.W., Fiedler F., Bröcker E.B., Kämpgen E., and Goebel W. 2000. *Listeria monocytogenes*-infected human dendritic cells: uptake and host cell response. Infect. Immun. **68**: 3680–3688.
11. Kolb-Mäurer A., Weissinger F., Kurzai O., Mäurer M., Wilhelm M., and Goebel W. 2004. Bacterial infection of human hematopoietic stem cells induces monocytic differentiation. FEMS Immunol. Med. Microbiol. **40**: 147–153.
12. Peng X., Hussain S.F., and Paterson Y. 2004. The ability of two *Listeria monocytogenes* vaccines targeting human papillomavirus-16 E7 to induce an antitumor response correlates with myeloid dendritic cell function. J. Immunol. **172**: 6030–6038.
13. Serbina N.V., Kuziel W., Flavell R., Akira S., Rollins B., and Pamer E.G. 2003. Sequential MyD88-independent and -dependent activation of innate immune responses to intracellular bacterial infection. Immunity. **19**: 891–901.
14. Serbina N.V., Salazar-Mather T.P., Biron C.A., Kuziel W.A., and Pamer E.G. 2003. TNF/iNOS-producing dendritic cells mediate innate immune defense against bacterial infection. Immunity. **19**: 59–70.
15. Tam M.A., and Wick M.J. 2006. Differential expansion, activation and effector functions of conventional and plasmacytoid dendritic cells in mouse tissues transiently infected with *Listeria monocytogenes*. Cell Microbiol. **8**: 1172–1187.
16. Wing E.J., Waheed A., and Shadduck R.K. 1984. Changes in serum colony-stimulating factor and monocytic progenitor cells during *Listeria monocytogenes* infection in mice. Infect. Immun. **45**: 180–184.
17. Scortti M., Monzó H.J., Lacharme-Lora L., Lewis D.A., and Vázquez-Boland J.A. 2007. The PrfA virulence regulon. Microbes Infect. **9**: 1196–1207.
18. Leimeister-Wächter M., Haffner C., Domann E., Goebel W., and Chakraborty T. 1990. Identification of a gene that positively regulates expression of listeriolysin, the major virulence factor of *Listeria monocytogenes*. Proc. Natl. Acad. Sci. U S A. **87**: 8336–8340.
19. Szalay G., Hess J., and Kaufmann S.H. 1994. Presentation of *Listeria monocytogenes* antigens by major histocompatibility complex class I molecules to CD8 cytotoxic T lymphocytes independent of listeriolysin secretion and virulence. Eur. J. Immunol. **24**: 1471–1477.
20. Cabanes D., Dussurget O., Dehoux P., and Cossart P. 2004. Auto, a surface associated autolysin of *Listeria monocytogenes* required for entry into eukaryotic cells and virulence. Mol. Microbiol. **51**: 1601–1614.

21. Cabanes D., Sousa S., Cebriá A., Lecuit M., García-del Portillo F., and Cossart P. 2005. Gp96 is a receptor for a novel *Listeria monocytogenes* virulence factor, Vip, a surface protein. EMBO J. **24**: 2827–2838.
22. Milohanic E., Jonquières R., Cossart P., Berche P., and Gaillard J.L. 2001. The autolysin Ami contributes to the adhesion of *Listeria monocytogenes* to eukaryotic cells via its cell wall anchor. Mol. Microbiol. **39**: 1212–1224.
23. Bierne H., and Cossart P. 2002. InlB, a surface protein of *Listeria monocytogenes* that behaves as an invasin and a growth factor. J. Cell Sci. **115**: 3357–3367.
24. Bonazzi M., Veiga E., Pizarro-Cerdá J., and Cossart P. 2008. Successive post-translational modifications of E-cadherin are required for InlA-mediated internalization of *Listeria monocytogenes*. Cell Microbiol. **10**: 2208–2222.
25. Cossart P., and Lecuit M. 1998. Interactions of *Listeria monocytogenes* with mammalian cells during entry and actin-based movement: bacterial factors, cellular ligands and signaling. EMBO J. **17**: 3797–3806.
26. Dramsi S., Dehoux P., Lebrun M., Goossens P.L., and Cossart P. 1997. Identification of four new members of the internalin multigene family of *Listeria monocytogenes* EGD. Infect. Immun. **65**: 1615–1625.
27. Greiffenberg L., Goebel W., Kim K.S., Weiglein I., Bubert A., Engelbrecht F., Stins M., and Kuhn M. 1998. Interaction of *Listeria monocytogenes* with human brain microvascular endothelial cells: InlB-dependent invasion, long-term intracellular growth, and spread from macrophages to endothelial cells. Infect. Immun. **66**: 5260–5267.
28. Khelef N., Lecuit M., Bierne H., and Cossart P. 2006. Species specificity of the *Listeria monocytogenes* InlB protein. Cell Microbiol. **8**: 457–470.
29. Lecuit M., Nelson D.M., Smith S.D., Khun H., Huerre M., Vacher-Lavenu M.C., Gordon J.I., and Cossart P. 2004. Targeting and crossing of the human maternofetal barrier by *Listeria monocytogenes*: role of internalin interaction with trophoblast E-cadherin. Proc. Natl. Acad. Sci. U S A. **101**: 6152–6157.
30. Lecuit M., Ohayon H., Braun L., Mengaud J., and Cossart P. 1997. Internalin of *Listeria monocytogenes* with an intact leucine-rich repeat region is sufficient to promote internalization. Infect. Immun. **65**: 5309–5319.
31. Lecuit M., Vandormael-Pournin S., Lefort J., Huerre M., Gounon P., Dupuy C., Babinet C., and Cossart P. 2001. A transgenic model for listeriosis: role of internalin in crossing the intestinal barrier. Science. **292**: 1722–1725.
32. Walz T. 2005. How cholesterol-dependent cytolysins bite holes into membranes. Mol. Cell. **18**: 393–394.
33. Portnoy D.A., Chakraborty T., Goebel W., and Cossart P. 1992. Molecular determinants of *Listeria monocytogenes* pathogenesis. Infect. Immun. **60**: 1263–1267.
34. Portnoy D.A., Tweten R.K., Kehoe M., and Bielecki J. 1992. Capacity of listeriolysin O, streptolysin O, and perfringolysin O to mediate growth of *Bacillus subtilis* within mammalian cells. Infect. Immun. **60**: 2710–2717.
35. Tweten R.K. 2005. Cholesterol-dependent cytolysins, a family of versatile pore-forming toxins. Infect. Immun. **73**: 6199–6209.
36. Alvarez-Dominguez C., Roberts R., and Stahl P.D. 1997. Internalized *Listeria monocytogenes* modulates intracellular trafficking and delays maturation of the phagosome. J. Cell Sci. **110**: 731–743.

37. Dussurget O., Pizarro-Cerda J., and Cossart P. 2004. Molecular determinants of *Listeria monocytogenes* virulence. Annu. Rev. Microbiol. **58**: 587–610.

38. Suárez M., González-Zorn B., Vega Y., Chico-Calero I., and Vázquez-Boland J.A. 2001. A role for ActA in epithelial cell invasion by *Listeria monocytogenes*. Cell Microbiol. **3**: 853–864.

39. Farber J.M., and Peterkin P.I. 1991. *Listeria monocytogenes*, a food-borne pathogen. Microbiol. Rev. **55**: 476–511.

40. Dalton C.B., Austin C.C., Sobel J., Hayes P.S., Bibb W.F., Graves L.M., Swaminathan B., Proctor M.E., and Griffin P.M. 1997. An outbreak of gastroenteritis and fever due to *Listeria monocytogenes* in milk. N. Engl. J. Med. **336**: 100–105.

41. Ewert D.P., Lieb L., Hayes P.S., Reeves M.W., and Mascola L. 1995. *Listeria monocytogenes* infection and serotype distribution among HIV-infected persons in Los Angeles County, 1985–1992. J. Acquir. Immune Defic. Syndr. Hum. Retrovirol. **8**: 461–465.

42. Fleming D.W., Cochi S.L., MacDonald K.L., Brondum J., Hayes P.S., Plikaytis B.D., Holmes M.B., Audurier A., Broome C.V., and Reingold A.L. 1985. Pasteurized milk as a vehicle of infection in an outbreak of listeriosis. N. Engl. J. Med. **312**: 404–407.

43. Jurado R.L., Farley M.M., Pereira E., Harvey R.C., Schuchat A., Wenger J.D., and Stephens D.S. 1993. Increased risk of meningitis and bacteremia due to *Listeria monocytogenes* in patients with human immunodeficiency virus infection. Clin. Infect. Dis. **17**: 224–227.

44. Linnan M.J., Mascola L., Lou X.D., Goulet V., May S., Salminen C., Hird D.W., Yonekura M.L., Hayes P., Weaver R., Audurier A., Plikaytis B.D., Fannin S.L., Kleks A., and Broome C.V. 1988. Epidemic listeriosis associated with Mexican-style cheese. N. Engl. J. Med. **319**: 823–828.

45. Salamina G., Dalle Donne E., Niccolini A., Poda G., Cesaroni D., Bucci M., Fini R., Maldini M., Schuchat A., Swaminathan B., Bibb W., Rocourt J., Binkin N., and Salmaso S. 1996. A foodborne outbreak of gastroenteritis involving *Listeria monocytogenes*. Epidemiol. Infect. **117**: 429–436.

46. Southwick F.S., and Purich D.L. 1996. Intracellular pathogenesis of listeriosis. N. Engl. J. Med. **334**: 770–776.

47. Stamm A.M., Dismukes W.E., Simmons B.P., Cobbs C.G., Elliott A., Budrich P., and Harmon J. 1982. Listeriosis in renal transplant recipients: report of an outbreak and review of 102 cases. Rev. Infect. Dis. **4**: 665–682.

48. Wing E.J., and Gregory S.H. 2002. *Listeria monocytogenes*: clinical and experimental update. J. Infect. Dis. **185**: S18–S24.

49. Aichele P., Zinke J., Grode L., Schwendener R.A., Kaufmann S.H., and Seiler P. 2003. Macrophages of the splenic marginal zone are essential for trapping of blood-borne particulate antigen but dispensable for induction of specific T cell responses. J. Immunol. **171**: 1148–1155.

50. Gregory S.H., Wing E.J., Danowski K.L., van Rooijen N., Dyer K.F., and Tweardy D.J. 1998. IL-6 produced by Kupffer cells induces STAT protein activation in hepatocytes early during the course of systemic listerial infections. J. Immunol. **160**: 6056–6061.

51. Mandel T.E., and Cheers C. 1980. Resistance and susceptibility of mice to bacterial infection: histopathology of listeriosis in resistant and susceptible strains. Infect. Immun. **30**: 851–861.

52. Arnold R., and Konig W., 1998. Interleukin-8 release from human neutrophils after phagocytosis of *Listeria monocytogenes* and *Yersinia enterocolitica*. J. Med. Microbiol. **47**: 55–62.
53. Conlan J.W., Dunn P.L., and North R.J. 1993. Leukocyte-mediated lysis of infected hepatocytes during listeriosis occurs in mice depleted of NK cells or CD4+ CD8+ Thy1.2+ T cells. Infect. Immun. **61**: 2703–2707.
54. Conlan J.W., and North R.J. 1994. Neutrophils are essential for early anti-*Listeria* defense in the liver, but not in the spleen or peritoneal cavity, as revealed by a granulocyte-depleting monoclonal antibody. J. Exp. Med. **179**: 259–268.
55. Rogers H.W., and Unanue E.R. 1993. Neutrophils are involved in acute, nonspecific resistance to *Listeria monocytogenes* in mice. Infect. Immun. **61**: 5090–5096.
56. Czuprynski C.J., Brown J.F., Maroushek N., Wagner R.D., and Steinberg H. 1994. Administration of anti-granulocyte mAb RB6-8C5 impairs the resistance of mice to *Listeria monocytogenes* infection. J. Immunol. **152**: 1836–1846.
57. Gekara N.O., Groebe L., Viegas N., and Weiss S. 2008. *Listeria monocytogenes* desensitizes immune cells to subsequent Ca^{2+} signaling via listeriolysin O-induced depletion of intracellular Ca^{2+} stores. Infect. Immun. **76**: 857–862.
58. Tsuchiya K., Kawamura I., Takahashi A., Nomura T., Kohda C., and Mitsuyama M. 2005. Listeriolysin O-induced membrane permeation mediates persistent interleukin-6 production in Caco-2 cells during *Listeria monocytogenes* infection in vitro. Infect. Immun. **73**: 3869–3877.
59. Ouadrhiri Y., Sibille Y., and Tulkens P.M. 1999. Modulation of intracellular growth of *Listeria monocytogenes* in human enterocyte Caco-2 cells by interferon-gamma and interleukin-6: role of nitric oxide and cooperation with antibiotics. J. Infect. Dis. **180**: 1195–1204.
60. Beckerman K.P., Rogers H.W., Corbett J.A., Schreiber R.D., McDaniel M.L., and Unanue E.R. 1993. Release of nitric oxide during the T cell-independent pathway of macrophage activation. Its role in resistance to *Listeria monocytogenes*. J. Immunol. **150**: 888–895.
61. Shiloh M.U., MacMicking J.D., Nicholson S., Brause J.E., Potter S., Marino M., Fang F., Dinauer M., and Nathan C. 1999. Phenotype of mice and macrophages deficient in both phagocyte oxidase and inducible nitric oxide synthase. Immunity. **10**: 29–38.
62. Rosen H., Gordon S., and North R.J. 1989. Exacerbation of murine listeriosis by a monoclonal antibody specific for the type 3 complement receptor of myelomonocytic cells. Absence of monocytes at infective foci allows *Listeria* to multiply in nonphagocytic cells. J. Exp. Med. **170**: 27–37.
63. Kurihara T., Warr G., Loy J., and Bravo R. 1997. Defects in macrophage recruitment and host defense in mice lacking the CCR2 chemokine receptor. J. Exp. Med. **186**: 1757–1762.
64. Miyamoto M., Emoto M., Emoto Y., Brinkmann V., Yoshizawa I., Seiler P., Aichele P., Kita E., and Kaufmann S.H. 2003. Neutrophilia in LFA-1-deficient mice confers resistance to listeriosis: possible contribution of granulocyte-colony-stimulating factor and IL-17. J. Immunol. **170**: 5228–5234.
65. Wu, H., Prince J.E., Brayton C.F., Shah C., Zeve D., Gregory S.H., Smith C.W., and Ballantyne C.M. 2003. Host resistance of CD18 knockout mice against systemic infection with *Listeria monocytogenes*. Infect. Immun. **71**: 5986–5993.

66. Tripp C.S., Wolf S.F., and Unanue E.R. 1993. Interleukin 12 and tumor necrosis factor alpha are costimulators of interferon gamma production by natural killer cells in severe combined immunodeficiency mice with listeriosis, and interleukin 10 is a physiologic antagonist. Proc. Natl. Acad. Sci. U S A. **90**: 3725–3729.
67. Tripp, C.S., Gately M.K., Hakimi J., Ling P., and Unanue E.R. 1994. Neutralization of IL-12 decreases resistance to *Listeria* in SCID and C.B-17 mice. Reversal by IFN-gamma. J. Immunol. **152**: 1883–1887.
68. Brombacher, F., Dorfmuller A., Magram J., Dai W.J., Kohler G., Wunderlin A., Palmer-Lehmann K., Gately M.K., and Alber G. 1999. IL-12 is dispensable for innate and adaptive immunity against low doses of *Listeria monocytogenes*. Int. Immunol. **11**: 325–332.
69. Buchmeier N.A., and Schreiber R.D. 1985. Requirement of endogenous interferon-gamma production for resolution of *Listeria monocytogenes* infection. Proc. Natl. Acad. Sci. U S A. **82**: 7404–7408.
70. Havell E.A. 1989. Evidence that tumor necrosis factor has an important role in antibacterial resistance. J. Immunol. **143**: 2894–2899.
71. Harty J.T., and Bevan M.J. 1995. Specific immunity to *Listeria monocytogenes* in the absence of IFN gamma. Immunity. **3**: 109–117.
72. Andersson A., Dai W.J., Di Santo J.P., and Brombacher F. 1998. Early IFN-gamma production and innate immunity during *Listeria monocytogenes* infection in the absence of NK cells. J. Immunol. **161**: 5600–5606.
73. Lertmemongkolchai G., Cai G., Hunter C.A., and Bancroft G.J. 2001. Bystander activation of CD8+ T cells contributes to the rapid production of IFN-gamma in response to bacterial pathogens. J. Immunol. **166**: 1097–1105.
74. Berg R.E., Crossley E., Murray S., and Forman J. 2005. Relative contributions of NK and CD8 T cells to IFN-gamma mediated innate immune protection against *Listeria monocytogenes*. J. Immunol. **175**: 1751–1757.
75. D'Orazio S.E., Troese M.J., and Starnbach M.N. 2006. Cytosolic localization of *Listeria monocytogenes* triggers an early IFN-gamma response by CD8+ T cells that correlates with innate resistance to infection. J. Immunol. **177**: 7146–7154.
76. Beatty G.L., and Paterson Y. 2001. Regulation of tumor growth by IFN-gamma in cancer immunotherapy. Immunol. Res. **24**: 201–210.
77. Dominiecki M.E., Beatty G.L., Pan Z.K., Neeson P., and Paterson Y. 2005. Tumor sensitivity to IFN-gamma is required for successful antigen-specific immunotherapy of a transplantable mouse tumor model for HPV-transformed tumors. Cancer Immunol. Immunother. **54**: 477–488.
78. Neighbors M., Xu X., Barrat F.J., Ruuls S.R., Churakova T., Debets R., Bazan J.F., Kastelein R.A., Abrams J.S., and O'Garra A. 2001. A critical role for interleukin 18 in primary and memory effector responses to *Listeria monocytogenes* that extends beyond its effects on interferon gamma production. J. Exp. Med. **194**: 343–354.
79. Tsuji N.M., Tsutsui H., Seki E., Kuida K., Okamura H., Nakanishi K., and Flavell R.A. 2004. Roles of caspase-1 in *Listeria* infection in mice. Int. Immunol. **16**: 335–343.
80. Cervantes J., Nagata T., Uchijima M., Shibata K., and Koide Y. 2008. Intracytosolic *Listeria monocytogenes* induces cell death through caspase-1 activation in murine macrophages. Cell Microbiol. **10**: 41–52.
81. McCaffrey R.L., Fawcett P., O'Riordan M., Lee K.D., Havell E.A., Brown P.O., and Portnoy D.A. 2004. A specific gene expression program triggered by

Gram-positive bacteria in the cytosol. Proc. Natl. Acad. Sci. U S A. **101**: 11386–11391.

82. Stetson D.B., and Medzhitov R. 2006. Type I interferons in host defense. Immunity. **25**: 373–381.
83. O'Connell R.M., Saha S.K., Vaidya S.A., Bruhn K.W., Miranda G.A., Zarnegar B., Perry A.K., Nguyen B.O., Lane T.F., Taniguchi T., Miller J.F., and Cheng G. 2004. Type I interferon production enhances susceptibility to *Listeria monocytogenes* infection. J. Exp. Med. **200**: 437–445.
84. Herskovits A.A., Auerbuch V., and Portnoy D.A. 2007. Bacterial ligands generated in a phagosome are targets of the cytosolic innate immune system. PLoS Pathog. **3**: e51.
85. Schmeck B., Beermann W., van Laak V., Zahlten J., Opitz B., Witzenrath M., Hocke A.C., Chakraborty T., Kracht M., Rosseau S., Suttorp N., and Hippenstiel S. 2006. *Listeria monocytogenes* activated p38 MAPK and induced IL-8 secretion in a nucleotide-binding oligomerization domain 1-dependent manner in endothelial cells. J. Immunol. **176**: 484–490.
86. Torres D., Barrier M., Bihl F., Quesniaux V.J., Maillet I., Akira S., Ryffel B., and Erard F. 2004. Toll-like receptor 2 is required for optimal control of *Listeria monocytogenes* infection. Infect. Immun. **72**: 2131–2139.
87. Way S.S., Thompson L.J., Lopes J.E., Hajjar A.M., Kollmann T.R., Freitag N.E., and Wilson C.B. 2004. Characterization of flagellin expression and its role in *Listeria monocytogenes* infection and immunity. Cell Microbiol. **6**: 235–242.
88. Chin A.I., Dempsey P.W., Bruhn K., Miller J.F., Xu Y., and Cheng G. 2002. Involvement of receptor-interacting protein 2 in innate and adaptive immune responses. Nature. **416**: 190–194.
89. Sha W.C., Liou H.C., Tuomanen E.I., and Baltimore D. 1995. Targeted disruption of the p50 subunit of NF-kappa B leads to multifocal defects in immune responses. Cell. **80**: 321–330.
90. Ballas Z.K., Rasmussen W.L., and Krieg A.M. 1996. Induction of NK activity in murine and human cells by CpG motifs in oligodeoxynucleotides and bacterial DNA. J. Immunol. **157**: 1840–1845.
91. Krieg A.M., and Vollmer J. 2007. Toll-like receptors 7, 8, and 9: linking innate immunity to autoimmunity. Immunol. Rev. **220**: 251–269.
92. Hemmi H., Takeuchi O., Kawai T., Kaisho T., Sato S., Sanjo H., Matsumoto M., Hoshino K., Wagner H., Takeda K., and Akira S. 2000. A Toll-like receptor recognizes bacterial DNA. Nature. **408**: 740–745.
93. Ishii K.J., Takeshita F., Gursel I., Gursel M., Conover J., Nussenzweig A., and Klinman D.M. 2002. Potential role of phosphatidylinositol 3 kinase, rather than DNA-dependent protein kinase, in CpG DNA-induced immune activation. J. Exp. Med. **196**: 269–274.
94. Tsujimura H., Tamura T., Kong H.J., Nishiyama A., Ishii K.J., Klinman D.M., and Ozato K. 2004. Toll-like receptor 9 signaling activates NF-kappaB through IFN regulatory factor-8/IFN consensus sequence binding protein in dendritic cells. J. Immunol. **172**: 6820–6827.
95. Ito S., Ishii K.J., Gursel M., Shirotra H., Ihata A., and Klinman D.M. 2005. CpG oligodeoxynucleotides enhance neonatal resistance to *Listeria* infection. J. Immunol. **174**: 777–782.

96. Ito S., Ishii K.J., Shirota H., and Klinman D.M. 2004. CpG oligodeoxynucleotides improve the survival of pregnant and fetal mice following *Listeria monocytogenes* infection. Infect. Immun. **72**: 3543–3548.

97. Finelli A., Kerksiek K.M., Allen S.E., Marshall N., Mercado R., Pilip I., Busch D.H., and Pamer E.G. 1999. MHC class I restricted T cell responses to *Listeria monocytogenes*, an intracellular bacterial pathogen. Immunol. Res. **19**: 211–223.

98. Villanueva M.S., Fischer P., Feen K., and Pamer E.G. 1994. Efficiency of MHC class I antigen processing: a quantitative analysis. Immunity. **1**: 479–489.

99. Schnupf P., Zhou J., Varshavsky A., and Portnoy D.A. 2007. Listeriolysin O secreted by *Listeria monocytogenes* into the host cell cytosol is degraded by the N-end rule pathway. Infect. Immun. **75**: 5135–5147.

100. Sijts A.J., Pilip I., and Pamer E.G. 1997. The *Listeria monocytogenes*-secreted p60 protein is an N-end rule substrate in the cytosol of infected cells. Implications for major histocompatibility complex class I antigen processing of bacterial proteins. J. Biol. Chem. **272**: 19261–19268.

101. Decatur A.L., and Portnoy D.A. 2000. A PEST-like sequence in listeriolysin O essential for *Listeria monocytogenes* pathogenicity. Science. **290**: 992–995.

102. Busch D.H., Pilip I.M., Vijh S., and Pamer E.G. 1998. Coordinate regulation of complex T cell populations responding to bacterial infection. Immunity. **8**: 353–362.

103. Badovinac V.P., and Harty J.T. 2000. Adaptive immunity and enhanced CD8+ T cell response to *Listeria monocytogenes* in the absence of perforin and IFN-gamma. J. Immunol. **164**: 6444–6452.

104. Badovinac V.P., and Harty J.T. 2007. Manipulating the rate of memory CD8+ T cell generation after acute infection. J. Immunol. **179**: 53–63.

105. Badovinac V.P., Porter B.B., and Harty J.T. 2002. Programmed contraction of CD8(+) T cells after infection. Nat. Immunol. **3**: 619–626.

106. Corbin G.A., and Harty J.T. 2004. Duration of infection and antigen display have minimal influence on the kinetics of the CD4+ T cell response to *Listeria monocytogenes* infection. J. Immunol. **173**: 5679–5687.

107. Mercado R., Vijh S., Allen S.E., Kerksiek K., Pilip I.M., and Pamer E.G. 2000. Early programming of T cell populations responding to bacterial infection. J. Immunol. **165**: 6833–6839.

108. Starks H., Bruhn K.W., Shen H., Barry R.A., Dubensky T.W., Brockstedt D., Hinrichs D.J., Higgins D.E., Miller J.F., Giedlin M., and Bouwer H.G. 2004. *Listeria monocytogenes* as a vaccine vector: virulence attenuation or existing antivector immunity does not diminish therapeutic efficacy. J. Immunol. **173**: 420–427.

109. Wong P., and Pamer E.G. 2001. Cutting edge: antigen-independent CD8 T cell proliferation. J. Immunol. **166**: 5864–5868.

110. Lauvau G., Vijh S., Kong P., Horng T., Kerksiek K., Serbina N., Tuma R.A., and Pamer E.G. 2001. Priming of memory but not effector CD8 T cells by a killed bacterial vaccine. Science. **294**: 1735–1739.

111. Datta S.K., Okamoto S., Hayashi T., Shin S.S., Mihajlov I., Fermin A., Guiney D.G., Fierer J., and Raz E. 2006. Vaccination with irradiated *Listeria* induces protective T cell immunity. Immunity. **25**: 143–152.

112. Mata M., and Paterson Y. 1999. Th1 T cell responses to HIV-1 Gag protein delivered by a *Listeria monocytogenes* vaccine are similar to those induced by endogenous listerial antigens. J. Immunol. **163**: 1449–1456.
113. Yamamoto K., Kawamura I., Tominaga T., Nomura T., Ito J., and Mitsuyama M. 2006. Listeriolysin O derived from *Listeria monocytogenes* inhibits the effector phase of an experimental allergic rhinitis induced by ovalbumin in mice. Clin. Exp. Immunol. **144**: 475–484.
114. Yamamoto K., Kawamura I., Tominaga T., Nomura T., Kohda C., Ito J., and Mitsuyama M. 2005. Listeriolysin O, a cytolysin derived from *Listeria monocytogenes*, inhibits generation of ovalbumin-specific Th2 immune response by skewing maturation of antigen-specific T cells into Th1 cells. Clin. Exp. Immunol. **142**: 268–274.
115. Janda J., Schöneberger P., Skoberne M., Messerle M., Rüssmann H., and Geginat G. 2002. Cross-presentation of *Listeria monocytogenes*-derived CD4 T cell epitopes. J. Immunol. **169**: 1410–1418.
116. Freeman M.M., and Ziegler H.K. 2005. Simultaneous Th1-type cytokine expression is a signature of peritoneal CD4+ lymphocytes responding to infection with *Listeria monocytogenes*. J. Immunol. **175**: 394–403.
117. Harty J.T., Schreiber R.D., and Bevan M.J. 1992. CD8 T cells can protect against an intracellular bacterium in an interferon gamma-independent fashion. Proc. Natl. Acad. Sci. U S A. **89**: 11612–11616.
118. Portnoy D.A., Schreiber R.D., Connelly P., and Tilney L.G. 1989. Gamma interferon limits access of *Listeria monocytogenes* to the macrophage cytoplasm. J. Exp. Med. **170**: 2141–2146.
119. Shedlock D.J., and Shen H. 2003. Requirement for CD4 T cell help in generating functional CD8 T cell memory. Science. **300**: 337–339.
120. Janssen E.M., Lemmens E.E., Wolfe T., Christen U., von Herrath M.G., and Schoenberger S.P. 2003. CD4+ T cells are required for secondary expansion and memory in CD8+ T lymphocytes. Nature. **421**: 852–856.
121. Sun J.C., and Bevan M.J. 2003. Defective CD8 T cell memory following acute infection without CD4 T cell help. Science. **300**: 339–342.
122. Bevan, M.J. 2004. Helping the CD8(+) T-cell response. Nat. Rev. Immunol. **4**: 595–602.
123. Stevenson P.G., Belz G.T., Altman J.D., and Doherty P.C. 1998. Virus-specific CD8(+) T cell numbers are maintained during gamma-herpesvirus reactivation in CD4-deficient mice. Proc. Natl. Acad. Sci. U S A. **95**: 15565–15570.
124. Zajac A.J., Blattman J.N., Murali-Krishna K., Sourdive D.J., Suresh M., Altman J.D., and Ahmed R. 1998. Viral immune evasion due to persistence of activated T cells without effector function. J. Exp. Med. **188**: 2205–2213.
125. Skeen M.J., and Ziegler H.K. 1993. Induction of murine peritoneal gamma/delta T cells and their role in resistance to bacterial infection. J. Exp. Med. **178**: 971–984.
126. Shum D.T., and Galsworthy S.B. 1979. Stimulation of monocyte precursors in vivo by an extract from *Listeria monocytogenes*. Can. J. Microbiol. **25**: 698–705.
127. Brzoza K.L., Rockel A.B., and Hiltbold E.M. 2004. Cytoplasmic entry of *Listeria monocytogenes* enhances dendritic cell maturation and T cell differentiation and function. J. Immunol. **173**: 2641–2651.

128. Bergmann B., Raffelsbauer D., Kuhn M., Goetz M., Hom S., and Goebel W. 2002. InlA- but not InlB-mediated internalization of *Listeria monocytogenes* by non-phagocytic mammalian cells needs the support of other internalins. Mol. Microbiol. **43**: 557–570.

129. Drevets D.A., Sawyer R.T., Potter T.A., and Campbell P.A. 1995. *Listeria monocytogenes* infects human endothelial cells by two distinct mechanisms. Infect. Immun. **63**: 4268–4276.

130. Parida S.K., Domann E., Rohde M., Müller S., Darji A., Hain T., Wehland J., and Chakraborty T. 1998. Internalin B is essential for adhesion and mediates the invasion of *Listeria monocytogenes* into human endothelial cells. Mol. Microbiol. **28**: 81–93.

131. Greiffenberg L., Sokolovic Z., Schnittler H.J., Spory A., Böckmann R., Goebel W., and Kuhn M. 1997. *Listeria monocytogenes*-infected human umbilical vein endothelial cells: internalin-independent invasion, intracellular growth, movement, and host cell responses. FEMS Microbiol. Lett. **157**: 163–170.

132. Krüll M., Nöst R., Hippenstiel S., Domann E., Chakraborty T., and Suttorp N. 1997. *Listeria monocytogenes* potently induces up-regulation of endothelial adhesion molecules and neutrophil adhesion to cultured human endothelial cells. J. Immunol. **159**: 1970–1976.

133. Drevets D.A. 1997. *Listeria monocytogenes* infection of cultured endothelial cells stimulates neutrophil adhesion and adhesion molecule expression. J. Immunol. **158**: 5305–5313.

134. Drevets, D.A. 1998. *Listeria monocytogenes* virulence factors that stimulate endothelial cells. Infect. Immun. **66**: 232–238.

135. Kayal S., Lilienbaum A., Poyart C., Memet S., Israel A., and Berche P. 1999. Listeriolysin O-dependent activation of endothelial cells during infection with *Listeria monocytogenes*: activation of NF-kappa B and upregulation of adhesion molecules and chemokines. Mol. Microbiol. **31**: 1709–1722.

136. Dramsi S., and Cossart P. 2003. Listeriolysin O-mediated calcium influx potentiates entry of *Listeria monocytogenes* into the human Hep-2 epithelial cell line. Infect. Immun. **71**: 3614–3618.

137. Repp H., Pamukçi Z., Koschinski A., Domann E., Darji A., Birringer J., Brockmeier D., Chakraborty T., and Dreyer F. 2002. Listeriolysin of *Listeria monocytogenes* forms Ca^{2+}-permeable pores leading to intracellular Ca^{2+} oscillations. Cell Microbiol. **4**: 483–491.

138. Gunn G.R., Zubair A., Peters C., Pan Z.K., Wu T.C., and Paterson Y. 2001. Two *Listeria monocytogenes* vaccine vectors that express different molecular forms of human papilloma virus-16 (HPV-16) E7 induce qualitatively different T cell immunity that correlates with their ability to induce regression of established tumors immortalized by HPV-16. J. Immunol. **167**: 6471–6479.

139. Seavey M.M., Pan Z.K., Maciag P.C., Wallecha A., Rivera S., Paterson Y., and Shahabi V. 2009. A novel human Her-2/neu chimeric molecule expressed by *Listeria monocytogenes* can elicit potent HLA-A2 restricted CD8-positive T cell responses and impact the growth and spread of Her-2/neu-positive breast tumors. Clin. Cancer Res. **15**: 924–932.

140. Singh R., Dominiecki M.E., Jaffee E.M., and Paterson Y. 2005. Fusion to listeriolysin O and delivery by *Listeria monocytogenes* enhances the immunogenicity of

HER-2/neu and reveals subdominant epitopes in the FVB/N mouse. J. Immunol. **175**: 3663–3673.

141. Maciag P.C., Seavey M.M., Pan Z.K., Ferrone S., and Paterson Y. 2008. Cancer immunotherapy targeting the high molecular weight melanoma-associated antigen protein results in a broad antitumor response and reduction of pericytes in the tumor vasculature. Cancer Res. **68**: 8066–8075.
142. Ikonomidis G., Paterson Y., Kos F.J., and Portnoy D.A. 1994. Delivery of a viral antigen to the class I processing and presentation pathway by *Listeria monocytogenes*. J. Exp. Med. **180**: 2209–2218.
143. Shahabi V., Reyes-Reyes M., Wallecha A., Rivera S., Paterson Y., and Maciag P. 2008. Development of a *Listeria monocytogenes* based vaccine against prostate cancer. Cancer Immunol. Immunother. **57**: 1301–1310.
144. Wallecha A., Maciag P.C., Rivera S., Paterson Y., and Shahabi V. 2009. Construction and characterization of an attenuated *Listeria monocytogenes* strain for clinical use in cancer immunotherapy. Clin. Vaccine Immunol. **16**: 96–103.
145. Pan Z.K., Ikonomidis G., Lazenby A., Pardoll D., and Paterson Y. 1995. A recombinant *Listeria monocytogenes* vaccine expressing a model tumour antigen protects mice against lethal tumour cell challenge and causes regression of established tumours. Nat. Med. **1**: 471–477.
146. Pan Z.K., Ikonomidis G., Pardoll D., and Paterson Y. 1995. Regression of established tumors in mice mediated by the oral administration of a recombinant *Listeria monocytogenes* vaccine. Cancer Res. **55**: 4776–4779.
147. Verch T., Pan Z.K., and Paterson Y. 2004. *Listeria monocytogenes*-based antibiotic resistance gene-free antigen delivery system applicable to other bacterial vectors and DNA vaccines. Infect. Immun. **72**: 6418–6425.
148. Lauer P., Chow M.Y., Loessner M.J., Portnoy D.A., and Calendar R. 2002. Construction, characterization, and use of two *Listeria monocytogenes* site-specific phage integration vectors. J. Bacteriol. **184**: 4177–4186.
149. Mata M., Yao Z.J., Zubair A., Syres K., and Paterson Y. 2001. Evaluation of a recombinant *Listeria monocytogenes* expressing an HIV protein that protects mice against viral challenge. Vaccine. **19**: 1435–1445.
150. Brockstedt D.G., Giedlin M.A., Leong M.L., Bahjat K.S., Gao Y., Luckett W., Liu W., Cook D.N., Portnoy D.A., and Dubensky T.W., Jr. 2004. *Listeria*-based cancer vaccines that segregate immunogenicity from toxicity. Proc. Natl. Acad. Sci. U S A. **101**: 13832–13837.
151. Brockstedt D.G., Bahjat K.S., Giedlin M.A., Liu W., Leong M., Luckett W., Gao Y., Schnupf P., Kapadia D., Castro G., Lim J.Y., Sampson-Johannes A., Herskovits A.A., Stassinopoulos A., Bouwer H.G., Hearst J.E., Portnoy D.A., Cook D.N., and Dubensky T.W., Jr. 2005. Killed but metabolically active microbes: a new vaccine paradigm for eliciting effector T-cell responses and protective immunity. Nat. Med. **11**: 853–860.
152. Sewell D.A., Shahabi V., Gunn G.R., 3rd, Pan Z.K., Dominiecki M.E., and Paterson Y. 2004. Recombinant *Listeria* vaccines containing PEST sequences are potent immune adjuvants for the tumor-associated antigen human papillomavirus-16 E7. Cancer Res. **64**: 8821–8825.
153. Lamikanra A., Pan Z.K., Isaacs S.N., Wu T.C., and Paterson Y. 2001. Regression of established human papillomavirus type 16 (HPV-16) immortalized tumors in

vivo by vaccinia viruses expressing different forms of HPV-16 E7 correlates with enhanced CD8(+) T-cell responses that home to the tumor site. J. Virol. **75**: 9654–9664.

154. Peng X., Treml J., and Paterson Y. 2007. Adjuvant properties of listeriolysin O protein in a DNA vaccination strategy. Cancer Immunol. Immunother. **56**: 797–806.

155. Neeson P., Pan Z.K., and Paterson Y. 2008. Listeriolysin O is an improved protein carrier for lymphoma immunoglobulin idiotype and provides systemic protection against 38C13 lymphoma. Cancer Immunol. Immunother. **57**: 493–505.

156. Nomura T., Kawamura I., Tsuchiya K., Kohda C., Baba H., Ito Y., Kimoto T., Watanabe I., and Mitsuyama M. 2002. Essential role of interleukin-12 (IL-12) and IL-18 for gamma interferon production induced by listeriolysin O in mouse spleen cells. Infect. Immun. **70**: 1049–1055.

157. Rose F., Zeller S.A., Chakraborty T., Domann E., Machleidt T., Kronke M., Seeger W., Grimminger F., and Sibelius U. 2001. Human endothelial cell activation and mediator release in response to *Listeria monocytogenes* virulence factors. Infect. Immun. **69**: 897–905.

158. Park J.M., Ng V.H., Maeda S., Rest R.F., and Karin M. 2004. Anthrolysin O and other gram-positive cytolysins are Toll-like receptor 4 agonists. J. Exp. Med. **200**: 1647–1655.

159. Tanabe Y., Xiong H., Nomura T., Arakawa M., and Mitsuyama M. 1999. Induction of protective T cells against *Listeria monocytogenes* in mice by immunization with a listeriolysin O-negative avirulent strain of bacteria and liposome-encapsulated listeriolysin O. Infect. Immun. **67**: 568–575.

160. Hamon M.A., Batsché E., Régnault B., Tham T.N., Seveau S., Muchardt C., and Cossart P. 2007. Histone modifications induced by a family of bacterial toxins. Proc. Natl. Acad. Sci. U S A. **104**: 13467–13472.

161. Tang P., Rosenshine I., Cossart P., and Finlay B.B. 1996. Listeriolysin O activates mitogen-activated protein kinase in eucaryotic cells. Infect. Immun. **64**: 2359–2361.

162. Singh R., and Paterson Y. 2007. In the FVB/N HER-2/neu transgenic mouse both peripheral and central tolerance limit the immune response targeting HER-2/neu induced by *Listeria monocytogenes*-based vaccines. Cancer Immunol. Immunother. **56**: 927–938.

163. Singh R., and Paterson Y. 2007. Immunoediting sculpts tumor epitopes during immunotherapy. Cancer Res. **67**: 1887–1892.

164. Bruhn K.W., Craft N., Nguyen B.D., Yip J., and Miller J.F. 2005. Characterization of anti-self CD8 T-cell responses stimulated by recombinant *Listeria monocytogenes* expressing the melanoma antigen TRP-2. Vaccine. **23**: 4263–4272.

165. Sewell D.A., Pan Z.K., and Paterson Y. 2008. *Listeria*-based HPV-16 E7 vaccines limit autochthonous tumor growth in a transgenic mouse model for HPV-16 transformed tumors. Vaccine. **26**: 5315–5320.

166. Souders N.C., Sewell D.A., Pan Z.K., Hussain S.F., Rodriguez A., Wallecha A., and Paterson Y. 2007. *Listeria*-based vaccines can overcome tolerance by expanding low avidity CD8+ T cells capable of eradicating a solid tumor in a transgenic mouse model of cancer. Cancer Immun. **7**: 2.

167. Liau L.M., Jensen E.R., Kremen T.J., Odesa S.K., Sykes S.N., Soung M.C., Miller J.F., and Bronstein J.M. 2002. Tumor immunity within the central nervous system stimulated by recombinant *Listeria monocytogenes* vaccination. Cancer Res. **62**: 2287–2293.

168. Souders N.C., Verch T., and Paterson Y. 2006. In vivo bactofection: listeria can function as a DNA-cancer vaccine. DNA Cell Biol. **25**: 142–151.

169. Jensen E.R., Selvakumar R., Shen H., Ahmed R., Wettstein F.O., and Miller J.F. 1997. Recombinant *Listeria monocytogenes* vaccination eliminates papillomavirus-induced tumors and prevents papilloma formation from viral DNA. J. Virol. **71**: 8467–8474.

170. Paglia P., Arioli I., Frahm N., Chakraborty T., Colombo M.P., and Guzmàn CA. 1997. The defined attenuated *Listeria monocytogenes* delta mp12 mutant is an effective oral vaccine carrier to trigger a long-lasting immune response against a mouse fibrosarcoma. Eur. J. Immunol. **27**: 1570–1575.

171. Kim S.H., Castro F., Gonzalez D., Maciag P.C., Paterson Y., and Gravekamp C. 2008. Mage-b vaccine delivered by recombinant *Listeria monocytogenes* is highly effective against breast cancer metastases. Br. J. Cancer. **99**: 741–749.

172. Seavey M.M., Maciag P.C., Al-Rawi N., Sewell D., and Paterson Y. 2009. An anti-vascular endothelial growth factor receptor 2/fetal liver kinase-1 *Listeria monocytogenes* anti-angiogenesis cancer vaccine for the treatment of primary and metastatic Her-2/neu+ breast tumors in a mouse model. J. Immunol. **182**: 5537–5546.

173. Rayevskaya M., Kushnir N., and Frankel F.R. 2002. Safety and immunogenicity in neonatal mice of a hyperattenuated *Listeria* vaccine directed against human immunodeficiency virus. J. Virol. **76**: 918–922.

174. Mustafa W., Maciag P.C., Pan Z.-K., Weaver J.R., Xiao Y., Isaacs S.N., and Paterson Y. 2009. *Listeria monocytogenes* delivery of HPV-16 major capsid protein, L1, induces systemic and mucosal cell-mediated $CD4^+$ and $CD8^+$ T cell responses after oral immunization. Viral Immunol. **22**: 195–204.

175. Soussi N., Saklani-Jusforgues H., Colle J.H., Milon G., Glaichenhaus N., and Goossens P.L. 2002. Effect of intragastric and intraperitoneal immunisation with attenuated and wild-type LACK-expressing *Listeria monocytogenes* on control of murine *Leishmania* major infection. Vaccine. 2002. **20**: 2702–2712.

176. Jia Q., Lee B.Y., Clemens D.L., Bowen R.A., and Horwitz M.A. 2009. Recombinant attenuated *Listeria monocytogenes* vaccine expressing *Francisella tularensis* IglC induces protection in mice against aerosolized Type A *F. tularensis*. Vaccine. **27**: 1216–1229.

177. Orr M.T., Orgun N.N., Wilson C.B., and Way S.S. 2007. Cutting edge: recombinant *Listeria monocytogenes* expressing a single immune-dominant peptide confers protective immunity to herpes simplex virus-1 infection. J. Immunol. **178**: 4731–4735.

178. Angelakopoulos H., Loock K., Sisul D.M., Jensen E.R., Miller J.F., and Hohmann E.L. 2002. Safety and shedding of an attenuated strain of *Listeria monocytogenes* with a deletion of actA/plcB in adult volunteers: a dose escalation study of oral inoculation. Infect. Immun. **70**: 3592–3601.

179. Thompson R.J., Bouwer H.G., Portnoy D.A., and Frankel F.R. 1998. Pathogenicity and immunogenicity of a *Listeria monocytogenes* strain that requires D-alanine for growth. Infect. Immun. **66**: 3552–3561.

180. Li Z., Zhao X., Higgins D.E., and Frankel F.R. 2005. Conditional lethality yields a new vaccine strain of *Listeria monocytogenes* for the induction of cell-mediated immunity. Infect. Immun. **73**: 5065–5073.

181. Hussain S.F., and Paterson Y. 2005. What is needed for effective antitumor immunotherapy? Lessons learned using *Listeria monocytogenes* as a live vector for HPV-associated tumors. Cancer Immunol. Immunother. **54**: 577–586.

182. Beatty G., and Paterson Y. 2001. IFN-gamma dependent inhibition of tumor angiogenesis by tumor infiltrating $CD4^+$ T cells requires tumor responsiveness to IFN-gamma. J. Immunol. **166**: 2276–2282.

183. Ozaki S., York-Jolley J., Kawamura H., and Berzofsky J.A. 1987. Cloned protein antigen-specific, Ia-restricted T cells with both helper and cytolytic activities: mechanisms of activation and killing. Cell Immunol. **105**: 301–316.

184. Yoshimura A., Shiku H., and Nakayama E. 1993. Rejection of an IA+ variant line of FBL-3 leukemia by cytotoxic T lymphocytes with CD4+ and CD4-CD8- T cell receptor-alpha beta phenotypes generated in CD8-depleted C57BL/6 mice. J. Immunol. **150**: 4900–4910.

185. Echchakir H., Bagot M., Dorothée G., Martinvalet D., Le Gouvello S., Boumsell L., Chouaib S., Bensussan A., and Mami-Chouaib F. 2000. Cutaneous T cell lymphoma reactive CD4+ cytotoxic T lymphocyte clones display a Th1 cytokine profile and use a fas-independent pathway for specific tumor cell lysis. J. Invest. Dermatol. **115**: 74–80.

186. Greenberg P.D. 1991. Adoptive T cell therapy of tumors: mechanisms operative in the recognition and elimination of tumor cells. Adv. Immunol. **49**: 281–355.

187. Liyanage U.K., Moore T.T., Joo H.G., Tanaka Y., Herrmann V., Doherty G., Drebin J.A., Strasberg S.M., Eberlein T.J., Goedegebuure P.S., and Linehan D.C. 2002. Prevalence of regulatory T cells is increased in peripheral blood and tumor microenvironment of patients with pancreas or breast adenocarcinoma. J. Immunol. **169**: 2756–2761.

188. Hussain S.F., and Paterson Y. 2004. CD4+CD25+ regulatory T cells that secrete TGFbeta and IL-10 are preferentially induced by a vaccine vector. J. Immunother. **27**: 339–346.

189. Nitcheu-Tefit J., Dai M.S., Critchley-Thorne R.J., Ramirez-Jimenez F., Xu M., Conchon S., Ferry N., Stauss H.J., and Vassaux G. 2007. Listeriolysin O expressed in a bacterial vaccine suppresses CD4+CD25 high regulatory T cell function in vivo. J. Immunol. **179**: 1532–1541.

190. Curtis M.M., and Way S.S. 2009. Interleukin-17 in host defence against bacterial, mycobacterial and fungal pathogens. Immunology. **126**: 177–185.

191. Xu L., Kitani A., Fuss I., and Strober W. 2007. Cutting edge: regulatory T cells induce CD4+CD25-Foxp3- T cells or are self-induced to become Th17 cells in the absence of exogenous TGF-beta. J. Immunol. **178**: 6725–6729.

192. Yu Y.A., Shabahang S., Timiryasova T.M., Zhang Q., Beltz R., Gentschev I., Goebel W., and Szalay A.A. 2004. Visualization of tumors and metastases in live animals

with bacteria and vaccinia virus encoding light-emitting proteins. Nat. Biotechnol. **22**: 313–320.

193. Yu Y.A., Zhang Q., and Szalay A.A. 2008. Establishment and characterization of conditions required for tumor colonization by intravenously delivered bacteria. Biotechnol. Bioeng. **100**: 567–578.

194. Huang B., Zhao J., Shen S., Li H., He K.L., Shen G.X., Mayer L., Unkeless J., Li D., Yuan Y., Zhang G.M., Xiong H., and Feng Z.H. 2007. *Listeria monocytogenes* promotes tumor growth via tumor cell toll-like receptor 2 signaling. Cancer Res. **67**: 4346–4352.

195. Maciag P.C., Radulovic S., and Rothman J. 2009. The first clinical use of a live-attenuated *Listeria monocytogenes* vaccine: a phase I safety study of Lm-LLO-E7 in patients with advanced carcinoma of the cervix. Vaccine. **27**: 3975–3983.

196. Moore D.H., Tian C., Monk B.J., Long H.J., Omura G.A., and Bloss J.D. 2010. Prognostic factors for response to cisplatin-based chemotherapy in advanced cervical carcinoma: a Gynecologic Oncology Group study. Gynecol. Oncol. **116**: 44–49.

197. Monk B.J., Sill M.W., Burger R.A., Gray H.J., Buekers T.E., and Roman L.D. 2009. Phase II trial of bevacizumab in the treatment of persistent or recurrent squamous cell carcinoma of the cervix: a Gynecologic Oncology Group study. J. Clin. Oncol. **27**: 1069–1074.

3

BACILLUS CALMETTE–GUERIN (BCG) FOR UROTHELIAL CARCINOMA OF THE BLADDER

Timothy P. Kresowik and Thomas S. Griffith
Department of Urology, University of Iowa, Iowa City, IA

THE BURDEN OF BLADDER CANCER AND BACILLUS CALMETTE–GUERIN (BCG) IMMUNOTHERAPY

Bladder cancer is one of the more prevalent cancers in the world. An estimated 357,000 patients were diagnosed with bladder cancer in 2002 worldwide, making it the ninth most commonly diagnosed malignancy (1). An estimated 145,000 deaths worldwide were attributed to bladder cancer in 2002, which ranks thirteenth among cancers by overall death rate. Bladder cancer is more common among men than women, with 77% of cases occurring in men worldwide. Smoking is the most common cause of bladder cancer and clear epidemiologic and mechanistic links have been identified. In a pooled analysis of several European case control studies of male smokers, 66% of the bladder cancer cases were directly attributed to a history of cigarette smoking (2).

Bladder cancer varies significantly in incidence throughout the world. Relative rates are high in the Middle East, especially Egypt, where chronic infection with *Schistosoma hematobium* leads to chronic inflammation and squamous cell carcinoma of the bladder (3). Industrialized countries with high

Emerging Cancer Therapy: Microbial Approaches and Biotechnological Tools, Edited by Arsénio M. Fialho and Ananda M. Chakrabarty

rates of cigarette smoking, such as most of Europe and the United States, also have high rates. In the United States, bladder cancer is the sixth most common cancer cause of death, and was estimated to be diagnosed in 68,810 people and caused 14,100 deaths in 2008 (4). In addition, bladder cancer is the most expensive malignancy to treat in the United States, in terms of cost per case from diagnosis to death. The direct costs of treatment are estimated at $3.7 billion per year (5). The cost is disproportionate to its incidence due to the frequent need for follow-up procedures, such as cystoscopy, and long-term treatment.

One of the most successful uses of bacteria for cancer therapy is the use of *Mycobacterium bovis* BCG to treat superficial urothelial carcinoma of the bladder. First used for that indication in 1976 (6), it has since become the most commonly used treatment for superficial urothelial carcinoma in the United States and in many other countries throughout the world.

DIAGNOSIS OF BLADDER CANCER

The most common presenting symptom of bladder cancer is hematuria. This can be macroscopic (gross blood in voided urine) or microscopic (red blood cells [RBC]) seen on urine microscopy or dipstick. Hematuria associated with bladder cancer is usually painless, although urinary symptoms, usually irritative symptoms such as frequency and urgency, can occur. Any patient with an episode of gross hematuria should be evaluated by a urologist, as rates of bladder cancer range from 13% to 34.5% (7). Patients with risk factors for bladder cancer (cigarette smoking, occupational exposure, older ages, voiding symptoms) and microhematuria should also have a urologic workup. Current American Urological Association (AUA) guidelines recommend urologic evaluation for asymptomatic patients with repeated microscopic hematuria (defined as >3 RBCs/hpf on 2 of 3 midstream urine collections).

Current evaluation for bladder cancer is by some combination of two components—cystoscopy and cytology. Cystoscopy (direct visual examination of the bladder) is most commonly done in an office setting with a flexible scope. The scope is inserted through the urethra and the bladder distended to show the entire wall. Any abnormal areas or lesions can be biopsied at that time with or without local anesthetic bladder instillation. Significantly abnormal areas or obvious tumors require transurethral resection of the bladder tumor (TURBT) done under general or spinal anesthesia. Figure 3.1 shows the cystoscopic appearance of urothelial carcinoma. Cytologic examination of urine, either voided or obtained through the cystoscope is a useful adjunct to cystoscopy. A positive cytologic examination is very specific for malignancy and accurately diagnoses carcinoma-*in-situ* (CIS)-flat tumors which can often be missed on cystoscopy alone.

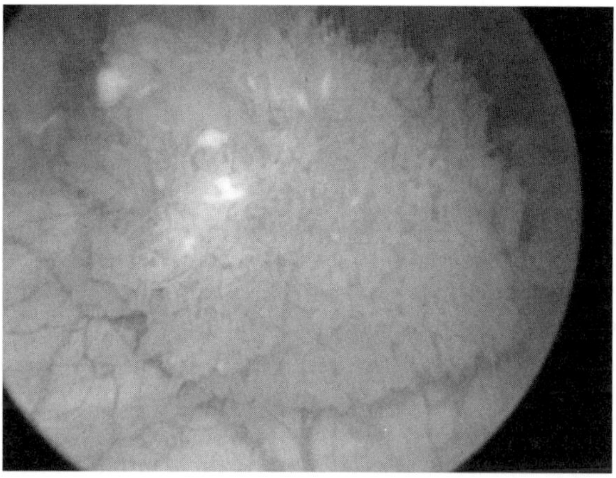

Figure 3.1 Cystoscopic appearance of urothelial carcinoma *in situ*.

STAGING OF BLADDER CANCER

Accurate staging is critically important to the successful management of bladder cancer, as optimal treatment varies based on the grade and depth of invasion of the tumor. Staging of bladder masses is generally accomplished by TURBT done under anesthesia with the goal of eradicating any visible tumor and providing sufficient tissue for pathologic diagnosis. With large masses or tumors that appear to be high grade, obtaining bladder muscle layers to determine the depth of invasion is critical. Random or targeted biopsies can be done on other areas in the bladder to determine whether any associated CIS is present, as this can influence optimal management.

Innovations in the diagnosis of bladder cancer include novel urinary screening markers such as bladder tumor antigen (BTA), nuclear matrix protein (NMP)-22, and fluorescence *in situ* hybridization (FISH) analysis for chromosomal abnormalities (8–11). These tests currently have their greatest utility for detecting recurrence as an adjunct to cytology, and optimal clinical uses are still being determined. Fluorescence cystoscopy, the procedure whereby fluorophores are instilled into the bladder that preferentially bind to neoplastic cells followed by cystoscopy with blue light, is another innovation (12). While this procedure is not in routine clinical use, some promise has been shown, especially for detecting CIS, which can be missed by standard cystoscopy (13).

Bladder cancer is staged based on the deepest layer of the bladder with the presence of malignant cells. The grade (degree of cellular abnormality) of the bladder tumor cells is critical, as high-grade lesions have a much higher risk of progressing to invasive disease than low grade. Knowing the approximate size of the largest tumor, number of tumors if multiple, and the presence of associated CIS with a papillary tumor is important for designing an ideal treatment regimen for an individual patient.

TREATMENT OF UROTHELIAL CARCINOMA

Following TURBT, there are several pathways for treatment. At initial diagnosis, 20–25% of urothelial carcinoma cases are muscle invasive (stage T2 or higher). Patients with muscle-invasive bladder cancer are optimally treated by radical cystectomy (bladder removal) and urinary diversion. The remaining patients have stage Ta (low-grade noninvasive), T1 (invasive into lamina propria), or CIS. These patients are generally treated by resection for diagnosis and local control with subsequent intravesical therapy for elimination of residual tumor and prevention of recurrence. In the United States, greater than 90% of bladder cancer is urothelial carcinoma. Other less common subtypes include squamous cell carcinoma and adenocarcinoma. BCG immunotherapy is used for the urothelial subtype and does not have demonstrated efficacy against the others. This chapter will focus primarily on the majority of patients with superficial (Ta, CIS, and T1) urothelial carcinoma. In the absence of contraindications, mitomycin C (MMC) (DNA-cross-linking chemotherapeutic agent) is instilled into the bladder following TURBT with resection of all visible tumor. For low-risk patients, periodic surveillance consisting only of cystoscopy is appropriate. For patients with higher risk disease, by comparison, treatment by intravesical immunotherapy with BCG is the recommended treatment. Several organizations have published guidelines for risk stratification and appropriate treatment of non-muscle-invasive bladder cancer (14, 15). These are summarized in Table 3.1. The European Association of Urology (EAU) guidelines reflect the European Organization for the Research and Treatment of Cancer (EORTC) risk tables for recurrence and progression, while the AUA guidelines are focused on specific patient situations.

HISTORY OF *M. BOVIS* BCG

M. bovis BCG is a live, attenuated mycobacterium most commonly used in oncology for the treatment of superficial urothelial carcinoma of the bladder. BCG was initially developed as a vaccine against tuberculosis (TB). TB is caused primarily by *Mycobacterium tuberculosis*, but other *Mycobacterium* sp. are pathogenic as well. TB is a common and often deadly infectious disease with a long latency, which affects the lungs as well as many other organs. Although TB has been present with humans since antiquity, awareness increased in nineteenth-century Europe. The squalid living conditions of the Industrial Revolution led to an increase in prevalence and diagnosis of the disease. The development of diagnostic instruments, such as the stethoscope, led to increased recognition and attempts at treatment.

TABLE 3.1. Risk Stratification and Treatment Guidelines

Patient	EAU	AUA
Single <3 cm low-grade Ta (low risk)	One immediate instillation of chemotherapy is strongly recommended as the complete adjuvant treatment.	Recommendation. MMC post-op
Multifocal, >3 cm, or recurrent low-grade Ta (intermediate risk)	Further instillations of chemotherapy or a minimum of 1 year of BCG.	Recommendation: induction BCG or MMC Option: maintenance BCG or MMC
High-grade Ta, T1, >3 cm tumors, or CIS (high risk)	>1 year BCG treatment. Consider cystectomy for highest-risk patients.	Recommendation: induction and maintenance BCG Option: cystectomy for selected patients
Recurrent high grade or CIS	High-risk group—>1 year BCG treatment. Consider cystectomy for highest-risk patients.	Recommendation: consider cystectomy for selected patients Option: further intravesical therapy (BCG/IFN-α, docetaxel/doxorubicin)

Following the discovery by Edward Jenner in 1796 that cowpox could immunize against smallpox, researchers attempted to translate that work to TB. Robert Koch identified mycobacteria as the causative agent of TB in 1882, and his work was immediately recognized as a breakthrough. His postulates laid the foundation for the explosion of research into infectious disease. Albert Calmette and Camille Guerin, working in Lille, France at the Pasteur Institute, provided the next breakthrough. They applied the principles of the successful smallpox vaccine by attempting to immunize patients with an animal pathogen. While direct immunization with isolated animal *Mycobacterium* was pathogenic, they continued to work on a cure.

Calmette and Guerin were working on a virulent *M. bovis* isolated from an infected cow by Nocard (16). They transferred the bacteria 231 times over 13 years in a medium of cow bile, potatoes, and glycerin. Experiments in cattle and guinea pigs demonstrated that a stable but nonvirulent form of the bacteria had developed (17). In 1921, they conducted their first experiments in humans, successfully immunizing Parisian children against *M. tuberculosis*. Subsequent experiments were published in 1924, and cultures were distributed around the world to at least 60 countries in 1924–1927. Enthusiasm for the BCG vaccine was quickly dampened by the Lubeck incident in 1930, where a laboratory error resulted in the inoculation of children with virulent mycobacteria rather than the attenuated BCG vaccine. Routine vaccination with BCG for TB did not become common practice until after World War II, although use was never widespread in some countries, including the United States. Over 3 billion individuals have been vaccinated, and the vaccine has proved safe and effective over time. It is very effective for the prevention of extrapulmonary TB in children and is administered at an early age. Its efficacy against adult pulmonary TB is mixed, and numerous factors affect the risk of acquiring this disease.

While studies exploring the oncologic efficacy of bacterial products were done starting in the late eighteenth century by Coley, the first landmark studies on BCG for cancer were not done until the 1950s. Intravenous injection of BCG into mice led to regression of several transplanted tumors (18). Mice that received BCG and the tumor at the same time showed no protection, while mice that received tumor implants a week after vaccination rejected their tumors. Encouraging early reports of therapy with combinations of BCG and chemotherapy for leukemia and intralesional injection for melanoma were published in 1969–1970. Further clinical studies did not bear out these results. Investigators found delayed hypersensitivity-type reactions when tumor implantation was attempted at the sites of previous BCG infection.

The first published report of intravesical use for bladder cancer was in 1976 by Alvaro Morales, a Canadian urologist (6). Morales' protocol involved the administration of one vial of the Montreal BCG strain at weekly intervals through a urethral catheter for six doses. This dosing interval was chosen based on the knowledge that delayed-type hypersensitivity reactions required 3 weeks or more to develop. In addition, the vials came in packs of six (17). An additional part of the protocol was intradermal BCG vaccination. Based on promising results in seven patients with recurrent bladder cancer despite resection and intravesical chemotherapy, two larger clinical trials were designed. These studies demonstrated the efficacy of a course of BCG following resection when compared to a control group of TURBT alone, and BCG became a major part of the therapeutic armamentarium for non-muscle-invasive urothelial carcinoma. Landmarks in the development of BCG for bladder cancer are summarized in Figure 3.2.

GENETIC VARIABILITY OF BCG

BCG has shown variable efficacy as a vaccine against TB. As BCG became used for vaccination against TB, the original BCG strain was distributed to other locations. Until the development of lyophilized forms for inert storage in the 1960s, these daughter strains were all transferred repeatedly and evolved in isolation, often in divergent conditions. BCG vaccination for the prevention of TB has shown variable results in different countries. Multiple explanations have been proposed for these differences, one of which is the regional differences between strains.

Landmarks in BCG development for urothelial carcinoma

- 1882: Robert Koch identifies *Mycobacteria* as the causative agent of tuberculosis.
- 1890s: William Coley's experiments with killed bacteria set the stage for modern cancer immunotherapy.
- 1921: Albert Calmette and Camille Guerin sequentially passage *Mycobacterium bovis* and successfully immunize children against tuberculosis.
- 1950s–1960s: Preclinical research by Old and others leads to administration of BCG for leukemia and melanoma.
- 1976: Alberto Morales publishes the first series of bladder cancer patients treated with intravesical BCG.
- 1980s–2000s: Subsequent randomized trials confirm BCG efficacy, need for maintenance BCG, and superiority over MMC for high-risk superficial bladder cancer and CIS.

Figure 3.2 Landmarks in the development of BCG as an immunotherapeutic agent for bladder cancer.

Regions of difference (RD) between the pathogenic strains of *Mycobacteria* and BCG have been identified by comparative genomic analysis. Deletion of RD1 in BCG may be the initial mutation that rendered the strain nonpathogenic in normal humans, as deletion of RD1 in pathogenic *Mycobacteria* leads to attenuation resembling that of BCG (19). Modern analysis has revealed significant genetic differences between strains of BCG related to time of initial dissemination from the Pasteur Institute (20). Clinical differences between strains administered to humans for protection against TB have been described in some studies, while others have not shown differences (21). It is clear that differences exist between BCG strains that may have clinical implications, but given the many different strains used and the difficulty of comparing the results between strains, no strong evidence exists for the use of one individual strain versus another. Studies comparing strains of BCG for use against bladder cancer have not shown significant differences between the commonly used strains, although few directly comparative trials have been performed (22). While differences between strains may exist, there is insufficient evidence to support the use of a particular variety of BCG as being more effective for the treatment of bladder cancer.

CLINICAL USE OF BCG

Currently, BCG is supplied as a lyophilized powder of live mycobacteria by several manufacturers for intravesical use. It is rehydrated immediately before use, and administered through a urethral catheter directly into an emptied

Fundamental principles of BCG therapy
• All visible tumor must be resected • No efficacy for deeply invasive tumors • Intact immune system needed as multiple different components play a role in response • Multiple instillations required for therapeutic effect • Improved results with periodic maintenance therapy • Side effects can lead to problems with treatment compliance • Induction of immune response correlates with success

Figure 3.3 Fundamental concepts relating to the use of BCG as an immunotherapeutic agent for superficial bladder cancer.

bladder. The catheter is then removed, leaving the live bacteria in the bladder. After holding the fluid inside the bladder for as long as the patient can tolerate (but less than 2 h), the patient voids to complete the treatment. It is important that antibiotics not be administered concurrently with BCG as they can reduce the efficacy.

Several important studies have confirmed the efficacy of BCG in various types of superficial bladder cancer. Meta-analysis of several trials comparing BCG to alternatives, most commonly MMC, have shown superiority in patients with high-risk tumors and CIS (23). The most commonly used dosing schedule for BCG consists of 6 weekly administrations of BCG, beginning 2–6 weeks following initial tumor resection. This is referred to as BCG induction. Patients with CIS or high-risk superficial tumors benefit from additional courses of BCG immunotherapy, referred to as maintenance treatment. Meta-analysis of several large controlled trials has shown that maintenance treatment is superior to induction-only for high-risk urothelial carcinoma and CIS (22). In addition, the benefit of BCG in preventing bladder cancer progression appears to only be present with maintenance therapy. Various permutations of maintenance courses have been used, mostly consisting of cycles of 3 weekly instillations at 3- to 6-month intervals. Currently at the University of Iowa, patients receive three cycles of BCG maintenance at 6-month intervals with surveillance at 3-month intervals starting 6 weeks after the previous maintenance cycle. Patients are monitored by cystoscopy, conventional cytology, and FISH analysis. Fundamental concepts relating to the use of BCG are summarized in Figure 3.3.

BCG SIDE EFFECTS

BCG immunotherapy for bladder cancer is associated with a variety of complications, ranging from minor cystitis to life-threatening BCG sepsis. Contraindications to BCG instillation include immunosuppression, history of

BCG sepsis, gross hematuria or traumatic catheterization, and active urinary tract infection. Side effects of intravesical BCG immunotherapy are common and affect treatment in up to 30% of patients (24). Mild cystitis with urgency, frequency, malaise, and low-grade fever can occur in up to 90% of patients (25). More severe reactions are uncommon (<5%), but can include mycobacterial sepsis that requires long-term treatment with anti-TB medications and can lead to death (26).

Supportive therapy with analgesics, such as acetaminophen and anticholinergic medications, is adequate for patients with minor reactions to BCG immunotherapy. Risk factors for more severe reactions to BCG instillation include immunosuppressed status and gross hematuria. Patients who are immunosuppressed may have difficulty mounting a sufficient immune response even against the attenuated BCG strain, whereas the presence of gross hematuria indicates a breakdown in the urothelial lining that may predispose patients to bacterial dissemination and systemic reactions. Patients who develop prolonged fever or severe side effects require urine culture, further evaluation, and possibly treatment with rifampin and isoniazid.

Strategies have evolved to preserve the antineoplastic efficacy of BCG while minimizing the side effects that cause many patients to modify or discontinue their treatment course. Dose reduction to 50% or 33% of the standard dose has been investigated as a way to reduce side effects while maintaining effect. Based on several small controlled trials, a one-third dose was as effective as the standard dose for low and intermediate risk tumors although the data are mixed for higher-risk (CIS, T1 high grade) tumors (27). Further dose reduction to a one-sixth dose was not beneficial for decreasing side effects, but it did decrease clinical efficacy (28).

Antibiotics can affect the viability of BCG, and for this reason, antibiotic prophylaxis is not generally administered prior to instillation. BCG is especially susceptible to fluoroquinolones (29). Live BCG is thought to be required for optimal therapeutic effect; however, if a patient develops a severe reaction, administration of a fluoroquinolone or other antibiotic can be helpful.

One of the debated issues in the BCG literature is whether side effects are an inevitable part of clinical effectiveness. The mechanism of action is the massive release of potent inflammatory compounds, and minor side effects/cystitis may be a "necessary evil" required for effect. Several studies have looked at whether the side effects of urinary cytokines are associated with therapeutic success, and most have found that pyuria and increased cytokine levels are indeed associated with decreased tumor recurrence rates (30, 31). Several authors have also demonstrated a clear association between local side effects and improved tumor response; however, the cause and effect relationship of this association is disputed (24).

Investigation into ways to reduce side effects is essential, as a significant portion of patients modify their treatment course because of BCG cystitis. However, the prevention of tumor recurrence and progression is the overarching goal, and significant evidence associates cystitis, pyuria, and cytokine release with improved therapeutic effect. The more serious side effects associated with systemic spread are not associated with efficacy, and reduction of these side effects is not only desirable but is also being actively pursued by a number of strategies.

BCG MECHANISM OF ACTION

BCG works by inducing an immune reaction in the bladder that has tumoricidal activity. While the exact mechanism of effect on bladder cancer remains unknown, many different cytokines and immune cells have been implicated in the response. It is clear that the response of bladder to BCG is multifactorial and is dependent on a functional immune system. It has also been clearly established that the mechanism of action is related to the massive release of cytokines from infiltrating inflammatory cells induced by BCG. The specific pathways by which BCG exerts its effects have not been as clearly defined, and there is certainly some overlap between "BCG effects" and the effects of urinary tract infection or cystitis.

Several insights from basic science research are applicable to clinical administration of BCG. An intact immune system is required for successful BCG therapy, and depletion of several immune components seems to have a negative effect on BCG efficacy. Depletion of CD4 or CD8 T cells in murine models abrogates BCG effectiveness, suggesting both cell types are necessary for optimal responses (32). Other authors have reported that natural killer (NK) cells and neutrophils are also essential components of the immune response to BCG, and the depletion of either of these populations in murine models eliminates BCG effectiveness (33, 34). The monocyte/macrophage lineage also plays a significant role and may be required for induction of NK responses (35).

BCG requires urine contact with tumor cells for therapeutic effect and does not have substantial efficacy against deeply invasive diseases. Original studies demonstrated that live BCG was required for the therapeutic effect (36). Subsequently, this observation has been challenged by murine models, which suggest that the initial instillation with live BCG is the critical step and that subsequent instillations with killed BCG also induce immune responses (37). While BCG has direct effects on some cell lines grown *in vitro* (38, 39), direct cytostatic effect is not a major contributor to the overall effectiveness of the therapy.

BCG binds to the bladder lining by fibronectin attachment to FAP (fibronectin attachment protein) on BCG (40). BCG is internalized by both urothelial cells and inflammatory cells and triggers a massive cascade of cytokine release and immune cell recruitment. Few cells infiltrate the bladder lumen 2h after BCG instillation, but urine analyzed by flow cytometry at 4h shows greatly increased levels of neutrophils and other inflammatory cells (41). Studies show that BCG persists in the bladder for several days following instillation, with 67.9% of voided urine samples containing BCG by culture at 24 h and 27.1% were positive at 1 week (42). Interestingly, some patients have BCG-positive cultures up to 16.5 months following intravesical instillation (43), suggesting that live bacteria persist and can lead to continuously elevated levels of inflammatory cytokines.

Other inflammatory cells also respond to BCG treatments and have been implicated in tumor responses. Monocyte/macrophages are the second most common cell found in the urine of patients treated with intravesical BCG and are a common immune cell found in the bladder wall. Granuloma formation is common after instillation, mediated by mononuclear lymphocytes as well as macrophages. Granuloma formation is an event more specific to BCG than generalized bladder infection or inflammation. However, most studies that examined bladder biopsies or clinical appearance of the bladder after BCG therapy have not shown a correlation with recurrence or progression risk (44).

Numerous studies have examined the cytokine environment in the bladder after BCG instillation. Elevated levels of many cytokines have been seen in the bladder; yet, establishing any correlation with outcome and finding cytokines specific to BCG rather than generalized inflammation has been challenging. Cytokines reported to be increased in the bladders of patients receiving BCG immunotherapy include interleukin (IL)-2, IL-6, IL-8, IL-10, IL-12, tumor necrosis factor (TNF), interferon (IFN)-α, and IFN-γ, among others (45).

The immune response to BCG is amplified with repeated instillations. Significantly increased levels of inflammatory cells and cytokines are seen after later instillations compared to earlier instillations. Levels of granulocytes 4 h after the third weekly instillation increased by 203 times compared to before instillation treatment, with other cells showing significant infiltration into the urine as well (41). This suggests the existence of a memory response, and provides a rationale for maintenance BCG treatment, as the immune response generated by successive BCG instillation dwarfs that of the initial instillation.

The most prominent and best-studied cytokines include IL-2, IL-8, and the IFNs. IFN-γ is a potent effector cytokine associated with clinical responses to BCG. IL-8 is a proinflammatory chemokine that is one of the earliest detected after BCG treatment. IL-8 is chemotactic for both neutrophils and lymphocytes. Lower IL-8 levels 4 h after BCG treatment have predictive value for recurrence (46). IL-2 levels in both the blood and urine have also been associated with responsiveness to BCG (47, 48). IL-2 serves as a potent stimulator of T cell proliferation, as well as NK cells. It has diverse effects on the immune system, but in many cases is thought to serve as an activator of immunity against malignant cells. It is also used as an agent against renal cell carcinoma and has been investigated in other cancers.

The concerted efforts of several immune system components work to create a response to BCG instillation that serves to eliminate small foci of residual

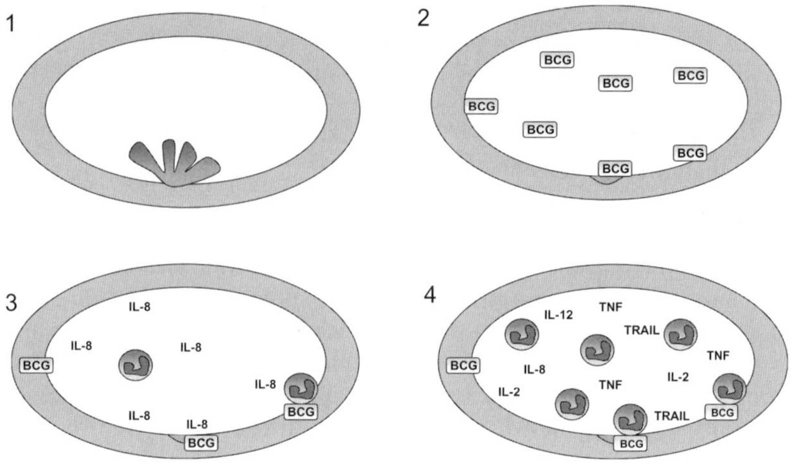

1. Superficial bladder tumor
2. Following resection of visible bladder tumor, BCG is instilled into the bladder, binds to and is internalized by both normal and malignant cells
3. BCG binding and internalization leads to the early cytokine response of IL-8 and others
4. Neutrophils respond, amplify the immune response, and secrete TRAIL

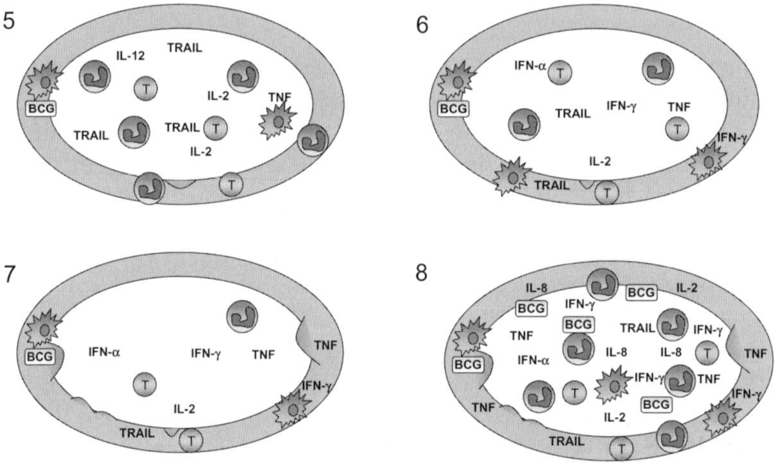

5. Macrophages and lymphocytes enter the bladder lumen and wall in response to cytokines
6. Residual tumor cell death from TRAIL, cytotoxic lymphocytes/NK cells, and others
7. Granuloma formation and cystitis in the bladder are sequelae of the massive immune response
8. When BCG is reintroduced it serves as a trigger for the inflammatory cells that are in place and the reaction is increased

Figure 3.4 Anti-tumor immune response in the bladder elicited by BCG immunotherapy. (1) A patient comes into the clinic with superficial bladder cancer, and (2) following resection of the visible tumor, intravesical BCG immunotherapy commences. The BCG binds to and is internalized by both normal urothelium and any remaining malignant cells. (3) The binding and internalization of BCG leads to an early proinflammatory cytokine response, with IL-8 being among the cytokines produced. (4) IL-8 initially recruits neutrophils into the bladder, which amplify the immune response via release of multiple molecules. In addition, neutrophil recognition of BCG cell wall components induces TRAIL release. (5) A second wave of immune cells enters the bladder, as lymphocytes and macrophages respond to the proinflammatory cytokines and chemokines present in the bladder, further amplifying the immune response. (6) The presence of soluble TRAIL, cytotoxic lymphocytes, and NK cells in the bladder lumen helps to induce residual tumor cell death. (7) Granuloma formation and cystitis are sequelae of the massive immune response induced by BCG instillation. (8) Subsequent BCG instillations trigger a more vigorous immune response within the bladder.

tumor cells and, therefore, prevent recurrence. Following BCG instillation, early cytokines such as IL-8 are released, which lead to a massive influx of neutrophils. Cytokines released from neutrophils attract lymphocytes and monocyte/macrophages, in addition to being cytotoxic to tumor cells. This initial response is necessary for the success of therapy, and the magnitude of the initial response measured by pyuria and cytokine release may correlate with therapeutic effect.

Lymphocytes, macrophages, and NK cells may play a role in bladder tumor eradication and immunity. Subsequent instillations lead to an amplification of this response and the presence of inflammatory cells and granulomas in the bladder wall. Many of the components of the BCG response are typical of any bladder insult or infection, but basic science and clinical research has helped elucidate which are required for BCG effect and which may be nonspecific components. A visual diagram of the bladder response to BCG is provided as Figure 3.4.

NEUTROPHILS, TRAIL, AND BCG

Neutrophils are central mediators of innate immunity and recognize bacterial antigens by several mechanisms. Release of inflammatory mediators by urothelial cells and responding inflammatory cells leads to immune cell recruitment. Neutrophils are the primary cells present in the urine following BCG instillation (comprising >75% of the cells present in the urine after BCG intravesical instillation) (49), and were originally viewed as nonspecific activators of the inflammatory cascade. However, elevated levels of neutrophils and mononuclear cells are seen in the bladder wall following BCG instillation. In a mouse model of urothelial carcinoma, neutrophils are essential for antitumor responses and their depletion leads to a reduced BCG responsiveness (34). More recent data demonstrate that neutrophils are the primary producers of TNF-related apoptosis-inducing ligand (TRAIL) after BCG stimulation, which suggest a role for neutrophils in the therapeutic effectiveness of BCG in urothelial carcinoma. TRAIL is a member of the TNF family that induces apoptosis in neoplastic cells, but not in normal cells and tissues (50). TRAIL expression can be induced in many inflammatory cell populations following activation or stimulation with cytokines, especially type I and II IFN (51). Patients who respond to BCG have significantly higher amount of urinary TRAIL than BCG nonresponders (52). In addition, responders to therapy had increasing amounts of TRAIL in the urine with each subsequent BCG treatment, whereas nonresponders had lower amounts of TRAIL for all treatments.

Neutrophils are the primary cells responsible for TRAIL secretion with BCG stimulation. Supernatants of BCG-stimulated neutrophils were able to kill bladder cancer cell lines *in vitro* in a TRAIL-dependent process (52). Analysis of urine from patients with traditional (*Escherichia coli*-based) urinary tract infections found low TRAIL levels when compared to BCG stimulation, suggesting that the TRAIL response is specific to BCG rather than nonspecific immune activation. Stimulation of peripheral blood neutrophils with various bacterial pathogens showed that BCG was the only bacterial species of those tested able to stimulate TRAIL release (53). Thus, TRAIL appears to be a contributor to the antineoplastic effect seen with BCG, and strategies aimed at increasing TRAIL release may have clinical effectiveness.

GENETIC DETERMINANTS OF THE BCG RESPONSE

While attention has focused at genetic differences between BCG strains, some investigators have studied genetic polymorphisms within humans that may affect certain individuals' ability to respond to BCG. Knowledge of this kind could better target BCG immunotherapy to those most likely to benefit from the therapy, while identifying alternatives for those with a higher potential risk for BCG failure. This is an extension of the idea that urinary cytokine levels can be used to predict which patients will respond to or fail BCG treatment. Several studies have identified increased urinary cytokine levels, such as IL-2 and TNF, as predictors of response to BCG; however, none have been translated into routine clinical use (44).

Analysis of variations in genes coding for cytokines have led to some interesting conclusions about the identification of BCG responders. Single nucleotide polymorphisms (SNPs) are one method of variation that can lead to differences in cytokine production or function between individuals. Analysis of several cytokines known to be induced by BCG showed that SNPs present in genes coding for production or regulation of IL-10, transforming growth factor (TGF)-β, and IL-4 were associated with disease progression despite BCG therapy (54). Polymorphism in *NRAMP1*, which has been implicated in human and murine responses to mycobacteria, was found in 2–3% of normal controls and in 12% of patients at high risk of recurrence (55). All of the patients with urothelial carcinoma and this polymorphism developed disease recurrence despite BCG therapy. Several other polymorphisms have been associated with low risk of recurrence after BCG therapy (56). Continued investigation into genetic determinants of host response to BCG could provide additional insights into the mechanism of BCG action. Further studies are needed in this area before translation into routine clinical use.

GENETIC MODIFICATION OF BCG

One strategy proposed to increase the potency of BCG immunotherapy or reduce the incidence of side effects is the modification of BCG. Given the known clinical effectiveness of BCG combined with IFN-α, the creation of recombinant BCG expressing IFN may be effective. Preclinical work with recombinant BCG expressing IFN-α has shown some success (57), where peripheral blood mononuclear cells (PBMC) treated *in vitro* with recombinant BCG-IFN-α had increased cytotoxic activity against bladder tumor cell lines and secreted more IFN-γ and IL-2 than controls.

Several investigators have modified BCG to secrete other cytokines important in BCG effect. Macrophages treated with BCG expressing IL-2 exhibited improved killing against a bladder tumor cell line (58). IL-18 has also been studied as a synergistic cytokine with BCG, and BCG expressing IL-18 has shown effect in murine cell lines (59). Recombinant BCG expressing IFN-γ has also been created and used in murine systems with some success (60). Other researchers have used recombinant BCG expressing pertussis toxin to further stimulate the immune system (61). This led to increased TNF expression and reduced tumor growth in a mouse model. Despite results in murine and *in vitro* human models, no human clinical trials have been published using recombinant BCG. While genetic modification of BCG has some promise to improve therapy, it remains experimental.

CELLULAR COMPONENT THERAPY

Cellular component therapy is another promising approach to reduce BCG toxicity and improve efficacy. Several mycobacterial components have been shown to play specific roles in the immune reaction to BCG, and administration of these components in the absence of live bacteria may avoid the most devastating side effects of BCG such as sepsis. The mycobacterial cell wall has been the best-characterized component. In studies performed in our laboratory, mycobacterial cell wall components were responsible for stimulating the majority of TRAIL released from neutrophils (62).

Clinical trials of mycobacterial cell wall extract given intravesically with a similar treatment schedule to BCG showed that the extract had a favorable safety profile with mild side effects comparable to BCG, but no serious side effects. In contrast to viable BCG, the cell wall extract can be given to patients with a disrupted urothelium and gross hematuria. This would enable the compound to be administered immediately after TURBT rather than waiting several weeks, as is the current practice. In addition, intravesical administration of the mycobacterial cell wall complex was able to rescue several BCG nonresponders (63). However, these clinical studies have not been controlled with a traditional BCG immunotherapy arm preventing a direct comparison of effectiveness. Analysis of the cell wall extract found it contained mycobacterial DNA, and has shown efficacy at inducing apoptosis, as well as leading to clinical responses in *in vitro* and *in vivo* models (64). Clinical trials in patients previously treated with BCG demonstrated some complete responses, dose responsiveness, and a favorable side-effect profile (65).

Arguments against the cellular component therapy as a replacement for BCG include evidence linking side effects to therapeutic effect. This suggests that while the massive influx of cytokines and inflammatory cells to the bladder may cause cystitis, these are also the engines of destruction for tumor cells. Earlier studies had shown that live BCG is necessary for therapeutic effect. These studies were comparing live BCG to killed BCG rather than cellular component therapy. Subsequent animal studies, however, have questioned the need for continuous live therapy and have suggested that induction with live BCG and maintenance with nonviable BCG is comparable to treatment with live BCG (37). While component therapy is a promising investigational approach to improving BCG therapy, additional clinical trials are needed before widespread adoption.

COMBINATION WITH CYTOKINES

Because of the elevated levels of several cytokines seen in the bladders of patients treated with BCG, administration of presumed effector cytokines alone has been investigated as a way to reduce side-effect risk from the instillation of viable bacteria. Cytokines such as IL-2, IL-12, and IFN-α lead to increased release of IFN-γ from human peripheral blood mononuclear cells alone or in combination with BCG (66). Experiments showing responses with cytokines only provided the rationale for clinical examination of cytokines alone or in combination with BCG for treatment of urothelial carcinoma. However, most clinical trials of single cytokine therapy have not shown improved efficacy (67, 68). The complicated immunologic cascade of BCG may be difficult or impossible to replicate with cytokine therapy.

Combinations of BCG and cytokines have been more effective. Combination of BCG with IFN-α has been the most widely studied and used clinically. Increased release of the cytotoxic effector IFN-γ has been seen with administration of IFN-α in addition to BCG (69). Another rationale for combination cytokine therapy is the synergy seen with the administration of TH1 cytokines such as IL-2, IFN-α, and IL-12 with BCG in increasing release of IFN–γ (70). Clinical trials have confirmed that BCG combined with IFN-α may have promise for patients who relapse after BCG therapy, as well as BCG naïve patients (71, 72). While the evidence for adding IFN-α is the strongest of the cytokines, others have been studied as adjuvants to BCG alone. Studies are ongoing examining combination with other cytokines such as IL-2 and granulocyte monocyte-colony stimulating factor (GM-CSF), among others.

CONCLUSION

BCG is the prototypic example of successful translation of a microbial product from its original use as a vaccine to a successful therapy in a different disease setting. The development of BCG for bladder cancer required background in microbiology and immunology to recognize an application in the arena of oncology. Further investigation into the mechanism of BCG activity should lead to improved safety and efficacy in the future and may spur the creation of next-generation therapeutics for intravesical treatment of bladder cancer.

REFERENCES

1. Parkin D.M., Bray F., Ferlay J., and Pisani P. 2005. Global cancer statistics, 2002. CA Cancer J. Clin. **55**: 74–108.
2. Brennan P., Bogillot O., Greiser E., Chang-Claude J., Wahrendorf J., Cordier S., Jöckel K.H., Lopez-Abente G., Tzonou A., Vineis P., Donato F., Hours M., Serra C., Bolm-Audorff U., Schill W., Kogevinas M., and Boffetta P. 2001. The contribution of cigarette smoking to bladder cancer in women (pooled European data). Cancer Causes Control. **12**: 411–417.
3. Ploeg M., Aben K.K., and Kiemeney L.A. 2009. The present and future burden of urinary bladder cancer in the world. World J. Urol. **27**: 289–293.
4. Jemal A., Siegel R., Ward E., Hao Y., Xu J., Murray T., Thun M.J., and Ward E. 2008. Cancer statistics 2008. CA Cancer J. Clin. **58**: 71–96.
5. Botteman M.F., Pashos C.L., Redaelli A., Laskin B., and Hauser R. 2003. The health economics of bladder cancer: a comprehensive review of the published literature. Pharmacoeconomics. **21**: 1315–1330.
6. Morales A., Eidinger D., and Bruce A.W. 1976. Intracavitary Bacillus Calmette-Guerin in the treatment of superficial bladder tumors. J. Urol. **116**: 180–183.
7. Carmack A.J., and Soloway M.S. 2006. The diagnosis and staging of bladder cancer: from RBCs to TURs. Urology. **67**: 3–8.

8. Priolo G., Gontero P., Martinasso G., Mengozzi G., Formiconi A., Pelucelli G., Zitella A., Casetta G., Viberti L., Aimo G., and Tizzani A. 2001. Bladder tumor antigen assay as compared to voided urine cytology in the diagnosis of bladder cancer. Clin. Chim. Acta. **305**: 47–53.
9. Savic S., Zlobec I., Thalmann G.N., Engeler D., Schmauss M., Lehmann K., Mattarelli G., Eichenberger T., Dalquen P., Spieler P., Schoenegg R., Gasser T.C., Sulser T., Forster T., Zellweger T., Casella R., and Bubendorf L. 2009. The prognostic value of cytology and fluorescence in situ hybridization in the follow-up of nonmuscle-invasive bladder cancer after intravesical Bacillus Calmette-Guerin therapy. Int. J. Cancer. **12**: 2899–2904.
10. Mengual L., Marín-Aguilera M., Ribal M.J., Burset M., Villavicencio H., Oliver A., Alcaraz A. 2007. Clinical utility of fluorescent in situ hybridization for the surveillance of bladder cancer patients treated with Bacillus Calmette-Guerin therapy. Eur. Urol. **52**: 752–759.
11. Gupta N.P., Sharma N., and Kumar R. 2009. Nuclear matrix protein 22 as adjunct to urine cytology and cystoscopy in follow-up of superficial TCC of urinary bladder. Urology. **73**: 592–596.
12. Jichlinski P., Guillou L., Karlsen S.J., Malmström P.U., Jocham D., Brennhovd B., Johansson E., Gärtner T., Lange N., van den Bergh H., and Leisinger H.J. Hexyl aminolevulinate fluorescence cystoscopy: new diagnostic tool for photodiagnosis of superficial bladder cancer—a multicenter study. J. Urol. **170**: 226–229.
13. Schmidbauer J., Witjes F., Schmeller N., Donat R., Susani M., and Marberger M. 2004. Improved detection of urothelial carcinoma in situ with hexaminolevulinate fluorescence cystoscopy. J. Urol. **171**: 135–138.
14. Babjuk M., Oosterlinck W., Sylvester R., Kaasinen E., Bohle A., and Palou-Redorta J. 2008. EAU guidelines on non-muscle-invasive urothelial carcinoma of the bladder. Eur. Urol. **54**: 303–314.
15. Hall M.C., Chang S.S., Dalbagni G., Pruthi R.S., Seigne J.D., Skinner E.C., Wolf J.S. Jr, and Schellhammer P.F. 2007. Guideline for the management of nonmuscle invasive bladder cancer (stages Ta, T1, and Tis): 2007 update. J. Urol. **178**: 2314–2330.
16. Oettinger T., Jorgensen M., Ladefoged A., Haslov K., and Andersen P. Development of the *Mycobacterium bovis* BCG vaccine: review of the historical and biochemical evidence for a genealogical tree. Tuber. Lung Dis. **79**: 243–250.
17. Herr H.W., and Morales A. 2008. History of bacillus Calmette-Guerin and bladder cancer: an immunotherapy success story. J. Urol. **179**: 53–56.
18. Old L.J., Clarke D.A., and Benacerraf B. 1959. Effect of Bacillus Calmette-Guerin infection on transplanted tumours in the mouse. Nature. **184**: 291–292.
19. Lewis K.N., Liao R., Guinn K.M., Hickey M.J., Smith S., Behr M.A., and Sherman D.R. 2003. Deletion of RD1 from *Mycobacterium tuberculosis* mimics bacille Calmette-Guerin attenuation. J. Infect. Dis. **187**: 117–123.
20. Brosch R., Gordon S.V., Garnier T., Eiglmeier K., Frigui W., Valenti P., Dos Santos S., Duthoy S., Lacroix C., Garcia-Pelayo C., Inwald J.K., Golby P., Garcia J.N., Hewinson R.G., Behr M.A., Quail M.A., Churcher C., Barrell B.G., Parkhill J., and Cole S.T. 2007. Genome plasticity of BCG and impact on vaccine efficacy. Proc. Natl. Acad. Sci. U S A. **104**: 5596–5601.

21. Ritz N., Hanekom W.A., Robins-Browne R., Britton W.J., and Curtis N. 2008. Influence of BCG vaccine strain on the immune response and protection against tuberculosis. FEMS Microbiol. Rev. **32**: 821–841.
22. Sylvester R.J., van der Meijden A.P., and Lamm D.L. 2002. Intravesical bacillus Calmette-Guerin reduces the risk of progression in patients with superficial bladder cancer: a meta-analysis of the published results of randomized clinical trials. J. Urol. **168**: 1964–1970.
23. Bohle A., and Bock P.R. 2004. Intravesical bacille Calmette-Guerin versus mitomycin C in superficial bladder cancer: formal meta-analysis of comparative studies on tumor progression. Urology. **63**: 682–686.
24. Sylvester R.J., van der Meijden A.P., Oosterlinck W., Hoeltl W., and Bono A.V. 2003. The side effects of Bacillus Calmette-Guerin in the treatment of Ta T1 bladder cancer do not predict its efficacy: results from a European Organisation for Research and Treatment of Cancer Genito-Urinary Group Phase III Trial. Eur. Urol. **44**: 423–428.
25. Bohle A., Jocham D., and Bock P.R. 2003. Intravesical bacillus Calmette-Guerin versus mitomycin C for superficial bladder cancer: a formal meta-analysis of comparative studies on recurrence and toxicity. J. Urol. **169**: 90–95.
26. Rawls W.H., Lamm D.L., Lowe B.A., Crawford E.D., Sarosdy M.F., Montie J.E., Grossman H.B., and Scardino P.T. 1990. Fatal sepsis following intravesical bacillus Calmette-Guerin administration for bladder cancer. J. Urol. **144**: 1328–1330.
27. Martínez-Piñeiro J.A., Martínez-Piñeiro L., Solsona E., Rodríguez R.H., Gómez J.M., Martín M.G., Molina J.R., Collado A.G., Flores N., Isorna S., Pertusa C., Rabadán M., Astobieta A., Camacho J.E., Arribas S., and Madero R. 2005. Has a 3-fold decreased dose of bacillus Calmette-Guerin the same efficacy against recurrences and progression of T1G3 and Tis bladder tumors than the standard dose? Results of a prospective randomized trial. J. Urol. **174**: 1242–1247.
28. Ojea A., Nogueira J.L., Solsona E., Flores N., Gómez J.M., Molina J.R., Chantada V., Camacho J.E., Piñeiro L.M., Rodríguez R.H., Isorna S., Blas M., Martínez-Piñeiro J.A., and Madero R. 2007. A multicentre, randomised prospective trial comparing three intravesical adjuvant therapies for intermediate-risk superficial bladder cancer: low-dose bacillus Calmette-Guerin (27 mg) versus very low-dose bacillus Calmette-Guerin (13.5 mg) versus mitomycin C. Eur. Urol. **52**: 1398–1406.
29. Durek C., Rusch-Gerdes S., Jocham D., and Bohle A. 2000. Sensitivity of BCG to modern antibiotics. Eur. Urol. **37**: 21–25.
30. Saint F., Patard J.J., Irani J., Salomon L., Hoznek A., Legrand P., Debois H., Abbou C.C., and Chopin D.K. 2001. Leukocyturia as a predictor of tolerance and efficacy of intravesical BCG maintenance therapy for superficial bladder cancer. Urology. **57**: 617–621.
31. Saint F., Patard J.J., Maille P., Soyeux P., Hoznek A., Salomon L., De La Taille A., Abbou C.C., and Chopin DK. 2001. T helper 1/2 lymphocyte urinary cytokine profiles in responding and nonresponding patients after 1 and 2 courses of bacillus Calmette-Guerin for superficial bladder cancer. J. Urol. **166**: 2142–2147.
32. Ratliff T.L., Ritchey J.K., Yuan J.J., Andriole G.L., and Catalona W.J. 1993. T-cell subsets required for intravesical BCG immunotherapy for bladder cancer. J. Urol. **150**: 1018–1023.

33. Brandau S., Riemensberger J., Jacobsen M., Kemp D., Zhao W., Zhao X., Jocham D., Ratliff T.L., and Böhle A. 2001. NK cells are essential for effective BCG immunotherapy. Int. J. Cancer. **92**: 697–702.

34. Suttmann H., Riemensberger J., Bentien G., Schmaltz D., Stöckle M., Jocham D., Böhle A., and Brandau S. 2006. Neutrophil granulocytes are required for effective Bacillus Calmette-Guerin immunotherapy of bladder cancer and orchestrate local immune responses. Cancer Res. **66**: 8250–8257.

35. Suttmann H., Jacobsen M., Reiss K., Jocham D., Böhle A., and Brandau S. 2004. Mechanisms of bacillus Calmette-Guerin mediated natural killer cell activation. J. Urol. **172**: 1490–1495.

36. Kelley D.R., Ratliff T.L., Catalona W.J., Shapiro A., Lage J.M., Bauer W.C., Haaff E.O., and Dresner S.M. 1985. Intravesical bacillus Calmette-Guerin therapy for superficial bladder cancer: effect of bacillus Calmette-Guerin viability on treatment results. J. Urol. **134**: 48–53.

37. De Boer E.C., Rooijakkers S.J., Schamhart D.H., and Kurth K.H. 2003. Cytokine gene expression in a mouse model: the first instillations with viable bacillus Calmette-Guerin determine the succeeding Th1 response. J. Urol. **170**: 2004–2008.

38. Pryor K., Stricker P., Russell P., Golovsky D., and Penny R. 1995. Antiproliferative effects of bacillus Calmette-Guerin and interferon alpha 2b on human bladder cancer cells in vitro. Cancer Immunol. Immunother. **41**: 309–316.

39. Chen F., Zhang G., Iwamoto Y., and See W.A. 2005. BCG directly induces cell cycle arrest in human transitional carcinoma cell lines as a consequence of integrin cross-linking. BMC Urol. **5**: 8.

40. Kavoussi L.R., Brown E.J., Ritchey J.K., and Ratliff T.L. 1990. Fibronectin-mediated Calmette-Guerin bacillus attachment to murine bladder mucosa. Requirement for the expression of an antitumor response. J. Clin. Invest. **85**: 62–67.

41. Bisiaux A., Thiounn N., Timsit M.O., Eladaoui A., Chang H.H., Mapes J., Mogenet A., Bresson J.L., Prié D., Béchet S., Baron C., Sadorge C., Thomas S., Albert E.B., Albert P.S., and Albert M.L. 2009. Molecular analyte profiling of the early events and tissue conditioning following intravesical bacillus Calmette-Guerin therapy in patients with superficial bladder cancer. J. Urol. **181**: 1571–1580.

42. Durek C., Richter E., Basteck A., Rüsch-Gerdes S., Gerdes J., Jocham D., and Böhle A. 2001. The fate of bacillus Calmette-Guerin after intravesical instillation. J. Urol. **165**: 1765–1768.

43. Bowyer L., Hall R.R., Reading J., Marsh M.M. 1995. The persistence of bacille Calmette-Guerin in the bladder after intravesical treatment for bladder cancer. Br. J. Urol. **75**: 188–192.

44. Saint F., Salomon L., Quintela R., Cicco A., Hoznek A., Abbou C.C., and Chopin D.K. 2003. Do prognostic parameters of remission versus relapse after Bacillus Calmette-Guerin (BCG) immunotherapy exist? Analysis of a quarter century of literature. Eur. Urol. **43**: 351–360.

45. Shintani Y., Sawada Y., Inagaki T., Kohjimoto Y., Uekado Y., and Shinka T. 2007. Intravesical instillation therapy with bacillus Calmette-Guerin for superficial bladder cancer: study of the mechanism of bacillus Calmette-Guerin immunotherapy. Int. J. Urol. **14**: 140–146.

46. Kumar A., Dubey D., Bansal P., Mandhani A., and Naik S. 2002. Urinary interleukin-8 predicts the response of standard and low dose intravesical bacillus

Calmette-Guerin (modified Danish 1331 strain) for superficial bladder cancer. J. Urol. **168**: 2232–2235.

47. Magno C., Melloni D., Galì A., Mucciardi G., Nicocia G., Morandi B., Melioli G., and Ferlazzo G. 2002. The anti-tumor activity of bacillus Calmette-Guerin in bladder cancer is associated with an increase in the circulating level of interleukin-2. Immunol. Lett. **81**: 235–238.
48. Watanabe E., Matsuyama H., Matsuda K., Ohmi C., Tei Y., Yoshihiro S., Ohmoto Y., and Naito K. 2003. Urinary interleukin-2 may predict clinical outcome of intravesical bacillus Calmette-Guerin immunotherapy for carcinoma in situ of the bladder. Cancer Immunol. Immunother. **52**: 481–486.
49. Siracusano S., Vita F., Abbate R., Ciciliato S., Borelli V., Bernabei M., and Zabucchi G. 2007. The role of granulocytes following intravesical BCG prophylaxis. Eur. Urol. **51**: 1589–1597.
50. Wiley S.R., Schooley K., Smolak P.J., Din W.S., Huang C.P., Nicholl J.K., Sutherland G.R., Smith T.D., Rauch C., Smith C.A., and Goodwin R.G. 1995. Identification and characterization of a new member of the TNF family that induces apoptosis. Immunity. **3**: 673–682.
51. Kayagaki N., Yamaguchi N., Nakayama M., Eto H., Okumura K., and Yagita H. 1999. Type I interferons (IFNs) regulate tumor necrosis factor-related apoptosis-inducing ligand (TRAIL) expression on human T cells: a novel mechanism for the antitumor effects of type I IFNs. J. Exp. Med. **189**: 1451–1460.
52. Ludwig A.T., Moore J.M., Luo Y., Chen X., Saltsgaver N.A., O'Donnell M.A., and Griffith T.S. 2004. Tumor necrosis factor-related apoptosis-inducing ligand: a novel mechanism for Bacillus Calmette-Guerin-induced antitumor activity. Cancer Res. **64**: 3386–3390.
53. Simons M.P., Nauseef W.M., and Griffith T.S. 2007. Neutrophils and TRAIL: insights into BCG immunotherapy for bladder cancer. Immunol. Res. **39**: 79–93.
54. Basturk B., Yavascaoglu I., Oral B., Goral G., and Oktay B. 2006. Cytokine gene polymorphisms can alter the effect of Bacillus Calmette-Guerin (BCG) immunotherapy. Cytokine. **35**: 1–5.
55. Decobert M., Larue H., Bergeron A., Harel F., Pfister C., Rousseau F., Lacombe L., and Fradet Y. 2006. Polymorphisms of the human NRAMP1 gene are associated with response to bacillus Calmette Guerin immunotherapy for superficial bladder cancer. J. Urol. **175**: 1506–1511.
56. Ahirwar D., Kesarwani P., Manchanda P.K., Mandhani A., and Mittal R.D. 2008. Anti- and proinflammatory cytokine gene polymorphism and genetic predisposition: association with smoking, tumor stage and grade, and bacillus Calmette-Guerin immunotherapy in bladder cancer. Cancer Genet. Cytogenet. **184**: 1–8.
57. Liu W., O'Donnell M.A., Chen X., Han R., and Luo Y. 2009. Recombinant bacillus Calmette-Guerin (BCG) expressing interferon-alpha 2B enhances human mononuclear cell cytotoxicity against bladder cancer cell lines in vitro. Cancer Immunol. Immunother. **58**: 1647–1655.
58. Yamada H., Matsumoto S., Matsumoto T., Yamada T., and Yamashita U. 2000. Murine IL-2 secreting recombinant Bacillus Calmette-Guerin augments macrophage-mediated cytotoxicity against murine bladder cancer MBT-2. J. Urol. **164**: 526–531.

59. Luo Y., Yamada H., Chen X., Ryan A.A., Evanoff D.P., Triccas J.A., and O'Donnell M.A. 2004. Recombinant *Mycobacterium bovis* bacillus Calmette-Guerin (BCG) expressing mouse IL-18 augments Th1 immunity and macrophage cytotoxicity. Clin. Exp. Immunol. **137**: 24–34.
60. Arnold J., de Boer E.C., O'Donnell M.A., Bohle A., and Brandau S. 2004. Immunotherapy of experimental bladder cancer with recombinant BCG expressing interferon-gamma. J. Immunother. **27**: 116–123.
61. Chade D.C., Borra R.C., Nascimento I.P., Villanova F.E., Leite L.C., Andrade E., Srougi M., Ramos K.L., and Andrade P.M. 2008. Immunomodulatory effects of recombinant BCG expressing pertussis toxin on TNF-alpha and IL-10 in a bladder cancer model. J. Exp. Clin. Cancer Res. **27**: 78.
62. Kemp T.J., Ludwig A.T., Earel J.K., Moore J.M., Vanoosten R.L., Moses B., Leidal K., Nauseef W.M., and Griffith T.S. 2005. Neutrophil stimulation with *Mycobacterium bovis* bacillus Calmette-Guerin (BCG) results in the release of functional soluble TRAIL/Apo-2L. Blood. **106**: 3474–3482.
63. Morales A., Chin J.L., and Ramsey E.W. 2001. Mycobacterial cell wall extract for treatment of carcinoma in situ of the bladder. J. Urol. **166**: 1633–1637.
64. Morales A. 2008. Evolution of intravesical immunotherapy for bladder cancer: mycobacterial cell wall preparation as a promising agent. Expert Opin. Investig. Drugs. **17**: 1067–1073.
65. Morales A., Phadke K., and Steinhoff G. 2009. Intravesical mycobacterial cell wall-DNA complex in the treatment of carcinoma in situ of the bladder after standard intravesical therapy has failed. J. Urol. **181**: 1040–1045.
66. Luo Y., Chen X., and O'Donnell M.A. 2003. Role of Th1 and Th2 cytokines in BCG-induced IFN-gamma production: cytokine promotion and simulation of BCG effect. Cytokine. **21**: 17–26.
67. Belldegrun A.S., Franklin J.R., O'Donnell M.A., Gomella L.G., Klein E., Neri R., Nseyo U.O., Ratliff T.L., and Williams R.D. 1998. Superficial bladder cancer: the role of interferon-alpha. J. Urol. **159**: 1793–1801.
68. Weiss G.R., O'Donnell M.A., Loughlin K., Zonno K., Laliberte R.J., and Sherman M.L. 2003. Phase 1 study of the intravesical administration of recombinant human interleukin-12 in patients with recurrent superficial transitional cell carcinoma of the bladder. J. Immunother. **26**: 343–348.
69. Luo Y., Chen X., Downs T.M., DeWolf W.C., and O'Donnell M.A. 1999. IFN-alpha 2B enhances Th1 cytokine responses in bladder cancer patients receiving *Mycobacterium bovis* bacillus Calmette-Guerin immunotherapy. J. Immunol. **162**: 2399–2405.
70. Chen X., O'Donnell M.A., and Luo Y. 2007. Dose-dependent synergy of Th1-stimulating cytokines on bacille Calmette-Guerin-induced interferon-gamma production by human mononuclear cells. Clin. Exp. Immunol. **149**: 178–185.
71. Joudi F.N., Smith B.J., and O'Donnell M.A. 2006. Final results from a national multicenter phase II trial of combination bacillus Calmette-Guerin plus interferon alpha-2B for reducing recurrence of superficial bladder cancer. Urol. Oncol. **24**: 344–348.
72. Gallagher B.L., Joudi F.N., Maymi J.L., and O'Donnell M.A. 2008. Impact of previous bacille Calmette-Guerin failure pattern on subsequent response to bacille Calmette-Guerin plus interferon intravesical therapy. Urology. **71**: 297–301.

4

LIVE *CLOSTRIDIA*: A POWERFUL TOOL IN TUMOR BIOTHERAPY

Lieve Van Mellaert,[1] Ming Q Wei,[2] and Jozef Anné[1]

[1]*Laboratory of Bacteriology, Rega Institute for Medical Research, Katholieke Universiteit Leuven, Leuven, Belgium*

[2]*Division of Molecular and Gene Therapies, Griffith Institute for Health and Medical Research, School of Medical Science, Griffith University, Queensland, Australia*

THE GENUS *CLOSTRIDIUM* IN A NUTSHELL

Clostridium is phylogenetically an extremely heterogeneous genus consisting of more than 80 species of strictly anaerobic, gram-positive, endospore-forming bacteria. Cells are rod-shaped with peritrichous flagella, and are mostly motile. Oxygen tolerance and biochemical properties vary greatly among different species. Several *Clostridium* spp. naturally inhabit the intestinal tract of animals and humans. The genus has a bad reputation as a consequence of *Clostridium tetani* and *Clostridium botulinum*, the causative agent of tetanus and botulism, respectively. Other species within this genus-causing disease are, for example, *Clostridium perfringens* which, besides being a source for enteritis, can infect wounds and forms as such the basis for subsequent gas gangrene. On antibiotic treatment, *Clostridium difficile* can overgrow the affected gut microflora, consequently causing antibiotic-associated diarrhea and pseudomembraneous enterocolitis. Several other *Clostridium* spp. including *Clostridium novyi*, *Clostridium sordellii*, and *Clostridium histolyticum* can also be associated with illness and even death among intravenous drug users as it may also be the case following infection with *C. botulinum* and *C. tetani* when

Emerging Cancer Therapy: Microbial Approaches and Biotechnological Tools, Edited by Arsénio M. Fialho and Ananda M. Chakrabarty

injecting heroin (1, 2). Nevertheless, the vast majority of species in this genus is entirely benign and because of their interesting fermentation properties, some species, including the solvent-producing *Clostridium acetobutylicum, Clostridium beijerinckii*, and *Clostridium saccharobutylicum*, are of interest in industrial biofuel production (3, 4).

THE TUMOR MICROENVIRONMENT: RESISTANT TO CONVENTIONAL TUMOR THERAPEUTICS BUT A HAVEN FOR *CLOSTRIDIA*

As early as 1813, a link was made between solid tumors and clostridial growth by reporting tumor regression in patients who suffered gas gangrene. Later on, it became clear that the unique physiology of solid tumors creates an exclusive niche in which proliferation of anaerobic bacteria can occur. The high demand for oxygen and nutrients by the fast-proliferating tumor cells triggers neo-angiogenesis, but the newly formed tumor vasculature is chaotic and disturbed as a result of which it becomes inadequate to provide all tumor cells of due oxygen and nutrients. As a consequence of the aberrant capillary network with a lot of shunts and torsions, a considerable part of the tumor becomes hypoxic. Severe hypoxia is not found in nonpathological, normal tissues (5–7). Those hypoxic/necrotic regions within a tumor are prognostic indicators of poor treatment outcome as hypoxia causes intrinsic therapy resistance: hypoxic cells require three times more irradiation for efficient killing than oxygenized cells, and the aberrant capillary network hinders the chemotherapeutic agents to readily reach the hypoxic cells. Moreover, under hypoxic stress conditions, cells undergo genetic and molecular changes in a way that some of them gain molecular resistance. Cells in the quiescent and necrotizing tumor regions become hypoxia tolerant, restart growth, and acquire a more malignant phenotype that promotes metastasis, while their apoptosis potential diminish (8). Additionally, hypoxia plays a key role in new blood vessel formation during tumor progression (9).

Precisely these therapeutically problematic regions provide ideal conditions for clostridial growth. The lack of oxygen concentration allows growth of anaerobes, and the necrotic tumor mass delivers the required nutrients. Altogether, the fact that solid tumors are hypoxic together with the prior observations that certain *Clostridium* strains exhibit innate oncolytic properties makes *Clostridium* an interesting tool for tumor therapy.

THE FIRST ERA OF TUMOR BIOTHERAPY USING *CLOSTRIDIUM* SPORES

The Onset of the *Clostridium*-Mediated Tumor Therapy

Upon the observations of clinical improvements of cancer treatment outcome following the administration of sterile filtrates of *C. histolyticum*, Parker and colleagues in 1947 injected *C. histolyticum* spores into transplanted mouse sarcomas, which resulted in spore germination and vegetative growth followed by "liquefaction" and regression of the tumor (10). To demonstrate the exquisite tumor selectivity of *Clostridium* spores upon systemic injection, Malmgren and Flanigan intravenously administered *C. tetani* spores to tumor-bearing and to healthy mice. In contrast to the healthy controls, the mice with tumors all succumbed to tetanus poisoning within 48h since the spores could only germinate within the anaerobic conditions provided by the tumor (11). Upon this proof of tumor specificity and tumor therapeutic efficacy of *Clostridium*, the issue of using pathogenic strains for therapeutic purposes was overcome by the isolation of the nonpathogenic *Clostridium butyricum* M-55 strain (later named *Clostridium oncolyticum* and finally reclassified as *Clostridium sporogenes* ATCC13732). This strain showed an extraordinary oncolytic capability and was proven to be safe and without adverse effects upon intravenous administration as shown by Möse and Möse who self-injected a spore suspension (cited in 12).

Clostridial Oncolysis Studies and Trials

Over the next 20 years, *C. butyricum* M-55 was used as reference strain for further studies carried out with a variety of *Clostridium* spp. against a variety of transplantable tumors in mice, rat, rabbits, and hamsters (13–16). These animal studies established critical parameters for *Clostridium*-mediated oncolysis including a tumor threshold size ($3\,cm^2$ or circa 2g of tumor weight), and a 10^{6-9} spore dose. Besides the fact that not all tumors were responsive to clostridial spore treatment, the mode of spore administration, the time–dose relationship between the phases of clostridial propagation, the germinative quality of spores, and of course the host-determined immune responses, all affected colliquative necrosis and resporulation. Although those studies showed that clostridial tumor colonization and proliferation could destruct a significant portion of the tumor, the oncolysis abruptly stopped at the outer rim of the tumor. Moreover, the extensive oncolysis mostly killed the animals in consequence of the tumor lysis syndrome (17) and if not, a high recurrence rate was observed.

The animal studies were rapidly followed by clinical tests of *Clostridium*-mediated tumor biotherapy in man. Single doses of 10^{6-10} intravenously administered *C. oncolyticum* spores were indeed well-tolerated, and in cases where oncolysis occurred, an increase of temperature and leukocytosis was observed in patients within 4–8 days accompanied with clostridial propagation and progressive oncolysis. As in animal studies, these clinical studies revealed that large portions of tumor (including brain tumors) could be liquefied, but the viable outer rim allowed regrowth of the tumor. Consequently, recurrence rate remained uninfluenced and no beneficial effect was obtained with *Clostridium* biotherapy (18, 19).

Meanwhile, basic studies on the oncolytic effect of *C. butyricum* M-55 using *in vitro* cell cultures revealed that the presence of both germinated spore suspensions or extracts, and vegetative clostridial growth had a cytotoxic and inhibitory effect on heteroploid tumor cells, whereas diploid cells remained unaffected (20–22). Furthermore, vegetative rods were reported to be closely associated with and inside tumor cells (22) whereas no penetration of the spores into cells was observed. In this respect, it was suggested that clostridial neuraminidase binds to neuraminic acid, the parent acid of the sialic acids which are abundantly present in cell membranes of malignant tumor cells (23). This neuraminidase activity was further exploited in tumor diagnosis upon clostridial oncolysis (*infra*).

Since spore suspensions, spore extracts as well as clostridial culture filtrates but not vegetative cell extracts reduced malignant cell growth, it was concluded that the inhibitory activity was exerted by soluble substances acting at a distance (20, 21). Several exoenzymes such as proteases, nucleases, and phospholipase A secreted by *C. oncolyticum* were shown to be of importance in the oncolysis process.

A Search for More Effective Combination Therapies

In order to enhance the oncolytic efficacy of clostridial biotherapy and to attack the remaining viable rim and thwart neoplastic resistance, several combination therapies were explored. While a series of antimetabolites, quinones, and peroxides did not improve the therapeutic effect of the spore therapy, 5-fluorodeoxyuridine and alkylating agents of the ethyleneimino type significantly enhanced sarcoma regression in mice. The toxicity observed with some of the latter products could be anticipated through adjustment of the drug therapy (15). Also, the combined administration of cyclophosphamide and *C. butyricum* M-55 spores resulted in a significant improvement on survival time. However, this effect could not be confirmed using other animal models and other *Clostridium* strains (18).

Another approach aimed to promote necrosis within the tumor applying high-frequency hyperthermia using microwaves as such creating a more favorable microenvironment for clostridial spore germination. In all tested tumor models, *C. butyricum* M-55-mediated oncolysis, but not survival rate, was improved when combined with microwave-induced hyperthermia, and a clear dependence of the extent of oncolysis on thermal dose and on time interval between hyperthermia treatment and *Clostridium* spore therapy was observed (24, 25). Additional irradiation to the combined hyperthermia and *Clostridium*-mediated therapy resulted for the first time in a cure rate of 20%. Upon repetition of the threefold-combined therapy for a second or third time, tumor relapse in the mouse tumor model could drastically be decreased (26).

Reduction of the oxygen concentration of air inhaled by tumor-bearing animals to 11–12% was an alternative approach to enhance clostridial oncolysis. This combination led to nearly 30% tumor cure (27). The positive results with these combination therapies in animal experiments could, however, not be transferred into clinical settings. In a review of brain glioma treatment,

F. Heppner reported in 1986 that none of the treatment modalities upon operative removal were really effective, but he implied that *Clostridium*-mediated oncolysis in combination with periodic postoperative generation of heat locally in the excision cavity of the tumor might justify cautious optimism about future developments in glioma treatment (28). An overview of strategies and results with *Clostridium*-mediated tumor therapy in this first era is given in Table 4.1.

Clostridium as a Tool for Tumor Serodiagnosis

The exquisite selectivity of clostridial growth for tumor tissue conceived the idea to use clostridia not only as a tool in tumor therapy, but also for diagnostic purposes. The prominent difference of target recognition of antibodies raised against spores or vegetative cells made it possible to clearly distinguish spores from vegetative cells in the serum upon germination of *Clostridium* spores and vegetative growth within a tumor. This made it possible to serologically diagnose the presence of a tumor even before it became clinically detectable (19). Alternatively, upon tumor colonization with *C. oncolyticum*, a more than 50% increase in the serum concentration of the tumor marker sialic acid was observed. This was a consequence of the high neuraminidase activity of *C. oncolyticum,* which hydrolyzes sialic acid containing substrates abundantly present in tumor cell membranes (23).

However, as *C. oncolyticum* caused a non-deniable local pathological effect, non-oncolytic *Clostridium* strains needed to be looked for, when aiming to use clostridia for tumor diagnostics. This search resulted in the selection of the saccharolytic *C. butyricum* CNZR 528 (later on reclassified as *Clostridium beijerinckii*). Using this strain in an animal trial including 251 dogs (with a malignant tumor, a benign tumor, or tumor-free) revealed that serodiagnosis of *Clostridium*-specific antibodies based on the combined evaluation of a complement fixation and passive hemagglutination test resulted in 70–80% test efficiency depending on the type of tumor. The suitability of the serologic tumor diagnosis based on *C. butyricum* CNRZ 528 was further investigated on bovine leukosis virus-infected cattle developing tumorogenic lymph nodes and with a small number of horses and green monkeys. Those trials supported the outcome of the dog studies (19).

Upon approval of *C. butyricum* CNRZ spores as live diagnostic agent in humans, clinical Phase I and Phase II studies were carried out, and final assessment of the tumor-*Clostridium* diagnostic test was based on 55 patients with known malignant or benign tumors. Anti-spore and anti-rod antibody responses were analyzed via a combination of complement fixation, passive hemagglutination, and via the use of enzyme-linked immunosorbent assay (ELISA) for IgM antibody detection and quantification. The serodiagnostic test achieved 86% sensitivity and 97% specificity. A positive tumor-*Clostridium* test on a patient was thus indicative for tumor growth. The tumor had to be sufficiently large and well vascularized and the test additionally required an amply large spore dose for triggering intratumoral vegetative growth (19).

TABLE 4.1. Overview of First-Generation *Clostridium* spp. with Respect to Tumor Therapy and Diagnosis

Clostridium Species	Year	Tumor Model	Strategy/Results	Reference
C. histolyticum	1947	Mice with transplanted sarcomas	Intratumoral spore injection leads to liquefaction and regression of the tumor.	10
C. tetani	1955	Mice with transplanted mammary tumors, fibrosarcomas, or hepatomas	Upon spore injection, tumor-bearing mice succumb to tetanus poisoning in contrast to healthy mice.	11
C. butyricum M-55 and other nonpathogenic clostridia	1964	Mice with Ehrlich ascites tumors	Among the different species, the M-55 strain shows the most extensive tumor lysis upon intravenous injection.	13
C. butyricum M-55	1964	Different models in mice, rats, rabbits, and hamsters	Clostridial oncolysis is dependent on the tumor model, the age and size of the tumor, and the animal species.	14, 16
C. butyricum M-55	1964	Mice bearing Ehrlich carcinomas, fibrosarcomas, or spontaneous tumors	Parental or local administration of metals, in particular iron dextran complex, enhances the tumor lysis effect induced by clostridial growth.	14
C. butyricum M-55	1964	Mice with Sarcoma 180	Combined therapy with 5-fluorodeoxyuridine and alkylating agents of the ethyleneimino type significantly improves clostridial oncolysis.	15
C. butyricum M-55	1967	Human	Lack of human pathogenicity demonstrated by self-injection of spore suspension and start of clinical trials.	12

C. butyricum M-55	From 1970	*In vitro* cell lines	Germinated spore suspensions or extracts and vegetative growth have a cytotoxic and inhibitory effect on heteroploid but not on diploid cells.	20–22
C. butyricum M-55	From 1976	Mice bearing neck cancers	Local tumor hyperthermy intensifies clostridial oncolysis.	24, 25
C. butyricum M-55	1979	Mice bearing Harding-Passey melanoma	Combined tumor treatment with local hyperthermy, irradiation, and clostridial oncolysis results in 20% cure rate; threefold repetition of this treatment drastically decreases tumor relapse.	26
C. butyricum M-55	1979	Mice with Ehrlich tumors or Harding-Passey melanoma	Reduction of oxygen concentration in inhaled air enhances clostridial oncolysis and leads to nearly 30% cure.	27
C. butyricum M-55	1986	Human glioblastomas	Cautious optimism about combined therapy with locally generated heat treatment in excision cavity upon tumor removal and clostridial oncolysis.	28
C. butyricum M-55	1988	–	Attempt to generate a recombinant strain producing colicin E3 compound, a cancerostatic compound derived from *E. coli*.	29

RECOMBINANT *CLOSTRIDIA* AS TOOL IN TUMOR BIOTHERAPY

Clinical outcome of intravenous administration of *C. oncolyticum* M-55 spores to tumor patients remained disappointing: the treatment was well-tolerated and resulted in tumor oncolysis but did not improve tumor cure nor recurrence rates. As a consequence, the *Clostridium*-mediated tumor therapy idea came to a halt; however, not for long. Keeping in mind the inherent tumor specificity of *Clostridium*, other approaches were looked at increasing tumor cure rate, with the main obstacle being the lack of killing tumor cells in the outer rim. The construction of a *C. butyricum* M-55 recombinant strain that produced the *Escherichia coli*-derived colicin E3, a known canceriostatic compound, was reported (29). However, the results obtained were not convincing. Moreover, the nonpathogenic character of the proteolytic *C. butyricum* M-55 (*C. oncolyticum*) was disputable and, in addition, faced the problem of extensive tumor oncolysis, which all too often resulted in toxemia. As a result, around 1995, the application of strains without a known association with disease, but engineered to increase antitumor effects, came into focus. For an overview, see Table 4.2. More details are described in the text below.

Saccharolytic *Clostridia*

The considerable advances in recombinant DNA technology and the increased availability of cloning tools for solvent-producing, saccharolytic clostridia stimulated the idea to genetically engineer these strains in order to endow them with therapeutic features and to apply them in tumor biotherapy. Therefore, the colonization capacity of nonpathogenic saccharolytic clostridia strains was reassessed. These studies clearly confirmed previous data concerning the exquisite tumor selectivity of clostridial growth. Twenty-four hours after intravenous injection of 10^8 *C. beijerinckii* NCIMB 8052 spores into EMT6 tumor-bearing mice, vegetative cells were present in tumor tissue, specifically in and surrounding the region of necrosis, while spores did not germinate in normal tissues (30). Similarly, systemic administration of at least 10^7 spores of *C. beijerinckii* ATCC 17778, *Clostridium limosum* DSM1400, *C. acetobutylicum* ATCC 824, and NI-4082 (later reclassified as *Clostridium saccharoperbutylacetonicum*) to WAG/Rij rats with syngeneic rhabdomyosarcomas, led in all cases to tumor colonization, but no bacterial growth occurred within other tissues. The NI-4082 strain showed a superior infiltration capacity compared to other tested strains, which was comparable to that of *C. oncolyticum* DSM754 (31).

In a next step, saccharolytic clostridia were engineered to produce antitumor compounds, either prodrug-converting enzymes or cytotoxic/tumor-inhibiting factors. The *Clostridium*-directed enzyme prodrug therapy (CDEPT) was introduced in 1995 (32). By endowing *Clostridium* with a gene encoding a prodrug-converting enzyme, prodrugs delivered to the tumor can be converted to highly cytotoxic compounds. *C. beijerinckii* NCIMB8052 recombinants were constructed to produce *E. coli*-derived cytosine deaminase (CDase) or

TABLE 4.2. Overview of "Second-Generation" Genetically Modified *Clostridium* Strains Investigated as Antitumor Tools in the Past 15 Years

Group	Species	Genetic Modifications/Adaptations	Results	References
Saccharolytic clostridia	*C. beijerinckii*	Intracellular recombinant production of the prodrug-converting enzymes *E. coli* NTR *E. coli* CDase	Killing effect on EMT6 cells *in vitro* in combination with appropriate prodrug. Demonstrated enzyme activity upon intratumoral spore injection. However, no antitumor effect upon intravenously spore administration.	32
	C. acetobutylicum	Extracellular recombinant production of *E. coli* CDase mouse TNF-α rat IL-2	Intratumoral occurrence of recombinant prodrug-converting enzyme and cytokines upon systemic spore administration to tumor-bearing rats, but only minor antitumor effects.	34, 37, 47
Proteolytic clostridia	*C. sporogenes*	Intracellular recombinant production of *E. coli* CDase Extracellular recombinant production of *H. influenzae* NTR Optimized intracellular production of *E. coli* NTR using an alternative promoter and a synthesized, codon-adapted gene	Significant antitumor efficacy and tumor growth delay observed in tumor-bearing mice when combined with current or newly developed prodrugs.	40–42
		Secretory production of anti-HIF-1α single chain VHH	First step to CDAT or *Clostridium* - directed antibody therapy.	65

TABLE 4.2. *Continued*

Group	Species	Genetic Modifications/Adaptations	Results	References
Oncolytic clostridia	*C. novyi-NT*	Attenuated strain of *C. novyi* via deletion of lethal α-toxin gene Recombinant production of *H. influenzae* NTR and anti-HIF-1α single chain VHHs (see *C. sporogenes*)	In combination with chemotherapeutics, a hemorrhagic tumor necrosis or a slow tumor regression was observed. Cure possible in immunocompetent animals with *C. novyi-NT* treatment alone due to immunoadjuvant effect. Liposomase activity of *C. novyi-NT* enhances outcome of liposomal drug delivery in combined treatments.	41, 52–57, 59, 65
	C. perfringens	*Sod^-/PVL* strain with a deleted superoxide dismutase gene and an inserted inflammation-suppressing gene *Plc^-/sod^-/PVL* strain in which the phospholipase C or α-toxin gene was additionally knocked out	Promising agents to induce necrosis in pancreatic cancer.	61, 62

nitroreductase (NTR). Notwithstanding CDase or NTR were not secreted from the cells, the amounts present in the culture supernatants were seemingly sufficient to kill EMT6 cells *in vitro* in the presence of the appropriate prodrugs 5-fluorocytosine (5-FC) and CB1954 (2,4-dinitro-5-aziridinylbenzamide), respectively (30, 33). *C. acetobutylicum* DSM792 and *C. saccharoperbutylacetonicum* containing a construct with a translational fusion of the *CDase* gene to the signal sequence derived from the *Clostridium* clostripain or gluthamine synthetase gene efficiently secreted functional CDase (34). Upon intratumoral administration of spores of these recombinant strains, prodrug-converting activity was detected within the tumor homogenates (30, 34). In contrast, intravenous administration of the recombinant *Clostridium* spores to tumor-bearing animals in combination with the prodrug did not result in any therapeutic effect. Alternatively, cDNAs of cytokines such as murine tumor necrosis factor α (mTNF-α) or rat interleukin 2 (rIL-2), which are known to have direct or indirect antitumor activity, were successfully introduced into *C. acetobutylicum* in such a way that the corresponding recombinant proteins were secreted in a bioactive form (35, 36). Upon injection of recombinant *Clostridium* spores to rhabdomyosarcoma-bearing rats, the intratumoral production of mTNF-α or IL-2 could be detected. Nevertheless, tumor colonization by the recombinant clones gave only minor tumor growth retardation (37). A possible explanation for the lack of antitumor activity is the insufficient colonization capacity of the used saccharolytic clostridia compared to proteolytic clostridia.

Saccharolytic *Clostridia* in Combination with Radiotherapy

In an attempt to enhance the antitumor effect of recombinant clostridia, a combination of intratumoral injection of IL-2-producing *C. acetobutylicum* with fractionated radiotherapy was set up because IL-2 therapy effectiveness could be improved by means of local tumor irradiation as earlier described (38). Combined therapy resulted in a minor but significant tumor growth delay compared to radiotherapy or treatment with *Clostridium* lacking the IL-2 cDNA alone. Surprisingly, growth retardation achieved in the combined therapy was not significantly different for the IL-2-recombinant strain or the "empty" *C. acetobutylicum* (Figure 4.1). Thus, the intratumoral presence of wild-type *C. acetobutylicum* was sufficient to elicit a clear radiation-adjuvant effect. It is known that radiation therapy has the capability to modify the tumor microenvironment and to generate inflammation (39). This radiation-induced immune response might be enhanced through the simultaneous immune stimulation by *C. acetobutylicum* growth (unpublished data). Although the obtained antitumoral effect was not very impressive, these results might be very important with respect to further exploitation of nonpathogenic bacteria in combined tumor therapy. Therefore, the observed adjuvant effect of *Clostridium* treatment surely deserves to be further investigated. Better insight in the adjuvant role of *Clostridium* will indeed lead to the development of new recombinant clostridia endowed with certain immunomodulating factors that will further improve the adjuvant effect in combined antitumor therapy.

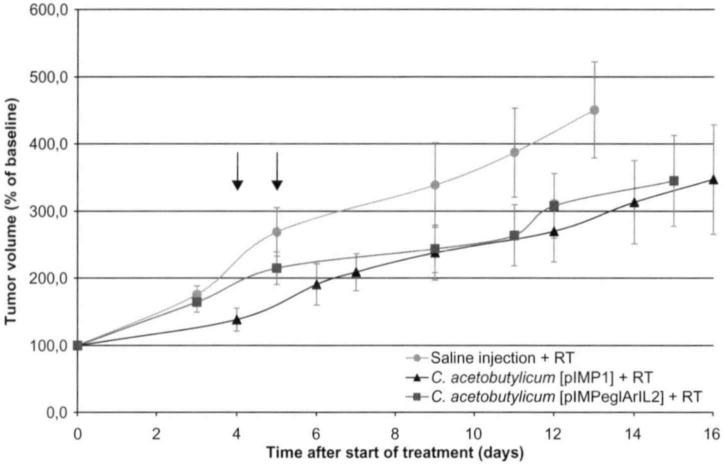

Figure 4.1 For evaluation of a combined therapy, animals were intratumorally injected with *C. acetobutylicum* [pIMP1] (▲) and *C. acetobutylicum* [pIMP1eglArIL-2] (■) spores, respectively, at day 0 while the control group (●) received a saline injection instead. The three groups received a 3-Gy irradiation dose at days 4 and 5 (↓).

Proteolytic Clostridia with Improved Antitumor Efficacy

The failure to transform particular clostridia delayed the use of recombinant proteolytic strains. A breakthrough came when J.M. Brown's group developed an electroporation protocol to transform proteolytic *C. sporogenes* with superior tumor colonization capacity compared to the saccharolytic strains under investigation. For the first time, intravenous injection of a recombinant *Clostridium* strain endowed with the gene-encoding CDase in combination with the administered prodrug 5-FC led to a significant antitumor effect which was greater than the effect obtained with maximally tolerated doses of 5-fluorouracil (5-FU) (40). Unfortunately, as in the case of daily 5-FU treatment, the *Clostridium* spore treatment combined with 5-FC was not able to sustain tumor growth delay after 1 week, notwithstanding the colony-forming unit number of the clostridia at the tumor site did not decrease.

As the described transformation procedure failed to be reproducible in other groups, Theys et al. (41) adapted a conjugation protocol originally set up for *C. difficile* to be used for *C. sporogenes* NCIMB 10696. With this protocol, a *C. sporogenes* strain which produced the *Haemophilus influenzae* NTR was developed. Systemic injection of this recombinant strain to nude mice bearing subcutaneously implanted HCT116 carcinomas led to a significant antitumor effect when combined with the CB1954 prodrug.

Very recently, Liu et al. (42) reported on the further optimization of genetically modified *C. sporogenes* in view of tumor therapy. By replacing the previously used *C. perfringens* ferrodoxin promoter (*fd*P) with very effective

C. acetobutylicum promoters from the thiolase gene (*thl*P) and the transcription regulator AbrB (*abrB*P), which both trigger constitutive gene expression, the prodrug-activating gene expression could be increased two- to threefold in comparison to *fd*P-controlled expression. To further improve NTR production, codons of the original encoding gene were adapted to the codon usage of *C. sporogenes*. With this artificially synthesized gene, NTR production in *C. sporogenes* was enhanced circa 20-fold. The *in vivo* antitumor effect of the newly constructed *C. sporogenes* strains on subcutaneous SiHa human xenografts in nude mice was tested in combination with two dinitrobenzamide prodrugs: the earlier used CB1954 and the novel hypoxia-activated prodrug PR-104 (43) which is currently in clinical trial (44). The combined effect of a stronger promoter activity, a codon-adapted *NTR* gene and the prodrug PR-104 resulted in an increased antitumor effect (42). As these experiments were done in immunodeficient mice, a possible adjuvant immunotherapeutic effect of recombinant *C. sporogenes* treatment as observed with *C. novyi-NT* could not be assessed.

Improvement of Intratumoral Targeting

To deal with the tumor threshold size in which the hypoxia level is sufficient to enable *Clostridium* germination, targeting the tumor vasculature was proposed as an opportunity to increase tumor hypoxia and, consequently, to improve *Clostridium* colonization of small tumors. Combretastatin A4 phosphate (CombreAp), for example, a prodrug dephosphorylated *in vivo* to its active metabolite, combretastatin A4, binds to tubulin resulting in morphological changes in proliferating endothelial cells. As a result, the permeability of tumor vasculature is increased, leading to reduced tumor blood flow and ischemic necrosis of tumor tissue. It was shown that CombreAp, when systemically administrated at nontoxic doses to Wag/Rij rats with rhabdomyosarcomas, triggered rapid and severe vascular shutdown within 3–6 h after injection, followed by necrosis within 1–3 days without side effects (45). Next, the effect of the vascular targeting compound CombreAp on *C. sporogenes* tumor colonization was assessed. A striking difference in colonization of small tumors ($<3\,cm^3$) was observed with and without vascular targeting. In the latter case, the amount of vegetative bacteria was less than 10^3 cfu/g tissue, while in combination with CombreAp, this number mounted up to $>10^7$ cfu/g tissue (46). These observations led to the conclusion that a strong relationship exists between the necrosis induced by CombreAp and the increased intratumoral *Clostridium* growth.

The positive effect of concomitant administration of CombreAp on tumor colonization was also observed using a recombinant CDase-secreting *C. acetobutylicum* strain. In presence of CombreAp, recombinant spore injection resulted in 100% CDase-positive tumors while this incidence was only 55% in absence of the vascular targeting compound. In addition, the recombinant CDase level was higher in the CombreAp-treated tumors than in tumors with an unaffected vasculature which resulted in a better conversion efficiency of 5-FC to 5-FU (~11% vs. ~3%) (34). Since 5-FU is also known as an effective radiosensitizer and a 1–3% conversion from 5-FC to 5-FU was calculated to be sufficient for radiosensitization (47), the positive effect of CombreAp on the outcome of CDase-recombinant *Clostridium* treatment would still be further enhanced when combined with radiotherapy. However, this triple treatment has not been explored so far. Importantly, this vascular targeting approach abated the critical parameter of tumor threshold size for *Clostridium*-mediated tumor therapy by improvement of intratumoral targeting which allows also smaller tumors to be colonized by *Clostridium* spores.

Temporally Controlled Delivery of Therapeutic Compounds

As 75% of all cancer patients are treated with radiotherapy during the course of their disease, "genetic radiotherapy" (48) or radiation-mediated activation of therapeutic genes in genetically modified clostridia is an obvious next step. While irradiation damages the oxygenized cells, the expression of the therapeutic gene in *Clostridium* cells in the vicinity of the less radiosensitive hypoxic cells could be induced if it can be linked to radiotherapy. As such, the spatial control of gene expression realized by tumor hypoxia could in addition be temporally controlled. This strategy might be useful to avoid accidently therapeutic gene expression in nontumor hypoxic environments.

Bacteria are known to be very radioresistant because of the existence of an efficient DNA repair mechanism in which *recA* plays a central role. Using a reporter gene, the radio-inducibility of the *C. acetobutylicum recA* promoter at a clinically relevant dose of 2 Gy was clearly demonstrated (49). This *recA* promoter was subsequently used to induce secretory mTNF-α production in *Clostridium*, which resulted in a 44% increase of mTNF-α in the culture supernatant at 3.5 h after a 2-Gy single-dose irradiation in comparison with nonirradiated controls. Furthermore, it was shown that the promoter could be reinduced using fractionated radiotherapy (50). However, the leakiness of the *recA* promoter resulted in a basal expression level of the therapeutic gene mTNF-α. This could be diminished by means of insertion of a second radio-responsive element, the Cheo box, which binds the DinR repressor (a LexA homolog) in non-induced conditions. The addition of an extra Cheo box to the *recA* promoter region resulted in a 412% increase in mTNF-α secretion at 2.5 h after a 2-Gy irradiation of the culture compared to nonirradiated cultures. Moreover, it was shown that the incorporation of the radio-responsive Cheo sequence in the promoter region rendered the constitutive *eglA* promoter into a radio-inducible one (51). In conclusion, this radio-responsive Cheo box looks promising for the temporal control of therapeutic gene expression in genetically modified clostridia when combined with radiotherapy. It would be worthwhile to see if this temporal control could also be obtained *in vivo*.

C. NOVYI-NT: THE REVIVAL OF NATIVE ONCOLYTIC *CLOSTRIDIA* FOR BIOTHERAPY PURPOSES

Convinced of the clinical potential of bacterial mediated tumor therapy, B. Vogelstein's group once more explored and selected anaerobes, which could be valuable for tumor treatment. Accordingly, at the turn of this century, they assessed the tumor colonization capacity of 26 common anaerobes belonging to the genera *Bifidobacterium*, *Lactobacillus*, and *Clostridium* (52). Of these bacteria, only *C. novyi* ATCC 19402 and *C. sordellii* ATCC 9714 exhibited extensive spreading throughout the poorly vascularized tumor regions. However, intratumoral vegetative growth of both strains was detrimental for the B16 tumor-bearing mice within 16–18 h as a consequence of the production of a lethal toxin. The researchers succeeded to "disarm" *C. novyi* by deleting the gene which encodes the α-toxin. They named this strain *C. novyi-NT*. Injection of *C. novyi-NT* spores into the animal tumor model was without side effects, while it caused hemorrhagic necrosis exclusively within tumor tissue, and tumor regression following the abundant intratumoral growth of *C. novyi-NT* (52). The vast tumor infiltration capacity of *C. novyi-NT*, which was comparable to that of the wild-type strain, was explained by the high bacterial mobility and the *C. novyi*-mediated destruction of viable tumor cells at the border of necrotic regions releasing metabolites which were favorable to the bacterial growth. Nevertheless, the extreme sensitivity of *C. novyi-NT* to oxygen halted bacterial spreading when they reached the well-oxygenated tumor rim. Hence, the issue of tumor recurrence from a viable rim formerly encountered with *C. oncolyticum* M-55 spore administration apparently still remained and combined therapies seemed inevitable.

COBALT: Combined Bacteriolytic Therapy

In consequence of the remaining viable tumor cells upon *C. novyi-NT* treatment of nude and C57BL/6 mice, combination with conventional chemotherapeutics was proposed in order to attack the tumor from both the inside and outside. In the first instance, different classes of compounds were evaluated: DNA-damaging compounds such as mitomycin C that selectively kills tumor cells, and antivascular agents such as dolastatin-10 (D10), a microtubule-binding agent. Intriguingly, *C. novyi-NT* treatment added to D10 or mitomycin C administration in mice tumor models resulted in a drastic antineoplastic effect and tumor shrinkage, phenomena that were not observed using the chemotherapeutic drugs alone. Dependent on the tumor model, even complete tumor cure was obtained in a considerable number of mice. This COBALT treatment, however, was accompanied with severe toxicity as up to 30% of the animals died as a consequence of the release of toxic products from the tumor or the tumor lysis syndrome (52). In an attempt to reduce this toxicity, the therapeutic effect of other microtubule inhibitors and stabilizers in combination with *C. novyi-NT* were evaluated on xenografts in nude or C57BL/6 mice.

Compounds of the first group, for example HTI-286 and vinorelbine, caused massive hemorrhagic necrosis as the blood flow was radically shut down, thereby enlarging the hypoxic regions in which *C. novyi-NT* vegetative cells thrive well. In contrast, treatment with microtubule stabilizers such as docetaxel and MAC-321 resulted in slow tumor regression, but neoplastic cells remained in poorly perfused tumor regions which then could be eradicated by *C. novyi-NT*. Although the long-term responses were highly tumor-dependent, these newly tested combinations revealed the wanted decrease in toxicity as compared to the use of D10. While still 4–8% of mice undergoing COBALT with microtubule destabilizers died, animal death was not observed upon *C. novyi-NT* treatment combined with microtubule stabilizers (53). This study further evidenced that the combined treatment of vascular-targeting agents and *Clostridium* can reduce the critical tumor threshold size as complete tumor regressions could also be observed in tumors as small as $100\,mm^3$. Another promising microtubule-stabilizing agent used in the COBALT approach was (+)-2,3-anhydrodiscodermolide. A single systemic administration of *C. novyi-NT* in combination with this drug led to rapid and complete regression of subcutaneous HCT116 tumors in mice (54). Remarkably, this microtubule-stabilizing compound induced a massive hemorrhagic necrosis within the tumor, an effect which was only achieved when using microtubule-destabilizing agents. Further investigations are required to unravel the mechanisms of actions of the observed antitumor responses.

C. novyi-NT in Combination with Radiotherapy

Conventional radiotherapy was likewise combined with the *C. novyi-NT*-mediated tumor therapy in order to kill the tumor cells escaping the bacterial antitumor effect and to achieve total tumor cure. Using three modes of radiotherapy, being external beam radiation, brachytherapy, and radioimmunotherapy, the additive effect of *C. novyi-NT* treatment to radiotherapy on tumor regression and long-term remission in several mouse models was demonstrated (55). The radio-enhancing effect of *C. novyi-NT* is assumed to be caused by the bacterial attack of cells that are the least sensitive to irradiation. Importantly, the diminished radiotherapy efficiency experienced with larger tumors disappeared when *C. novyi-NT* treatment was added to radiotherapy as the antitumor effect of the combined therapy was independent of tumor volume. A vast advantage of the combined therapy lies within the fact that the effectual radiation dose could be lowered compared to radiotherapy alone. This is of importance for tumor therapy at sites in which the commonly used dose would be injurious and for the reduction of damage to healthy tissues (55).

C. novyi-NT and Liposomal Drug Delivery

A third, innovative combination therapy links *C. novyi-NT* treatment to liposomal drug delivery. The development of sterically stabilized liposomes (SSLs) as drug delivery system for cancer therapy has dramatically reduced the systemic toxicity encountered with cytotoxic drugs. Furthermore, since the robust, drug-loaded SSLs that have a unique lipid formulation cannot pass the gaps found in normal endothelium, but can enter the significantly larger pores in the aberrant tumor vessels, these liposomes selectively infiltrate tumor tissues. The drawback of the physical robustness of SSLs is the slow degradation at the tumor site that impedes the high drug concentration needed for provoking a therapeutic response (56). Because of the hemolytic properties of *C. novyi-NT*, it was hypothesized that the responsible membrane-disrupting enzymes could be exploited to enhance the release of liposome-encapsulated drugs within the tumor. Indeed, while treatment of tumor-bearing mice with the DNA-damaging doxorubicin or the liposome-formulated variant Doxil on its own did not result in prolonged therapeutic effects, combined treatment of Doxil with *C. novyi-NT* led to complete tumor regression in 100% of mice and 65% of them survived a period of 90 days. The applicability of this system for any chemotherapeutic agent that can be encapsulated in a liposome was proven by using liposomes loaded with the topoisomerase inhibitor CPT-11, extensively used in cancer therapy. Combining these liposomes with *C. novyi-NT* treatment provided similar tumor therapeutic effects as observed with Doxil (57). Further research identified the liposome-disrupting factor as a new lipase, called liposomase, encoded by gene NT01CX2047 (GeneID: 4539629). Surprisingly, the substitution of the Ser residue by Gly in the highly conserved GXSXG motif abolished the lipase activity of the secreted liposomase but retained substantial liposome-disrupting activity (57). The secreted liposomase was proven to act through physical perturbation of the liposome membrane lipid layer, but the real liposome disruption mechanism still needs to be elucidated.

Interestingly, transcriptome analyses carried out on *C. novyi-NT* infected tumors revealed that the liposomase gene together with a number of genes involved in fatty acid and lipid metabolism were expressed at higher amount in tumors compared to *in vitro* (58). This *C. novyi*-encoded liposomase opens new promising possibilities for multimodality therapy of solid tumors, not only for the liposomal delivery of toxic compounds to tumors but possibly also in combination with genetically modified clostridia producing prodrug-converting enzymes or combined with immunotherapeutic approaches.

Long-Term Tumor Cure upon a Single *C. novyi*-NT Treatment

Further investigations, focused on the importance of the immune system on the tumor therapeutic effects provided by combined treatments described above, surprisingly revealed that up to 30% of immunocompetent animals carrying syngeneic tumors—mice as well as rabbits—exhibited complete tumor regression upon administration of *C. novyi-NT* spores without any additional treatment, resulting in long-term tumor cure (59). As total cure was not obtained using immunodeficient mice with human xenografts, these results clearly indicated that total tumor cure upon *Clostridium* treatment is due to an additional immune-mediated killing of the tumor cells. The proof that vegetative *C. novyi-NT* cells within a tumor also triggers a tumor cell-directed immune response was convincingly given by the rejection of a subsequent challenge to the same tumor cells in the animals cured following *C. novyi-NT* treatment, whereas this was not the case for surgically cured animals. This anti-tumor immune response is probably due to an adjuvant effect of the local *C. novyi-NT* infection as *C. novyi-NT* germination was indeed quickly followed by the appearance of cytokines and chemokines, such as interleukin-6 (IL-6), granulocyte-colony stimulating factor (G-CSF), and the keratinocyte-derived chemokine (KC) which stimulate the infiltration of inflammatory cells. This massive influx of immune cells gives an opportunity to activate an immune response against tumor antigens derived from the destroyed tumor cells and thus to burst through the tumor self-tolerance barrier. The only 30% cure rate in immunocompetent animals was suggested to be the outcome of the balance between bacteriolysis, angiogenesis, regrowth of the residual tumor cells, and the rate of immune response development (59). Nevertheless, this adjuvant effect of *Clostridium* has surely to be taken into account and exploited in the further development of *Clostridium* strains as antitumor tools.

C. novyi-NT Spores on the Road to Clinical Application

In view of the clinical application of *C. novyi-NT* spores as antitumor bugs, extensive pharmacological and toxicological studies with this oncolytic strain were carried out on tumor-bearing mice (60). In the first instance, these experiments revealed that systemically injected spores were quickly cleared from the circulation, with 95% spore fading within 1 day, and at that time the majority of viable spores were present in liver and spleen. Moreover, by means of three different models, it was proven that *C. novyi-NT* does not germinate in hypoxic non-neoplastic tissues such as myocardial infarcts, a concern often expressed. Spore injection caused no toxicity in animals without tumors but toxicity was clearly related to bacterial growth in tumor tissue. However, this toxicity, which appeared to be more severe when larger tumors and higher spore doses were employed, could easily be controlled by simply systemic hydration (60). In 2006, a Phase I clinical trial has been started to evaluate the safety of *C. novyi-NT* spore treatment in humans (NCT00358397, www.clinicaltrials.gov); however, meanwhile, this study was suspended for unknown reasons.

MODULATION OF *C. PERFRINGENS* TO AN ONCOTHERAPEUTIC MICROBE

Besides *C. novyi-NT*, very recently, *C. perfringens* has been considered as an antitumor bug for its ability to selectively colonize and induce necrosis in advanced pancreatic cancer, for which conventional treatment modalities were insufficient. In order to use it in tumor biotherapy, *C. perfringens* mutant strains were developed by means of double homologous recombination techniques. By deleting the superoxide dismutase (*sod*) gene, the oxygen tolerance of the wild-type strain was lowered, thereby elevating the maximal tolerated dose of spores with 1 log and enhancing the tumor selectivity. In addition, to evade the severe inflammatory cellular responses, the inflammation-suppressing Panton-Valentine Leukocidin (*PVL*) gene derived from *Staphylococcus aureus* was inserted into the chromosome. Upon intravenous injection into immunocompetent pancreas tumor-bearing mice, this newly constructed $Cp/sod^{-}/PVL$ strain showed considerably enhanced tumor selectivity and oncotherapeutic potency compared to the wild type leading to a prolonged survival, but this treatment did not destroy all tumor cells (61).

In a next step, the role of α-toxin, the major virulence factor of *C. perfringens*, for its oncotherapeutic effect was investigated (62). Therefore, the phospholipase C (*plc*) gene encoding the α-toxin, which is the most important toxin for gas gangrene development, was knocked out in the $Cp/sod^{-}/PVL$ strain with the aim to considerably enhance the safety profile of the bacterium. The $Cp/plc^{-}/sod^{-}/PVL$ mutant was indeed unable to induce gas gangrene in mice whereas it kept the same growth rate, spore yield, germination efficiency and, more importantly, an equally oncotherapeutic effect as the parental mutant. Conclusively, the oncopathic capacity of *C. perfringens* was independent of its α-toxin. Moreover, it was shown that the presence of PVL in the mutant strains did not compromise the host immune response to other pathogens (61).

CDAT: *CLOSTRIDIUM*-DIRECTED ANTIBODY THERAPY

Immunotherapy has become an additional, very attractive modality for many malignant diseases that face shortcomings in conventional treatments. Monoclonal antibodies directed against tumor-associated antigens are now commonly used therapeutic agents that mediate antibody-dependent cellular toxicity, complement-dependent lysis, and activation of tumor antigen-specific T cells. Moreover, these antibodies can affect crucial signaling pathways which sustain the malignant phenotype (63, 64). With the aim to transfer tumor therapeutic antibodies to the hypoxic, therapy-resistant tumor regions, Groot et al. (65) conceived the idea to use the exquisitely tumor-targeting clostridia as antibody carriers. To proof the concept of antibody production by clostridia, they constructed recombinant *C. novyi-NT* and *C. sporogenes* strains that secrete small, soluble single-domain fragments derived from the unique variable region of dromedary heavy-chain antibodies (VHHs) directed against the hypoxia-inducible factor 1 alpha (HIF-1α), a key regulator of the hypoxia-induced response. HIF is frequently overexpressed in solid tumors, and the HIF-upregulated gene transcription can promote survival, tumor angiogenesis and a malignant phenotype, and has been correlated with poor prognosis (9, 66). Hence, interrupting the hypoxia-induced pathways would have a dramatic impact on tumor progression. The *Clostridium*-produced, purified VHH (AG2) antibodies were shown to be functional in HIF-1α binding (65). This study shows that CDAT offers promising prospects for the use of clostridia as antitumor tools.

ADVANTAGES OF CLOSTRIDIAL SPORES AS ANTITUMOR TOOLS

Despite the many objections raised against *Clostridium*-mediated tumor therapy, this biotherapy has a great potential and deals with many drawbacks of conventional and recently developed cancer therapies. The major benefits of using *Clostridium* in tumor therapy can be summarized as follows. (a) It has exquisite, unequalled tumor selectivity as it thrives only in the unique tumor environment that provides the required hypoxic conditions and metabolic needs, and not in oxygenized, non-neoplastic regions. (b) In contrast to viral vectors, *Clostridium* acts as an extracellular therapeutic agent that does not need to be integrated in the genome of neoplastic cells nor does its replication depend directly on the neoplastic cell itself. This consequently avoids the risk of unwanted insertions or mutations in the eukaryotic genome. (c) *Clostridium*-mediated tumor killing is often not only based on the action of the recombinant compound produced by the genetically modified clostridia but may be also a consequence of the innate antitumor capacity of the bacterium. It competes with the cancer cells for the limited nutrients and additionally produces extracellular, hydrolytic enzymes attacking both tumor cells and many nonmalignant stromal cells, such as fibroblasts, endothelial cells, and inflammatory cells. Besides these direct killing effects, the intratumoral, vegetative *Clostridium* growth may act as an immunoadjuvant that evokes an indirect antitumor effect by means of immune cell activation thereby breaking up tumor self-tolerance. This versatility of killing effects is not found in any other cancer therapy. (d) The clostridial spores can be simply and economically produced. The obtained spore suspensions are quite stable and can be stored at room temperature for at least 3–6 months. Subsequent spore administration only requires an intravenous injection which is sufficient for tumor colonization, far most the easiest way of application. (e) *Clostridium*-based antitumor therapy on its own will probably be not sufficient to obtain complete tumor control, but it could readily be combined with other therapy modalities lowering the doses of the toxic agents and diminishing possible harmful side effects. (f) Last but not least, objections have been made concerning the safety of the *Clostridium*-mediated tumor therapy. Nevertheless, a lot of experiments have shown that this vector system can be safely used. *Clostridium* spore injections are well tolerated in animals as well as in human beings, which was first demonstrated by the experiments of self-injection by Möse and Möse and later on with the clinical trials using *C. butyricum* M-55. The toxicity linked to *C. novyi-NT* tumor colonization could be effectively controlled by systemic hydration. With respect to safety, a major advantage of using a bacterial system is the possibility to easily eliminate the vector from the body by means of antibiotics in case of an adverse effect.

This was proven via experiments in which metronidazole was administered to rats bearing *C. sporogenes*-colonized tumors. Results illustrated that the tumor and the body could be cleared from *C. sporogenes* within 4 days, demonstrating that the antibiotic can indeed reach the hypoxic, badly perfused tumor regions (41, 46). Upon elimination of the clostridia, it was yet possible to repeat the *Clostridium* treatment in rhabdomyosarcoma-bearing rats, indicating that the immune response against *Clostridium* spores or vegetative cells does not hinder repeated tumor colonization (46). Moreover, repeated cycles of combined treatment of hyperthermia, irradiation, and *C. oncolyticum* spore administration to Harding-Passey melanoma-bearing mice revealed formerly a higher percentage of cure compared to only one treatment cycle (26).

FURTHER IMPROVEMENTS AND FUTURE PERSPECTIVES

Genetic modification of clostridia offers a plenitude of possibilities to construct a variety of *Clostridium*-based antitumor tools. In theory, there are no limitations in therapeutically active genes that can be cloned in *Clostridium*. However, introduction of the genes of interest in clostridia has been obtained so far by plasmid-based transformation or conjugation and subsequent selection of transformants or conjugants using plasmid-encoded antibiotic resistance markers. For evident reasons, absence of such markers is a prerequisite for clinical application. Furthermore, plasmid instability would also be problematic, since this would lead to loss of the gene needed for therapeutic action. Therefore, chromosomal integration of the therapeutic gene under control of a strong (inducible) promoter is needed. Since single crossover integration mutants are segregationally unstable through the presence of a directly repeated DNA sequence flanking the integrated sequence, host-dependent double homologous recombination or site-specific integration is preferred. However, such recombinational events have been only rarely reported in *Clostridium* spp. except for *C. perfringens*. The recently developed ClosTron, a *Clostridium*-specific targetron (67), might offer a solution. This insertional system is based on the mobile group II intron from *Lactococcus lactis*, which has a broad host range as its mobility is intron-encoded and essentially independent of host factors. Using this intron integration, the α-toxin gene of *C. perfringens* was inactivated (68) and insertional mutants were achieved from *C. difficile* and *C. acetobutylicum*, and for the first time also from *C. botulinum* and *C. sporogenes* (67). Importantly, the group II intron-based insertion system enables "knock-out" as well as "knock-in" modifications, which provides the possibility to also introduce the genes of interest into the genomes of clostridia applied as antitumor bugs. Moreover, the intron insertion is stable, and in principle, quite large DNA fragments can be used for genomic integration (67). Together with the increasing number of sequenced genomes from clostridia, including *C. acetobutylicum* (69), *C. perfringens* (70), and *C. novyi-NT* (58), the newly available sequence modification system opens a spectrum of possibilities to further adapt the clostridia in view of their antitumor capacity. Therapeutically interesting clostridial genes (e.g., the *C. novyi* liposomase gene or particular *C. sporogenes* protease genes) can easily be exchanged from one species to another so that the most effective antitumor *Clostridium* strain can be engineered.

Another issue of which the importance has already been demonstrated but definitely deserves more fundamental research, is the immune response evoked by the presence of vegetative clostridial cells within a tumor. As it is already known that this immune response has an impact on the final outcome of the therapy (59), a better insight in the cytokine/chemokine profile of *Clostridium*-colonized tumors makes a targeted enhancement of the antitumoral immune response feasible. Hence, genetically modified *Clostridium* strains that produce *in situ* a specific cytokine or chemokine at an early phase of tumor growth, might increase infiltration and/or activation of immune cells finally overruling the tumor self-tolerance and thus improving the antitumoral immune activity. As such, the restriction of *Clostridium* to solely destroy tumor cells within the hypoxic regions might be taken away, and *Clostridium*-mediated killing might be broadened to all cancer cells, even metastatic ones.

As mentioned earlier, *Clostridium* colonization of a tumor had previously also been used for tumor serodiagnosis. Due to recently developed genetic manipulation tools and tumor imaging technologies, the idea of using genetically modified clostridia to generate noninvasive tumor diagnostics will become a reality. If the introduction of a gene encoding a light-emitting or fluorescent protein into the clostridial genome and a bioactive protein expression would be achieved, these genetically modified strains could be useful to estimate the degree of the tumor colonization, a prerequisite for prodrug administration, or to monitor tumor reduction. They could also be applied as diagnostic imaging markers for the presence of a primary tumor and metastases.

Taken together, the potential of *Clostridium*-mediated tumor therapy has clearly been demonstrated and the first clinical trials with the "second-generation" antitumor clostridia are bound to start. Given that antitumor *Clostridium* strains preferably attack the tumor cells within the therapy-resistant hypoxic regions, they can be extremely suitable for combined therapies directed against the oxygenated tumor areas. Further fine-tuning of relevant bacteria becomes feasible via recently developed, *Clostridium*-adapted genetic modification procedures. This will certainly allow accelerating the clinical implementation of clostridia as antitumor and maybe tumor diagnostic tools in a near future.

REFERENCES

1. Brazier J.S., Duerden B.I., Hall V., Salmon J.E., Hood J., Brett M.M., McLauchlin J., and George R.C. 2002. Isolation and identification of *Clostridium* spp. from infections associated with the injection of drugs: experiences of a microbiological investigation team. J. Med. Microbiol. **51**: 985–989.

2. Brett M.M., Hood J., Brazier J.S., Duerden B.I., and Hahné S.J. 2005. Soft tissue infections caused by spore-forming bacteria in injecting drug users in the United Kingdom. Epidemiol. Infect. **133**: 575–582.
3. Dürre P. 2007. Biobutanol: an attractive biofuel. Biotechnol. J. **2**: 1525–1534.
4. Dürre P. 2008. Fermentative butanol production, bulk chemical and biofuel. Ann. N Y Acad. Sci. **1125**: 353–362.
5. Lartigau E., Le Ridant A.M., Lambin P., Weeger P., Martin L., Sigal R., Lusinchi A., Luboinski B., Eschwege F., and Guichard M. 1993. Oxygenation of head and neck tumors. Cancer. **71**: 2319–2324.
6. Vermeulen P.B., Verhoeven D., Hubens G., Van Marck E., Goovaerts G., Huyghe M., De Bruijn E.A., Van Oosterom A.T., and Dirix L.Y. 1995. Microvessel density, endothelial cell proliferation and tumour cell proliferation in human colorectal adenocarcinomas. Annals Oncology. **6**: 59–64.
7. Hermans R., Lambin P., Van den Bogaert W., Haustermans K., Van der Goten A., and Baert A.L. 1997. Non-invasive tumor perfusion measurement by dynamic CT: preliminary results. Radiother. Oncol. **44**: 159–162.
8. Wouters B.G., van den Beucken T., Magagnin M.G., Lambin P., and Koumenis C. 2004. Targeting hypoxia tolerance in cancer. Drug Resist. Updates. **7**: 25–40.
9. Liao D., and Johnson R.S. 2007. Hypoxia: a key regulator of angiogenesis in cancer. Cancer Metastasis Rev. **26**: 281–290.
10. Parker R.C., Plummer H.C., Siebenmann C.O., and Chapman M.G. 1947. Effect of histolyticus infection and toxin on transplantable mouse tumours. Proc. Soc. Exp. Biol. Med. **66**: 461–465.
11. Malmgren R.A., and Flanigan C.C. 1955. Localization of the vegetative form of *Clostridium tetani* in mouse tumors following intravenous spore administration. Cancer Res. **15**: 473–478.
12. Carey R.W., Holland F.W., Whang H.Y., Neter E., and Bryant B. 1967. Clostridial oncolysis in man. Eur. J. Cancer. **3**: 37–46.
13. Möse J.R., and Möse G. 1964. Oncolysis by clostridia. I. Activity of *Clostridium butyricum* (M-55) and other nonpathogenic clostridia against the Ehrlich carcinoma. Cancer Res. **24**: 212–216.
14. Gericke D., and Engelbart K. 1964. Oncolysis by clostridia. II. Experiments on a tumor spectrum with a variety of clostridia in combination with heavy metal. Cancer Res. **24**: 217–221.
15. Thiele E.H., Arison R.N., and Boxer G.E. 1964. Oncolysis by clostridia. III. Effects of clostridia and chemotherapeutic agents on rodent tumors. Cancer Res. **24**: 222–233.
16. Engelbart K., and Gericke D. 1964. Oncolysis by clostridia. V. Transplanted tumors of the hamster. Cancer Res. **24**: 239–243.
17. Cairo M.S., and Bishop M. 2004. Tumour lysis syndrome: new therapeutic strategies and classification. Br. J. Hematol. **127**: 3–11.
18. Minton N.P., Brown J.M., Lambin P., and Anné J. 2001. Clostridia in cancer therapy. In: Bahl H., Dürre P., eds. Clostridia—Biotechnology and Medical Applications. Weinheim: Wiley-VCH Verlag GmbH, pp. 251–270.
19. Schmidt W., Fabricius E.M., and Schneeweiss U. 2006. The tumour-*Clostridium* phenomenon: 50 years of developmental research. Int. J. Oncol. **29**: 1479–1492.

20. Rousseau P., Chagnon A., and Fredette V. 1970. Effect of oncolytic anaerobic spores on animal cell cultures. Cancer Res. **30**: 849–854.
21. Chagnon A., Hudon C., McSween G., Vinet G., and Fredette V. 1972. Cytotoxicity and reduction of animal cell growth by *Clostridium* M-55 spores and their extracts. Cancer. **29**: 431–434.
22. Schlechte H., Baumbach L., and Elbe B. 1981. Cocultivation of *Clostridium oncolyticum* with normal and tumour cell lines. Arch. Geschwulstforsch. **51**: 51–57.
23. Marth E., and Möse J.R. 1987. Oncolysis by *Clostridium oncolyticum* M55 and subsequent enzymatic determination of sialic acid in serum. Zentralbl. Bakteriol. Mikrobiol. Hyg. [A]. **265**: 33–44.
24. Dietzel F., Gericke D., and König W. 1976. Tumor hyperthermia using high frequency for increase of oncolysis by *Clostridium butyricum* (M 55). Strahlentherapie. **152**: 537–541.
25. Dietzel F., and Gericke D. 1977. Intensification of the oncolysis by clostridia by means of radio-frequency hyperthermy in experiments on animals—dependence on dosage and on intervals. Strahlentherapie. **153**: 263–266.
26. Gericke D., Dietzel F., König W., Rüster I., and Schumacher L. 1979. Further progress with oncolysis due to apathogenic clostridia. Zentralbl. Bakteriol. [Orig A]. **243**: 102–112.
27. Möse J.R. 1979. Experiments to improve the oncolysis effect of clostridial-strain M55. Zentralbl. Bakteriol. [Orig A]. **244**: 541–545.
28. Heppner F. 1986. The glioblastoma multiforme: a lifelong challenge to the neurosurgeon. Neurochirurgia (Stuttg). **29**: 9–14.
29. Schlechte H., and Elbe B. 1988. Recombinant plasmid DNA variation of *Clostridium oncolyticum*—model experiments of cancerostatic gene transfer. Zentralbl. Bakteriol. Mikrobiol. Hyg. [A]. **268**: 347–356.
30. Lemmon M.J., van Zijl P., Fox M.E., Mauchline M.L., Giaccia A.J., Minton N.P., and Brown J.M. 1997. Anaerobic bacteria as a gene delivery system that is controlled by the tumor microenvironment. Gene Ther. **4**: 791–796.
31. Lambin P., Theys J., Landuyt W., Rijken P., van der Kogel A., van der Schueren E., Hodgkiss R., Fowler J., Nuyts S., de Bruijn E., Van Mellaert L., and Anné J. 1998. Colonisation of *Clostridium* in the body is restricted to hypoxic and necrotic areas of tumours. Anaerobe. **4**: 183–188.
32. Minton N.P., Mauchline M.L., Lemmon M.J., Brehm J.K., Fox M., Michael N.P., Giaccia A., and Brown J.M. 1995. Chemotherapeutic tumour targeting using clostridial spores. FEMS Microbiol. Rev. **17**: 357–364.
33. Fox M.E., Lemmon M.J., Mauchline M.L., Davis T.O., Giaccia A.J., Minton N.P., and Brown J.M. 1996. Anaerobic bacteria as a delivery system for cancer gene therapy: *in vitro* activation of 5-fluorocytosine by genetically engineered clostridia. Gene Ther. **3**: 173–178.
34. Theys J., Landuyt W., Nuyts S., Van Mellaert L., van Oosterom A., Lambin P., and Anné J. 2001. Specific targeting of cytosine deaminase to solid tumors by engineered *Clostridium acetobutylicum*. Cancer Gene Ther. **8**: 294–297.
35. Theys J., Nuyts S., Landuyt W., Van Mellaert L., Dillen C., Böhringer M., Dürre P., Lambin P., and Anné J. 1999. Stable *Escherichia coli-Clostridium acetobutylicum* shuttle vector for secretion of murine tumor necrosis factor alpha. Appl. Environ. Microbiol. **65**: 4295–4300.

36. Barbé S., Van Mellaert L., Theys J., Geukens N., Lammertyn E., Lambin P., and Anné J. 2005. Secretory production of biologically active rat interleukin-2 by *Clostridium acetobutylicum* DSM792 as a tool for anti-tumor treatment. FEMS Microbiol. Lett. **246**: 67–73.

37. Barbé S. 2005. Optimization and evaluation of the *Clostridium*-mediated transfer system of therapeutic proteins to solid tumours. PhD dissertation, K.U. Leuven.

38. Younes E., Haas G.P., Dezso B., Ali E., Maughan R.L., Kukuruga M.A., Montecillo E., Pontes J.E., and Hillman G.G. 1995. Local tumour irradiation augments the response to IL-2 therapy in a murine renal adenocarcinoma. Cell Immunol. **165**: 243–251.

39. Damaria S., Bhardwaj N., McBride W.H., and Formenti S.C. 2005. Combining radiotherapy and immunotherapy: a revived partnership. Int. J. Radiat. Oncol. Biol. Phys. **63**: 655–666.

40. Liu S.C., Minton N.P., Giaccia A.J., and Brown J.M. 2002. Anticancer efficacy of systemically delivered anaerobic bacteria as gene therapy vectors targeting tumor hypoxia/necrosis. Gene Ther. **9**: 291–296.

41. Theys J., Pennington O., Dubois L., Anlezark G., Vaughan T., Mengesha A., Landuyt W., Anné J., Burke P.J., Dûrre P., Wouters B.G., Minton N.P., and Lambin P. 2006. Repeated cycles of *Clostridium*-directed enzyme prodrug therapy result in sustained antitumour effects *in vivo*. Br. J. Cancer. **95**: 1212–1219.

42. Liu S.C., Ahn G.O., Kioi M., Dorie M.J., Patterson A.V., and Brown JM. 2008. Optimized *Clostridium*-directed enzyme prodrug therapy improves the antitumor activity of the novel DNA cross-linking agent PR-104. Cancer Res. **68**: 7995–8003.

43. Patterson A.V., Ferry D.M., Edmunds S.J., Gu Y., Singleton R.S., Patel K., Pullen S.M., Hicks K.O., Syddall S.P., Atwell G.J., Yang S., Denny W.A., and Wilson W.R. 2007. Mechanism of action and preclinical antitumor activity of the novel hypoxia-activated DNA cross-linking agent PR-104. Clin. Cancer Res. **13**: 3922–3932.

44. Singleton R.S., Guise C.P., Ferry D.M., Pullen S.M., Dorie M.J., Brown J.M., Patterson A.V., and Wilson W.R. 2009. DNA cross-links in human tumor cells exposed to the prodrug PR-104A: relationships to hypoxia, bioreductive metabolism, and cytotoxicity. Cancer Res. **69**: 3884–3891.

45. Landuyt W., Verdoes O., Darius D.O., Drijkoningen M., Nuyts S., Theys J., Stockx L., Wynendaele W., Fowler J.F., Maleux G., Van den Bogaert W., Anné J., van Oosterom A., and Lambin P. 2000. Vascular targeting of solid tumours: a major "inverse" volume-response relationship following combretastatin A-4 phosphate treatment of rat rhabdomyosarcomas. Eur. J. Cancer. **36**: 1833–1843.

46. Theys J., Landuyt W., Nuyts S., Van Mellaert L., Bosmans E., Rijnders A., Van Den Bogaert W., van Oosterom A., Anné J., and Lambin P. 2001. Improvement of *Clostridium* tumour targeting vectors evaluated in rat rhabdomyosarcomas. FEMS Immunol. Med. Microbiol. **30**: 37–41.

47. Lambin P., Theys J., Nuyts S., Landuyt W., Van Mellaert L., and Anné J. 2002. *Clostridium*-mediated transfer of therapeutic proteins to solid tumors. In: Curiel D.T., Douglas J.T., eds. Vector Targeting for Therapeutic Gene Delivery. New York: Wiley-Liss, pp. 527–546.

48. Kufe D., and Weichselbaum R. 2003. Radiation therapy: activation for gene transcription and the development of genetic radiotherapy—therapeutic strategies in oncology. Cancer Biol. Ther. **2**: 326–329.
49. Nuyts S., Van Mellaert L., Theys J., Landuyt W., Lambin P., and Anné J. 2001. The use of radio-induced bacterial promoters in anaerobic conditions: a means to control gene expression in *Clostridium* mediated therapy for cancer. Radiat. Res. **155**: 716–723.
50. Nuyts S., Van Mellaert L., Theys J., Landuyt W., Bosmans E., Anné J., and Lambin P. 2001. Radio-responsive recA promoter significantly increases TNFα production in recombinant clostridia after 2 Gy irradiation. Gene Ther. **8**: 1197–1201.
51. Nuyts S., Van Mellaert L., Barbé S., Lammertyn E., Theys J., Landuyt W., Bosmans E., Lambin P., and Anné J. 2001. Insertion or deletion of the Cheo box modifies radiation inducibility of *Clostridium* promoters. Appl. Environ. Microbiol. **67**: 4464–4470.
52. Dang L.H., Bettegowda C., Huso D.L., Kinzler K.W., and Vogelstein B. 2001. Combination bacteriolytic therapy for the treatment of experimental tumors. Proc. Natl. Acad. Sci. U S A. **98**: 15155–15160.
53. Dang L.H., Bettegowda C., Agrawal N., Cheong I., Huso D., Frost P., Loganzo F., Greenberger L., Barkoczy J., Pettit G.R., Smith A.B. III, Gurulingappa H., Khan S., Parmigiani G., Kinzler K.W., Zhou S., and Vogelstein B. 2004. Targeting vascular and avascular compartments of tumors with *C. novyi-NT* and anti-microtubuli agents. Cancer Biol. Ther. **3**: 326–337.
54. Smith A.B. III, Freeze B.S., LaMarche M.J., Sager J., Kinzler K.W., and Vogelstein B. 2005. Discodermolide analogues as the chemical component of combination bacteriolytic therapy. Bioorg. Med. Chem. Lett. **15**: 3623–3626.
55. Bettegowda C., Dang L.H., Abrams R., Huso D.L., Dillehay L., Cheong I., Agrawal N., Borzillary S., McCaffery J.M., Watson E.L., Lin K.S., Bunz F., Baidoo K., Pomper M.G., Kinzler K.W., Vogelstein B., and Zhou S. 2003. Overcoming the hypoxic barrier to radiation therapy with anaerobic bacteria. Proc. Natl. Acad. Sci. U S A. **100**: 15083–15088.
56. Cheong I., Huang X., Thornton K., Diaz L.A. Jr., and Zhou S. 2007. Targeting cancer with bugs and liposomes: ready, aim, fire. Cancer Res. **67**: 9605–9608.
57. Cheong I., Huang X., Bettegowda C., Diaz L.A. Jr., Kinzler K.W., Zhou S., and Vogelstein B. 2006. A bacterial protein enhances the release and efficiency of liposomal cancer drugs. Science. **314**: 1308–1311.
58. Bettegowda C., Huang X., Lin J., Cheong I., Kohli M., Szabo S.A., Zhang X., Diaz L.A. Jr., Velculescu V.E., Parmigiani G., Kinzler K.W., Vogelstein B., and Zhou S. 2006. The genome and transcriptomes of the anti-tumor agent *Clostridium novyi-NT*. Nat. Biotechnol. **24**: 1573–1580.
59. Agrawal N., Bettegowda C., Cheong I., Geschwind J.F., Drake C.G., Hipkiss E.L., Tatsumi M., Dang L.H., Diaz L.A. Jr., Pomper M., Abusedera M., Wahl R.L., Kinzler K.W., Zhou S., Huso D.L., and Vogelstein B. 2004. Bacteriolytic therapy can generate a potent immune response against experimental tumors. Proc. Natl. Acad. Sci. U S A. **101**: 15172–15177.
60. Diaz L.A. Jr., Cheong I., Foss C.A., Zhang X., Peters B.A., Agrawal N., Bettegowda C., Karim B., Liu G., Khan K., Huang X., Kohli M., Dang L.H., Hwang P., Vogelstein A., Garrett-Mayer E., Kobrin B., Pomper M., Zhou S., Kinzler

K.W., Vogelstein B., and Huso D.L. 2005. Pharmacologic and toxicologic evaluation of *C. novyi-NT* spores. Toxicol. Sci. **88**: 562–575.

61. Li Z., Fallon J., Mandeli J., Wetmur J., and Woo S.L.C. 2008. A genetically enhanced anaerobic bacterium for oncopathic therapy of pancreatic cancer. J. Natl. Cancer Inst. **100**: 1389–1400.
62. Li Z., Fallon J., Mandeli J., Wetmur J., and Woo S.L.C. 2009. The oncopathic potency of *Clostridium perfringens* is independent of its α-toxin gene. Hum. Gene Ther. **20**: 1–8.
63. Lee S.C., López-Albaitero A., and Ferris RL. 2009. Immunotherapy of head and neck cancer using tumor antigen-specific monoclonal antibodies. Curr. Oncol. Rep. **11**: 156–162.
64. Weiner L.M., Dhodapkar M.V., and Ferrone S. 2009. Monoclonal antibodies for cancer immunotherapy. Lancet. **373**: 1033–1040.
65. Groot A.J., Mengesha A., van der Wall E., van Diest P.J., Theys J., and Vooijs M. 2007. Functional antibodies produced by oncolytic clostridia. Biochem. Biophys. Res. Commun. **364**: 985–989.
66. Magagnin M.G., Koritzinsky M., and Wouters B.G. 2006. Patterns of tumor oxygenation and their influence on the cellular hypoxic response and hypoxia-directed therapies. Drug Resist. Update. **9**: 185–197.
67. Heap J.T., Pennington O.J., Cartman S.T., Carter G.P., and Minton NP. 2007. The ClosTron: a universal knock-out system for the genus *Clostridium*. J. Microbiol. Methods. **70**: 452–464.
68. Chen Y., McClane B.A., Fisher D.J., Rood J.I., and Gupta P. 2005. Construction of an alpha toxin gene knockout mutant of *Clostridium perfringens* type A by use of a mobile group II intron. Appl. Environ. Microbiol. **71**: 7542–7547.
69. Nölling J., Breton G., Omelchenko M.V., Makarova K.S., Zeng Q., Gibson R., Lee H.M., Dubois J., Qiu D., Hitti J., Wolf Y.I., Tatusov R.L., Sabathe F., Doucette-Stamm L., Soucaille P., Daly M.J., Bennett G.N., Koonin E.V., and Smith D.R. 2001. Genome sequence and comparative analysis of the solvent-producing bacterium *Clostridium acetobutylicum*. J. Bacteriol. **183**: 4823–4838.
70. Shimizu T., Ohshima S., Ohtani K., Shimizu T., and Hayashi H. 2001. Genomic map of *Clostridium perfringens* strain 13. Microbiol. Immunol. **45**: 179–189.

5

BIFIDOBACTERIUM AS A DELIVERY SYSTEM OF FUNCTIONAL GENES FOR CANCER GENE THERAPY

GENG-FENG FU,[1] YAN YIN,[2] BI HU,[2] AND GEN-XING XU[2,3]

[1]*Jiangsu Provincial Center for Disease Prevention and Control, Nanjing, China*
[2]*Center for Public Health Research, Medical School, Nanjing University, Nanjing, China*
[3]*Jiangsu Research Center for Gene Pharmaceutical Engineering and Technology, Suzhou, China*

EXPRESSION PLASMIDS IN *BIFIDOBACTERIUM* AS A DELIVERY SYSTEM OF FUNCTIONAL GENES

Endogenous Plasmids and Cloning Vectors in *Bifidobacterium*

To develop a cloning vector for modification of *Bifidobacterium*, a comprehensive understanding of the replication mechanism and characterization of bifidobacterial plasmids is necessary. Plasmids in *Bifidobacterium* attracted lots of interests because they encode many special characters as well as they play an important role in the research on genetics and construction of genetic engineering strains for *Bifidobacterium*. *Bifidobacterium* with related plasmids not only showed its own characteristics but also gained characters encoded by the plasmids, such as lactose catabolism, generation of bacteriocin, drug resistance, and antibiotic resistance (1–3).

Emerging Cancer Therapy: Microbial Approaches and Biotechnological Tools, Edited by Arsénio M. Fialho and Ananda M. Chakrabarty

Only a few *Bifidobacterium* species have been demonstrated to harbor extrachromosomal DNA molecules, and these species are *B. asteroides*, *B. breve*, *B. globosum*, *B. indicum*, *B. longum*, *B. pseudolongum* subsp. *globosum*, *B. catenulatum*, and *B. pseudocatenulatum*. Fourteen plasmids from *B. longum* have been completely sequenced: pMB1 (4), pMG1 (GenBank Accession No. NC_006997), pKJ36 (GenBank Accession No. NC_002635), pKJ50 (5), pBLO1 (GenBank Accession No. NC_004943), pNAC1, pNAC2 (GenBank Accession No. NC_004769), pNAC3 (6), pBIFA24 (7), pTB6 (GenBank Accession No. NC_006843), pDOJH10L, pDOJH10S (8), pNAL8 (9), and pB44 (GenBank accession No. NC_004443). Furthermore, plasmids from other *Bifidobacterium* species have been sequenced as well: pVS809 from *B. globosum* (10), pCIBb1 and pNBb1 (GenBank accession No. E17316) from *B. breve* (11), pBI from *B. indicum* (12), pBC1 from *B. catenulatum* (13), pAP1 from *B. asteroides* (GenBank accession No. Y11549), pASV479 from *B. pseudolongum* subsp. *globosum* (14), pB80 from *B. bifidum*, pB21a from *B. breve* (15), and p4M from *B. pseudocatenulatum* (GenBank accession No. NC003527). A number of cloning vectors have been constructed with plasmids from *Bifidobacterium* and *Escherichia coli*, and transformed into both of them by electroporation. In all cases, electroporation efficiency in *Bifidobacterium* was very low, and attempts have been made to optimize it.

The nucleotide sequence of the *B. longum* B2577 cryptic plasmid pMB1 (1.8 kb) was determined (4). The plasmid had a G + C content of 62% and contained two open reading frames, orf1 and orf2, likely arranged in an operon. Recombinant plasmids containing the pMB1 replicon were able to replicate in *B. animalis* MB209 (4). Another pMB1-based vector pNC7 (4.9 kb) was used to optimize the electroporation protocol and successfully transformed into 10 different *Bifidobacterium* species. Transformation efficiencies ranged from 3.6×10^1 to 1.2×10^5 transformants per microgram DNA (16). Plasmid pKJ50 was isolated from *B. longum* KJ and analyzed to be 4.9 kb in size with a G + C content of 61.7 mol%. A shuttle vector constructed by cloning pKJ50 and a chloramphenicol resistance gene into pBR322 was constructed and was capable of transforming *B. animalis* MB209 (5).

pDOJH10L and pDOJH10S were two cryptic plasmids isolated from *B. longum* DJO10A. Lee et al. (8) conducted sequence analysis of these two plasmids and found that plasmid pDOJH10L was a cointegrate plasmid consisting of DNA regions exhibiting high sequence identity to two other *B. longum* plasmids, pNAC2 (98%) and pKJ50 (96%). The smaller pDOJH10S had no sequence similarity to any other characterized plasmid from Bifidobacteria and did not contain any features consistent with rolling circular replication (RCR). An *E. coli–B. longum* shuttle cloning vector pDOJHR was constructed from pDOJH10S and the *E. coli ori* region of p15A, a *lacZ* gene with a multiple cloning site of pUC18. A chloramphenicol resistance gene of pCI372 was transformed successfully into *E. coli* and *B. longum*, and stably maintained in *B. longum* without antibiotic pressure for 92 generations. pDOJHR was the first cloning vector for *Bifidobacteria* that did not utilize

RCR and should be useful for the stable introduction of heterologous genes into these dominant inhabitants of the large intestine (8). Corneau et al. (6) determined the complete nucleotide sequences for pNAC1 (3.5 kb) from strain RW048 as well as for pNAC2 (3.7 kb) and pNAC3 (10.2 kb) from strain RW041 of *B. longum* and analyzed their molecular characteristics. The putative *Rep*B gene product of pNAC2 was found to be similar to the replication protein of pDOJH10L and pKJ36. Bifidobacterial plasmids were divided into five groups based on Rep amino acid sequence homology, and the results suggest a new plasmid family for *B. longum* (6).

Three plasmids, pB44 (3.6 kb) from *B. longum*, pB80 (4.9 kb) from *B. bifidum*, and pB21a (5.2 kb) from *B. breve*, have been sequenced. While the former two plasmids were found to be highly similar to previously characterized rolling circle replicating pKJ36 and pKJ50, respectively, the third plasmid, pB21a, does not share significant nucleotide homology with known plasmids. Two sets of *E. coli–Bifidobacteria* shuttle vectors, pSUW64/123 and pEESH80, constructed based on pB44 and pB80 replicons were capable of transforming *B. bifidum* and *B. breve* strains with efficiency up to 3×10^4 cfu/μg DNA. Additionally, an attempt was made to employ a broad host range conjugation element, RP4, in developing an efficient *E. coli–Bifidobacterium* gene transfer system (15).

Expression Plasmids in *Bifidobacterium* for Cancer Gene Therapy

Hypoxic regions are characteristic of solid tumors. Tissue oxygen electrode measurements taken in cancer patients show a median range of oxygen partial pressure of 10–30 mmHg in tumors, with a significant proportion of readings below 2.5 mmHg, whereas those in normal tissues range from 24 to 66 mmHg (17, 18). Kimura et al. (19) demonstrated that anaerobic bacteria of the genus *Bifidobacterium* could selectively germinate and grow in the hypoxic regions of solid tumors after intravenous injection (19). Based on the research of the endogenous plasmids and cloning vectors in *Bifidobacterium*, many shuttle vectors encoding target genes were constructed, and *Bifidobacterium* was used as a delivery system for cancer gene therapy. The main expression plasmids used in cancer gene therapy are as follows.

Plasmid pBLES100 pBLES100, which was constructed by cloning with a *B. longum* plasmid, pTB6, and an *E. coli* vector, pBR322, has been used as an expression vector for different genes in cancer gene therapy. Yazawa et al. (20) cloned a gene encoding spectinomycin adenyltransferase (AAD) into pBLES100 and transferred it into *B. longum* 105-A or 108-A by electroporation. When these genetically engineered Bifidobacteria were introduced into tumor-bearing mice, bacteria were found only in the tumors. The result strongly suggests that obligate anaerobic bacteria such as *Bifidobacterium* can be used as highly specific gene delivery vectors for cancer gene therapy (18). *B. longum* 105-A carrying pBLES100-AAD was also used in the gene therapy of chemically induced rat mammary tumors, where transformed *B. longum* selectively germinated in tumor was observed in an autochthonous tumor system using 7,12-dimethylbenz[a]anthracene (DMBA)-induced mammary carcinoma in rats. When these genetically engineered Bifidobacteria were introduced systemically into tumor-bearing mice, bacteria were found only in the tumors, presumable due to the hypoxic environment required for the growth of these bacteria (19, 20).

Nakamura et al. (21) constructed a plasmid pBLES100-S-eCD, which included the histone-like protein (HU) gene promoter and the gene encoding the cytosine deaminase of *E. coli* in the pBLES100 vector. The cytosine deaminase enzyme converts 5-fluorocytosine (5-FC) to 5-fluorouracil (5-FU). The expression of cytosine deaminase in *B. longum* successfully converted 5-FC to 5-FU. These findings suggest *B. longum* as an excellent gene delivery system and candidate for enzyme/prodrug therapy, which was considered to be safe and effective (21). These findings suggested that the genus *B. longum* was an attractive and safe tumor-targeting vector, and transformed *B. longum* was a potential anticancer agent that could effectively and specifically treat solid tumors (22).

Plasmid pGEX-1LamdaT Yi et al. (23) successfully constructed a *B. infantis*/CD targeting gene therapy system with a recombinant CD/pGEX-1LamdaT plasmid. The recombinant plasmid was used in the inhibition of melanoma *in vitro* and *in vivo*. Generally, *B. infantis* transfected by recombinant CD/pGEX-1LamdaT plasmid was incubated with 5-FC anaerobically, and the supernatant fluid was collected and added to melanoma B16-F10 cells to observe the killing effect for B16-F10 cells. Tumor-bearing mice were injected with 5-FC and transfected *B. infantis* to examine the antitumor effect. *In vitro*, B16-F10 cells treated by supernatant fluid were remarkably damaged morphologically, and the cell growth was significantly inhibited. Experiments on the mice melanoma model showed that the tumor volume was significantly inhibited compared with controls after treatment with a combination of transfected *B. infantis* and 5-FC (24).

Plasmids pBV220 and pBV22210 In our lab, a shuttle pBV220 was used to construct pBV220-endostatin and transform the recombinant plasmid into *B. adolescentis* and *B. longum*. *B. adolescentis* with endostatin gene was injected into mice bearing Heps liver cancer. After determination of the expression of endostatin, the distribution and antitumor effects of transfected *B. adolescentis* were examined. At 168 h after the third injection of *B. adolescentis* with endostatin gene, *B. adolescentis* cells were only found in the tumors, and no bacilli were found in other normal tissues (Figure 5.1). The growth of primary tumors was inhibited by 69.9% by systemic therapy with *B. adolescentis*-carrying endostatin gene as compared to control mice treated with 5% glucose in 0.9% NaCl. The inhibitory role of wild-type *B. adolescentis* (23.1%) was rather weak than that of *B. adolescentis*-carrying endostatin gene (25). Fu et al. (26) used *B. longum*-carrying shuttle vector pBV220 (Amp^+) encoding human

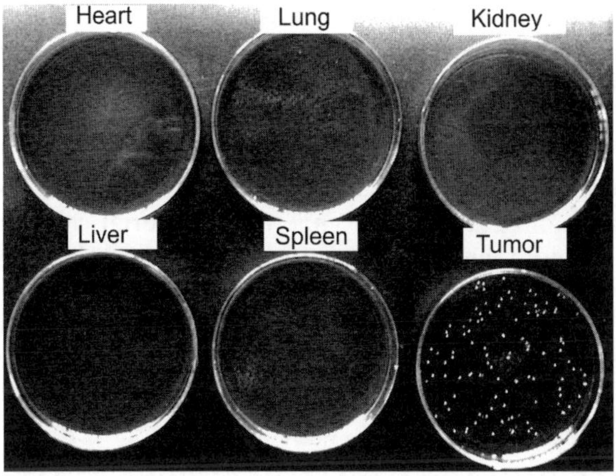

Figure 5.1 Comparison of the number of *B. adolescentis* carrying pBV220-endostatin plasmid in both tumors and normal tissues, after 168 h of the third administration of 1×10^8 viable bacilli into tumor-bearing mice through tail vein for each time. After 72 h of anaerobic cultivation, many colonies of *B. adolescentis* carrying pBV220-endostatin plasmid were observed in the tumor, but no colonies were found in normal tissues.

endostatin as an oral delivery system and succeeded in selective localization within solid tumors and inhibiting growth of HepG2 liver tumor xenografts in mice. In addition, enrichment of selenium to transformed *B. longum* was proved to show stronger efficacy in enhancement of host immunity and tumor inhibition (26). However, since ampicillin is considered to be harmful to bacterial cytoderm synthesis after electroporation, Xu et al. (27) constructed a new vector pBV22210-endostatin combining a chloramphenicol resistance gene and a cryptic plasmid pMB1 from the wild-type *B. longum* strain. *B. longum*-pBV22210-endostatin exhibited higher stability and stronger inhibitory effect on H22 liver tumor growth in xenograft models than *B. longum*-pBV220-endostatin. Our results also indicated that the plasmid electroporated into *B. longum* was maintained stably in the absence of selective antibiotics and did not significantly affect biological characteristics of *B. longum*. These results suggested that pBV22210 may be a stable vector in *B. longum* for transporting anticancer genes in cancer gene therapy (27).

Other genes were also successfully expressed by shuttle vectors in *Bifidobacterium*. Park et al. (28) constructed a constitutive high-level expression vector pBES16PR-CHOL with the structural gene for cholesterol oxidase (CHOL) under the control of the 16S rRNA promoter and transformed it to *B. longum*. The gene was successfully expressed, and high level of cholesterol oxidase activity was obtained in *B. longum* (28). Reyes Escogido et al. (29) used a novel expression vector (pLR) driven by *hup* promoter for the expression of the optimized human interleukin (IL)-10 synthetic gene in *E. coli* and *B. longum*; recombinant human (rh)IL-10 was obtained in a soluble form in total extract cells in both microorganisms (29). In another report, Guglielmetti et al. (30) transformed *B. longum* biovar longum with a vector (pGBL8b) containing the insect luciferase gene. The bioluminescent *B. longum* was employed for a quick test of the efficacy of different carbohydrates to preserve cell physiology under acidic conditions. The bioluminescent *B. longum* transformed with the pGBL8b plasmid is a potentially valuable tool for rapidly studying the physiological state of anaerobic bacterial cells under different environmental conditions (30). Though no anticancer gene was inserted in these cases, the successful expression of other genes suggested the stability and expression level of these shuttle vectors, and the vectors were useful for the further application of *Bifidobacterium* in cancer gene therapy.

BIFIDOBACTERIUM AS AN ORAL DELIVERY SYSTEM OF FUNCTIONAL GENES FOR CANCER GENE THERAPY

Oral Administration of *Bifidobacterium* Affects Cancer Growth and the Immune System

Human colon is a complex microbial ecosystem, comprising several hundred bacterial species. Some of these enteric bacteria are beneficial to the host and have been shown to exert antimutagenic and anticarcinogenic properties (31, 32). *Bifidobacterium* is the most important enteric bacterium in human colon with anticarcinogenic properties. In 1978, Kohwi et al. (33) had shown that *B. infantis* could inhibit cancer growth significantly in mice, and the inhibitory effects were related to the amounts of the Bifidobacteria. *Bifidobacterium* could enhance the anticancer function in cancer-bearing mice. Not only did *Bifidobacterium* inhibit cancer growth directly, but *Bifidobacterium* could also enhance the killing activity of natural killer (NK) cells and increase the activity of IL-2, interferon (IFN)-γ, and tumor necrosis factor (TNF)-α in blood (34). In the cell wall preparation (WPG), the ultrasonicated WPG and framework of WPG of *B. infant* were administrated to tumor-bearing mice; cancer cell growth was reported to be inhibited by 70%, 40%, and 20%, respectively. Moreover, WPG was the main component in anticancer effect, and the mRNA expression of several cytokines (IL-1β, IL-6, IL-10, IFN-α, and TNF-α) was induced in mice after administering WPG (35). There are some effective components in *Bifidobacterium* for anticancer function besides WPG of *Bifidobacterium*. Takahashi et al. (36) evaluated the efficiency of immunostimulatory DNA sequence from *B. longum* in preventing allergic responses by oral administration. Oral administration of the immunostimulatory DNA sequence suppressed serum ovalbumin (OVA)-specific immunoglobulin (Ig) E levels and improved the OVA-specific IgG2a/IgG1 ratio. The immunostimulatory DNA sequence increased Th-1 cytokine and decreased Th-2 cytokine production in splenocytes. The former such as IFN-γ and IL-12 p40 increased significantly, while the latter such as IL-4, IL-5, IL-10, and TGF-β levels decreased. These results appear to indicate that oral administration of immunostimulatory DNA sequence of *B. longum* improved Th1/Th2 balance (36). *Bifidobacterium* not only can inhibit cancer growth in animal models, but also can exert the anticancer function in clinic. Kubota (37) checked the fecal intestinal flora in patients with colon adenoma; the results showed significant decrease of *Bifidobacterium* in patients with colon adenoma (37). However, *Bifidobacterium* has been reported to exert satisfactory therapeutic effects in melanoma from clinic. Cell wall preparations of *B. longum*, IL-2, IFN-γ, and TNF-α were used to treat melanoma. WPG was shown to inhibit melanoma growth, and the inhibitory effect of WPG was the same as IFN-γ. Furthermore, WPG had no side effect on the melanoma patients. Some components of *Bifidobacterium* were shown to be beneficial to the host. Moreover, some of them had anticancer effects, such as cell wall preparation, nucleic acid, and whole peptidoglycan from *Bifidobacterium* (38, 39).

Oral Administration of *Bifidobacterium* Inhibits Cancer Growth Through Cell Signal Modulation

Many growth factors can regulate cancer cell growth and differentiation, through activation of the ligands on the surface of the cells. These factors include platelet-derived growth factor (PDGF), epidermal growth factor (EGF), fibroblast growth factor (FGF), hepatocyte growth factor (HGF), and insulin growth factor (IGF) (40–42). Tang et al. (43, 44) reported that *Bifidobacterium* and the culture fluid would regulate the contents of cyclic adenosine monophosphate (cAMP), cyclic guanosine monophosphate (cGMP), and diacylglycerol (DAG) in ccL-187 cancer cells (43, 44); Jiang et al. (45) had found that the WPG of *Bifidobacterium* could regulate the concentration of Ca^{2+} in Lovo cancer cells, inhibiting cancer cell growth and inducing cancer cell differentiation to mature cells (45). Some reports indicate that the cell growth and differentiation are related to the signal transmissions of protein tyrosine kinase (PTK), nuclear factor (NF)-κB, and protein kinase C (PKC) (46). Wang et al. (47) reported that *Bifidobacterium* would inhibit the activities of PTK, NF-κB, and PKC in animal models bearing large bowel carcinoma. Dai et al. (48) explored the antitumor mechanisms of *B. adolescence in vivo* and found that the levels of extracellular signal-regulated protein (ERK), c-jun, p38, and c-fos in nude mouse transplanted with large bowel carcinoma were reduced compared with controls.

BIFIDOBACTERIUM IS A GOOD VECTOR FOR ORAL CANCER GENE THERAPY

The concept of probiotics was established near the beginning of the last century by Metchnikoff, who hypothesized that the ingestion of fermented milk products containing lactic acid-producing bacteria had a beneficial impact on health and human longevity. *Bifidobacterium* is one of the most important lactic acid-producing bacteria (31), and so, it is a logical choice to be a vector bearing any anticancer gene. *Bifidobacterium* can selectively germinate and proliferate in the hypoxic regions of solid tumors after intravenous injection. Fujimori (49) constructed a plasmid containing the cytosine deaminase gene and transferred the recombinant plasmid into a strain of *Bifidobacterium*. The *Bifidobacterium* carrying the recombinant plasmid was used to treat animals bearing Lewis lung cancer via injection in tail vein of the animals. The results suggested that *Bifidobacterium* selectively germinated and proliferated in tumor tissue, inhibiting tumor growth (49).

Bifidobacterium can inhibit cancer growth and regulate immune action in animals bearing tumors; moreover, *Bifidobacterium* can directly grow in the host intestine carrying anticancer functional genes. The live *Bifidobacterium* carrying the anticancer gene in the host intestine becomes the "reservoir" of anticancer medicine. Our research group utilized a strain of *B. longum* as a delivery system to transport an endostatin gene that can inhibit growth of tumor. The *B. longum* strain with the endostatin gene (*B. longum*-endostatin) was given orally to tumor-bearing nude mice through drencher preparation. The results showed that *B. longum*-endostatin could strongly inhibit the growth of solid liver tumor in nude mice and prolong the survival time of tumor-bearing nude mice. Furthermore, tumor growth was inhibited more efficiently when the *B. longum*-endostatin treatment included selenium (Se-*B. longum*-En) (Figure 5.2). Se-*B. longum*-En also could improve the activities of NK and T cells, and stimulate the activity of IL-2 and TNF-α in BALB/c mice (26). Nakamura et al. (21) constructed pBLES100-S-eCD plasmid containing the cytosine deaminase gene and was transfected into a strain of *Bifidobacterium*. The recombinant *Bifidobacterium* produced cytosine deaminase that converted 5-FC into 5-FU. The result indicated that *Bifidobacterium* could be useful for enzyme/prodrug therapy of hypoxic solid cancer (21).

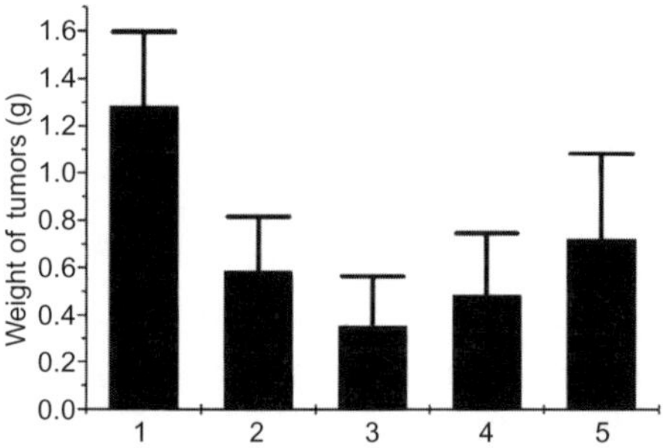

Figure 5.2 Average tumor weights in different treatment groups. Bar 1 was weight of control group; bars 2, 3, 4, and 5 were weight of tumors orally treated with *B. longum*-endostatin (middle dose), Se-*B. longum*-endostatin (high dose, $p < 0.001$), Se-*B. longum*-endostatin (middle dose, $p < 0.01$), and Se-*B. longum*-endostatin (low dose, $p < 0.01$).

Bifidobacterium is a good vector for oral cancer gene therapy; the reasons are the following: (1) *Bifidobacterium* is the most important and prevalent enteric bacterium in human colon, and it can inhibit cancer occurrence and development; (2) oral administration of recombinant *Bifidobacterium* shows no side effects to host, and it can increase the immune response; (3) *Bifidobacterium* can selectively germinate and proliferate in the host colon, and any anticancer gene can be expressed in the host. The anticancer reactions can be persistent if *Bifidobacterium*-carrying anticancer gene is supplied continuously; (4) oral administration of recombinant *Bifidobacterium* is more convenient, and it can avoid harmful effects of venous injection to the host.

BIFIDOBACTERIUM AS A DELIVERY SYSTEM OF FUNCTIONAL GENES FOR CANCER GENE THERAPY AND ITS APPLICATION

Bifidobacterium as an Agent for Cancer Gene Therapy

The World Health Organization (WHO) and the Food and Agriculture Organization (FAO) have recognized *Bifidobacterium* as one of the most important probiotic bacteria. Many kinds of dairy and pharmaceutical products containing *Bifidobacterium* have been proposed as dietary adjuncts (28). Accordingly, several kinds of nonpathogenic anaerobic bacteria became a research tool recently. Among all of these bacteria, *Bifidobacterium* has particular predominance in cancer gene therapy. Anaerobic bacteria themselves exhibit desirable anticancer effect. Several mechanisms could be responsible for the broad antitumor activity of anaerobic bacteria, and consideration of these mechanisms is important for optimization of antitumor activity during clinical development. First, anaerobic bacteria may induce local inflammatory responses, perhaps by virtue of their capacity to accumulate to high levels in tumors and to reside there for long periods. Second, anaerobic bacteria can compete with tumor cells for nutrients and growth factors, and they may also secrete toxins into the extracellular environment. In addition, anaerobic bacteria may induce cytokine within the tumor, consequently resulting in selective intratumoral and intravascular coagulation, and enhanced tumor encapsulation (50).

Many studies have shown that *Bifidobacterium* has tumor-targeting property. Kimura et al. (19) demonstrated that *B. bifidum* (Lac B) can selectively localize and proliferate in several types of mouse tumors following intravenous administration. None of the same bacilli could be detected in the tissues of healthy organs, such as the liver, spleen, kidney, lung, blood, bone marrow, and muscle, 48 or 96h after intravenous administration into tumor-bearing mice (19). *Bifidobacterium* as a strain of representative anaerobic bacteria also has antitumor effect itself, and the mechanism may be relevant to the content mentioned above. Furthermore, studies have been confirmed that *Bifidobacterium* could stimulate the proliferation of B cells and augment production of antibody against pathogens or food allergens, and prevent the penetration of the pathogen or allergen into the body (51). The enhancement of immune response of *Bifidobacterium* may be relevant to its anticancer property. On all accounts, *Bifidobacterium* can selectively localize to and proliferate in hypoxic areas within tumors, and it has antitumor effect itself.

Bifidobacterium as a Delivery System of Functional Genes for Cancer Gene Therapy

In recent years, gene therapy in solid tumors that targets gene expression to hypoxic tumor cells is being investigated (52). However, a crucial obstacle in cancer gene therapy is the specific targeting of therapy directly to a solid tumor and lack of specificity of current delivery systems. As already mentioned, *Bifidobacterium* can selectively germinate and grow in the hypoxic regions of solid tumors after intravenous injection. In view of the tumor specificity of *Bifidobacterium*, it could be used as a gene delivery vector for anticancer gene therapy. There are several advantages of selecting *Bifidobacterium* as a gene delivery vector for cancer gene therapy: (1) *Bifidobacterium* is a domestic anaerobic bacterium in the human body that does not produce toxins; (2) *Bifidobacterium* increases immune response and inhibits many tumor growth *in vivo*, such as liver cancer, breast cancer, and so on (53); (3) *Bifidobacterium* can be killed easily by antibiotics or in oxygen environment *in vitro* or *in vivo*. Both wild-type and genetically engineered *Bifidobacterium* were killed easily with kanamycin, cefoperazone, and penicillin *in vitro* or *in vivo*; and (4) when *Bifidobacterium* was injected intravenously, it only germinated and proliferated in solid tumor but not in normal tissues. Therefore, *Bifidobacterium* could be used as a highly specific gene delivery vector for anticancer gene therapy, and its inherent role as an antitumor agent makes it more effective in the inhibition of tumor growth. So *Bifidobacterium* is a good gene delivery system, which is not only tumor specific but also nontoxic for gene therapy of tumors.

TNF-related apoptosis-inducing ligand (TRAIL) was identified as a member of the TNF superfamily. In our lab, pBV22210 encoding the extracellular domain of TRAIL was introduced into *B. longum* (*B. longum*-TRAIL). The growth curve of *B. longum*-TRAIL and wide-type *B. longum* cells was similar in TPY medium without selective pressure, while the lag phase of *B. longum*-TRAIL in selective medium was statistically longer than that in nonselective medium as shown in Figure 5.3. Both *B. longum* in TPY medium without chloramphenicol grew to an exponential phase after 6 h and stationary phase (optical density $[OD]_{600} = 1.1$) after 10 h of incubation. However, *B. longum*-TRAIL cells in TPY medium with chloramphenicol (5 μg/mL) grew to an exponential phase after 15 h and stationary phase after 18 h of incubation (Figure 5.3).

After 72 h of subcutaneous inoculation with S180 cells, tumor-bearing mice were intravenously injected with *B. longum*-TRAIL cells. Tumor-bearing mice were sacrificed at 1, 24, 48, and 96 h after the third injection of *B. longum*-

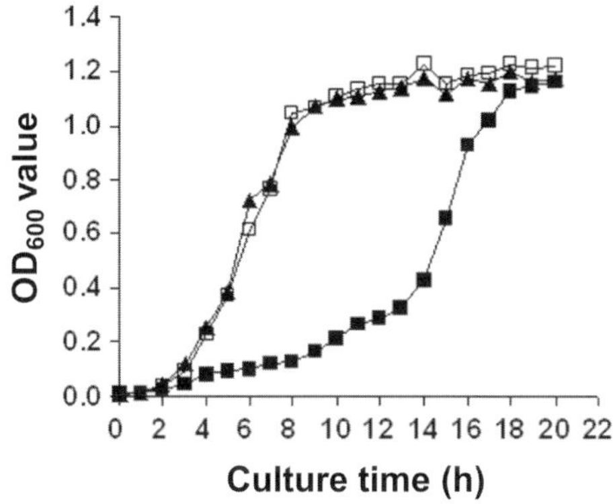

Figure 5.3 The growth curves of *B. longum*-TRAIL and wide-type *B. longum* cells. *B. longum*-TRAIL cells were incubated anaerobically at 37°C in TPY medium with 5 μg/mL chloramphenicol (filled squares) or without chloramphenicol (filled triangles). Wide-type *B. longum* was incubated anaerobically in TPY medium without selective pressure (open squares). OD_{600}, optical density at 600 nm.

TRAIL cells, and the location of transformed *B. longum* in tumors and several normal tissues were determined. Other mice were sacrificed at 24, 48, and 96 h after an additional injection of ampicillin (Amp) at 96 h following the treatment of *B. longum*-TRAIL cells, and the presence of transformed *B. longum* in tumors was detected. At 96 h, about 1.55×10^7 bacilli/g tumor tissue was found, but few bacilli were detected in normal tissues, such as the heart, liver, spleen, lung, and kidney, from the tumor-bearing mice (Figure 5.4). The increasing number of bacilli in tumors suggested that the transformed *B. longum* selectively proliferated in the tumor tissue. In contrast, the number of transformed *B. longum* in normal tissues decreased rapidly 24 h after injection, indicating that transformed *B. longum* did not germinate in normal tissues. Interestingly, the number of transformed *B. longum* located in tumors decreased rapidly after an additional treatment with ampicillin; few viable bacilli were detected in tumors 96 h after the administration of ampicillin (Figure 5.4) (54).

Compared to dextrose–saline solution group, combined treatment with *B. longum*-TRAIL and *B. longum*-endostatin significantly suppressed tumor in weight by 79.6% ($p = 0.0034$) and in volume by 82.6% ($p = 0.001$), *B. longum*-TRAIL alone by 58.0% ($p = 0.006$) and by 60.9% ($p = 0.0038$), *B. longum*-endostatin alone by 55.3% ($p = 0.0074$) and by 58.1% ($p = 0.0037$), and cyclophosphamide (CTX) by 63.4% ($p = 0.0051$) and by 65.1% ($p = 0.0026$), respectively. Compared with wide-type *B. longum*, combination group inhibited the tumor growth in tumor weight by 73.6% ($p = 0.0077$) and in tumor volume by 76.7% ($p = 0.0006$), while *B. longum*-TRAIL alone, respectively, by 45.6% ($p = 0.0229$) and by 47.7% ($p = 0.0059$), and *B. longum*-endostatin alone, respectively, by 42.1% ($p = 0.0299$) and by 40.0% ($p = 0.0074$). Based on these results, we could conclude that combination of *B. longum*-TRAIL and

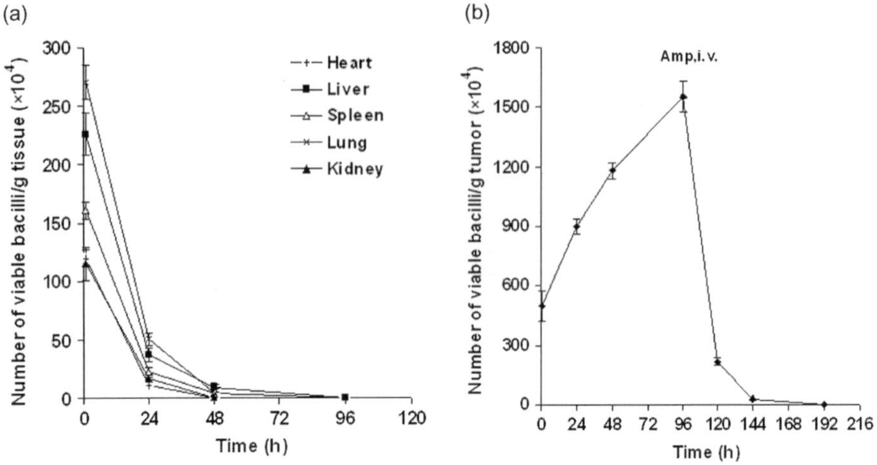

Figure 5.4 Viable bacilli number of *B. longum*-TRAIL in tumors and normal organs of treated mice. (a) Shows the distribution of *B. longum*-TRAIL in different normal organs at different times after the third injection of 1×10^8 viable bacilli. (b) Shows the viable bacilli number in tumors after administration of ampicillin (50 mg/kg) following the treatment of *B. longum*-TRAIL cells.

B. longum-endostatin showed synergistic interactions and had a stronger inhibitory effect on tumor growth than either *B. longum*-TRAIL or *B. longum*-endostatin alone (Figure 5.5). In addition, when low dosage of Adriamycin (Haizheng Pharmaceutical Enterprise, Taizhou, China) (5 mg/kg) or *B. longum*-endostatin was combined, the antitumor effect was significantly enhanced. The successful inhibition of S180 tumor growth suggests that a stable vector in *B. longum* for transporting anticancer genes combined with low-dose chemotherapeutic drugs or other target genes is a promising approach in cancer gene therapy (54).

Moreover, *Bifidobacterium* can be used as delivery system of many other antitumor genes, for example, granulocyte colony-stimulating factor (GCSF), IL-2, and so on, and can be introduced into *Bifidobacterium* for tumor therapy (55). Thus, *Bifidobacterium* can be used as a safe delivery vector of many interesting genes for potential cancer therapy.

BIFIDOBACTERIUM COMBINATION WITH OTHER FACTORS

Cancer is a complex disease where many factors can influence; so, the therapy of cancer can not depend on a single method. Though *Bifidobacterium* is effective for tumor therapy, it can only slow the progress of cancer yet can not cure the tumor completely. So, in order to improve curative effect, *Bifidobacterium* should be combined with other factors to treat the tumor. Many combined strategies can be adopted.

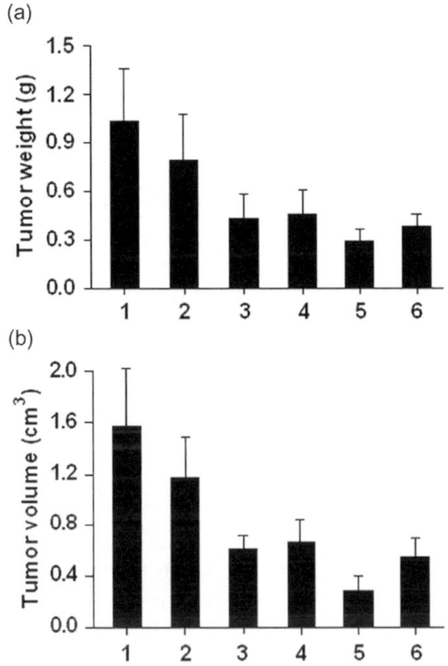

Figure 5.5 The inhibition effects on S180 tumor growth by *B. longum*-TRAIL cells and/or *B. longum*-endostatin cells in tumor-bearing mice. The tumor weights and tumor volumes were measured for each mouse. (a) Shows tumor weights in different treatment groups. Row 1, dextrose–saline solution group; row 2, wide-type *B. longum* cells group; row 3, *B. longum*-endostatin cells group; row 4, *B. longum*-TRAIL cells group; row 5, *B. longum*-TRAIL cells combined with *B. longum*-endostatin cells group; row 6, cyclophosphamide (CTX) group. (b) Shows the average tumor volume. Both tumor weights and tumor volumes were significantly reduced in *B. longum*-TRAIL cells combined with *B. longum*-endostatin cells group.

Combination with Radiation and Chemotherapeutic Drugs

Presently, the surgical treatment combined with radiation and chemotherapy is still the major approach to treat malignancies. But the presence of necrotic and hypoxic areas within solid tumors contributes to the treatment failure of conventional chemotherapy. Chemotherapeutic approaches for cancer are, in part, limited by the inability of drugs to destroy neoplastic cells within poorly vascularized compartments of tumors. The poorly vascularized regions are less sensitive to ionizing radiation because its cell-killing effects depend on oxygen, and they are less sensitive to chemotherapeutic drugs because drug delivery to these regions is suboptimal (56). These conditions reduce the effectiveness of radiation and some chemotherapeutic agents. Radiation and chemotherapeutic drugs can kill tumor cells around the necrotic and hypoxic areas, which the application of *Bifidobacterium* can not help. Therefore, the combination of *Bifidobacterium* with the conventional chemotherapy drugs and radiation can destroy tumor cells more drastically.

Furthermore, using high dosage of many antitumor agents would bring side effects. *Bifidobacterium* with cloned genes can reduce the side effect and is relevant to cancer therapy. For example, CTX is in the nitrogen mustard group of alkylating antineoplastic chemotherapeutic agents. High doses of CTX often lead to severe neutropenia, which greatly affect the quality of patients' life (57). *B. longum* with GCSF can effectively antagonize the effect of myelosuppression induced by chemotherapy agents, effectively antagonize the myelosupression caused by CTX, and increase the number of peripheral leukocytes of the tumor-bearing mice. Combination of *Bifidobacterium*-GCSF with the conventional chemotherapy drugs and radiation might be of value for the therapy of tumors (55).

Strategies for Enzyme–Prodrug Cancer Therapy

The success of systemic application of chemotherapy drugs is limited by several drawbacks, including insufficient drug concentrations in tumors, sytemic toxicity, and lack of selectivity for tumor cells over normal cells and of drug resistance of tumor cells. Enzyme–prodrug cancer therapy can use *Bifidobacterium* to deliver drug-activating enzyme gene or functional protein to tumor tissues, followed by systemic administration of a prodrug. Then, a systemically administered prodrug can be converted to high local concentration of an active anticancer drug only in tumors. So, enzyme prodrug therapy is one promising area for improving tumor selectivity.

Combination with Some Chemical Compounds

Some chemical compounds were shown to have antitumor effect. For instance, selenium compounds have been found to inhibit carcinogenesis in a variety of animal models. Many reports have shown that selenium supplementation can enhance lymphocyte functions and inhibit solid tumor growth in mice (58, 59). Fu et al. (26) chose sodium selenite (Na_2SeO_3) to enrich selenium in the *B. longum*-endostatin strain. The result indicated that the high dose and the middle dose of Se-*B. longum*-endostatin can inhibit tumor growth by 72.5% and 62.2%, while the middle dose of *B. longum*-endostatin can inhibit tumor growth by 54.4% (26). This study demonstrated that Se-*B. longum*-endostatin can inhibit tumor growth as well as stimulation of the immune function more efficiently. Moreover, engineered *Bifidobacterium* with different genes can be used together for cancer treatment, and the inhibition of tumor can be more effective.

PROSPECT

Bifidobacterium has many advantages in cancer gene therapy as the vector for anticancer genes; however, it is difficult to commercially use *Bifidobacterium* as a vector because the basic studies on molecular biology of *Bifidobacterium* are in their infancy. Similarly, the plasmids in *Bifidobacterium* are few, molecular weights are high, and the plasmids are very difficult to isolate. These difficulties make it hard to construct shuttle plasmids through engineering of *Bifidobacterium*. Also, transfected *Bifidobacterium* will likely be subject to stomach acid digestion while orally administered. All of the above difficulties have limited the utilization of *Bifidobacterium* as the vector for anticancer gene therapy. However, many researchers are interested in *Bifidobacterium* as the vector for anticancer genes, and it is likely that *Bifidobacterium* will be a suitable and useful vector for anticancer genes in cancer gene therapy in the future.

To date, only a few plasmids were found to replicate stably and successfully expressed exogenous proteins in *Bifidobacterium*. In addition, transformation efficiency and expression level are low in all cases. So, there is a strong need for work with other vectors that can transfect *Bifidobacterium* with high efficiency and generate target proteins in *Bifidobacterium* at a high level. Thus, one avenue for future work is construction of new shuttle vectors with high stability and expression level in *Bifidobacterium*. As more and more advantages of *Bifidobacterium* have been discovered, the molecular characterization of different *Bifidobacterium* strains should be elucidated to select the most suitable delivery system.

Another avenue that should be explored is the combination of transformed *Bifidobacterium* carrying target genes with different anticancer mechanisms. These agents may have cross-talks and result in synergistic effect, causing massive vascular shutdown and tumor cell apoptosis because of the selective replication of *Bifidobacterium* in tumors. Further work is needed to explore the potential of such combinations.

The identification of suitable selection markers and *Bifidobacterium* host system should also be pursued. If the host vector system is to be used in the dairy industry, it must eventually have food-grade selection markers, such as genes related to lactose metabolism and/or bacteriocin resistance.

This chapter describes the power of genetically engineered *Bifidobacteria* in cancer therapy. Bacteria are not vectors for the introduction of genes into cancer cells. However, they can and do concentrate in tumors by various means and can carry genes of interest to produce proteins of choice in tumors. This can be a powerful adjunct to cancer therapy. We believe that this novel approach for tumor targeting using *Bifidobacteria* could be useful for gene therapy of solid tumors.

REFERENCES

1. Klun A., Mercenier A., and Arigoni F. 2005. Lemons from the genomes of *Bifidobacteria*. FEMS Microbiol. Rev. **29**: 491–509.
2. Mckay L.L., Baldwin K.A., and Zottola E.A. 1972. Loss of lactose metabolism in lactic streptococci. Appl. Environ. Microbiol. **23**: 1090–1096.

3. Sgorbati B., Scardovi V., and Leblanc D.J. 1982. Plasmids in the genus *Bifidobacterium*. J. Gen. Microbiol. **128**: 2121–2131.
4. Rossi M., Brigidi P., Gonzalez Vara y Rodriguez A., and Matteuzzi D. 1996. Characterization of the plasmid pMB1 from *Bifidobacterium longum* and its use for shuttle vector construction. Res. Microbiol. **147**: 133–143.
5. Park M.S., Shin D.W., Lee K.H., and Ji G.E. 1999. Sequence analysis of plasmid pKJ50 from *Bifidobacterium longum*. Microbiology. **145**: 585–592.
6. Corneau N., Émond E., and LaPointe G. 2004. Molecular characterization of three plasmids from *Bifidobacterium longum*. Plasmid. **51**: 87–100.
7. Park Y.S., Kim K.H., Park J.H., Oh I.K., and Yoon S.S. 2008. Isolation and molecular characterization of a cryptic plasmid from *Bifidobacterium longum*. Biotechnol. Lett. **30**: 145–151.
8. Lee J.H., and O'Sullivan D.J. 2006. Sequence analysis of two cryptic plasmids from *Bifidobacterium longum* DJO10A and construction of a shuttle cloning vector. Appl. Environ. Microbiol. **72**: 527–535.
9. Guglielmetti S., Karp M., Mora D., Tamagnini I., and Parini C. 2007. Molecular characterization of *Bifidobacterium longum* biovar longum NAL8 plasmids and construction of a novel replicon screening system. Appl. Microbiol. Biotechnol. **74**: 1053–1061.
10. Mattarelli P., Biavati B., Alessandrini A., Crociani F., and Scardovi V. 1994. Characterization of the plasmid pVS809 from *Bifidobacterium globosum*. Microbiology. **17**: 327–331.
11. O'Riordan K., and Fitzgerald G.F. 1999. Molecular characterisation of a 5.75-kb cryptic plasmid from *Bifidobacterium breve* NCFB2258 and determination of a mode of replication. FEMS Microbiol. Lett. **174**: 285–294.
12. Sgorbati B., Scardovi V., and LeBlanc D.J. 1982. Plasmids in the genus *Bifidobacterium*. J. Gen. Microbiol. **128**: 2121–2131.
13. Alvarez-Martín P., Florez A.B., and Mayo B. 2007. Screening for plasmids among human bifidobacteria species: sequencing and analysis of pBC1 from *Bifidobacterium catenulatum* L48. Plasmid. **57**: 165–174.
14. Sangrador-Vegas A., Stanton C., van Sinderen D., Fitzgerald G.F., and Ross R.P. 2007. Characterisation of plasmid pASV479 from *Bifidobacterium pseudolongum* subsp. *globosum* and its use for expression vector construction. Plasmid. **58**: 140–147.
15. Shkoporov A.N., Efimov B.A., Khokhlova E.V., Steele J.L., Kafarskaia L.I., and Smeianov V.V. 2008. Characterization of plasmids from human infant *Bifidobacterium* strains: sequence analysis and construction of *E. coli–Bifidobacterium* shuttle vectors. Plasmid. **60**: 136–148.
16. Rossi M., Brigidi P., and Matteuzzi D. 1997. An efficient transformation system for *Bifidobacterium* spp. Lett. Appl. Microbiol. **24**: 33–36.
17. Vaupel P.W. 1993. Oxygenation of solid tumors in drug resistance in oncology. In: Teicher B.A., ed., Drug Resistance in Oncology. New York: Marcel Dekker, pp. 53–85.
18. Yazawa K., Fujimori M., Amano J., Kano Y., and Taniguchi S. 2000. *Bifidobacterium longum* as a delivery system for cancer gene therapy: selective location and growth in hypoxic tumors. Cancer Gene Ther. **7**: 269–274.

19. Kimura N.T., Taniguchi S., and Aoki K. 1980. Selective localization and growth of *Bifidobacterium bifidum* in mouse tumors following intravenous administration. Cancer Res. **40**: 2061–2068.
20. Yazawa K., Fujimori M., Nakamura T., Sasaki T., Amano J., Kano Y., and Taniguchi S. 2001. *Bifidobacterium longum* as a delivery system for gene therapy of chemically induced rat mammary tumors. Breast Cancer Res. Treat. **66**: 165–170.
21. Nakamura T., Sasaki T., Fujimori M., Yazawa K., Kano Y., Amano J., and Taniguchi S. 2002. Cloned cytosine deaminase gene expression of *Bifidobacterium longum* and application to enzyme/pro-drug therapy of hypoxic solid tumors. Biosci. Biotechnol. Biochem. **66**: 2362–2366.
22. Sasaki T., Fujimori M., Hamaji Y., Amano J., and Taniguchi S. 2006. Genetically engineered *Bifidobacterium longum* for tumor-targeting enzyme-prodrug therapy of autochthonous mammary tumors in rats. Cancer Sci. **97**: 649–657.
23. Yi C., Huang Y., Guo Z., and Wang S. 2005. Construction of *Bifidobacterium infantis*/CD targeting gene therapy system. Chinese-German J. Clin. Oncol. **4**: 244–247.
24. Yi C., Huang Y., Guo Z.Y., and Wang S.R. 2005. Antitumor effect of cytosine deaminase/5-fluorocytosine suicide gene therapy system mediated by *Bifidobacterium infantis* on melanoma. Acta Pharmacol. Sin. **26**: 629–634.
25. Li X., Fu G.F., Fan Y.R., Liu W.H., Liu X.J., Wang J.J., and Xu G.X. 2003. *Bifidobacterium adolescentis* as a delivery system of endostatin for cancer gene therapy: selective inhibitor of angiogenesis and hypoxic tumor growth. Cancer Gene Ther. **10**: 105–111.
26. Fu G.F., Li X., Hou Y.Y., Fan Y.R., Liu W.H., and Xu G.X. 2005. *Bifidobacterium longum* as an oral delivery system of endostatin for gene therapy on solid liver cancer. Cancer Gene Ther. **12**: 133–140.
27. Xu Y.F., Zhu L.P., Hu B., Fu G.F., Zhang H.Y., Wang J.J., and Xu G.X. 2007. A new expression plasmid in *Bifidobacterium longum* as a delivery system of endostatin for cancer gene therapy. Cancer Gene Ther. **14**: 151–157.
28. Park M.S., Kwon B., Shim J.J., Huh C.S., and Ji G.E. 2008. Heterologous expression of cholesterol oxidase in *Bifidobacterium longum* under the control of 16S rRNA gene promoter of bifidobacteria. Biotechnol. Lett. **30**: 165–172.
29. Reyes Escogido M.L., De Leon Rodriguez A., and Barba de la Rosa A.P. 2007. A novel binary expression vector for production of human IL-10 in *Escherichia coli* and *Bifidobacterium longum*. Biotechnol. Lett. **29**: 1249–1253.
30. Guglielmetti S., Ciranna A., Mora D., Parini C., and Karp M. 2008. Construction, characterization and exemplificative application of bioluminescent *Bifidobacterium longum* biovar longum. Int. J. Food Microbiol. **124**: 285–290.
31. Singh J., Rivenson A., Tomita M., Shimamura S., Ishibashi N., and Reddy B.S. 1997. *Bifidobacterium longum*, a lactic acid-producing intestinal bacterium inhibits colon cancer and modulates the intermediate biomarkers of colon carcinogenesis. Carcinogenesis. **18**: 833–841.
32. Leng S., and Wang R. 2005. Experimental study on antitumor effect with *Bifidobacterium* and its broth. Chin. J. Microecol. **17**: 411–414.
33. Kohwi Y., Imai K., Tamura Z., and Hashimoto Y. 1978. Antitumor effect of *Bifidobaterium infants* in mice. Gann. **69**: 613–618.

34. Liu R.Z., and Zhang L.Z. 2001. Study on immune mechanism in anticancer of *Bifidobacterium*. Chin. J. Cancer Biother. **8**: 44–45
35. Sekine K., Ohta J., Onishi M., Tatsuki T., Shimokawa Y., Toida T., Kawashima T., and Hashimoto Y. 1995. Analysis of antitumor properties of effector cells stimulated with a cell wall preparation (WPG) of *Bifidobacterium infantis*. Biol. Pharm. Bull. **18**: 148–153.
36. Takahashi N., Kitazawa H., Iwabuchi N., Xiao J.Z., Miyaji K., Iwatsuki K., and Saito T. 2006. Oral administration of an immunostimulatory DNA sequence from *Bifidobacterium longum* improves Th1/Th2 balance in murine model. Biosci. Biotechnol. Biochem. **70**: 2013–2017.
37. Kubota Y. 1990. Fecal intestinal flora in patients with colon adenoma and colon cancer. Nippon Shokakibyo Gakkai Zasshi. **87**: 771–779.
38. Cai L.Z., Li S.J., and Guan C.N. 2005. In vitro and in vivo antitumor effects of whole peptidoglycan from *Bifidobacterium* on colon carcinoma. J. Guangdong Med. Coll. **23**: 500–502.
39. Goldin B.R., Adlercreutz H., Dwyer J.T., Swenson L., Warram J.H., and Gorbach S.L. 1980. Effect of diet on *Lactobacillus acidophilus* supplements on human fecal bacterial enzymes. J. Natl. Cancer Inst. **64**: 255–260.
40. Bianco R., Melisi D., Ciardiello F., and Tortora G. 2006. Key cancer cell signal transduction pathways as therapeutic targets. Eur. J. Cancer. **42**: 290–294.
41. Yarden Y., and Sliwkowski M.X. 2001. Untangling the ErbB signaling network. Nat. Rev. Mol. Cell Biol. **2**: 127–137.
42. Li Y., and Wang L. 2006. Study on anticancer signal mechanism of *Bifidobacterium*. Chin. J. Microecol. **18**: 497–498.
43. Tang L., Ran F., and Zhao B. 1998. Effect of *Bifidobacterium longum* on cAMP, cGMA in colon cancer cell lines. Chin. J. Microecol. **10**: 257–259.
44. Tang L., Zhang Y., and Zhao B. 2000. Regulatory effect of *Bifidobaterium* on DAG in human colon cancer cells. Chin. J. Microecol. **12**: 322–323.
45. Jiang H., and Hu H. 2001. Differentiation inducing effect of bifidobaterial surface molecules on human colonic cancer cells. Chin. J Microbiol. Immunol. **21**: 186–190.
46. Madhusudan S., and Ganesan T.S. 2004. Tyrosine kinase inhibitors in cancer therapy. Clin. Biochem. **37**: 618–635.
47. Wang L., Zhu H., and Pan L. 2003. Influence of *Bifidobacterium* on NF-κB and IκB in experimental large bowel carcinoma. Chin. J Pathophysiol. **19**: 1062–1064.
48. Dai J., Wang L., and Yang D. 2003. Effect of bifidobateria on MAPK and AP-1 of experimental large bowel carcinoma. Chin. J Exp. Surgery. **20**: 1119–1120.
49. Fujimori M. 2006. Genetically engineered *Bifidobacterium* as a drug delivery system therapy of metastatic breast cancer patients. Breast Cancer. **13**: 27–31.
50. Sznol M., Lin S.L., Bermudes D., Zheng L.M., and King I. 2000. Use of preferentially replicating bacteria for the treatment of cancer. J. Clin. Invest. **10**: 1027–1030.
51. Yasui H., and Ohwaki M. 1991. Enhancement of immune response in Peyer's patch cells cultured with *Bifidobacterium breve*. J. Dairy Sci. **74**: 1187–1195.

52. Dachs G.U., Patterson A.V., Firth J.D., Ratcliffe P.J., Townsend K.M., Stratford I.J., and Harris A.L. 1997. Targeting gene expression to hypoxic tumor cells. Nat. Med. **3**: 515–520.

53. Reddy B.S., and Rivenson A. 1993. Inhibitory effect of *Bifidobacterium longum* on colon, mammary and liver carcinogenesis induced by 2-amino-3-methylimidazo quinoline, a food mutagen. Cancer Res. **53**: 3914–3918.

54. Hu B., Kou L., Li C., Zhu L.P., Fan Y.R., Wu Z.W., Wang J.J., and Xu G.X. 2009. *Bifidobacterium longum* as a delivery system of TRAIL and endostatin cooperates with chemotherapeutic drugs to inhibit hypoxic tumor growth. Cancer Gene Ther. **16**: 655–663.

55. Zhu L.P., Yin Y., Xing J., Li C., Kou L., Hu B., Wu Z.W., Wang J.J., and Xu G.X. 2009. Therapeutic efficacy of *Bifidobacterium longum*-mediated human granulocyte colony-stimulating factor and/or endostatin combined with cyclophosphamide in mouse-transplanted tumors. Cancer Sci. **100** (10): 1986–1990.

56. Dang L.H., Bettegowda C., Huso D.L., Kinzler K.W., and Vogelstein B. 2001. Combination bacteriolytic therapy for the treatment of experimental tumors. Proc. Natl. Acad. Sci. U S A. **98**: 15155–15160.

57. Slavin R.E., Millan J.C., and Mullins G.M. 1975. Pathology of high dose intermittent cyclophosphamide therapy. Hum. Pathol. **6**: 693–699.

58. Cheng D.S., Deng Y.P., and Lu X.Y. 2000. Effect of selenium supplementation on transplantive tumor growth and lymphocyte functions in mice. Acta Univ. Sci. Med. Chongqing. **25**: 41–43.

59. Zhao R., Yu B.M., and Zhang G.C. 2000. Selenium enhancing the inhibitory effect of lymphocytes on the growth of colorectal neoplasm in nude mice. Chin. J. Exp. Surg. **17**: 302–303.

6

REPLICATION-SELECTIVE VIRUSES FOR THE TREATMENT OF CANCER

PADMA SAMPATH AND STEVE H. THORNE
Departments of Surgery and Immunology, University of Pittsburgh, Pittsburgh, PA

INTRODUCTION

The most basic requirement of an oncolytic virus is the ability to infect, replicate within, and subsequently destroy human tumor cells coupled to, at most, limited replication ability in essential normal tissues. A host of ingenious mechanisms have been incorporated into an extensive variety of different viruses in order to achieve these goals. Additional features that would also be desirable include genetic stability, both for ease of manufacture and safety, and the use of nonintegrating viruses, which have potential safety advantages as unpredicted events can result from genomic integration. Ideally, antiviral agents should be available to treat a patient in case of any adverse event, or some other mechanism for controlled viral inactivation incorporated into the vector itself. Conversely, it may also be necessary to increase certain aspects of the potency of the virus, such as the ability to reach the tumor following systemic delivery especially in the face of antiviral immunity. Finally, the virus needs to be amenable to high-titer production and purification.

Because oncolytic viruses typically target and kill tumor cells through distinct mechanisms of action from traditional therapies, they can often be successfully combined with other therapies. In particular, the cloning of exogenous and therapeutic genes into some viruses has allowed for their increased antitumoral effects both at the sites of replication and, in some cases,

Emerging Cancer Therapy: Microbial Approaches and Biotechnological Tools, Edited by Arsénio M. Fialho and Ananda M. Chakrabarty

at a distance, while the majority of the clinical benefits of oncolytic viruses have been seen when they are used in combination with chemotherapy.

ENGINEERING TUMOR SELECTIVITY INTO VIRUSES

Targeting the Biology of the Cancer Cell

Incorporating Tumor- or Tissue-Specific Expression of Early Viral Genes This approach, which involves engineering of viruses such that the expression of essential viral genes are placed under the control of tumor- or tissue-specific promoters (Table 6.1), has principally been applied to strains of adenovirus and herpes simplex virus (HSV). This results in replication of the viral constructs being limited to cells capable of expressing genes from these promoters. For example, the HSV essential immediate-early ICP4 gene product has been placed under the control of the albumin promoter/enhancer element, creating a hepatocellular carcinoma-specific virus (1), or under the control of the calponin promoter in order to target the virus to human sarcoma. Alternatively, the HSV γ-34.5 gene has been placed under the control of promoters for either the human carcinoembryonic antigen (CEA) or the MUC1/DF3 tumor-associated antigens (2). These viruses displayed preferential replication in cells overexpressing the corresponding tumor-associated antigen. However, the

TABLE 6.1. Tumor Selectivity and Killing with Oncolytic Viruses

Step in Viral Life Cycle	Mechanisms of Selectivity/Enhanced Killing
Delivery to target tissue	Intratumoral/local regional injections Viruses evolved to spread in blood Cell-based delivery (trafficking to tumor) Polymer coating (evasion of removal)
Cell recognition/entry	Natural/engineered recognition of receptors upregulated on tumor cells
Early gene expression	Place early viral genes under control of tissue/tumor specific promoter
Virulence gene function to evade antiviral response	Deletion of viral virulence genes that are redundant in tumor cells Use of viruses lacking virulence genes that cannot naturally replicate in normal tissues
Killing of cell targets	Sensitizing cells to chemotherapy and radiotherapy Expression of toxic transgenes from virus
Immune response to viral infection	Expression of cytokines to direct immune response Vascular collapse as result of immune infiltration

ICP4 gene regulated by the CEA promoter demonstrated overly attenuated replication even in CEA-positive cells, demonstrating the need to carefully choose and test the gene and promoter combination used.

A panel of tissue-specific promoters has also been utilized in adenovirus vectors. These include strains with the adenovirus E1A gene under control of the prostate-specific antigen (PSA) promoter/enhancer element (3) (to target prostate carcinomas) or the α-fetoprotein (AFP) antigen promoter (targeting hepatocellular carcinoma) (4). Alternatively, the use of two separate tissue-specific promoters, each independently controlling an essential early viral gene, has also been utilized (5) (the prostate-specific rat probasin promoter driving the adenovirus E1A gene, and the human prostate-specific enhancer/promoter, driving the E1B gene). Alternatively, the E1A and E1B genes may be separated by an internal ribosome entry site (IRES), and both are regulated by a single transcriptional response element (TRE), such as with the bladder-specific uroplakin II promoter (6). Further examples of TREs used in this context include that of the surfactant protein B (SPB) (7), the MUC1/DF3 promoter (8), the human IAI.3B promoter element (used to target ovarian cancers) (9), human telomerase reverse transcriptase (hTERT) (10, 11), nestin (used to target glioma) (12), and the tyrosinase promoter/enhancer (used to target melanoma cells) (13).

As an alternative to the use of tissue-specific promoters that may be abnormally overexpressed in cancers originating from that tissue, another approach has been to identify promoters that are generally overexpressed in different tumor types. For example, the HSV Myb34.5 strain contains the B-myb promoter driving the γ-34.5 gene (14). This promoter is E2F responsive, and so the virus replicates in proliferating cells where the E2F transcription factor is active (15). Additionally, several adenoviral strains have been reported that display more general promoter-driven selectivity. These include adenoviruses with E1A under the control of the E2F-1 promoter (16) and several hypoxia-specific constructs such as incorporation of promoters with a hypoxia response element (HRE) (17, 18) or with E1A under the control of an HRE and E4 under the control of the telomerase reverse transcriptase (TERT) promoter (19). However, the use of more general, tumor promoters raises the need to carefully assess what normal tissue these promoters may also be expressed in.

Although the use of tumor- and tissue-specific promoters is undoubtedly a promising area of research, and several of these strains have begun clinical evaluation (20), their use is often limited by the fact that the strength of the promoters can be significantly diminished relative to native viral promoters, and that a degree of leakiness, or readthrough, is often seen if the native viral promoter is not deleted. These issues are being addressed with the design of synthetic promoters incorporating combinations of transcription response elements from different sources.

Retargeting of Viruses for Tumor-Selective Uptake Efforts have also been made to engineer tumor selectivity of viral vectors at the level of cell recognition and entry (Table 6.1). This is typically achieved by coat protein alteration, masking of ligands or redirecting of the virus. Although most of the work to date in this area has been carried out with adenovirus and measles virus, this approach should be applicable to other viruses. Successful retargeting requires an understanding of viral entry into its target cell as both ablation of the natural viral-uptake mechanism and engineering of new "ligands" that target tumor-specific "receptors" into the viral coat, without disrupting viral integrity, are needed. Current efforts to retarget adenovirus can be grouped into three categories: (1) alteration of viral coat proteins to ablate natural targeting and/or insertion of new targeting moieties, (2) addition of bispecific binding molecules, or (3) polymer coating of the virus. Effective ablation of the natural tropism of Adenovirus serotype 5 has been achieved through removal of the viral coxsackie and adenovirus receptor (CAR) and integrin binding domains (21). Attempts have then been made to redirect adenovirus to different cell types, such as by incorporation of a hemagglutinin (HA) ligand into the HI loop of the fiber protein to enable the modified virus to transduce pseudoreceptor-expressing cells. Alternatively, it has been found that the switching of the fiber knob between adenovirus serotypes 5 and 35 (group B) redirects viral infection away from CAR, and allows the transfection of cell lines previously resistant to Ad5 infection. This is because Ad35 does not bind CAR, but CD46 (22), resulting in altered intracellular trafficking of the virus (23). However, it is again necessary to carefully determine how the interchanging of serotype factors will affect viral uptake into other tissues. An efficient system has also been devised for retargeting of measles virus using single chain antibodies (scFv) fused to the viral hemagglutinin (H) envelope protein, with anti-CD20 (24), anti-CD38 (25), anti-alpha-folate receptor (26), and anti-epidermal growth factor receptor (EGFR) (27) antibody fusions constructed to target non-Hodgkin's lymphoma, myeloma, ovarian cancer, and glioma cells, respectively. These viruses displayed appropriate *in vitro* targeting and *in vivo* efficacy. Similar approaches have been used to retarget oncolytic adenovirus strains with the addition of bispecific antibodies that bind both the virus and specific cellular receptors (such as EGFR) (28) or fusion of bispecific single chain antibody (scFv) fragments to the viral coat proteins (29). Alternatively, covalent polymer coating of adenovirus has been shown to ablate its natural tropism (30), while incorporation of EGF into the polymer coat results in retargeting to cells expressing EGFR (31).

Recently, several groups have also used microRNA sequences expressed from different viruses in order to either repress viral gene expression in specific normal tissues in which the virus may otherwise cause toxicity (such as coxsackievirus toxicity in muscle cells) (32), or else to more generally target cancers where suppression of translation by certain miRNAs may be relieved (demonstrated in both adenovirus [33] and vesicular stomatitis virus (VSV) [34]). Although still very early, this approach may add an additional level of safety to already attenuated viruses.

Tumor Selectivity through Targeting the Viral Life Cycle

RNA Viruses A large and growing panel of RNA viruses have been used as oncolytic vectors (Table 6.2); however, due to the inherent complexity in engineering these viruses, the majority of these are wild-type strains that are naturally tumor targeting. Because RNA viruses are often small viruses that do not encode extensive immune modulatory proteins, they rely on rapid replication and spread to escape immune clearance. The immune privileged status of the tumor microenvironment means that some viruses that typically cause little or no disease in humans can replicate for extended periods within the cancer. As a result, these viruses have had excellent safety profiles, but little scope for further modification. However, recent advances in techniques to engineer and modify these viruses have created the opportunity to further

TABLE 6.2. Viruses Used as Oncolytic Vector Backbones

Virus	Class	Natural Selectivity	Engineered Selectivity	Clinical Testing
Reovirus	Small RNA	Yes—IFN response	No	Phase II
NDV	Small RNA	Yes—IFN response	No	Phase I
VSV	Small RNA	Yes—IFN response	Yes IFN response	None
Semliki Forest	Small RNA	Yes—IFN response	No	None
Coxsackievirus	Small RNA	Yes—receptor mediated	Yes miRNA expression	Phase I
Retrovirus	Small RNA	Yes—dividing cells	No	None
Measles	RNA	Yes—Receptor mediated	Yes receptor targeting	Phase I/II
Poliovirus	RNA	No	Yes—reduced pathogenicity	None
Influenza	RNA	No	Yes—block PKR inhibition	None
Seneca Valley	RNA	Yes—receptor mediated	No	Phase I/II
Adenovirus	Small DNA	No	Yes—receptor targeting; viral gene expression; deletion of virulence genes	Phase III
Parvovirus	Small DNA	Yes—IFN response	No	None
HSV	Large DNA	No	Yes—viral gene expression; deletion of virulence genes	Phase II
Vaccinia	Large DNA	Yes—not known	Yes—deletion of virulence genes	Phase II

develop them as tumor-targeting agents. It has been observed for over 100 years that viral infections can have beneficial effects in cancer patients (35), and clinical trials with live wild-type viruses were first attempted over 50 years ago (36). Inherent tumor selectivity in humans is a characteristic found in a diverse range of RNA viruses including:

1. *Reovirus, VSV, Newcastle Disease Virus (NDV), Semliki Forest Virus:* A common mechanism behind the tumor selectivity of all of these viruses involves tumor-specific inactivation of protein kinase R (PKR) and interferon (IFN) response pathways. For example, it is believed reovirus infection leads to activation of dsRNA-activated protein kinase, PKR, resulting in termination of translation of viral transcripts (37). However, in most human cancers, where Ras-mediated signal transduction is activated, PKR kinase activity is impaired such that reovirus can replicate normally. VSV is also known to replicate selectively in tumors with defects in the IFN response pathway involving PKR (38), as does Semliki Forest virus (39), and NDV appears to do the same (40). Naturally occurring infections with these viruses are either asymptomatic or cause mild disease in humans, and so their safety profiles are attractive. However, several (such as VSV or NDV) cause severe disease in farm animals, and so their use as cancer therapies will need to be carefully controlled to prevent spread to agriculture. The fact that these viruses are normally simple to manufacture to high titers means that some have been extensively used in the clinic, notably NDV (41) and Reovirus (as Reolysin) (42, 43). While safety profiles have been excellent, their single agent efficacy has been limited. However, when used in combination with chemotherapy, encouraging antitumor effects have been witnessed. VSV, unlike many other small RNA viruses, is amenable to genetic alteration, with tumor-selective genetically modified strains described, and it has the capacity to encode foreign transgenes. However, surprisingly, to date, it has not been used in the clinic.
2. *Coxsackievirus A21:* This common cold-causing picornavirus has been found to preferentially target cancer cells that overexpress the viral receptors ICAM-1 and/or decay-accelerating factor (DAF) (44). Some potential safety concerns have been addressed with the insertion of miRNA sequences (see above). Phase I trials have been started.
3. *Retrovirus:* Replication-selective retroviral vectors based on murine leukemia virus (MLV) have been described. Selectivity appears to be dependent on dividing cells within the tumor and the immunosuppressive nature of the tumor environment (45). The virus itself possesses limited antitumor effect, and the fact that it stably integrates into the tumor genome may raise concerns. However, the expression of therapeutic transgenes (especially prodrug-converting enzymes) from the virus, coupled to the fact that rapid spread throughout the tumor has

been demonstrated, means that potent antitumor effects in preclinical models are possible. Phase I trials are planned.

4. *Poliovirus:* Obviously, a highly pathogenic virus like poliovirus will need to be significantly attenuated and undergo careful safety testing before clinical assessment can begin. Attenuation has been achieved by switching the poliovirus IRES with that of human rhinovirus type 2 (HRV2), producing the oncolytic strain PVS-RIPO. This virus retains the natural CNS targeting of poliovirus, and so is a natural target for CNS neoplasia, including glioblastoma multiforme, but has lost the ability to replicate in nonmalignant cells within the CNS (46). Demonstrable safety testing has meant that Phase I trials are also planned with this virus.
5. *Measles:* The observation that a Hodgkin's lymphoma patient displayed a dramatic response after measles vaccination (47) has led to the development of the Edmonton vaccine strain of measles as a leukemia and lymphoma therapy (48). The observation that the tumor targeting of this strain was primarily achieved through upregulation of the CD46 receptor on the tumor cells has led to its use against other tumor types that often display increased CD46 expression, including lymphoma (49), multiple myeloma (50), ovarian (51), and brain (52) tumors. Clinical trials have commenced targeting glioblastoma, ovarian cancer, and multiple myeloma. In addition, retargeting of the virus to other receptors has been used to target other cancer types (see above). The high percentage of cancer patients presenting with preexisting immunity to measles virus has led to the development of several strategies to try to overcome this (see below).
6. *Seneca Valley Virus:* Although only recently discovered, and still poorly understood, this virus has demonstrated the ability to infect a variety of neuroendocrine cancers, but not normal cells, and has the advantage that preexisting immunity is rare (53). It appears that selectivity is controlled at the level of infectivity, with several cell surface proteins, including integrin alpha 4, implicated in its tumor-targeting ability. Phase I trials have been completed, with safety and some potential activity demonstrated, and Phase II trials are planned.
7. *Influenza:* Another normally pathogenic virus that has been proposed as an oncolytic agent is influenza. A strain of Influenza A (delNS1), in which the NS1 gene was deleted, was found to be unable to block the antiviral effects of PKR, and so is only capable of replicating in cells expressing oncogenic ras (54). This represents an example of a viral gene deletion designed to increase tumor selectivity, as discussed in more detail below.

Small DNA Viruses In addition to the natural tumor tropism displayed by many viruses, a host of viral virulence genes that are essential for effective infection and replication in normal tissues have been shown to be redundant in certain tumor types. These may, for example, be involved in the synthesis of nucleotide pools, the induction of cellular proliferation, the blocking of cellular apoptosis or escape from immune-mediated clearance, such as through disruption of the IFN signaling pathway. This is because many cancer cells have undergone a series of adaptation steps (including uncontrolled entry into S-phase, disruption of apoptotic and p53 pathways, loss of the ability to produce and/or respond to innate immune effectors, and evasion of cell-mediated immunity) (55) during the progression to malignancy that make these gene products redundant. A large amount of work has therefore involved examining deletion mutants of these genes in different viruses in order to assess their effects on tumor targeting.

Perhaps the most extensively studied virus in this respect is the common cold virus adenovirus type 5. For example, one function lost in many tumor cells is the ability to undergo apoptosis. This is significant as one of the main cellular responses and host defense mechanisms against viral infection is also apoptosis, and many viruses carry genes whose products function to inhibit this process. Perhaps the most frequent genetic abnormality identified in tumors to date is mutation of the tumor suppressor p53 and/or inactivation of the p53 pathway. The p53 protein functions to both block cell cycling and/or to promote apoptosis in response to a number of stimuli. Other blocks in apoptosis in cancers may involve overexpression of anti-apoptotic proteins such as Bcl-2, or loss of Bax expression. Viruses target numerous points in apoptosis pathways. For example, the adenovirus E1B-55K protein acts to inactivate p53 function in combination with the E4ORF6 protein. The tumor-targeting oncolytic adenovirus *dl*1520 (aka Onyx-015, H101) contains a deletion in this gene and so should only be able to replicate in tumor cells with an inactivated p53 pathway (56, 57). Although the tumor selectivity of this virus has been demonstrated both in patients (58) and *in vitro* (57), the role of p53 during *dl*1520 infection and replication remains controversial, and an additional role for its interaction with heat shock proteins has been proposed (59, 60). The Onyx-015 virus represents the first rationally designed and engineered oncolytic virus to enter the clinic, and progressed to Phase III testing in the United States. In addition, a generic version of this virus (H101) represents the first oncolytic virus approved for cancer treatment (for SCCHN and some other injectable tumors in some Asian markets).

Alternatively, the loss of the retinoblastoma protein (pRB) tumor suppressor function in cancers has been targeted by a variety of adenoviral constructs. The pRB protein interacts with the cellular transcription factor E2F in order to prevent uncontrolled cellular proliferation and is inactivated in many tumor types. The adenovirus E1A protein interacts with pRB through the conserved region 2 (CR2) and this region of E1A has been deleted in several adenoviral mutants currently under investigation (61–64). Alternatively, the E1A-CR1 domain of adenovirus specifically targets the p300 protein (65). p300/CBP transcriptional coactivator proteins play a central role in coordinating and integrating multiple signal-dependent events, including cellular proliferation, differentiation, and apoptosis, and are involved in activating p53 by acetylation. Several oncolytic adenovirus strains with deletions in the E1A-CR1 domain have also been described (7, 64).

Many adenoviral strains under investigation contain deletions in all (e.g., 01/PEME) or part (e.g., *dl*1520) of the E3 region to allow increased cloning capacity for the insertion of exogenous genes. However, as the main function of this region is to protect infected cells from immune-mediated clearance (66), it would be expected that deletions in the E3 region would attenuate the virus *in* vivo and may contribute to tumor selectivity. However, the E3B region was found to increase tumor cell killing *in vivo*, even in human tumor xenografts in immunodeficient mice (67), and so the roles of the individual genes in this region need to be studied in more detail. The effects of inserting foreign genes into different locations within the E3 region have been studied (68) and in 01/PEME, the E3 gene encoding the adenovirus death protein (ADP, E3-11.6K) has been reinserted into the genome, because efficient cell lysis and release of adenovirus has been shown to be dependent on this protein (69). Finally, the adenovirus E1B-19K gene is known to disrupt tumor necrosis factor (TNF) signaling pathways, and deletions of this gene have also been found to produce tumor targeting due to TNF-dependent clearance from normal tissues (70, 71). Although oncolytic or conditionally replicating adenovirus strains (sometimes known as CRADs) represent the most extensively studied viruses preclinically, and have advanced furthest in clinical testing, there are inherent disadvantages to this background that may prove difficult to overcome. In particular, it spreads slowly in the tumor (meaning it is often cleared before complete tumor destruction is achieved) and has not evolved to spread through the blood (meaning systemic delivery and spread to metastatic tumors is often limited).

The autonomous parvovirus H-1 represents another small DNA virus that has been proposed for use as an oncolytic without further genetic modification. Like many RNA viruses, this rodent pathogen appears to rely on disrupted IFN response in tumor cells to selectively replicate.

Large DNA Viruses Several larger DNA viruses have also been proposed as oncolytic agents, notably HSV and poxvirus strains. Unlike adenovirus, where most gene products are multifunctional, these larger viruses tend to have evolved single function gene products, which means their deletion may produce a more predictable phenotype. A large number of virulence genes means that a variety of different tumor-selective agents with different phenotypic targets can be designed, with multiple deletions leading to increased safety. However, the large size of these viruses makes it harder to control for the appearance of spontaneous secondary mutations, and can make manufacture more complicated, especially when filter sterilization of viral preparations is not possible.

One set of viral genes that are not needed for replication in many tumor cells are those involved in induction of cellular proliferation in infected cells. The potential of this gene-deletion approach was first demonstrated with the restriction of HSV replication to proliferating cells (72). Thymidine kinase (U_L23)-negative deletion mutants such as *dl*sptk were found to display limited replication in nonproliferating normal cells but were able to replicate within and kill malignant glioma cells, resulting in a dose-dependent prolongation of survival among animals with brain tumors. The dependence on host cell nucleotides, and so actively proliferating cells, seen with viral tymidine kinase (TK) deletion has also been studied in vaccinia virus (73). In order to further limit the replication of TK-deleted vaccinia virus to proliferating tumor cells, a second mutation was introduced, in the vaccinia growth factor (VGF) gene, in vaccinia strain vvDD (74, 75). The double deletion was found to have markedly enhanced tumor specificity and reduced pathogenicity. An alternative to TK deletion has also been explored in HSV, with the deletion of the ICP6, ribonucleotide reductase gene (76). Through a similar mechanism to deletion of TK, loss of viral ribonucleotide reductase will restrict viral replication to proliferating cells where cellular ribonucleotide reductase is producing elevated pools of dNTPs. This deletion is also exquisitely sensitive to antivirals.

Alternatively, the HSV ICP10 gene contains a protein kinase domain that activates the Ras/MEK/MAPK mitogenic signaling pathway, and deletion of this domain has also been found to be tumor targeting (77). Similarly, as several inherently selective wild-type viruses target tumor cells with defective IFN pathways, including those driven by oncogenic ras, it is logical that deletion of viral genes involved with suppression of IFN production and/ or the IFN/PKR response may also produce tumor-selective vectors. This has been found to be the case in HSV with deletion of ICP34.5, an inhibitor of PKR function. Five different HSV strains with this deletion have entered clinical testing: G207 (78, 79), HSV1716 (80), NV1020 (81), JS1/34.5-/47- (82) and OncoVEXGM-CSF (83). Whereas HSV1716 has a single deletion of ICP34.5, G207 has an additional deletion in the UL39 locus, JS-1/34.5-/47- carries an additional deletion in ICP47 gene (known to downregulate MHC Class I) and is derived from a clinical HSV-1 isolate, and OncoVEXGM-CSF represents this virus additionally expressing granulocyte/macrophage-colony stimulating factor (GM-CSF). Deletion of ICP34.5 also removes section of the LAT's, which makes the virus unable to establish latency in the cells. G47Δ (84) carries deletions in ICP34.5, UL39, and ICP47 with Us11 under the control of the ICP47 promoter, which enhances the growth by preventing the shutoff of host protein synthesis. This vector demonstrates how third- and fourth-generation oncolytic viruses are continually becoming more distinct from their parental strains and designed specifically to target cancers. However, care is needed not to create viruses that are so attenuated that they are unable to efficiently destroy their tumor targets. A vaccinia virus strain with a deletion in the B18R gene, a secreted protein that acts as a type I IFN decoy receptor, has also been shown to target tumors through a loss of type I IFN response, and was also engineered to express IFN-β, to further restrict replication, and to provide additional antitumor effects (85). In addition, the host range of myxoma virus is limited by IFN-β pathways in species other than rabbit, and so human tumor cells, without an innate IFN response, are also capable of replicating this virus (86).

Vaccinia virus also contains several genes encoding serpins that are involved in preventing apoptosis, and the deletion of these genes was found to produce a tumor-targeting phenotype (87). Finally, although not directly tumor targeting, vaccinia mutations that produce a form of the virus that is wrapped in a host cell-derived envelope (the EEV form), were found to have improved delivery and spread within tumors, and so enhanced antitumor effects (88).

FURTHER ENHANCING ONCOLYTIC VIRUS THERAPIES

Interaction with the Host Immune Response

How best to integrate oncolytic vectors with the host immune response remains one of the most poorly understood components of oncolytic viral therapy, yet is likely to prove critical to the overall success of the platform. The host immune response undoubtedly represents a double-edged sword, as premature immune clearance of the virus prevents effective therapy, a problem compounded when preexisting antiviral immunity is present. However, it is also true that immune-mediated clearance of infected cancer cells represents an alternate tumor-killing mechanism mediated by oncolytic therapy, that immune clearance of the virus is necessary to prevent viral-mediated toxicity, and that the priming of an antitumor adaptive immune response as a result of viral infection within the tumor has been described for multiple oncolytic agents.

A variety of strategies have therefore been described that either suppress or enhance different components of the immune response. In general, these strategies look to evade or suppress the humoral immune response, while enhancing the cellular immune response. Normally, but not always, strategies that inhibit the innate immune response also appear beneficial. In this respect, the use of cyclophosphamide (CPA), cyclosporine, or B-cell depletion to prevent the upregulation or appearance of neutralizing antibodies after therapy has been found to allow for repeat dosing and/or improved therapeutic outcome with some viruses, including reovirus (89). The use of CPA as a means to block innate immune responses, and prevent NK cell infiltration into the tumor has also been found to enhance oncolytic HSV therapy (90), while CPA and rapamycin were both found to reduce innate immune responses and so enhance vaccinia therapy (91). Rapamycin has also been found to enhance myxoma therapy, but this is primarily due to its action on mammalian target of Rapamycin (mTOR) signaling pathways (92). Alternatively, the use of histone deacetylase inhibitors as a means of disrupting IFN signaling, and so blocking the innate immune response, has been found to enhance oncolytic viral therapy with VSV, vaccinia (93), and HSV (94).

The expression of cytokines from viral strains as a means to skew the balance of the Th1/Th2 arms of the immune response toward Th1 has also proven an effective way to create a more "beneficial" cellular immune response, often at the expense of the humoral response. As such, cytokines such as

GM-CSF, IL-2, IL-12, and IFN-γ have all demonstrated benefits when expressed from oncolytic agents. Interestingly however, some of these cytokines may actually also have immune suppressive effects as, for example, IL-2 can also induce proliferation of regulatory T cells, and GM-CSF has been shown to enhance proliferation of myeloid-derived suppressor cells. It remains to be determined whether possible oncolytic activity resulting from this immune suppression may actually be beneficial, or if they could instead enhance tumor progression. In addition, several reports have shown that type I IFN (a mediator primarily of the innate immune response) can enhance the therapeutic benefits of some oncolytic agents (85).

One major limitation of oncolytic viral therapies that is only poorly addressed with any of these approaches is that of overcoming preexisting antiviral immunity (either as a result of natural pre-exposure to wild-type virus, as part of a vaccination program, or as a result of prior therapy with the same agent) and, in particular, how to evade neutralizing antibody. Recently, several strategies have been proposed, and these include the use of polymer coating (see above) or the use of cell-based delivery mechanisms. Different cell types might be preinfected with the viral agent prior to delivery and have been shown to subsequently be effective as a means to deliver different viruses to the tumor (including tumor cells [95], immune cells [96–98], and stem cells [99]), and several of these have also been found to conceal the virus from circulating antibody. This cell-based delivery approach has also been found to provide other advantages in some situations, including improving biodistribution of viral infection within the tumor (due to the cell's ability to traffic within the tumor microenvironment) and even resulting in synergistic tumor cell killing if the cellular delivery agent is also a therapeutic (97).

One final immune-mediated, tumor-killing mechanisms induced by oncolytic viral therapy is that of vascular collapse. VSV and vaccinia therapy have both been shown to induce a rapid collapse of the tumor vasculature, leading to necrosis of large areas of the tumor that were not infected by the virus (85, 100). During VSV infection, this appears to be due to recruitment of neutrophils into the infected tumor, resulting in thrombosis. Vaccinia therapy may additionally lead to selective infection of tumor endothelial cells, leading to their destruction and resulting vascular collapse. Although this phenomenon has only recently been described, strategies to enhance it may lead to improved therapeutiuc benefits.

Combining Oncolytic Viral Therapy with Traditional Therapies

Frontline therapies for most cancers involve combinations of multiple therapeutics, reflecting the need to target the tumor through distinct mechanisms. This approach better destroys the many heterogeneous cell populations within the cancer, and better prevents the development of resistance. Because oncolytic viruses often recognize and destroy tumors through distinct mechanisms to traditional chemo and radiotherapies, they often display additive or synergistic antitumor effects, without increased toxicity, when used in combination with these agents. In some cases, therapeutic benefits of oncolytic viruses are only seen clinically when used in combination studies. As such, a variety of preclinical reports and clinical trials have described the successful combination of oncolytic viral therapies and either radiotherapy (101, 102) or chemotherapy (103). These include adenovirus with cisplatin and 5FU (104) or cisplatin and paclitaxel (105) and HSV with cisplatin (106) or mitomycin C (107). In some cases, mechanisms for synergy have been described, with the action of chemotherapies often enhancing viral replication in tumor cells. For example, taxanes were shown to stabilize microtubules and so enhance movement of oncolytic adenovirus strains within the cell during its replication cycle, or HSV was found to adjust cellular gene expression, so sensitizing cells to taxane therapies. Alternatively, it was found that pretreating solid tumors *in vivo* with paclitaxel sensitized them to subsequent HSV therapy, with apoptosis induced by the chemotherapy found to open up the tumor, creating channels of apoptosis, and so allowing better delivery of the viral agent (108). However, careful preclinical testing of oncolytic agent and chemotherapy combinations are needed, as in some cases, chemotherapies may directly inhibit the viral life cycle (109), or rapid destruction of infected tumor cells may prevent viral replication. In addition to applying combinations of therapies, it is possible to express different therapeutic transgenes from oncolytic viruses in order to incorporate an additional dimension to their tumor cell killing; these might include cytokines (as described above), prodrug-converting enzymes (that selectively convert a systemically delivered prodrug into a toxic drug within the infected tumor) (110), or anti-angiogenic factors (111), for example. Alternatively, the use of reporter genes to follow the biodistribution and replication patterns of the viral agents *in vivo* have been designed to better assess the early effectiveness of the therapy *in vivo* (112). Finally, it has also recently been demonstrated that in some cases, regulation of transgene expression may further enhance the therapeutic potential of the viral vectors, by allowing the virus to selectively replicate and spread within the tumor before a toxic or immune stimulatory tarnsgene function is activated (113).

SUMMARY

As our understanding of tumor biology and virology expands, so does our ability to manipulate viruses as therapies. It is clear that viruses can be used as more than simple gene expression systems and can be incorporated as therapies themselves, notably in the treatment of cancers, where a replication-selective virus can safely destroy tumor cells by multiple mechanisms, including direct cell lysis, immune targeting of infected cells, and vascular collapse. Because these viruses destroy tumors by a variety of distinct mechanisms, they can be safely combined with standard treatment modalities with resultant synergy. There is much still to be learned, including improving systemic delivery and balancing lytic potential against safety concerns, but great strides have been made. As different combinations of attenuating and tumor-targeting mutations are incorporated, these viruses will become continually distinct from their native wild-type parents, and more efficient at treating cancer patients.

REFERENCES

1. Miyatake S.I., Tani S., Feigenbaum F., Sundaresan P., Toda H., Narumi O., Kikuchi H., Hashimoto N., Hangai M., Martuza R.L., and Rabkin S.D. 1999. Hepatoma-specific antitumor activity of an albumin enhancer/promoter regulated herpes simplex virus in vivo. Gene Ther. **6**: 564–572.
2. Mullen J.T., and Tanabe K.K. 2002. Viral oncolysis. Oncologist. **7**: 106–119.
3. Chen Y., DeWeese T., Dilley J., Zhang Y., Li Y., Ramesh N., Lee J., Pennathur-Das R., Radzyminski J., Wypych J., Brignetti D., Scott S., Stephens J., Karpf D.B., Henderson D.R., and Yu D.C. 2001. CV706, a prostate cancer-specific adenovirus variant, in combination with radiotherapy produces synergistic antitumor efficacy without increasing toxicity. Cancer Res. **61**: 5453–5460.
4. Li Y., Yu D.C., Chen Y., Amin P., Zhang H., Nguyen N., and Henderson D.R. 2001. A hepatocellular carcinoma-specific adenovirus variant, CV890, eliminates distant human liver tumors in combination with doxorubicin. Cancer Res. **61**: 6428–6436.
5. Yu D.C., Chen Y., Dilley J., Li Y., Embry M., Zhang H., Nguyen N., Amin P., Oh J., and Henderson D.R. 2001. Antitumor synergy of CV787, a prostate cancer-specific adenovirus, and paclitaxel and docetaxel. Cancer Res. **61**: 517–525.
6. Zhang J., Ramesh N., Chen Y., Li Y., Dilley J., Working P., and Yu D.C. 2002. Identification of human uroplakin II promoter and its use in the construction of CG8840, a urothelium-specific adenovirus variant that eliminates established bladder tumors in combination with docetaxel. Cancer Res. **62**: 3743–3750.
7. Doronin K., Toth K., Kuppuswamy M., Ward P., Tollefson A.E., and Wold W.S. 2000. Tumor-specific, replication-competent adenovirus vectors overexpressing the adenovirus death protein. J. Virol. **74**: 6147–6155.
8. Kurihara T., Brough D.E., Kovesdi I., and Kufe D.W. 2000. Selectivity of a replication-competent adenovirus for human breast carcinoma cells expressing the MUC1 antigen. J. Clin. Invest. **106**: 763–771.
9. Hamada K., Kohno S., Iwamoto M., Yokota H., Okada M., Tagawa M., Hirose S., Yamasaki K., Shirakata Y., Hashimoto K., and Ito M. 2003. Identification of the human IAI.3B promoter element and its use in the construction of a replication-selective adenovirus for ovarian cancer therapy. Cancer Res. **63**: 2506–2512.
10. Taki M., Kagawa S., Nishizaki M., Mizuguchi H., Hayakawa T., Kyo S., Nagai K., Urata Y., Tanaka N., and Fujiwara T. 2005. Enhanced oncolysis by a tropism-modified telomerase-specific replication-selective adenoviral agent OBP-405 ("Telomelysin-RGD"). Oncogene. **24**: 3130–3140.
11. Huang Q., Zhang X., Wang H., Yan B., Kirkpatrick J., Dewhrist M.W., and Li C.Y. 2004. A novel conditionally replicative adenovirus vector targeting telomerase-positive tumor cells. Clin. Cancer Res. **10**: 1439–1445.

12. Kambara H., Okano H., Chiocca E.A., and Saeki Y. 2005. An oncolytic HSV-1 mutant expressing ICP34.5 under control of a nestin promoter increases survival of animals even when symptomatic from a brain tumor. Cancer Res. **65**: 2832–2839.

13. Peter I., Graf C., Dummer R., Schaffner W., Greber U.F., and Hemmi S. 2003. A novel attenuated replication-competent adenovirus for melanoma therapy. Gene Ther. **10**: 530–539.

14. Chung R.Y., Saeki Y., and Chiocca E.A. 1999. B-myb promoter retargeting of herpes simplex virus gamma34.5 gene-mediated virulence toward tumor and cycling cells. J. Virol. **73**: 7556–7564.

15. Nakamura H., Kasuya H., Mullen J.T., Yoon S.S., Pawlik T.M., Chandrasekhar S., Donahue J.M., Chiocca E.A., Chung R.Y., and Tanabe K.K. 2002. Regulation of herpes simplex virus gamma(1)34.5 expression and oncolysis of diffuse liver metastases by Myb34.5. J. Clin. Invest. **109**: 871–882.

16. Ramesh N., Ge Y., Ennist D.L., Zhu M., Mina M., Ganesh S., Reddy P.S., and Yu D.C. 2006. CG0070, a conditionally replicating granulocyte macrophage colony-stimulating factor—armed oncolytic adenovirus for the treatment of bladder cancer. Clin. Cancer Res. **12**: 305–313.

17. Binley K., Askham Z., Martin L., Spearman H., Day D., Kingsman S., and Naylor S. 2003. Hypoxia-mediated tumour targeting. Gene Ther. **10**: 540–549.

18. Post D.E., and Van Meir E.G. 2003. A novel hypoxia-inducible factor (HIF) activated oncolytic adenovirus for cancer therapy. Oncogene. **22**: 2065–2072.

19. Hernandez-Alcoceba R., Pihalja M., Qian D., and Clarke M.F. 2002. New oncolytic adenoviruses with hypoxia- and estrogen receptor-regulated replication. Hum. Gene Ther. **13**: 1737–1750.

20. Small E.J., Carducci M.A., Burke J.M., Rodriguez R., Fong L., van Ummersen L., Yu D.C., Aimi J., Ando D., Working P., Kirn D., and Wilding G. 2006. A phase I trial of intravenous CG7870, a replication-selective, prostate-specific antigen-targeted oncolytic adenovirus, for the treatment of hormone-refractory, metastatic prostate cancer. Mol. Ther. **14**: 107–117.

21. Akiyama M., Thorne S., Kirn D., Roelvink P.W., Einfeld D.A., King C.R., and Wickham T.J. 2004. Ablating CAR and integrin binding in adenovirus vectors reduces nontarget organ transduction and permits sustained bloodstream persistence following intraperitoneal administration. Mol. Ther. **9**: 218–230.

22. Gaggar A., Shayakhmetov D.M., Liszewski M.K., Atkinson J.P., and Lieber A. 2005. Localization of regions in CD46 that interact with adenovirus. J. Virol. **79**: 7503–7513.

23. Kawakami Y., Li H., Lam J.T., Krasnykh V., Curiel D.T., and Blackwell J.L. 2003. Substitution of the adenovirus serotype 5 knob with a serotype 3 knob enhances multiple steps in virus replication. Cancer Res. **63**: 1262–1269.

24. Bucheit A.D., Kumar S., Grote D.M., Lin Y., von Messling V., Cattaneo R.B., and Fielding A.K. 2003. An oncolytic measles virus engineered to enter cells through the CD20 antigen. Mol. Ther. **7**: 62–72.

25. Peng K.W., Donovan K.A., Schneider U., Cattaneo R., Lust J.A., and Russell SJ. 2003. Oncolytic measles viruses displaying a single-chain antibody against CD38, a myeloma cell marker. Blood. **101**: 2557–2562.

26. Hasegawa K., Nakamura T., Harvey M., Ikeda Y., Oberg A., Figini M., Canevari S., Hartmann L.C., and Peng K.W. 2006. The use of a tropism-modified measles virus in folate receptor-targeted virotherapy of ovarian cancer. Clin. Cancer Res. **12**: 6170–6178.

27. Allen C., Vongpunsawad S., Nakamura T., James C.D., Schroeder M., Cattaneo R., Giannini C., Krempski J., Peng K.W., Goble J.M., Uhm J.H., Russell S.J., and Galanis E. 2006. Retargeted oncolytic measles strains entering via the EGFRvIII receptor maintain significant antitumor activity against gliomas with increased tumor specificity. Cancer Res. **66**: 11840–11850.

28. Hemminki A., Wang M., Hakkarainen T., Desmond R.A., Wahlfors J., and Curiel D.T. 2003. Production of an EGFR targeting molecule from a conditionally replicating adenovirus impairs its oncolytic potential. Cancer Gene Ther. **10**: 583–588.

29. van der Poel H.G., Molenaar B., van Beusechem V.W., Haisma H.J., Rodriguez R., Curiel D.T., and Gerritsen W.R. 2002. Epidermal growth factor receptor targeting of replication competent adenovirus enhances cytotoxicity in bladder cancer. J. Urol. **168**: 266–272.

30. Fisher K.D., Stallwood Y., Green N.K., Ulbrich K., Mautner V., and Seymour L.W. 2001. Polymer-coated adenovirus permits efficient retargeting and evades neutralising antibodies. Gene Ther. **8**: 341–348.

31. Morrison J., Briggs S.S., Green N., Fisher K., Subr V., Ulbrich K., Kehoe S., and Seymour L.W. 2008. Virotherapy of ovarian cancer with polymer-cloaked adenovirus retargeted to the epidermal growth factor receptor. Mol. Ther. **16**: 244–251.

32. Kelly E.J., Hadac E.M., Greiner S., and Russell S.J. 2008. Engineering microRNA responsiveness to decrease virus pathogenicity. Nat. Med. **14**: 1278–1283.

33. Ylosmaki E., Hakkarainen T., Hemminki A., Visakorpi T., Andino R., and Saksela K. 2008. Generation of a conditionally replicating adenovirus based on targeted destruction of E1A mRNA by a cell type-specific MicroRNA. J. Virol. **82**: 11009–11015.

34. Edge R.E., Falls T.J., Brown C.W., Lichty B.D., Atkins H., and Bell J.C. 2008. A let-7 MicroRNA-sensitive vesicular stomatitis virus demonstrates tumor-specific replication. Mol. Ther. **16**: 1437–1443.

35. Dock G. 1904. Rabies virus vaccination in a patient with cervical carcinoma. American J. Med. Sci. **127**: 563.

36. Southam C.M., and Moore A.E. 1952. Clinical studies of viruses as antineoplastic agents with particular reference to Egypt 101 virus. Cancer. **5**: 1025–1034.

37. Strong J.E., Coffey M.C., Tang D., Sabinin P., and Lee P.W. 1998. The molecular basis of viral oncolysis: usurpation of the Ras signaling pathway by reovirus. EMBO J. **17**: 3351–3362.

38. Stojdl D.F., Lichty B., Knowles S., Marius R., Atkins H., Sonenberg N., and Bell J.C. 2000. Exploiting tumor-specific defects in the interferon pathway with a previously unknown oncolytic virus. Nat. Med. **6**: 821–825.

39. Vähä-Koskela M.J., Kallio J.P., Jansson L.C., Heikkilä J.E., Zakhartchenko V.A., Kallajoki M.A., Kähäri V.M., and Hinkkanen A.E. 2006. Oncolytic capacity of attenuated replicative semliki forest virus in human melanoma xenografts in severe combined immunodeficient mice. Cancer Res. **66**: 7185–7194.

40. Reichard K.W., Lorence R.M., Cascino C.J., Peeples M.E., Walter R.J., Fernando M.B., Reyes H.M., and Greager J.A. 1992. Newcastle disease virus selectively kills human tumor cells. J. Surg. Res. **52**: 448–453.

41. Lorence R.M., Roberts M.S., O'Neil J.D., Groene W.S., Miller J.A., Mueller S.N., and Bamat M.K. 2007. Phase 1 clinical experience using intravenous administration of PV701, an oncolytic Newcastle disease virus. Curr. Cancer Drug Targets. **7**: 157–167.

42. Forsyth P., Roldán G., George D., Wallace C., Palmer C.A., Morris D., Cairncross G., Matthews M.V., Markert J., Gillespie Y., Coffey M., Thompson B., and Hamilton M. 2008. A phase I trial of intratumoral administration of reovirus in patients with histologically confirmed recurrent malignant gliomas. Mol. Ther. **16**: 627–632.

43. Vidal L., Pandha H.S., Yap T.A., White C.L., Twigger K., Vile R.G., Melcher A., Coffey M., Harrington K.J., and DeBono J.S. 2008. A phase I study of intravenous oncolytic reovirus type 3 Dearing in patients with advanced cancer. Clin. Cancer Res. **14**: 7127–7137.

44. Shafren D.R., Au G.G., Nguyen T., Newcombe N.G., Haley E.S., Beagley L., Johansson E.S., Hersey P., and Barry R.D. 2004. Systemic therapy of malignant human melanoma tumors by a common cold-producing enterovirus, coxsackievirus a21. Clin. Cancer Res. **10**: 53–60.

45. Tai C.K., and Kasahara N. 2008. Replication-competent retrovirus vectors for cancer gene therapy. Front. Biosci. **13**: 3083–3095.

46. Dobrikova E.Y., Broadt T., Poiley-Nelson J., Yang X., Soman G., Giardina S., Harris R., and Gromeier M. 2008. Recombinant oncolytic poliovirus eliminates glioma in vivo without genetic adaptation to a pathogenic phenotype. Mol. Ther. **16**: 1865–1872.

47. Zygiert Z. 1971. Hodgkin's disease: remissions after measles. Lancet. **1**: 593.

48. Russell S.J., and Peng K.W. 2009. Measles virus for cancer therapy. Curr. Top. Microbiol. Immunol. **330**: 213–241.

49. Grote D., Russell S.J., Cornu T.I., Cattaneo R., Vile R., Poland G.A., and Fielding A.K. 2001. Live attenuated measles virus induces regression of human lymphoma xenografts in immunodeficient mice. Blood. **97**: 3746–3754.

50. Peng K.W., Ahmann G.J., Pham L., Greipp P.R., Cattaneo R., and Russell S.J. 2001. Systemic therapy of myeloma xenografts by an attenuated measles virus. Blood. **98**: 2002–2007.

51. Peng K.W., TenEyck C.J., Galanis E., Kalli K.R., Hartmann L.C., and Russell S.J. 2002. Intraperitoneal therapy of ovarian cancer using an engineered measles virus. Cancer Res. **62**: 4656–4662.

52. Phuong L.K., Allen C., Peng K.W., Giannini C., Greiner S., TenEyck C.J., Mishra P.K., Macura S.I., Russell S.J., and Galanis E.C. 2003. Use of a vaccine strain of measles virus genetically engineered to produce carcinoembryonic antigen as a novel therapeutic agent against glioblastoma multiforme. Cancer Res. **63**: 2462–2469.

53. Reddy P.S., Burroughs K.D., Hales L.M., Ganesh S., Jones B.H., Idamakanti N., Hay C., Li S.S., Skele K.L., Vasko A.J., Yang J., Watkins D.N., Rudin C.M., and Hallenbeck P.L. 2007. Seneca Valley virus, a systemically deliverable oncolytic picornavirus, and the treatment of neuroendocrine cancers. J. Natl. Cancer Inst. **99**: 1623–1633.

54. Bergmann M., Romirer I., Sachet M., Fleischhacker R., García-Sastre A., Palese P., Wolff K., Pehamberger H., Jakesz R., and Muster T. 2001. A genetically engineered influenza A virus with ras-dependent oncolytic properties. Cancer Res. **61**: 8188–8193.
55. Hanahan D., and Weinberg R.A. 2000. The hallmarks of cancer. Cell. **100**: 57–70.
56. Bischoff J.R., Kirn D.H., Williams A., Heise C., Horn S., Muna M., Ng L., Nye J.A., Sampson-Johannes A., Fattaey A., and McCormick F. 1996. An adenovirus mutant that replicates selectively in p53-deficient human tumor cells. Science. **274**: 373–376.
57. Heise C., Sampson J.A., Williams A., McCormick F., Von H.D., and Kirn D.H. 1997. ONYX-015, an E1B gene-attenuated adenovirus, causes tumor-specific cytolysis and antitumoral efficacy that can be augmented by standard chemotherapeutic agents. Nat. Med. **3**: 639–645.
58. Nemunaitis J., Ganly I., Khuri F., Arseneau J., Kuhn J., McCarty T., Landers S., Maples P., Romel L., Randlev B., Reid T., Kaye S., and Kirn D. 2000. Selective replication and oncolysis in p53 mutant tumors with Onyx-015, an E1B-55 kD gene-deleted adenovirus, in patients with advanced head and neck cancer: a phase II trial. Cancer Research **60**: 6359–6366.
59. O'Shea C.C., Johnson L., Bagus B., Choi S., Nicholas C., Shen A., Boyle L., Pandey K., Soria C., Kunich J., Shen Y., Habets G., Ginzinger D., and McCormick F. 2004. Late viral RNA export, rather than p53 inactivation, determines ONYX-015 tumor selectivity. Cancer Cell. **6**: 611–623.
60. Thorne S.H., Brooks G., Lee Y.L., Au T., Eng L.F., and Reid T. 2005. Effects of febrile temperature on adenoviral infection and replication: implications for viral therapy of cancer. J. Virol. **79**: 581–591.
61. Heise C., Hermiston T., Johnson L., Brooks G., Sampson-Johannes A., Williams A., Hawkins L., and Kirn D. 2000. An adenovirus E1A mutant that demonstrates potent and selective systemic anti-tumoral efficacy. Nat. Med. **6**: 1134–1139.
62. Fueyo J., Gomez-Manzano C., Alemany R., Lee P.S., McDonnell T.J., Mitlianga P., Shi Y.X., Levin V.A., Yung W.K., and Kyritsis A.P. 2000. A mutant oncolytic adenovirus targeting the Rb pathway produces anti-glioma effect *in vivo*. Oncogene **19**: 2–12.
63. Johnson L., Shen A., Boyle L., Kunich J., Pandey K., Lemmon M., Hermiston T., Giedlin M., McCormick F., and Fattaey A. 2002. Selectively replicating adenoviruses targeting deregulated E2F activity are potent, systemic antitumor agents. Cancer Cell. **1**: 325–337.
64. Balague C., Noya F., Alemany R., Chow L.T., and Curiel D.T. 2001. Human papillomavirus E6E7-mediated adenovirus cell killing: selectivity of mutant adenovirus replication in organotypic cultures of human keratinocytes. J. Virol. **75**: 7602–7611.
65. Arany Z., Sellers W.R., Livingston D.M., and Eckner R. 1994. E1A-associated p300 and CREB-associated CBP belong to a conserved family of coactivators. Cell. **77**: 799–800.
66. Horwitz M.S. 2001. Adenovirus immunoregulatory genes and their cellular targets. Virology. **279**: 1–8.

67. Zhu M., Bristol J.A., Xie Y., Mina M., Ji H., Forry-Schaudies S., and Ennist D.L. 2005. Linked tumor-selective virus replication and transgene expression from E3-containing oncolytic adenoviruses. J. Virol. **79**: 5455–5465.

68. Toth K., Spencer J.F., Tollefson A.E., Kuppuswamy M., Doronin K., Lichtenstein D.L., La Regina M.C., and Prince G.A. 2005. Wold WS Cotton rat tumor model for the evaluation of oncolytic adenoviruses. Hum. Gene Ther. **16**: 139–146.

69. Tollefson A.E., Ryerse J.S., Scaria A., Hermiston T.W., and Wold W.S. 1996. The E3-11.6-kDa adenovirus death protein (ADP) is required for efficient cell death: characterization of cells infected with ADP mutants. Virology. **220**: 152–162.

70. Liu T.C., Hallden G., Wang Y., Brooks G., Francis J., Lemoine N., and Kirn D. 2004. An E1B-19kDa gene deletion mutant adenovirus demonstrates tumor necrosis factor-enhanced cancer selectivity and enhanced oncolytic potency. Mol. Ther. **9**: 786–803.

71. Liu T.C., Wang Y., Hallden G., Brooks G., Francis J., Lemoine N.R., and Kirn D. 2005. Functional interactions of antiapoptotic proteins and tumor necrosis factor in the context of a replication-competent adenovirus. Gene Ther. **12**: 1333–1346.

72. Martuza R.L., Malick A., Markert J.M., Ruffner K.L., and Coen D.M. 1991. Experimental therapy of human glioma by means of a genetically engineered virus mutant. Science. **252**: 854–856.

73. Puhlmann M., Brown C.K., Gnant M., Huang J., Libutti S.K., Alexander H.R., and Bartlett D.L. 2000. Vaccinia as a vector for tumor-directed gene therapy: biodistribution of a thymidine kinase-deleted mutant. Cancer Gene Ther. **7**: 66–73.

74. McCart J.A., Ward J.M., Lee J., Hu Y., Alexander H.R., Libutti S.K., Moss B., and Bartlett D.L. 2001. Systemic cancer therapy with a tumor-selective vaccinia virus mutant lacking thymidine kinase and vaccinia growth factor genes. Cancer Res. **61**: 8751–8757.

75. Thorne S.H., Hwang T.H., O'Gorman W.E., Bartlett D.L., Sei S., Kanji F., Brown C., Werier J., Cho J.H., Lee D.E., Wang Y., Bell J., and Kirn D.H. 2007. Rational strain selection and engineering creates a broad-spectrum, systemically effective oncolytic poxvirus, JX-963. J. Clin. Invest. **117**: 3350–3358.

76. Mineta T., Rabkin S.D., and Martuza R.L. 1994. Treatment of malignant gliomas using ganciclovir-hypersensitive, ribonucleotide reductase-deficient herpes simplex viral mutant. Cancer Res. **54**: 3963–3966.

77. Fu X., Tao L., Cai R., Prigge J., and Zhang X. 2006. A mutant type 2 herpes simplex virus deleted for the protein kinase domain of the ICP10 gene is a potent oncolytic virus. Mol. Ther. **13**: 882–890.

78. Mineta T., Rabkin S.D., Yazaki T., Hunter W.D., and Martuza R.L. 1995. Attenuated multi-mutated herpes simplex virus-1 for the treatment of malignant gliomas. Nat. Med. **1**: 938–943.

79. Aghi M.K., and Chiocca E.A. 2009. Phase ib trial of oncolytic herpes virus G207 shows safety of multiple injections and documents viral replication. Mol. Ther. **17**: 8–9.

80. Mace A.T., Ganly I., Soutar D.S., and Brown S.M. 2008. Potential for efficacy of the oncolytic Herpes simplex virus 1716 in patients with oral squamous cell carcinoma. Head Neck. **30**: 1045–1051.

81. Fong Y., Kim T., Bhargava A., Schwartz L., Brown K., Brody L., Covey A., Karrasch M., Getrajdman G., Mescheder A., Jarnagin W., and Kemeny N. 2009. A herpes

oncolytic virus can be delivered via the vasculature to produce biologic changes in human colorectal cancer. Mol. Ther. **17**: 389–394.

82. Hu J.C., Booth M.J., Tripuraneni G., Davies D., Zaidi S.A., Tamburo de Bella M., Slade M.J., Marley S.B., Gordon M.Y., Coffin R.S., Coombes R.C., and Kamalati T. 2006. A novel HSV-1 virus, JS1/34.5-/47-, purges contaminating breast cancer cells from bone marrow. Clin. Cancer Res. **12**: 6853–6862.
83. Hu J.C., Coffin R.S., Davis C.J., Graham N.J., Groves N., Guest P.J., Harrington K.J., James N.D., Love C.A., McNeish I., Medley L.C., Michael A., Nutting C.M., Pandha H.S., Shorrock C.A., Simpson J., Steiner J., Steven N.M., Wright D., and Coombes R.C. 2006. A phase I study of OncoVEXGM-CSF, a second-generation oncolytic herpes simplex virus expressing granulocyte macrophage colony-stimulating factor. Clin. Cancer Res. **12**: 6737–6747.
84. Todo T., Martuza R.L., Rabkin S.D., and Johnson P.A. 2001. Oncolytic herpes simplex virus vector with enhanced MHC class I presentation and tumor cell killing. Proc. Natl. Acad. Sci. U S A. **98**: 6396–6401.
85. Kirn D.H., Wang Y., Le Boeuf F., Bell J., and Thorne S.H. 2007. Targeting of interferon-beta to produce a specific, multi-mechanistic oncolytic vaccinia virus. PLoS Med. **4**: e353.
86. Wang F., Ma Y., Barrett J.W., Gao X., Loh J., Barton E., Virgin H.W., and McFadden G. 2004. Disruption of Erk-dependent type I interferon induction breaks the myxoma virus species barrier. Nat. Immunol. **5**: 1266–1274.
87. Guo Z.S., Naik A., O'Malley M.E., Popovic P., Demarco R., Hu Y., Yin X., Yang S., Zeh H.J., Moss B., Lotze M.T., and Bartlett D.L. 2005. The enhanced tumor selectivity of an oncolytic vaccinia lacking the host range and antiapoptosis genes SPI-1 and SPI-2. Cancer Res. **65**: 9991–9998.
88. Kirn D.H., Wang Y., Liang W., Contag C.H., and Thorne S.H. 2008. Enhancing poxvirus oncolytic effects through increased spread and immune evasion. Cancer Res. **68**: 2071–2075.
89. Qiao J., Wang H., Kottke T., White C., Twigger K., Diaz R.M., Thompson J., Selby P., de Bono J., Melcher A., Pandha H., Coffey M., Vile R., and Harrington K. 2008. Cyclophosphamide facilitates antitumor efficacy against subcutaneous tumors following intravenous delivery of reovirus. Clin. Cancer Res. **14**: 259–269.
90. Fulci G., Breymann L., Gianni D., Kurozomi K., Rhee S.S., Yu J., Kaur B., Louis D.N., Weissleder R., Caligiuri M.A., and Chiocca E.A. 2006. Cyclophosphamide enhances glioma virotherapy by inhibiting innate immune responses. Proc. Natl. Acad. Sci. U S A. **103**: 12873–12878.
91. Lun X.Q., Jang J.H., Tang N., Deng H., Head R., Bell J.C., Stojdl D.F., Nutt C.L., Senger D.L., Forsyth P.A., and McCart JA. 2009. Efficacy of systemically administered oncolytic vaccinia virotherapy for malignant gliomas is enhanced by combination therapy with rapamycin or cyclophosphamide. Clin. Cancer Res. **15**: 2777–2788.
92. Lun X.Q., Zhou H., Alain T., Sun B., Wang L., Barrett J.W., Stanford M.M., McFadden G., Bell J., Senger D.L., and Forsyth PA. 2007. Targeting human medulloblastoma: oncolytic virotherapy with myxoma virus is enhanced by rapamycin. Cancer Res. **67**: 8818–8827.
93. Nguyên T.L., Abdelbary H., Arguello M., Breitbach C., Leveille S., Diallo J.S., Yasmeen A., Bismar T.A., Kirn D., Falls T., Snoulten V.E., Vanderhyden B.C.,

Werier J., Atkins H., Vähä-Koskela M.J., Stojdl D.F., Bell J.C., and Hiscott J. 2008. Chemical targeting of the innate antiviral response by histone deacetylase inhibitors renders refractory cancers sensitive to viral oncolysis. Proc. Natl. Acad. Sci. U S A. **105**: 14981–14986.

94. Liu T.C., Castelo-Branco P., Rabkin S.D., and Martuza R.L. 2008. Trichostatin A and oncolytic HSV combination therapy shows enhanced antitumoral and antiangiogenic effects. Mol. Ther. **16**: 1041–1047.
95. Power A.T., Wang J., Falls T.J., Paterson J.M., Parato K.A., Lichty B.D., Stojdl D.F., Forsyth P.A., Atkins H., and Bell J.C. 2007. Carrier cell-based delivery of an oncolytic virus circumvents antiviral immunity. Mol. Ther. **15**: 123–130.
96. Ong H.T., Hasegawa K., Dietz A.B., Russell S.J., and Peng K.W. 2006. Evaluation of T cells as carriers for systemic measles virotherapy in the presence of antiviral antibodies. Gene Ther. **14**: 324–333.
97. Thorne S.H., Negrin R.S., and Contag C.H. 2006. Synergistic antitumor effects of immune cell-viral biotherapy. Science. **311**: 1780–1784.
98. Cole C., Qiao J., Kottke T., Diaz R.M., Ahmed A., Sanchez-Perez L., Brunn G., Thompson J., Chester J., and Vile R.G. 2005. Tumor-targeted, systemic delivery of therapeutic viral vectors using hitchhiking on antigen-specific T cells. Nat. Med. **11**: 1073–1081.
99. Komarova S., Kawakami Y., Stoff-Khalili M.A., Curiel D.T., and Pereboeva L. 2006. Mesenchymal progenitor cells as cellular vehicles for delivery of oncolytic adenoviruses. Mol. Cancer Ther. **5**: 755–766.
100. Breitbach C.J., Paterson J.M., Lemay C.G., Falls T.J., McGuire A., Parato K.A., Stojdl D.F., Daneshmand M., Speth K., Kirn D., McCart J.A., Atkins H., and Bell J.C. 2007. Targeted inflammation during oncolytic virus therapy severely compromises tumor blood flow. Mol. Ther. **15**: 1686–1693.
101. Twigger K., Vidal L., White C.L., De Bono J.S., Bhide S., Coffey M., Thompson B., Vile R.G., Heinemann L., Pandha H.S., Errington F., Melcher A.A., and Harrington K.J. 2008. Enhanced in vitro and in vivo cytotoxicity of combined reovirus and radiotherapy. Clin. Cancer Res. **14**: 912–923.
102. Harrington K.J., Melcher A., Vassaux G., Pandha H.S., and Vile R.G. 2008. Exploiting synergies between radiation and oncolytic viruses. Curr. Opin. Mol. Ther. **10**: 362–370.
103. Kumar S., Gao L., Yeagy B., and Reid T. 2008. Virus combinations and chemotherapy for the treatment of human cancers. Curr. Opin. Mol. Ther. **10**: 371–379.
104. Khuri F.R., Nemunaitis J., Ganly I., Arseneau J., Tannock I.F., Romel L., Gore M., Ironside J., MacDougall R.H., Heise C., Randlev B., Gillenwater A.M., Bruso P., Kaye S.B., Hong W.K., and Kirn D.H. 2000. A controlled trial of Onyx-015, an E1B gene-deleted adenovirus, in combination with chemotherapy in patients with recurrent head and neck cancer. Nat. Med. **6**: 879–885.
105. Cheong S.C., Wang Y., Meng J.H., Hill R., Sweeney K., Kirn D., Lemoine N.R., and Halldén G. 2008. E1A-expressing adenoviral E3B mutants act synergistically with chemotherapeutics in immunocompetent tumor models. Cancer Gene Ther. **15**: 40–50.
106. Chahlavi A., Todo T., Martuza R.L., and Rabkin S.D. 1999. Replication-competent herpes simplex virus vector G207 and cisplatin combination therapy for head and neck squamous cell carcinoma. Neoplasia. **1**: 162–169.

107. Post D.E., Fulci G., Chiocca E.A., and Van Meir E.G. 2004. Replicative oncolytic herpes simplex viruses in combination cancer therapies. Curr. Gene Ther. **4**: 41–51.

108. Nagano S., Perentes J.Y., Jain R.K., and Boucher Y. 2008. Cancer cell death enhances the penetration and efficacy of oncolytic herpes simplex virus in tumors. Cancer Res. **68**: 3795–3802.

109. Reeves P.M., Bommarius B., Lebeis S., McNulty S., Christensen J., Swimm A., Chahroudi A., Chavan R., Feinberg M.B., Veach D., Bornmann W., Sherman M., and Kalman D. 2005. Disabling poxvirus pathogenesis by inhibition of Abl-family tyrosine kinases. Nat. Med. **11**: 731–739.

110. Barak Y., Thorne S.H., Ackerley D.F., Lynch S.V., Contag C.H., and Matin A. 2006. New enzyme for reductive cancer chemotherapy, YieF, and its improvement by directed evolution. Mol. Cancer Ther. **5**: 97–103.

111. Thorne S.H., Tam B.Y., Kirn D.H., Contag C.H., and Kuo C.J. 2006. Selective intratumoral amplification of an antiangiogenic vector by an oncolytic virus produces enhanced antivascular and anti-tumor efficacy. Mol. Ther. **13**: 938–946.

112. Peng K.W., Facteau S., Wegman T., O'Kane D., and Russell S.J. 2002. Non-invasive in vivo monitoring of trackable viruses expressing soluble marker peptides. Nat. Med. **8**: 527–531.

113. Banaszynski L.A., Sellmyer M.A., Contag C.H., Wandless T.J., and Thorne S.H. 2008. Chemical control of protein stability and function in living mice. Nat. Med. **14**: 1123–1127.

7

ENGINEERING HERPES SIMPLEX VIRUS FOR CANCER ONCOLYTIC VIROTHERAPY

JASON S. BUHRMAN,[1,2] TOOBA A. CHEEMA,[1] AND GIULIA FULCI[1]

[1]*Massachusetts General Hospital, Harvard Medical School, MA*
[2]*University of Illinois at Chicago School of Medicine, IL*

INTRODUCTION

The first evidence of viruses acting as anticancer agents was reported over 100 years ago when a woman with chronic myelogenous leukemia showed improved symptoms during an infection with the influenza virus (1). In 1911, an Italian physician published a report on a spontaneous regression of cervical carcinoma after injection with a live rabies virus vaccine (2). Over the next century, other isolated cases reported that infection from a virus resulted in improved symptoms from an invading cancer (3). However, it became evident only after the advent of molecular biology that there is a large similarity between the environment that a virus requires for replication and the intracellular environment in cancer cells. The discovery of such overlap in biological pathways made viral treatment of tumors a plausible reality, and in the 1980s, interest in the oncolytic effects of viruses reemerged.

Oncolytic virotherapy employs live, actively replicating viruses as therapeutic vehicles. These viruses have the capacity to selectively kill cancer cells by design, either inherent or engineered, to the intracellular environment

Emerging Cancer Therapy: Microbial Approaches and Biotechnological Tools, Edited by Arsénio M. Fialho and Ananda M. Chakrabarty

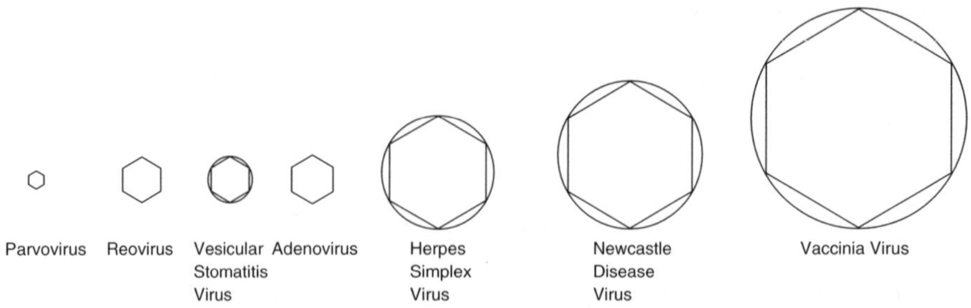

Virus	Abbrev.	Envelope	Genome	Genome Size	Virion Size (nm)
Parvovirus	APV	No	ss-DNA	5kb	20–25
Reovirus	RV	No	ds-RNA	24kb	60–80
Vesicular Stomatitis Virus	VSV	Yes	neg-RNA	11kb	45–100
Adenovirus	Ad	No	ds-DNA	36kb	70–90
Herpes Simplex Virus	HSV	Yes	ds-DNA	153kb	150–200
Newcastle Disease Virus	NDV	Yes	neg-RNA	150kb	150–300
Vaccinia Virus	W	Yes	ds-DNA	192kb	300–400

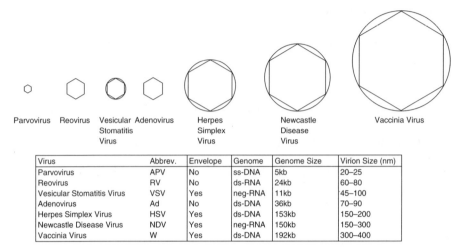

Figure 7.1 Summary of virology of selected oncolytic viruses. Top: respective sizes of different virus particles used for their oncolytic effects. Bottom: table summarizing the notable virology of each virus.

setup after transformation for their own replication. At the same time, viruses can be engineered to deliver recombinant transgenes to the cancer tissue. Moreover, the replicating capacity of oncolytic viruses (OVs) allows them to amplify *in situ* and to spread through the whole tissue. This auto-amplification and spreading capacity of OVs makes them the perfect tool to reach isolated cancer cells of invasive tumors and metastatic neoplastic formation.

Several viruses have been identified with the inherent capacity to specifically replicate in tumor cells and to have therapeutic benefits following such delivery. The selectivity is due to metabolic similarities between cancer cells and viruses, and the discovery of these metabolic similarities has allowed specific tumor targeting of viruses that are not endogenously oncolytic through genetic engineering (Figure 7.1). Numerous other virus mutants have been made in different backbones, such as herpes simplex virus (HSV), adenovirus, vaccinia virus (VV), and measles virus to enhance oncolytic effects (Figure 7.1). Thus, a wide variety of viruses are currently available for cancer treatment.

However, to provide an in-depth look of the versatility and broad application that one virus strain can have, we will focus in this chapter on oncolytic herpes simplex viruses (oHSVs). We will provide an overview of the different oHSVs that have been engineered, their application to cancer treatment, and the current problems remaining to be solved to make oncolytic virotherapy an effective anticancer therapeutic strategy.

ONCOLYTIC VIRUSES

Because oncolytic virotherapy employs live, actively replicating viruses as therapeutic vehicles, several key features are necessary in a virus safe for clinical use: (i) the virus must be able to replicate and destroy infected cells; (ii) it must be selective only for cancer cells; (iii) it must not undergo any form of lysogenic or dormant cycle as to persist after the tumor has been destroyed; (iv) it must be stable and not able to easily acquire mutations that lessen its replication specificity; (v) the virus should cause only mild, and well-characterized human diseases in its wild-type form; (vi) the genome of the virus should not integrate into the host one to avoid transformations associated with random integration; and (vii) high titers of the virus should be attainable for clinical use.

Given these characteristics, several inherently tumor-selective viruses have been identified as potential cancer therapeutic tools including reovirus, vesicular stomatitis virus (VSV), Newcastle disease virus (NDV), parvovirus, and VV (Figure 7.1). The incidental tumor specificity of these viruses can be attributed to their inability to counteract host defense responses that are active in normal but are disrupted in cancerous cells (4–9). However, viruses that are not endogenously tumor selective have also displayed desirable characteristics that could be exploited in cancer treatment. Genetic modification of these viruses has successfully limited their replication to cancer cells and has enabled them to be exploited for their desirable therapeutic traits.

Engineering viruses provides a versatile and powerful therapeutic approach because it allows scientists to engineer oncolytic mechanisms into more virulent viruses. Among the commonly engineered viruses are adenovirus (10), VV (11), measles virus (12), NDV (13), and HSV (14). All of these viruses, except NDV, contain double-stranded DNA genomes that simplify genetic alterations to optimize oncolytic effects. Adenovirus and HSV present several additional advantages: (i) their genomes contain multiple, "nonessential" genes that can be replaced or deleted to attenuate replication in normal cells and to insert transgenes with anticancer properties in a process known as "arming" the virus (15, 16); (ii) both genomes have also been fully sequenced, providing a convenient blueprint for these alterations; (iii) neither virus integrates into the host genome, and so, undesired effects of further transforming cells during treatment is unlikely; (iv) their receptors are widely distributed on many tumor cells, suggesting that both viruses can be versatile in the types of tumors they may be employed to treat (17–19); and (v) finally, both HSV1 and adenovirus are relatively common viruses that are tolerated by most humans. An additional advantage of using HSV is that effective antiviral compounds such as acyclovir (ACV) and ganciclovir (GCV) have been approved for treatment of HSV infection and can be used in case OV therapy becomes too aggressive. On the other hand, a number of studies have focused mainly on oncolytic adenovirus because of its smaller-sized genome and non-enveloped virion (18) (Figure 7.1).

TARGETING HSV TO CANCER CELLS

HSV Biology

HSV is a human neurotropic virus of the alpha-herpes subfamily. The viral particle, 150–200 nm in diameter, is composed of four components: (i) an electron dense core containing the viral genome (a linear double-stranded DNA molecule of about 153 kb), which is packed into (ii) an icosahedral nucleocapsid surrounded by (iii) the tegument, an amorphous proteinaceous layer, and (iv) a lipid envelope (Figure 7.1). The HSV genome consists of two segments: a unique L (long) and S (short) region bracketed by inverted repeats (Figure 7.2). Genes encoded within the inverted repeats (ICP0, ICP4, γ34.5) are diploid

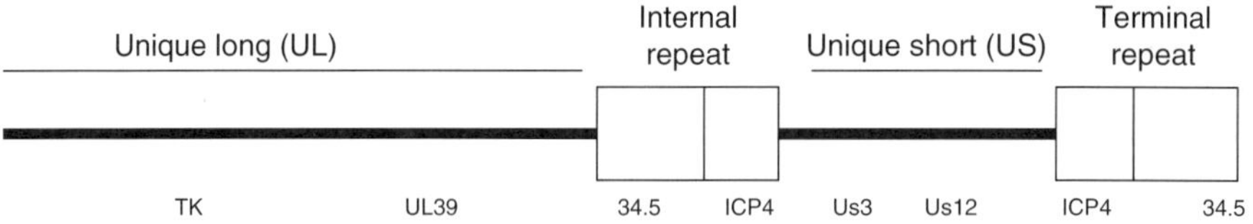

Virus	Strain	Deletions	Function	Duplications	Function
hrR3	Kos	UL39	Ribonucleotide reductase		
R3616	F	Gamma 34.5	Neurovirulence, PKR shutdown		
G207	F	Gamma 34.5	Neurovirulence, PKR shutdown		
		UL39	Ribonucleotide reductase		
MGH1	F	Gamma 34.5	Neurovirulence, PKR shutdown		
		UL39	Ribonucleotide reductase		
G47D	F	Gamma 34.5	Neurovirulence, PKR shutdown		
		UL39	Ribonucleotide reductase		
		US12	MHC class I loading		
rQNestin34.5****	F	Gamma 34.5	Ribonucleotide reductase		
		UL39	Ribonucleotide reductase		
NV1020*	F	Gamma 34.5 +/–	Neurovirulence, PKR shutdown	UL5	DNA replication
		ICP0 +/–	Transcription factor	UL6	DNA packing
		LAT +/–	Latency-associated transcript		Glycoproteins
		ICP4 +/–	Transcription factor		
		UL24–			
		UL56–	Unknown		
NV1066**	F	Gamma 34.5 +/–	Neurovirulence, PKR shutdown	UL5	DNA replication
		ICP0 +/–	Transcription factor	UL6	DNA packing
		LAT +/–	Latency-associated transcript		
		ICP4 +/–	Transcription factor		
		UL24			
		UL56	Unknown		
HF10***	HF	UL56	Unknown	UL53	gk
				UL54	Transcriptional regulation
				UL55	Unknown
R7041	F	US3	Anti-apoptosis		
HSV1716	17+	Gamma 34.5	Neurovirulence, PKR shutdown		
OncoVEX	JS-1	Gamma 34.5	Neurovirulence, PKR shutdown		
		US12	MHC class I loading		
dlsptk	Kos	TK	Nudeoside kinase		

* Contains insertions of HSV2 glycoproteins G, D, and I.
** Contains bacterial artificial chromosome (BAC) sequences for shuttling transgenes.
*** Carries additional mutations.
**** Has reinsertion of gamma 34.5 under Nestin1 promoter.

Figure 7.2 Summary of oHSV genome and additional parental OVs. Top: pictorial representation of herpes simplex virus (HSV) genome. Bottom: mutations in common parental oncolytic HSV.

and are present in two copies. Each end of the genome contains a direct repeat, the "a" sequence that serves as a DNA cleavage/packaging signal. The HSV genome is ideal for genetic manipulations as it contains only 80 genes and lacks overlapping intron-containing genes. Two to four hours after HSV enters the infected cells, immediate early genes are synthesized and initiate the temporally regulated expression of early (E or β) and late (L or γ) genes. Following DNA replication, approximately 6 to 10 hours after infection, the viral genome is packaged into the capsid and the cell is lysed for release of HSV progeny. The whole viral cycle is completed in approximately 12 to 15 hours (20).

Genetic Complementation

Genetic complementation of a virus to a host occurs when genes that are essential for viral metabolism are removed from the viral genome and the virus uses the host's products, complementary to viral deleted genes, to accomplish its cycle; that is, the virus uses host gene products to "complement" its lack of genes necessary to continue the replication cycle. Because cancer cells express products that are present only in dividing cells and are required for viral replication, it has been possible to engineer several HSV strains to specifically replicate in dividing cells through genetic complementation.

The first example of oHSV was provided in 1991 by Martuza et al. (14) when they engineered a virus containing a mutation in the thymidine kinase (*tk*) gene (183) (*dl*sptk, Figure 7.2). *Tk* is necessary for the conversion of deoxythymidine to thymidine monophosphate during DNA synthesis and is highly expressed in actively replicating cells but is absent in non-replicating cells (14). Therefore, this mutation targets oHSV to dividing cells because only there will endogenous *tk* be active enough to meet the demands of a replicating virus. Although the virus may infect these nondividing cells upon initial delivery, it will not have a sufficient resource of deoxythymidine triphosphate (dTTP) necessary for viral DNA replication, so it will not further infect nondividing cells. Martuza et al. (14) have demonstrated the ability of *dl*sptk viral treatment to significantly reduce tumor size and to improve survival in mice. However, there were two problems associated with *dl*sptk treatment. Evidence of encephalitis was observed upon sectioning of the orthotopic xenographs indicating that the virus retained its neurovirulant functions, and deletion of the *tk* gene caused resistance to the antiviral treatment of ACV and GCV.

To address these problems, two new OVs were engineered retaining the *tk* gene while still presenting significant attenuation in normal cells. The first virus, termed R3616 (created in the lab of B. Roizman), is a mutant virus created by deleting both copies of the diploid gene γ34.5 (Figure 7.2). This gene restricts viral replication to cells with an inactivated protein kinase R (PKR) viral defense pathway, that is, cancer cells. Moreover, it blocks cellular autophagy by sequestering the protein Beclin 1 which is thought to be critical for HSV1-induced encephalitis (21–23, 185). The second virus, hrR3, has a deletion in the UL39 (ICP6) gene encoding viral ribonucleotide reductase

(RR, Figure 7.2) (24, 184) that limits viral replication only to actively dividing cells with upregulated RR. It was observed through these studies that the γ34.5 mutants were indeed significantly neuro-attenuated and that the UL39 mutant maintained significant tumor specificity while remaining sensitive to GCV and ACV antiviral treatment. However, each of these viruses demonstrated respective problems either in specificity in the case of R3616, or in neurovirulence in the case of hrR3.

To generate a clinically useful virus, it became apparent that a single mutation would not be sufficient to satisfy the safety requirements necessary for clinical use. A second-generation oHSV became necessary harboring mutations in more than one gene to safely attenuate the virus for use in humans. An oHSV termed G207 (MGH1) containing deletions in both copies of the γ34.5 genes and a disruption of the UL39 RR was created for this purpose (25). Additional advantages to the G207 double-mutant OV include attenuated neurovirulence, hypersensitity to the antiherpetic drugs ACV and GCV, an easily detectable histochemical marker, and an increased unlikelihood that the virus would revert to its wild-type form. It was demonstrated in mice and in primates that G207 was safe when delivered intracranially, intraprostatically, and intravenously with no evidence of encephalitis in nervous tissue (25, 26). However, the replication efficacy of this virus was reduced compared to the single-mutant viruses.

A third-generation vector, G47Δ, was generated containing an additional deletion of the US12 (ICP47) gene, which functions as an inhibitor of major histocompatibility complex (MHC) class I molecule loading. G47Δ showed much higher replication potential than its parent, G207. It was then discovered that this deletion also places the late US11 gene, another inhibitor of the PKR viral defense pathway, under control of the US12 immediate-early promoter (27, 28). This reestablishes the virus' ability to shut down the host PKR viral defense pathway early after cell infection and allows G47Δ to replicate more efficiently than the parental G207 while retaining its avirulent characteristics due to deletion of the γ34.5 genes (Figures 7.2 and 7.3) (29).

Two other HSV genes that have been deleted to target HSV specifically to cancer cells are US3 and UL56. US3 codes for a protein kinase that blocks cellular apoptosis. Indeed, apoptosis is a common mechanism of cell defense against virus replication, and it is frequently disrupted in cancer cells (30). The function of UL56 in HSV1 is not completely understood, but it has been shown to be heavily involved in the virulent phenotype of HSV (31).

Tissue-Specific Promoters

Another mechanism to target viruses to cancer cells is to drive the expression of essential viral genes under promoters that are active in ectopic or cancer cells (17, 32). Examples of tumor promoters used for HSV retargeting are the liver-specific albumin promoter (33), the smooth muscle calponin promoter (34), the epithelial mucin 1 promoter and the carcinoembryonic antigen promoter (35), the stem cell promoters B-myb (36), Nestin (37), and Musashi1

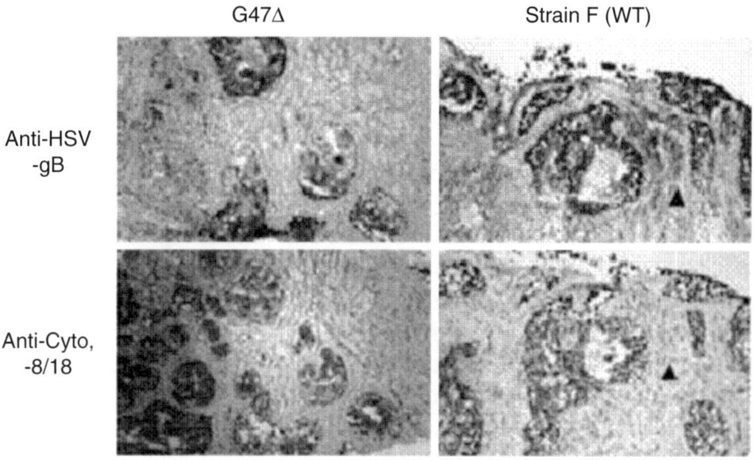

Figure 7.3 Analysis of oHSV infectivity in prostate cancer surgical samples. Tissue fragments derived from radial prostatectomies were incubated with G47Δ or strain F (wild type [WT]) at 1.5×10^6 pfu for 1 h and thereafter, placed on a semi-submersed collagen sponge for 3 days. Tissue specimens were sectioned and immunostained with anti-HSV-1 gB (upper panels) or anti-cytokeratin-8/18 (lower panels) antibodies. Note that G47Δ is confined to the prostatic glands, while strain F is present in both the prostatic glands and stroma (arrowheads). Sections were counterstained with hemotoxylin. Scale bar = 50 μM.

(38), and the oral cancer cell promoter from the human papillomavirus URR16 (39). Promoter-driven virus retargeting to cancer cells has a relatively restrictive use because such viruses can be used only in specific tumors. However, this retargeting strategy has the advantage of maintaining viruses closer in virulence to their wild-type parents as they need fewer deletions.

Interaction with Receptors of Cancer Cells

A third method for targeting viruses to cancer cells involves retargeting the virion to nonconventional receptors. Wild-type HSV infects cells through three known receptors: herpesvirus entry mediator (HVEM) (40), Nectin1 (41), and 3-O-sulfated sites on heparin sulfate (42) by interaction with glycoprotein D (gD) on the surface of the HSV virion. Retargeting of the HSV can be achieved by making chimeras of gD with ligands to other receptors. Such an approach has been used to retarget HSV to the IL-13 receptor (43), the urokinase plasminogen activator (44), and the human epidermal growth factor receptor 2 (45). Increased levels of these receptors are associated with cellular transformation.

Amplicons

A more recent strategy to target HSV to cancer cells is through the use of a combined system including an amplicon and a helper virus. An HSV amplicon is an enveloped, HSV capsid-containing exogenous DNA that has coding regions under the control of a tissue-specific promoter and a single essential viral gene, whereas the helper virus is a virus lacking this gene. The helper virus is then delivered to the tumor together with its corresponding amplicon coding for the lacking gene under a tumor-specific promoter. Following transduction of a mixed population of cells with the amplicon, only those cells with transcription factors able to activate the tissue-specific promoter will produce the essential viral gene, and only in those cells can a virus lacking that single essential gene replicate. The concept for this system is very similar to tissue-specific promoters engineered into the viral genome, but the amplicon method is thought to provide several treatment benefits. First, it is much easier to create amplicons than to engineer HSV. Amplicons move the regulatory elements of the tissue-specific promoter outside of the viral genome, thus eliminating any effect that the genomic structure may have in influencing gene expression. Finally, the amplicon targeting system allows for modification of the potency of OV treatment by varying the ratio of amplicon to helper virus (46).

COMBINATION THERAPY INVOLVING THE ONCOLYTIC HSV VIRUS

"Arming" Oncolytic HSV Vectors with Transgenes

OVs present the advantage to serve as both oncolytic bioreagents and as vectors that can provide amplified gene delivery within the tumor. One of the advantages of HSV-1 vectors is the capacity to incorporate large and/or multiple transgenes within the viral genome allowing for expression of factors that target cancers through multiple mechanisms. There are four main categories of therapeutic transgenes engineered into oHSV that have been explored in laboratory research and have been extensively reviewed. These are listed in Table 7.1 (47–50).

Immunomodulatory Transgenes Immunotherapy by stimulating the immune system to attack tumor cells has long been investigated as an adjuvant to conventional therapy. Cytokines produced by either malignant or immune cells can serve to activate antigen-presenting cells, NK, CD4, and CD8 T cells against the tumor (51). There are several immune-modulatory peptides reported to enhance tumor immunity including interleukins 4, 12, 18, B7-1 (CD80), granulocyte-macrophage colony-stimulating factor (GM-CSF), chemokine (C-C motif) ligand 2 (CCL2) and tumor necrosis factor α (TNFα) when "armed" with HSV (Table 7.1).

IL-12 is a proinflammatory cytokine secreted by activated macrophages and dendritic cells that has important stimulatory effects on T cells and natural killer cells and can mediate anti-angiogenesis effects via interferon (IFN)-γ

TABLE 7.1. Arming Oncolytic HSV Vectors with Transgenes

Categories	Oncolytic Vector	Transgene	Parent Virus	Backbone Wild-Type Strain	Mutated Viral Genes	References
Cytokines immuno modulatory peptides	R8306	Mouse IL-4	R3659	F	γ34.5–/–	(188)
	R8308	Mouse IL-10	R3659	F	γ34.5–/–	(188)
	NV1042	Mouse IL-12	NV1020 (R7020)	F	ICP6–/γ34.5+/–	(186, 189)
	HSV-IL12	Mouse IL-12	MGH1 (G207)	F	ICP6–/γ34.5–	(52)
	HSV-IL18	Mouse IL-18	MGH1 (G207)	F	ICP6–/γ34.5–	(52)
	HSV-B7.1-Ig	Mouse B7-1/human IgG Fc fusion	MGH1 (G207)	F	ICP6–/γ34.5–	(52)
	G47D-IL18/ B7-1-IgG	Mouse IL-18/mouse B7-1/ human IgG Fc fusion	G47D	F	ICP6–/γ34.5–/ICP47–	(56)
	M002	Mouse IL-12	R3659	F	γ34.5–	(190)
	M010	Mouse CCL2	R3659	F	γ34.5–	(57)
	Oncovex GM-CSF	Human GM-CSF	JS-1	JS-1	γ34.5–/ICP47–	(143)
	NV1034	Mouse GM-CSF	NV1023 (R7020)	F	γ34.5–/ICP47–	(54, 186)
	Oncovex-hTNF-alpha	Human TNF-a	JS-1	JS-1	γ34.5–/ICP47–	(143, 191)
ProDrug-converting enzymes	HSV1yCD	Yeast cytosine deaminase	KOS	KOS	ICP6–	(62)

TABLE 7.1. *Continued*

Categories	Oncolytic Vector	Transgene	Parent Virus	Backbone Wild-Type Strain	Mutated Viral Genes	References
	Onco-Vex CD	Yeast cytosine deaminase + uracil phophoribosyl transferase	JS-1	JS-1	γ34.5/ICP47–	(82)
	M012	Bacterial cytosine deaminase	R3659	F	γ34.5–	(63)
	MGH2	CYP2B1	MGH1	F	ICP6–/γ34.5–	(65)
		Secreted human intestinal carboxylesterase (shiCE)	MGH1	F	ICP6–/γ34.5–	(65)
	rRP450	Rat cytochrome P450 (CYP2B1)	hrR3	KOS	ICP6–	(64, 184, 192)
Anti-angiogenic peptides	AE618	Endostatin–angiostatin fusion protein	G207	F	ICP6–/γ34.5–	(68)
	HSV-endo	Mouse endostatin	KOS	KOS	ICP6–	(67)
	G47Δ-dnFGFR	Mouse dnFGFR	G47D	F	ICP6–/γ34.5–/ICP47–	(69)
	G47Δ-PF4	Human PF4	G47D	F	ICP6–/γ34.5–/ICP47–	(70)
	rQT3	TIMP-3	rHSVQ1	F	ICP6–/γ34.5–/–	(193)
Fusogenic protein	Synco2D	Hyperfusogenic glycoprotein gene of gibbon ape leukemia virus (GALV.fus)	Baco-1	17	γ34.5–	(81)
	OncoVEX GALV	GALV env R	JS-1	JS-1	γ34.5–/ICP47–	(82)
	OncoVEX GALV/CD	Yeast CD + Uracil phosporibosyl transferase	JS-1	JS-1	γ34.5–/ICP47–	(82)
	OncdSyn	gBsyn3 and gKsyn1	NV1020	F	γ34.5–	(83, 84)

The table summarizes therapeutic genes inserted into oHSV that are reported in the literature for the treatment of cancer.

and other downstream effectors. Different groups have demonstrated that IL-12 is the most potent immunostimulatory gene at inducing tumor immunity. For example, the oHSV MGH1 expressing IL-12 was significantly more efficacious when compared with those expressing IL-18 and soluble murine B7-1-immunoglobulin (B7-1-Ig) (52). GM-CSF is another cytokine secreted by activated lymphocytes, macrophages, and endothelial cells, and whose primary effects include recruitment and differentiation of antigen-presenting cells and production of marrow-derived myeloid cells. In metastatic colorectal and hepatic adenocarcinoma, GM-CSF expressing oHSV (NV1034, Table 7.1) had enhanced antitumor efficacy compared to the control NV1023 (see NV1020, Figure 7.2) (53, 54). As a result, an oHSV expressing human GM-CSF (OncoVEXGM-CSF, Table 7.1) is now being used in Phase II clinical trials on malignant melanoma and head and neck cancer (55). However, the most efficient response is observed with a combination of multiple immunogenic transgenes, such as the double-armed IL-18/B7-1 or the combination treatment with IL-12, IL-18, and B7-1 or CCL2 and IL-12 (Table 7.1) (52, 56, 57).

Even though most studies have been done with the idea of enhancing the immune system, caution is needed when "arming" OVs with immune-modulatory transgenes as stimulation of antitumor immunity may be accompanied by induction of antiviral immunity as well (58). In fact, studies with OVs engineered with anti-inflammatory factors to antagonize immune response have also demonstrated enhancement in oncolysis that is safe and effective (59–61). The dual role of the immune system, antiviral or antitumor, remains to be elucidated and more understanding of the specific immune mechanism need to be achieved for optimal immunomodulation.

Prodrug-Converting Enzymes Prodrugs are inactive precursor forms of active drugs that can be converted to their active forms by prodrug-converting enzymes. The rationale of using OVs engineered with prodrug-converting enzymes is to increase the concentration of an active, cytotoxic drug selectively in cancer cells in order to minimize its potential side effects. Several oHSVs have been engineered with prodrug-converting enzymes and have been tested in a number of preclinical studies. HSV-1yCD is an ICP6-inactivated vector carrying the yeast form of cytosine deaminase (CD) (Table 7.1). CD deaminates 5-fluorocytosine (5-FC) to 5-fluorouracil (5-FU), which is a well-known anticancer agent. Treatment with HSV-1yCD and 5-FC combination enhances efficacy in several cancer models compared to virus alone (62, 63).

The *CYP2B1* gene encodes the liver-specific enzyme cytochrome P450 that plays an important role in the metabolism of many drugs including cyclophosphamide (CPA) and ifosfamide prodrugs into anticancer agents. The oncolytic rRp450 was constructed by inserting the *CYP2B1* gene in the hrR3 oHSV (Figure 7.1), allowing cytochrome P450 2B1 expression under the viral ICP6 promoter. The combination of rRp450 and CPA has shown significantly better efficacy *in vitro* and *in vivo* models of gliomas (64). The combination of two prodrug-converting enzymes, CYP2B1 and secreted human intestinal carboxylesterase (shiCE), an enzyme that metabolizes CPT-11 topoisomerase I inhibitor into its active metabolite SN-38 suggests that these agents can increase oHSV's anticancer effect in a synergistic way (65).

Although it seems counterintuitive to attempt to kill the same cells with an activated, cytotoxic drug and viral oncolysis, it is the "bystander" effect where cells surrounding the enzyme producing cells are also killed that makes this arming strategy attractive. Indeed, it has been shown that only 2% of cells expressing a prodrug-converting enzyme are sufficient to induce a significant reduction in tumor volume (66).

Anti-Angiogenic Transgenes Angiogenesis is a critical process in the growth and metastasis of tumors and constitutes an important process for the control of cancer. As tumors evolve, they secrete a variety of angiogenic factors that enables proliferation of blood vessels and supports the proliferation of cancerous cells.

So far, the short list of anti-angiogenic factors that have been engineered into the oHSV genome includes endostatin and angiostatin (67, 68) (degradation products of collagen XVIII and plasminogen, respectively), dominant negative fibroblast growth factor (dnFGFR) (69), platelet factor 4 (PF4) (70), and tissue inhibitor of metalloproteinase-3 (TIMP-3) (71, 72, 193) (all vectors listed in Table 7.1). The latter one, known to inhibit all known matrix metalloproteinases (MMPs), is involved in tumor neovascularization as well as in migration and invasion, and resulted particularly efficient in preclinical studies on neuroblastoma and on a malignant peripheral nerve sheath tumor (MPNST) xenograft model. All together, these "anti-angiogenic" vectors show significant tumor suppression due to the fact that they can target both the cancer cells through the oncolytic activity of the virus and the vessel density through the anti-angiogenic activity of the transgene. The importance of using viruses targeting tumor angiogenesis is even more emphasized by recent discoveries showing that oHSV indirectly induce a tumor blood supply, as a result of host inflammatory responses, which results in infiltration of antiviral innate immune cells (73–75).

Fusogenic Glycoproteins Fusogenic membrane glycoproteins are a family of proteins used by enveloped viruses to gain entry into the cytoplasm (76–78). It has been shown previously that expression of the membrane glycoproteins of the measles virus and the gibbon ape leukemia virus (GALV) can induce membrane fusion between cells leading to multicellular synctia and cell death by apoptosis (79). For example, Synco-2D, created by the combination of inserting a hyperfusogenic glycoprotein from GALV into the spontaneously synctial mutant (80), replicated better than the parental and showed better efficacy on ovarian cancer metastases and prostate cancer (81). A triple combination of this transgenic oHSV with the expression of CD and with the *Escherichia coli* or yeast uracil phosphoribosyltransferase (UPRT) enzyme has also been shown to markedly increase the antitumor effects of 5-FC/CD compared to single transgenes (82). Mutations in oHSV glycoproteins gB and gK (OncSyn and OncdSyn) also enable the virus to become synctial and to spread among cells by virus-induced cell fusion while replicating more efficiently in cancer cells (83, 84). This intratumoral expression of viral fusogenic membrane proteins is a promising approach for cancer gene therapy because their expression in tumor cells is directly cytotoxic and associated with a local bystander effect (85) and can also stimulate an antitumor immunity (86).

Combination with Radiotherapy and Small Molecules

Another strategy to maximize the efficacy of OV therapy is by combining the virus with radiation and/or drugs before, after, or during treatment in order to enhance therapeutic effects in combination with each other. A combination of two different therapeutics can result in an antagonistic, synergistic, or additive therapeutic outcome. In the first two cases, the two drugs act on different parts of the same pathway and may result as inhibitory to another (antagonistic) or, on the contrary, stimulating each other (synergistic). When the combination of two drugs is additive, they usually act on two different pathways and do not influence each other. Synergy between drug and virus is achieved when the combined effect of drug and virus is greater than the additive effects of drug alone and virus alone, thus being preferred in clinical use. Ideal synergistic combinations will result in treatment regimes where less drug and virus will be necessary when used together to achieve similar results as either drug or virus alone. On the other hand, an antagonistic effect is when the combination of therapeutics is worse than the additive effect of each therapeutic alone. Clinically, this is the least useful outcome, but it still does imply a biological overlap in therapeutic pathways, making antagonistic effects interesting to study. It is important to note that the timing at which each therapeutic is delivered with respect to the other may influence the final therapeutic efficacy. Indeed, two drugs that are synergistic may become additive or even antagonistic if the order at which they are delivered is changed. There are several classes of drugs that have been found to be synergistic with OVs, and hypotheses have been put forth to explain these synergies (87). The combination of drugs with oHSV that have been studied so far is shown in Table 7.2.

Combination with Radiotherapy Combining OV therapy with radiation therapy (RT) has shown some complementarity; however, most of the effects have only been additive upon analysis (reviewed in Reference 88). In theory, irradiated cells can upregulate cellular GADD34 and RR that should complement oHSV deletion mutants lacking γ34.5 and UL39, respectively (89, 90) (Figure 7.2). Accordingly, a synergistic therapeutic effect was observed when combining radiation with the R3616 and NV1023 oHSV (Figure 7.2) in mouse glioma models (89, 91). However, synergism between RT and OV therapy seems to be tumor specific because cases of only enhanced or additive effects when combining RT with other γ34.5 mutants have been published (90, 92–94).

TABLE 7.2. Combination of oHSV and Small Molecules

Virus	Drug class	Drug	Effect	Tumor Type	*In Vitro/In Vivo*	References
R1716	DNA cross-linker	Cisplatin	Additive	NSCLC, HNSCC	*In vitro*	(185, 194, 195)
	Topoisomerase inhibitor	Doxorubicin	Additive	NSCLC	*In vitro*	(195)
	Antimetabolite	Methotrexate	Additive	NSCLC	*In vitro*	(195)
	Alkylating agent	Mitomycin C	Additive-synergistic	NSCLC	*In vitro/in vivo*	(195)
G207	DNA cross-linker	Cisplatin	Additive	HNSCC	*In vivo*	(196)
	Antimetabolite	5-Fluorouracil	Additive	Gastric/colon	*In vivo*	(197)
	Anti-angiogenic	Erlotinib	Additive	MPNST	*In vivo*	(128)
	Antimetabolite	Fluorodeoxyuridine	Synergistic	Colon carcinoma	*In vivo*	(198)
	Antiherpetic	Ganciclovir	No effect	Neuroblastoma	*In vitro/in vivo*	(130)
	Antitumor antibiotic	Mitomycin C	Synergistic	Gastric	*In vitro/in vivo*	(106)
	Taxol	Paclitaxol	Synergistic	Anaplastic thyroid	*In vitro/in vivo*	(102)
	DNA alkylating agent	Temazolamide/ O6-BG	Synergistic	Glioma	*In vitro/in vivo*	(96, 105)
	Vinca-alkyloid	Vincristine	CKE	Rhabdomyosarcoma	*In vitro/in vivo*	(118)
G470	Taxol	Docetaxel	Synergistic	Prostate	*In vitro*	(103)
	Antimetabolite	FOLFOX*	Prolong survival	Colorectal	*In vivo*	(79)
	HDAC inhibitor	Trichostatin A	Synergistic	Glioma, colorectal	*In vitro/in vivo*	(112)
hrR3	Anti-angiogenic	Erlotonib	Additive	MPNST	*In vivo*	(128)
	Anti-angiogenic	Cilengitide	CKE	Glioma	*In vivo*	(74)
	DNA alkylating agent	Cyclophosphamide	Additive	Glioma	*In vivo*	(124)
	Antiherpetic	Ganciclovir	Prolong survival	Gilosarcoma	*In vivo*	(95, 142, 199)
	Antimetabolite/RR inhibitor	Gemcitabine	Antagonistic	Pancreatic	*In vitro/in vivo*	(200)

rQNestin34.5	HDAC inhibitor	APAH compound 8	VRE	Melanoma, glioma	*In vitro*	(119)
	HDAC inhibitor	Sodium butyrate	VRE	Melanoma, glioma	*In vitro*	(119)
	HDAC inhibitor	Trichostatin A	VRE	Melanoma, glioma	*In vitro*	(119)
	HDAC inhibitor	Valproic acid	VRE	Melanoma, glioma	*In vitro/in vivo*	(119)
R7041	AKT inhibitor	LY294002	Synergistic	Glioma, lung adenocarcinoma	*In vitro/in vivo*	(30)
R3616/R849	Antimetabolite/RR inhibitor	Gemcitabine	No effect-additive	Pancreatic	*In vitro/in vivo*	(200)
	Inducer of apoptosis	Hexamethylene Bisacetamide	VRE	Oral SCC	*In vitro/in vivo*	(201)
	HDAC inhibitor	Triohostatin A	VRE	Oral SCC	*In vitro*	(202)
HF10	Antiherpetic	Gancyclovir	CKE	Human laryngeal epidermoid carcinoma	*In vitro*	(142)
NV1020	Antimetabolite	5-Flourouracil	Additive–synergistic	Colon carcinoma	*In vitro/in vivo*	(129, 186)
	Antimetabolite	Oxaliplatin	Additive–synergistic	Colon carcinoma	*In vitro/in vivo*	(129)
	Topoisomerase inhibitor	SN38 (irinotecan)	Additive–synergistic	Colon carcinoma	*In vitro/in vivo*	(129)
NV1066	Hormone	Estrogen	Additive	Breast	*In vitro*	(54)
	Antimetabolite	5-Fluorouracil	Synergistic	Human pancreatic	*In vitro*	(203)
	DNA cross-linker	Cisplatin	Synergistic	MPM	*In vitro*	(108, 182)
	Antimetabolite	Gemcitabine	Synergistic	Human pancreatic	*In vitro*	(203)
	Alkylating agent	Mitomycin C	Synergistic	Bladder	*In vitro*	(107)

The table summarizes combinations with oHSV and combinations with small molecules.

*5-Fluorouracil, leucovorin, oxaliplatin.

MPM, malignant plurel mesothelioma; SCC, squamous cell carcinoma; NSCLC, non-small cell lung carcinoma; VRE, viral replication enhancement; CKE, cell killing enhancement.

Since RT is very common in cancer treatment, it is important to note that antagonism with oHSV has not been reported.

Antiherpetic Drugs ACV and GCV are antiherpetic prodrugs that are activated to their cytotoxic form through phosphorylation by the HSV gene *tk*. Using antiherpetic drugs in combination with oHSV results in increased cytotoxicity mediated by oHSV. However, they will also destroy infected cells before the virus has completed its entire replication cycle and therefore will interfere with the course of viral treatment. Nevertheless, the combination of antiherpetic drugs and oHSV was reported to prolong survival in gliosarcoma models with hrR3 (Table 7.2) (95).

Chemotherapeutics There are several classes of chemotherapeutics that have been used in combination with oHSV therapy. These include antimetabolites, topoisomerase inhibitors, DNA alkylating agents, taxols, vinca-alkyloids, antitumor antibiotics, and anti-angiogenic compounds. The synergy between chemotherapeutics and viruses, in many cases, arises through secondary effects of the chemotherapeutics that may either enhance an active viral gene, enhance cellular complementation of a viral gene that was deleted to promote the oncolytic phenotype, or modify the cellular component of the tumor stroma. These effects cannot always be appreciated in the context of the primary mechanism of the drug. Instead, they involve complex mechanisms that regulate the interaction between a virus and its host cell and/or extracellular environment.

The first cellular response to virus infection includes activation of the RNA-dependent serine/threonine PKR which phosphorylates eIF2α resulting in a shutoff of all protein synthesis including viral proteins (96). HSV replication would be severely attenuated if it did not express the products of the γ34.5 and US11 genes that inactivate PKR. Additionally, the HSV immediate-early genes ICP27, ICP0, ICP4, and the tegument protein VP22 block cells in the G1 phase of the cell cycle, also possibly the G2 phase, in order to maximize the viral replication potential. Under normal situations, such a blockade would induce eukaryotic cells to "commit suicide" through the induction of apoptosis rather than allow the spread of the virus throughout the organism. However, HSV is able to continue its lytic cycle because the viral US3 protein kinase blocks cellular apoptosis by phosphorylating a variety of cellular targets including the cell survival PKA substrates. Finally, the cell can inhibit viral replication through the synthesis of cytokines, such as type 1 and type 2 interferons (97), and the induction of an intracellular oxygen burst (98). The activation of a cytokine response in virus-infected cells leads to an inflammatory response and a consequent establishment of antiviral adaptive immunity. However, the virion host shutoff (VHS) protein or UL41 inhibits protein synthesis and degrades most of the host cell's mRNA, thus preventing the synthesis of inflammatory factors and early viral inactivation (99). Moreover, through the viral α47 protein product of the US12 gene, HSV has developed a mechanism to avoid MHC class I binding in order to escape the adaptive anti-HSV immunity (29).

Synergies can arise when drugs help oHSV establish, or reestablish, blockades at these critical points of interaction between virus and host cell. For example, taxols (paclitaxel and docetaxel) stabilize microtubules and can synergize with oHSV because of their capacity to block cell cycle progression during mitosis, a mechanism shared by oHSV for replication (100–103). Similarly, synergy was found when combining oHSV with another small molecule (LY294002) that inhibits the AKT (protein kinase B) pathway essential for inducing apoptosis in infected cells. Synergy of the AKT inhibitor LY294002 and oHSV was observed particularly with oHSV viruses that are targeted to cancer cells through the deletion of the anti-apoptotic US3 gene (R7041, Figure 7.2 and Table 7.2) (30). This gene blocks the AKT pathway that is often altered in cancer cells (30, 104). Other common chemotherapeutic drugs such as temozolamide (TMZ) (96, 105), mitomycin C (MMC) (106, 107), and cisplatin (108) synergize with oHSV vectors deleted for γ34.5 or the UL39 RR (Figure 7.2) through the upregulation of cellular GADD34 and/or RR (see section on genetic complementation). Indeed, the product of the HSV γ34.5 gene presents a GADD34 domain that inactivates the cellular PKR antiviral defense pathway, and the UL39 gene codes for the viral RR necessary to generate nucleotides for viral replication.

Histone deacetylase (HDAC) is important in the regulation of genes whose products deacetylate the N-terminal of histones, a process that is associated with gene regulation and is out of balance in many types of cancer (109). HDAC inhibitors have been shown to synergize with various chemotherapeutics by sensitizing cells to apoptosis-inducing factors (109), inducing the cell cycle inhibitor p21(WAF1) (110), and downregulating cell survival genes such as AKT and ERK (111). Similarly, the synergy between the HDAC inhibitor trichostatin A (TSA) and HSV-G47Δ (Figure 7.2 and Table 7.2) is thought to arise from a joint suppression of cyclin D1 (112) resulting in the suppression of the cell cycle in the G1 phase and allowing efficient viral replication (113). In addition, HDAC inhibitors have been shown to possess anti-angiogenic qualities (114–116). These are qualities HDAC inhibitors share with oHSV (69, 73, 117, 118), and Liu et al. (69) hypothesized that these commonalities may have contributed to the enhanced tumor killing seen *in vivo*. In addition, HDAC inhibitors can also suppress the synthesis of cytokines such as type 1 interferons, and this phenomenon was shown to mediate synergy between the rQNestin34.5 oHSV (Figure 7.2) and the HDAC inhibitor valproic acid (VPA) (119).

Finally, CPA, an alkylating agent that efficiently kills replicating cells and presents a powerful immunosuppressive phenotype, was shown to increase oHSV replication *in vivo*. Even though CPA does not increase the intracellular replication of the virus, its immunosuppressive action leads to a strong induction of intratumoral oHSV replication, spread, and persistence (120–126). For example, Kambara et al. (123) has shown that in the presence of CPA, rQNestin34.5 (Figure 7.2) could be delivered at 10^2 plaque-forming units (pfu) less to obtain the same therapeutic effect in the absence of CPA. An advantage of CPA is also that it has a very broad application and is not restricted to specific oHSV vectors. Indeed, CPA was shown to increase replication of several OV strains in multiple tumors. As a result, a clinical trial testing CPA with OV has recently been funded (A. Dispenzieri, pers. comm.).

Increased intratumoral oHSV titers were also observed when specifically depleting peripheral macrophages with clodronated liposomes; however, such a selective treatment was not sufficient to induce survival (127). Indeed, the ideal scenario is to use a drug that has both anticancer and broad anti-inflammatory properties to observe the best effect. Putative drugs for this include anti-angiogenic agents, as most innate immune cells penetrate to the cancer through increased tumor perfusion mediated by the virus. Accordingly, a recent study with cilengitide has demonstrated induction of oHSV correlating with cilengitide-mediated decreased perfusion and intratumoral infiltration of innate immune cells (74). In addition, Mahller et al. (128) have shown that the angiogenesis inhibitor erolotinib in combination with oHSV increases survival in MPNSTs.

Other mechanisms of synergy between chemotherapeutics and oHSV occur when the virus sensitizes a cell to the drug. For example, 5-FU and its conversion product fluorodeoxyuridine (FUdR) are nucleotide analogues that induce DNA damage after being incorporated into growing nucleotide chains. It is reported that combination of NV1020 (Figure 7.2) with 5-FU or FUdR is highly additive to slightly synergistic (129) (Table 7.2). Gutermann et al. (129) speculated two reasons for this synergy. Both viral infection and 5-FU arrest cells in S-phase, and cells arrested in S-phase may be more susceptible to the DNA-damaging effects of 5-FU while arrested in S-phase. Another possible mechanism for this synergy speculates that cytokines secreted as a result of HSV infection may enhance cell-killing effects by inducing apoptosis independently in combination with DNA damage-induced apoptosis caused by the drug.

CLINICAL TRIALS WITH ONCOLYTIC HSV

The first clinical trials with oHSV were conducted simultaneously in the United States and in Scotland using G207 and HSV1716, respectively (Figure 7.2) (130, 131). Given the innate ability of HSV to infect neural tissue and the level of neuro-attenuation achieved of each of these two OVs, the first clinical trials enrolled patients with malignant gliomas (132). These trials (summarized in Table 7.3) were designed to test the safety profiles and efficacy of G207 and HSV1716. It was found that neither virus showed adverse effects when administered intracranially, but neither virus caused drastic improvements in the patient's measurable clinical parameters. However, since these initial two trials were successful in demonstrating the safety of modified HSV in the clinic, several other trials have taken place using oHSV in the treatment of gliomas

TABLE 7.3. Clinical Trials with oHSV

Virus	Sponsoring Company	No. of Patients	Phase	Tumors Treated	Virus Safety	Tumor Regression	References
HSV1716	Crusade Laboratories Ltd.	9	I	Glioma	NAE	22% by thallium SPEC	(131)
		12	II	Glioma	NAE	Not determined	(133)
		12	I	Glioma (intracavity)	NAE	Not determined	(134)
		5	I	Melanoma	NAE	Not determined	(136)
G207	Medigene Inc.	21	I	Glioma	NAE	Not determined	(130)
NV1020	Medigene Inc.	12	I	CRC liver metastases	Fever, nausea, headache	Reduced CEA	(139)
HF-10	M's Science Corp/ Nagoya University	3	I	Pancreatic	NAE	Between 30% and 100% regress	(137)
		6	I	Metastatic breast	NAE	Between 30% and 100% regress	(141)
		2	I	HNSCC	Slight fever	Not determined	(138)
OncoVex GM–CSF	BioVex Ltd.	30	I	Breast, head, neck, GI, melanoma	Flu-like symptoms	73% showed necrosis (19/26)	(135)
		60	II	Melanoma	Pending	Pending	(55)
		430	III	Melanoma	Pending	Pending	(55)

The table summarizes clinical trials with oHSV.
CRC, colorectal cancer; HNSCC, head and neck squamous cell carcinoma; GI, gastrointestinal; NAE, no adverse effect.

(133, 134), malignant melanomas (135, 136), pancreatic cancer (137), head and neck squamous cell carcinomas (HNSCCs) (135, 138), colorectal cancer (139), and breast cancer (140, 141) (Table 7.3).

Viruses taken to clinical trials have included the neuro-attenuated G207 and HSV1716 for brain tumors, multimutated but theoretically neurotoxic HF10 (Figure 7.2) for melanoma, pancreatic cancer, and HNSCC, and the neuro-attenuated but proinflammatory virus OncoVEXGM-CSF (Figure 7.2) for a plethora of malignant cancers including breast, head, neck, and melanoma (Table 7.3). The perceived clinical characteristics of each virus dictate which tumors it can be used to treat. For instance, HF10 might not be suitable in any tumor where nerve-sparing treatments are necessary as it maintains functional genes known to cause neurotoxcicity (142). Likewise, OncoVEXGM-CSF might not be suitable in any area where inflammation could be problematic as it carries a proinflammatory transgene (143).

Promising clinical results in the trials include tumor regression in 6 of 21 patients enrolled in the G207 trial (130) and tumor regression in one of the nine patients in the initial HSV1716 trial (131). A subsequent HSV1716 trial enrolling 12 patients had one patient live 22 months without tumor progression after surgical debulking and OV treatment, and another patient live 15 months without tumor progression (134). Fourteen of the nineteen tumor biopsies after treatment with OncoVEXGM-CSF showed areas of necrosis in tumor tissue with no evidence of necrosis in the normal surrounding tissue (135). In the dose escalation trial using HSV1716 on melanoma patients, the fifth patient receiving the highest dose showed tumor flattening in injected nodules (136). Finally, the HF10 trials in HNSCC and in pancreatic cancer showed between 30% and 100% areas of necrosis in biopsied tumors after treatment (137, 138). All of these trials report few, if any, adverse viral-induced side effects.

All except one of the clinical studies conducted use oHSV delivered directly into tumors, and though this is a viable treatment option, one of the advantages of OV therapy is the ability to systemically treat multiple tumors through intravenous delivery. The NV1020 study with colorectal cancer explores this option by delivering virus into the hepatic artery of 12 patients (139). The most common side effects of systemic delivery were fever, headache, and nausea. Only one patient experienced a transient increase in γ-glutamyl transpeptidase (GGT) levels 12 h postinjection thought to be due to NV1020 delivery. No virus was found in any patients who were delivered a dose less than 3×10^7 pfu. Sera tested positive for HSV in five patients for the first hour; however, only one demonstrated culturable virus. Only three patients showed disease progression at 28 days postinjection, while the other seven patients showed stable disease.

These studies demonstrate that the current oHSV treatment is safe, although not entirely efficient. Clinical outcomes were influenced in only a few of the many patients enrolled in these trials. Most of oHSV effects during these trials were assessed by tumor biopsy and histology that do show an effect, albeit not strong enough to correlate with improvement in clinical status. These effects are thought to result from inefficient viral replication and spread throughout the tumors. The one trial performed to assess the safety and efficacy of the intravenous delivery of NV1020 suggests that the virus is quickly cleared from the blood, and therefore the efficiency of this delivery method is questionable. The study does suggest effective treatment in 7 of the 12 enrolled patients (139). Overall, the clinical studies indicate the need for improvements in delivery, spread, and longevity of survival and replication for the future generations of oHSV vectors.

CURRENT LIMITATIONS OF ONCOLYTIC VIROTHERAPY

Despite the strong biological rationale for using OVs as anticancer agents, published reports of clinical trials indicate that the survival of patients was not significantly prolonged (131, 144). However, these trials demonstrated that no toxicity was associated with these viruses and the maximum tolerated dose was never reached. Together, these data indicate the need to enhance the therapeutic efficacy of OVs, and several researchers have been working to understand the factors that limit OV efficacy. Our current knowledge has identified the following limitations associated with the development of an effective oncolytic virotherapy:

1. *Limited systemic delivery:* Inhibition of viral trespassing through the blood–brain barrier (BBB), neutralization of the virus by innate and preexisting host immunity, and engulfment of the virus in the liver all limit the efficacy of the intravascular delivery of OV. Scientists have developed strategies using vehicles such as immune cells and stem cells to serve as carriers to deliver OVs to the tumor with the potential to bypass the preexisting antiviral immunity (145–148). The advantage of using immune cells as cellular vehicles is it may allow scientists to combine immune therapy with virotherapy (149). Additionally, selective intra-arterial instead of intravenous delivery may circumvent the first-pass effect of liver engulfment. Finally, osmotic and pharmacological agents can now be used to increase trespassing through BBB (150, 151).
2. *Insufficient dispersal through the matrix:* The compact network created by extracellular matrix (ECM) along with the heterogeneity of tumor cells restricts the intratumoral spread of OVs (152). Strategies utilizing pretreatment with matrix proteases (153–155) as well as reengineering OV with tropism to surface receptors (156) show an increased cancer cell transduction of OVs. Also, in the brain, the increased interstitial pressure from the skull created by the presence of a tumor limits the distribution of the OV that is directly injected into the tumor bed. One potential way to enhance this spread is by creating a convection-enhanced flow (CED) that results in distal diffusion of OVs or liposomes carrying OVs through the tumor (150, 157–159).

3. *Antiviral innate immune responses:* Three mathematical models have been established predicting that efficacy of oncolytic virotherapy is given by the ratio of virus efficacy (or OV intratumoral burst) and efficacy of the antiviral innate immune system (160–162). The predictions of these models have been widely confirmed in experimental models emphasizing the need to achieve a balance between repressing and enhancing the immune system away from the virus and toward tumor cells.
4. *Lack of syngeneic models:* Most research studies use human-derived tumors implanted in the brain or flanks of immunocompromised mice. These xenografts in nude mice do not represent a realistic host–tumor interaction as they lack the ability to wage adaptive immune responses (144, 163). Existing syngeneic tumor models tend not to be permissive for human OV replication and are phenotypically different from the human tumors that they are intended to model (164–166). To translate the therapeutic efficacy of OVs from rodents to humans, models that more accurately reproduce human–tumor environments are needed (17, 144, 167, 168).
5. *In vivo imaging of OVs:* The lack of effective systems to image viral intratumoral spread in a noninvasive fashion limits the therapeutic efficacy and outcome of such treatment. One clinical trial tested the efficacy of using positron electron tomography (PET) to image the HSV *tk* gene by comparing different modes of delivery. Results indicated a direct relation between smaller tumor size and a broad tumor distribution of the *tk* gene (169). Schellingerhout et al. (151) and Rehemtulla et al. (170) attempted to image OV vectors *in vivo* using HSV and adenovirus, respectively, radiolabeled with ^{111}In-Oxine or expressing the *E. coli LacZ* gene. Each was found to distribute broadly in their respective glioma models. These imaging systems are not yet efficiently applicable to the clinical setting. The recent discovery of an artificial reporter gene coding for a lysine-rich protein (LRP) that provides MRI enhancement through the phenomenon of chemical exchange saturation transfer (CEST) might be pivotal in overcoming the problem of OV imaging *in vivo* (171).

CONCLUSIONS AND FUTURE PERSPECTIVES

Considering the recent evidence from human genomic analysis of various tumors, multiple gene mutations and therefore multiple pathways are responsible for tumor development. With this, it is clear that a single virus with one transgene or one drug will not be sufficient to treat such complex pathways. Multi-transgenic OVs or a combination of OVs bearing multiple transgenes in combination with drugs are more likely to be effective against these complex tumors. High-throughput screening to identify genes and biomolecules that may enhance OV replication is currently under way and will provide novel ways to enhance OV therapy for cancer.

The field of small interfering RNA (siRNA)-based therapy offers a great promise for personalized treatment of cancer, including targets such as oncogenes and genes that are involved in angiogenesis, metastasis, survival, antiapoptosis, and resistance to chemotherapy. Combining the siRNA platform with OV may demonstrate superior efficacy over other armed transgenes since it will allow for specific knockdown of genes of interest that may be overexpressed in tumors (172–174). Another newly emerged field in OV combination therapy is microRNA (miRNA). These molecules inhibit RNA translation and seem to regulate all aspects of cellular biology including tumor biogenesis and cell–virus interaction (175–178). Because a single miRNA can repress the transcription of multiple mRNAs belonging to the same pathway, it is considered an optimal target for modulating and manipulating the complex mechanisms regulating cell–virus interaction. Accordingly, three recent works have reported the generation of more selective and less toxic OVs through the analysis and utilization of miRNA expression that differs between cancer cells and other cell types (179–181).

The next challenge for OV research is to translate these findings to the clinic and to bridge the gap between our current knowledge and our ability to treat cancer.

ACKNOWLEDGMENTS

We would like to acknowledge Dr. Brent Passer (Mass General Hospital) for providing us with Figure 7.3.

REFERENCES

1. Dock G. 1904. Influence of complicating diseases upon leukemia. Am. J. Med. Sci. **127**: 563–592.
2. DePace N. 1912. Sulla Scomparsa di un enorme cancro begetante del callo dell'utero senza cura chirurgica [Italian]. Ginecolgia. **9**: 82.
3. Liu T.C., Galanis E., and Kirn D. 2007. Clinical trial results with oncolytic virotherapy: a century of promise, a decade of progress. Nat. Clin. Pract. Oncol. **4**: 101–117.
4. Mundschau L.J., and Faller D.V. 1992. Oncogenic ras induces an inhibitor of double-stranded RNA-dependent eukaryotic initiation factor 2 alpha-kinase activation. J. Biol. Chem. **267**: 23092–23098.
5. Telerman A., Tuynder M., Dupressoir T., Robaye B., Sigaux F., Shaulian E., Oren M., Rommelaere J., and Amson R. 1993. A model for tumor suppression using H-1 parvovirus. Proc. Natl. Acad. Sci. U.S.A. **90**: 8702–8706.

6. Stojdl D.F., Lichty B., Knowles S., Marius R., Atkins H., Sonenberg N., and Bell J.C. 2000. Exploiting tumor-specific defects in the interferon pathway with a previously unknown oncolytic virus. Nat. Med. **6**: 821–825.
7. McCart J.A., Ward J.M., Lee J., Hu Y., Alexander H.R., Libutti S.K., Moss B., and Bartlett D.L. 2001. Systemic cancer therapy with a tumor-selective vaccinia virus mutant lacking thymidine kinase and vaccinia growth factor genes. Cancer Res. **61**: 8751–8757.
8. Moehler M., Blechacz B., Weiskopf N., Zeidler M., Stremmel W., Rommelaere J., Galle P.R., and Cornelis J.J. 2001. Effective infection, apoptotic cell killing and gene transfer of human hepatoma cells but not primary hepatocytes by parvovirus H1 and derived vectors. Cancer Gene Ther. **8**: 158–167.
9. Herold-Mende C., Karcher J., Dyckhoff G., and Schirrmacher V. 2005. Antitumor immunization of head and neck squamous cell carcinoma patients with a virus-modified autologous tumor cell vaccine. Adv. Otorhinolaryngol. **62**: 173–183.
10. Bischoff J.R., Kirn D.H., Williams A., Heise C., Horn S., Muna M., Ng L., Nye J.A., Sampson-Johannes A., Fattaey A., and McCormick F. 1996. An adenovirus mutant that replicates selectively in p53-deficient human tumor cells. Science. **274**: 373–376.
11. Whitman E.D., Tsung K., Paxson J., and Norton J.A. 1994. *In vitro* and *in vivo* kinetics of recombinant vaccinia virus cancer-gene therapy. Surgery. **116**: 183–188.
12. Schneider U., Bullough F., Vongpunsawad S., Russell S.J., and Cattaneo R. 2000. Recombinant measles viruses efficiently entering cells through targeted receptors. J. Virol. **74**: 9928–9936.
13. Vigil A., Park M.S., Martinez O., Chua M.A., Xiao S., Cros J.F., Martínez-Sobrido L., Woo S.L., and García-Sastre A. 2007. Use of reverse genetics to enhance the oncolytic properties of Newcastle disease virus. Cancer Res. **67**: 8285–8292.
14. Martuza R.L., Malick A., Markert J.M., Ruffner K.L., and Coen D.M. 1991. Experimental therapy of human glioma by means of a genetically engineered virus mutant. Science **252**: 854–856.
15. Kieff E.D., Bachenheimer S.L., and Roizman B. 1971. Size, composition, and structure of the deoxyribonucleic acid of herpes simplex virus subtypes 1 and 2. J. Virol. **8**: 125–132.
16. Nettelbeck D.M., 2008. Cellular genetic tools to control oncolytic adenoviruses for virotherapy of cancer. J. Mol. Med. **86**: 363–377.
17. Fulci G., and Chiocca E.A. 2003. Oncolytic viruses for the therapy of brain tumors and other solid malignancies: a review. Front. Biosci. **8**: e346–360.
18. Mizuguchi H., and Hayakawa T. 2004. Targeted adenovirus vectors. Hum. Gene Ther. **15**: 1034–1044.
19. Guzman G., Oh S., Shukla D., Engelhard H.H., and Valyi-Nagy T. 2006. Expression of entry receptor nectin-1 of herpes simplex virus 1 and/or herpes simplex virus 2 in normal and neoplastic human nervous system tissues. Acta Virol. **50**: 59–66.
20. Driever P.H., and Rabkin S.D., eds. 2001. Replication-Competent Viruses for Cancer Therapy. Basel: Karger.
21. Markert J.M., Malick A., Coen D.M., and Martuza R.L. 1993. Reduction and elimination of encephalitis in an experimental glioma therapy model with attenuated herpes simplex mutants that retain susceptibility to acyclovir. Neurosurgery **32**: 597–603.

22. Brown S.M., MacLean A.R., McKie E.A., and Harland J. 1997. The herpes simplex virus virulence factor ICP34.5 and the cellular protein MyD116 complex with proliferating cell nuclear antigen through the 63-amino-acid domain conserved in ICP34.5, MyD116, and GADD34. J. Virol. **71**: 9442–9449.

23. Orvedahl A., Alexander D., Tallóczy Z., Sun Q., Wei Y., Zhang W., Burns D., Leib D.A., and Levine B. 2007. HSV-1 ICP34.5 confers neurovirulence by targeting the Beclin 1 autophagy protein. Cell Host Microbe. **1**: 23–35.

24. Mineta T., Rabkin S.D., and Martuza R.L. 1994. Treatment of malignant gliomas using ganciclovir-hypersensitive, ribonucleotide reductase-deficient herpes simplex viral mutant. Cancer Res. **54**: 3963–3966.

25. Mineta T., Rabkin S.D., Yazaki T., Hunter W.D., and Martuza R.L. 1995. Attenuated multi-mutated herpes simplex virus-1 for the treatment of malignant gliomas. Nat. Med. **1**: 938–943.

26. Varghese S., Newsome J.T., Rabkin S.D., McGeagh K., Mahoney D., Nielsen P., Todo T., and Martuza R.L. 2001. Preclinical safety evaluation of G207, a replication-competent herpes simplex virus type 1, inoculated intraprostatically in mice and nonhuman primates. Hum. Gene Ther. **12**: 999–1010.

27. He B., Chou J., Brandimarti R., Mohr I., Gluzman Y., and Roizman B. 1997. Suppression of the phenotype of gamma(1)34.5- herpes simplex virus 1: failure of activated RNA-dependent protein kinase to shut off protein synthesis is associated with a deletion in the domain of the alpha 47 gene. J. Virol. **71**: 6049–6054.

28. He B., Gross M., and Roizman B. 1997. The gamma(1)34.5 protein of herpes simplex virus 1 complexes with protein phosphatase 1alpha to dephosphorylate the alpha subunit of the eukaryotic translation initiation factor 2 and preclude the shutoff of protein synthesis by double-stranded RNA-activated protein kinase. Proc. Natl. Acad. Sci. U.S.A. **94**: 843–848.

29. Todo T., Martuza R.L., Rabkin S.D., and Johnson P.A. 2001. Oncolytic herpes simplex virus vector with enhanced MHC class I presentation and tumor cell killing. Proc. Natl. Acad. Sci. U.S.A. **98**: 6396–6401.

30. Liu T.C., Wakimoto H., Martuza R.L., and Rabkin S.D. 2007. Herpes simplex virus Us3(-) mutant as oncolytic strategy and synergizes with phosphatidylinositol 3-kinase-Akt targeting molecular therapeutics. Clin. Cancer Res. **13**: 5897–5902.

31. Kehm R., Rosen-Wolff A., and Darai G. 1996. Restitution of the UL56 gene expression of HSV-1 HFEM led to restoration of virulent phenotype; deletion of the amino acids 217 to 234 of the UL56 protein abrogates the virulent phenotype. Virus Res. **40**: 17–31.

32. Hardcastle J., Kurozumi K., Chiocca E.A., and Kaur B. 2007. Oncolytic viruses driven by tumor-specific promoters. Curr. Cancer Drug Targets. **7**: 181–189.

33. Miyatake S., Iyer A., Martuza R.L., and Rabkin S.D. 1997. Transcriptional targeting of herpes simplex virus for cell-specific replication. J. Virol. **71**: 5124–5132.

34. Yamamura H., Hashio M., Noguchi M., Sugenoya Y., Osakada M., Hirano N., Sasaki Y., Yoden T., Awata N., Araki N., Tatsuta M., Miyatake S.I., and Takahashi K. 2001. Identification of the transcriptional regulatory sequences of human calponin promoter and their use in targeting a conditionally replicating herpes vector to malignant human soft tissue and bone tumors. Cancer Res. **61**: 3969–3977.

35. Mullen J.T., Kasuya H., Yoon S.S., Carroll N.M., Pawlik T.M., Chandrasekhar S., Nakamura H., Donahue J.M., and Tanabe K.K. 2002. Regulation of herpes simplex virus 1 replication using tumor-associated promoters. Ann. Surg. **236**: 502–512.
36. Chung R.Y., Saeki Y., and Chiocca E.A. 1999. B-myb promoter retargeting of herpes simplex virus gamma 34.5 gene-mediated virulence toward tumor and cycling cells. J. Virol. **73**: 7556–7564.
37. Kambara H., Okano H., Chiocca E.A., and Saeki Y. 2005. An oncolytic HSV-1 mutant expressing ICP34.5 under control of a nestin promoter increases survival of animals even when symptomatic from a brain tumor. Cancer Res. **65**: 2832–2839.
38. Kanai R., Tomita H., Shinoda A., Takahashi M., Goldman S., Okano H., Kawase T., and Yazaki T. 2006. Enhanced therapeutic efficacy of G207 for the treatment of glioma through Musashi1 promoter retargeting of gamma34.5-mediated virulence. Gene Ther. **13**: 106–116.
39. Griffith C, Noonan S, Lou E, and Shillitoe EJ. 2007. An oncolytic mutant of herpes simplex virus type-1 in which replication is governed by a promoter/enhancer of human papillomavirus type-16. Cancer Gene Ther. **14**: 985–993.
40. Montgomery R.I., Warner M.S., Lum B.J., and Spear P.G. 1996. Herpes simplex virus-1 entry into cells mediated by a novel member of the TNF/NGF receptor family. Cell **87**: 427–436.
41. Geraghty R.J., Krummenacher C., Cohen G.H., Eisenberg R.J., and Spear P.G. 1998. Entry of alpha herpes viruses mediated by poliovirus receptor-related protein 1 and poliovirus receptor. Science **280**: 1618–1620.
42. Shukla D., Liu J., Blaiklock P., Shworak N.W., Bai X., Esko J.D., Cohen G.H., Eisenberg R.J., Rosenberg R.D., and Spear P.G. 1999. A novel role for 3-O-sulfated heparan sulfate in herpes simplex virus 1 entry. Cell **99**: 13–22.
43. Zhou G., Ye G.J., Debinski W., and Roizman B. 2002. Engineered herpes simplex virus 1 is dependent on IL13Ralpha 2 receptor for cell entry and independent of glycoprotein D receptor interaction. Proc. Natl. Acad. Sci. U.S.A. **99**: 15124–15129.
44. Kamiyama H., Zhou G., and Roizman B. 2006. Herpes simplex virus 1 recombinant virions exhibiting the amino terminal fragment of urokinase-type plasminogen activator can enter cells via the cognate receptor. Gene Ther. **13**: 621–629.
45. Menotti L., Cerretani A., Hengel H., and Campadelli-Fiume G. 2008. Construction of a fully retargeted herpes simplex virus 1 recombinant capable of entering cells solely via human epidermal growth factor receptor 2. J. Virol. **82**: 10153–10161.
46. Lee C.Y., Bu L.X., Rennie P.S., and Jia W.W. 2007. An HSV-1 amplicon system for prostate-specific expression of ICP4 to complement oncolytic viral replication for in vitro and in vivo treatment of prostate cancer cells. Cancer Gene Ther. **14**: 652–660.
47. Lam P.Y., and Breakefield X.O. 2001. Potential of gene therapy for brain tumors. Hum. Mol. Genet. **10**: 777–787.
48. Aghi, M., and Rabkin S. 2005. Viral vectors as therapeutic agents for glioblastoma. Curr. Opin. Mol. Ther. **7**: 419–430.
49. Post D.E., Shim H., Toussaint-Smith E., and Van Meir E.G. 2005. Cancer scene investigation: how a cold virus became a tumor killer. Future Oncol. **1**: 247–258.

50. Jeyaretna D.S., and Kuroda T. 2007. Recent advances in the development of oncolytic HSV-1 vectors: 'arming' of HSV-1 vectors and application of bacterial artificial chromosome technology for their construction. Curr. Opin. Mol. Ther. **9**: 447–466.
51. Smyth M.J., Swann J., Kelly J.M., Cretney E., Yokoyama W.M., Diefenbach A., Sayers T.J., and Hayakawa Y. 2004. NKG2D recognition and perforin effector function mediate effective cytokine immunotherapy of cancer. J. Exp. Med. **200**: 1325–1335.
52. Ino Y., Saeki Y., Fukuhara H., and Todo T. 2006. Triple combination of oncolytic herpes simplex virus-1 vectors armed with interleukin-12, interleukin-18, or soluble B7-1 results in enhanced antitumor efficacy. Clin. Cancer Res. **12**: 643–652.
53. Malhotra S., Kim T., Zager J., Bennett J., Ebright M., D'Angelica M., and Fong Y. 2007. Use of an oncolytic virus secreting GM-CSF as combined oncolytic and immunotherapy for treatment of colorectal and hepatic adenocarcinomas. Surgery. **141**: 520–529.
54. Derubertis B.G., Stiles B.M., Bhargava A., Gusani N.J., Hezel M., D'Angelica M., and Fong Y. 2007. Cytokine-secreting herpes viral mutants effectively treat tumor in a murine metastatic colorectal liver model by oncolytic and T-cell-dependent mechanisms. Cancer Gene Ther. **14**: 590–597.
55. U.S. National Institutes of Health, OncoVEX Clinical Trials. http://clinicaltrials.gov/ct2/results?term=OncoVEX.
56. Fukuhara H., Ino Y., Kuroda T., Martuza R.L., and Todo T. 2005. Triple gene-deleted oncolytic herpes simplex virus vector double-armed with interleukin 18 and soluble B7-1 constructed by bacterial artificial chromosome-mediated system. Cancer Res. **65**: 10663–10668.
57. Parker J.N., Meleth S., Hughes K.B., Gillespie G.Y., Whitley R.J., and Markert J.M. 2005. Enhanced inhibition of syngeneic murine tumors by combinatorial therapy with genetically engineered HSV-1 expressing CCL2 and IL-12. Cancer Gene Ther. **12**: 359–368.
58. Osorio Y., Cai S., and Ghiasi H. 2005. Treatment of mice with anti-CD86 mAb reduces CD8+ T cell-mediated CTL activity and enhances ocular viral replication in HSV-1-infected mice. Ocul. Immunol. Inflamm. **13**: 159–167.
59. Altomonte J., Wu L., Chen L., Meseck M., Ebert O., García-Sastre A., Fallon J., and Woo S.L. 2008. Exponential enhancement of oncolytic vesicular stomatitis virus potency by vector-mediated suppression of inflammatory responses in vivo. Mol. Ther. **16**: 146–153.
60. Altomonte J., Wu L., Meseck M., Chen L., Ebert O., Garcia-Sastre A., Fallon J., Mandeli J., and Woo S.L. 2009. Enhanced oncolytic potency of vesicular stomatitis virus through vector-mediated inhibition of NK and NKT cells. Cancer Gene Ther. **16**: 266–278.
61. Zamarin D., Martínez-Sobrido L., Kelly K., Mansour M., Sheng G., Vigil A., García-Sastre A., Palese P., and Fong Y. 2009. Enhancement of oncolytic properties of recombinant Newcastle disease virus through antagonism of cellular innate immune responses. Mol Ther. **17**: 697–706.
62. Nakamura H., Mullen J.T., Chandrasekhar S., Pawlik T.M., Yoon S.S., and Tanabe K.K. 2001. Multimodality therapy with a replication-conditional herpes simplex virus 1 mutant that expresses yeast cytosine deaminase for intratumoral conversion of 5-fluorocytosine to 5-fluorouracil. Cancer Res. **61**: 5447–5452.

63. Guffey M.B., Parker J.N., Luckett W.S., Jr., Gillespie G.Y., Meleth S., Whitley R.J., and Markert J.M. 2007. Engineered herpes simplex virus expressing bacterial cytosine deaminase for experimental therapy of brain tumors. Cancer Gene Ther. **14**: 45–56.

64. Chase M., Chung R.Y., and Chiocca E.A. 1998. An oncolytic viral mutant that delivers the CYP2B1 transgene and augments cyclophosphamide chemotherapy. Nat. Biotechnol. **16**: 444–448.

65. Tyminski E., Leroy S., Terada K., Finkelstein D.M., Hyatt J.L., Danks M.K., Potter P.M., Saeki Y., and Chiocca E.A. 2005. Brain tumor oncolysis with replication-conditional herpes simplex virus type 1 expressing the prodrug-activating genes, CYP2B1 and secreted human intestinal carboxylesterase, in combination with cyclophosphamide and irinotecan. Cancer Res. **65**: 6850–6857.

66. Huber B.E. 1994. Gene therapy strategies for treating neoplastic disease. Ann. N. Y. Acad. Sci. **716**: 6–11.

67. Mullen J.T., Donahue J.M., Chandrasekhar S., Yoon S.S., Liu W., Ellis L.M., Nakamura H., Kasuya H., Pawlik T.M., and Tanabe K.K. 2004. Oncolysis by viral replication and inhibition of angiogenesis by a replication-conditional herpes simplex virus that expresses mouse endostatin. Cancer. **101**: 869–877.

68. Yang C.T., Lin Y.C., Lin C.L., Lu J., Bu X., Tsai Y.H., and Jia W.W. 2005. Oncolytic herpes virus with secretable angiostatic proteins in the treatment of human lung cancer cells. Anticancer Res. **25**: 2049–2054.

69. Liu T.C., Zhang T., Fukuhara H., Kuroda T., Todo T., Canron X., Bikfalvi A., Martuza R.L., Kurtz A., and Rabkin S.D. 2006. Dominant-negative fibroblast growth factor receptor expression enhances antitumoral potency of oncolytic herpes simplex virus in neural tumors. Clin. Cancer Res. **12**: 6791–6799.

70. Liu T.C., Zhang T., Fukuhara H., Kuroda T., Todo T., Martuza R.L., Rabkin S.D., and Kurtz A. 2006. Oncolytic HSV armed with platelet factor 4, an antiangiogenic agent, shows enhanced efficacy. Mol. Ther. **14**: 789–797.

71. Jodele S., Chantrain C.F., Blavier L., Lutzko C., Crooks G.M., Shimada H., Coussens L.M., and Declerck Y.A. 2005. The contribution of bone marrow-derived cells to the tumor vasculature in neuroblastoma is matrix metalloproteinase-9 dependent. Cancer Res. **65**: 3200–3208.

72. Saunders W.B., Bohnsack B.L., Faske J.B., Anthis N.J., Bayless K.J., Hirschi K.K., and Davis G.E. 2006. Coregulation of vascular tube stabilization by endothelial cell TIMP-2 and pericyte TIMP-3. J. Cell Biol. **175**: 179–191.

73. Aghi M., Rabkin S.D., and Martuza R.L. 2007. Angiogenic response caused by oncolytic herpes simplex virus-induced reduced thrombospondin expression can be prevented by specific viral mutations or by administering a thrombospondin-derived peptide. Cancer Res. **67**: 440–444.

74. Kurozumi K., Hardcastle J., Thakur R., Yang M., Christoforidis G., Fulci G., Hochberg F.H., Weissleder R., Carson W., Chiocca E.A., and Kaur B. 2007. Effect of tumor microenvironment modulation on the efficacy of oncolytic virus therapy. J. Natl. Cancer Inst. **99**: 1768–1781.

75. Kurozumi K., Hardcastle J., Thakur R., Shroll J., Nowicki M., Otsuki A., Chiocca E.A., and Kaur B. 2008. Oncolytic HSV-1 infection of tumors induces angiogenesis and upregulates CYR61. Mol. Ther. **16**: 1382–1391.

76. Lanzrein M., Schlegel A., and Kempf C. 1994. Entry and uncoating of enveloped viruses. Biochem. J. **302**: 313–320.

77. Hernandez L.D., Hoffman L.R., Wolfsberg T.G., and White J.M. 1996. Virus-cell and cell-cell fusion. Annu. Rev. Cell Dev. Biol. **12**: 627–661.

78. Weissenhorn W., Dessen A., Calder L.J., Harrison S.C., Skehel J.J., and Wiley D.C. 1999. Structural basis for membrane fusion by enveloped viruses. Mol. Membr. Biol. **16**: 3–9.

79. Hoffmann D., Bayer W., and Wildner O. 2007. In situ tumor vaccination with adenovirus vectors encoding measles virus fusogenic membrane proteins and cytokines. World J. Gastroenterol. **13**: 3063–3070.

80. Fu X., and Zhang X. 2002. Potent systemic antitumor activity from an oncolytic herpes simplex virus of syncytial phenotype. Cancer Res. **62**: 2306–2312.

81. Nakamori M., Fu X., Meng F., Jin A., Tao L., Bast R.C., Jr., and Zhang X. 2003. Effective therapy of metastatic ovarian cancer with an oncolytic herpes simplex virus incorporating two membrane fusion mechanisms. Clin. Cancer Res. **9**: 2727–2733.

82. Simpson G.R., Han Z., Liu B., Wang Y., Campbell G., and Coffin R.S. 2006. Combination of a fusogenic glycoprotein, prodrug activation, and oncolytic herpes simplex virus for enhanced local tumor control. Cancer Res. **66**: 4835–4842.

83. Israyelyan A.H., Melancon J.M., Lomax L.G., Sehgal I., Leuschner C., Kearney M.T., Chouljenko V.N., Baghian A., and Kousoulas K.G. 2007. Effective treatment of human breast tumor in a mouse xenograft model with herpes simplex virus type 1 specifying the NV1020 genomic deletion and the gBsyn3 syncytial mutation enabling high viral replication and spread in breast cancer cells. Hum. Gene Ther. **18**: 457–473.

84. Israyelyan A., Chouljenko V.N., Baghian A., David A.T., Kearney M.T., and Kousoulas K.G. 2008. Herpes simplex virus type-1 (HSV-1) oncolytic and highly fusogenic mutants carrying the NV1020 genomic deletion effectively inhibit primary and metastatic tumors in mice. Virol. J. **5**: 68.

85. Bateman A., Bullough F., Murphy S., Emiliusen L., Lavillette D., Cosset F.L., Cattaneo R., Russell S.J., and Vile R.G. 2000. Fusogenic membrane glycoproteins as a novel class of genes for the local and immune-mediated control of tumor growth. Cancer Res. **60**: 1492–1497.

86. Errington F., Bateman A., Kottke T., Thompson J., Harrington K., Merrick A., Hatfield P., Selby P., Vile R., and Melcher A. 2006. Allogeneic tumor cells expressing fusogenic membrane glycoproteins as a platform for clinical cancer immunotherapy. Clin. Cancer Res. **12**: 1333–1341.

87. Chou T.C., and Talalay P. 1984. Quantitative analysis of dose-effect relationships: the combined effects of multiple drugs or enzyme inhibitors. Adv. Enzyme Regul. **22**: 27–55.

88. Advani S.J., Mezhir J.J., Roizman B., and Weichselbaum R.R. 2006. ReVOLT: radiation-enhanced viral oncolytic therapy. Int. J. Radiat. Oncol. Biol. Phys. **66**: 637–646.

89. Jarnagin W.R., Zager J.S., Hezel M., Stanziale S.F., Adusumilli P.S., Gonen M., Ebright M.I., Culliford A., Gusani N.J., and Fong Y. 2006. Treatment of cholangiocarcinoma with oncolytic herpes simplex virus combined with external beam radiation therapy. Cancer Gene Ther, **13**: 326–334.

90. Stanziale S.F., Petrowsky H., Joe J.K., Roberts G.D., Zager J.S., Gusani N.J., Ben-Porat L., Gonen M., and Fong Y. 2002. Ionizing radiation potentiates the antitumor efficacy of oncolytic herpes simplex virus G207 by upregulating ribonucleotide reductase. Surgery **132**: 353–359.
91. Bradley J.D., Kataoka Y., Advani S., Chung S.M., Arani R.B., Gillespie G.Y., Whitley R.J., Markert J.M., Roizman B., and Weichselbaum R.R. 1999. Ionizing radiation improves survival in mice bearing intracranial high-grade gliomas injected with genetically modified herpes simplex virus. Clin. Cancer Res. **5**: 1517–1522.
92. Blank S.V., Rubin S.C., Coukos G., Amin K.M., Albelda S.M., and Molnar-Kimber K.L. 2002. Replication-selective herpes simplex virus type 1 mutant therapy of cervical cancer is enhanced by low-dose radiation. Hum. Gene Ther. **13**: 627–639.
93. Advani S.J., Sibley G.S., Song P.Y., Hallahan D.E., Kataoka Y., Roizman B., and Weichselbaum R.R. 1998. Enhancement of replication of genetically engineered herpes simplex viruses by ionizing radiation: a new paradigm for destruction of therapeutically intractable tumors. Gene Ther. **5**: 160–165.
94. Jorgensen T.J., Katz S., Wittmack E.K., Varghese S., Todo T., Rabkin S.D., and Martuza R.L. 2001. Ionizing radiation does not alter the antitumor activity of herpes simplex virus vector G207 in subcutaneous tumor models of human and murine prostate cancer. Neoplasia. **3**: 451–456.
95. Boviatsis E.J., Park J.S., Sena-Esteves M., Kramm C.M., Chase M., Efird J.T., Wei M.X., Breakefield X.O., and Chiocca E.A. 1994. Long-term survival of rats harboring brain neoplasms treated with ganciclovir and a herpes simplex virus vector that retains an intact thymidine kinase gene. Cancer Res. **54**: 5745–5751.
96. Aghi M., Rabkin S., and Martuza R.L. 2006. Effect of chemotherapy-induced DNA repair on oncolytic herpes simplex viral replication. J. Natl. Cancer Inst. **98**: 38–50.
97. Garcia-Sastre A., and Biron C.A. 2006. Type 1 interferons and the virus-host relationship: a lesson in detente. Science **312**: 879–882.
98. Kodukula P., Liu T., Rooijen N.V., Jager M.J., and Hendricks R.L. 1999. Macrophage control of herpes simplex virus type 1 replication in the peripheral nervous system. J. Immunol. **162**: 2895–2905.
99. Murphy J.A., Duerst R.J., Smith T.J., and Morrison LA. 2003. Herpes simplex virus type 2 virion host shutoff protein regulates alpha/beta interferon but not adaptive immune responses during primary infection in vivo. J. Virol. **77**: 9337–9345.
100. Chen J.G., and Horwitz S.B. 2002. Differential mitotic responses to microtubule-stabilizing and -destabilizing drugs. Cancer Res. **62**: 1935–1938.
101. Elliott G. and O'Hare P. 1998. Herpes simplex virus type 1 tegument protein VP22 induces the stabilization and hyperacetylation of microtubules. J. Virol. **72**: 6448–6455.
102. Lin S.F., Gao S.P., Price D.L., Li S., Chou T.C., Singh P., Huang Y.Y., Fong Y., and Wong R.J. 2008. Synergy of a herpes oncolytic virus and paclitaxel for anaplastic thyroid cancer. Clin. Cancer Res. **14**: 1519–1528.
103. Passer B.J., Castelo-Branco P., Buhrman J.S., Varghese S., Rabkin S.D., and Martuza R.L. 2009. Oncolytic herpes simplex virus vectors and taxanes synergize to promote killing of prostate cancer cells. Cancer Gene Ther. **16**: 551–560.

104. Benetti L., and Roizman B. 2006. Protein kinase B/Akt is present in activated form throughout the entire replicative cycle of deltaU(S)3 mutant virus but only at early times after infection with wild-type herpes simplex virus 1. J. Virol. **80**: 3341–3348.

105. Aghi M., Rabkin S., and Martuza R.L. 2006. Oncolytic herpes simplex virus mutants exhibit enhanced replication in glioma cells evading temozolomide chemotherapy through deoxyribonucleic acid repair. Clin. Neurosurg. **53**: 65–76.

106. Bennett J.J., Adusumilli P., Petrowsky H., Burt B.M., Roberts G., Delman K.A., Zager J.S., Chou T.C., and Fong Y. 2004. Up-regulation of GADD34 mediates the synergistic anticancer activity of mitomycin C and a gamma134.5 deleted oncolytic herpes virus (G207). FASEB J. **18**: 1001–1003.

107. Mullerad M., Bochner B.H., Adusumilli P.S., Bhargava A., Kikuchi E., Hui-Ni C., Kattan M.W., Chou T.C., and Fong Y. 2005. Herpes simplex virus based gene therapy enhances the efficacy of mitomycin C for the treatment of human bladder transitional cell carcinoma. J. Urol. **174**: 741–746.

108. Adusumilli P.S., Chan M.K., Chun Y.S., Hezel M., Chou T.C., Rusch V.W., and Fong Y. 2006. Cisplatin-induced GADD34 upregulation potentiates oncolytic viral therapy in the treatment of malignant pleural mesothelioma. Cancer Biol. Ther. **5**: 48–53.

109. Marks P., Rifkind R.A., Richon V.M., Breslow R., Miller T., and Kelly W.K. 2001. Histone deacetylases and cancer: causes and therapies. Nat. Rev. Cancer. **1**: 194–202.

110. Archer S.Y., Meng S., Shei A., and Hodin R.A. 1998. p21(WAF1) is required for butyrate-mediated growth inhibition of human colon cancer cells. Proc. Natl. Acad. Sci. U.S.A. **95**: 6791–6796.

111. Rahmani M., Reese E., Dai Y., Bauer C., Payne S.G., Dent P., Spiegel S., and Grant S. 2005. Coadministration of histone deacetylase inhibitors and perifosine synergistically induces apoptosis in human leukemia cells through Akt and ERK1/2 inactivation and the generation of ceramide and reactive oxygen species. Cancer Res. **65**: 2422–2432.

112. Liu T.C., Castelo-Branco P., Rabkin S.D., and Martuza R.L. 2008. Trichostatin A and oncolytic HSV combination therapy shows enhanced antitumoral and antiangiogenic effects. Mol. Ther. **16**: 1041–1047.

113. Song B., Liu J.J., Yeh K.C., and Knipe D.M. 2000. Herpes simplex virus infection blocks events in the G1 phase of the cell cycle. Virology. **267**: 326–334.

114. Kim M.S., Kwon H.J., Lee Y.M., Baek J.H., Jang J.E., Lee S.W., Moon E.J., Kim H.S., Lee S.K., Chung H.Y., Kim C.W., and Kim K.W. 2001. Histone deacetylases induce angiogenesis by negative regulation of tumor suppressor genes. Nat. Med. **7**: 437–443.

115. Deroanne C.F., Bonjean K., Servotte S., Devy L., Colige A., Clausse N., Blacher S., Verdin E., Foidart J.M., Nusgens B.V., and Castronovo V. 2002. Histone deacetylases inhibitors as anti-angiogenic agents altering vascular endothelial growth factor signaling. Oncogene. **21**: 427–436.

116. Qian D.Z., Kato Y., Shabbeer S., Wei Y., Verheul H.M., Salumbides B., Sanni T., Atadja P., and Pili R. 2006. Targeting tumor angiogenesis with histone deacetylase inhibitors: the hydroxamic acid derivative LBH589. Clin. Cancer Res. **12**: 634–642.

117. Benencia F., Courreges M.C., Conejo-García J.R., Buckanovich R.J., Zhang L., Carroll R.H., Morgan M.A., and Coukos G. 2005. Oncolytic HSV exerts direct antiangiogenic activity in ovarian carcinoma. Hum. Gene Ther. **16**: 765–778.
118. Cinatl J., Jr., Michaelis M., Driever P.H., Cinatl J., Hrabeta J., Suhan T., Doerr H.W., and Vogel J.U. 2004. Multimutated herpes simplex virus g207 is a potent inhibitor of angiogenesis. Neoplasia. **6**: 725–735.
119. Otsuki A., Patel A., Kasai K., Suzuki M., Kurozumi K., Chiocca E.A., and Saeki Y. 2008. Histone deacetylase inhibitors augment antitumor efficacy of herpes-based oncolytic viruses. Mol. Ther. **16**: 1546–1555.
120. Ikeda K., Ichikawa T., Wakimoto H., Silver J.S., Deisboeck T.S., Finkelstein D., Harsh G.R., IV, Louis D.N., Bartus R.T., Hochberg F.H., and Chiocca EA. 1999. Oncolytic virus therapy of multiple tumors in the brain requires suppression of innate and elicited antiviral responses. Nat. Med. **5**: 881–887.
121. Ikeda K., Wakimoto H., Ichikawa T., Jhung S., Hochberg F.H., Louis D.N., and Chiocca EA. 2000. Complement depletion facilitates the infection of multiple brain tumors by an intravascular, replication-conditional herpes simplex virus mutant. J. Virol. **74**: 4765–4775.
122. Wakimoto H., Fulci G., Tyminski E., and Chiocca E.A. 2004. Altered expression of antiviral cytokine mRNAs associated with cyclophosphamide's enhancement of viral oncolysis. Gene Ther. **11**: 214–223.
123. Kambara H., Saeki Y., and Chiocca E.A. 2005. Cyclophosphamide allows for in vivo dose reduction of a potent oncolytic virus. Cancer Res. **65**: 11255–11258.
124. Fulci G., Breymann L., Gianni D., Kurozomi K., Rhee S.S., Yu J., Kaur B., Louis D.N. Weissleder R., Caligiuri M.A., and Chiocca E.A. 2006. Cyclophosphamide enhances glioma virotherapy by inhibiting innate immune responses. Proc. Natl. Acad. Sci. U.S.A. **103**: 12873–12878.
125. Currier M.A., Gillespie R.A., Sawtell N.M., Mahller Y.Y., Stroup G., Collins M.H., Kambara H., Chiocca E.A., and Cripe T.P. 2008. Efficacy and safety of the oncolytic herpes simplex virus rRp450 alone and combined with cyclophosphamide. Mol. Ther. **16**: 879–885.
126. Thomas M.A., Spencer J.F., Toth K., Sagartz J.E., Phillips N.J., and Wold W.S. 2008. Immunosuppression enhances oncolytic adenovirus replication and antitumor efficacy in the Syrian hamster model. Mol. Ther. **16**: 1665–1673.
127. Fulci G., Dmitrieva N., Gianni D., Fontana E.J., Pan X., Lu Y., Kaufman C.S., Kaur B., Lawler S.E., Lee R.J., Marsh C.B., Brat D.J., van Rooijen N., Stemmer-Rachamimov A.O., Hochberg F.H., Weissleder R., Martuza R.L., and Chiocca E.A. 2007. Depletion of peripheral macrophages and brain microglia increases brain tumor titers of oncolytic viruses. Cancer Res. **67**: 9398–9406.
128. Mahller Y.Y., Vaikunth S.S., Currier M.A., Miller S.J., Ripberger M.C., Hsu Y.H., Mehrian-Shai R., Collins M.H., Crombleholme T.M., Ratner N., and Cripe T.P. 2007. Oncolytic HSV and erlotinib inhibit tumor growth and angiogenesis in a novel malignant peripheral nerve sheath tumor xenograft model. Mol. Ther. **15**: 279–286.
129. Gutermann A., Mayer E., von Dehn-Rothfelser K., Breidenstein C., Weber M., Muench M., Gungor D., Suehnel J., Moebius U., and Lechmann M. 2006. Efficacy of oncolytic herpes virus NV1020 can be enhanced by combination with chemotherapeutics in colon carcinoma cells. Hum. Gene Ther. **17**: 1241–1253.

130. Markert J.M., Medlock M.D., Rabkin S.D., Gillespie G.Y., Todo T., Hunter W.D., Palmer C.A., Feigenbaum F., Tornatore C., Tufaro F., and Martuza R.L. 2000. Conditionally replicating herpes simplex virus mutant, G207 for the treatment of malignant glioma: results of a phase I trial. Gene Ther. **7**: 867–874.

131. Rampling R., Cruickshank G., Papanastassiou V., Nicoll J., Hadley D., Brennan D., Petty R., MacLean A., Harland J., McKie E., Mabbs R., and Brown M. 2000. Toxicity evaluation of replication-competent herpes simplex virus (ICP 34.5 null mutant 1716) in patients with recurrent malignant glioma. Gene Ther. **7**: 859–866.

132. Chou J., Kern E.R., Whitley R.J., and Roizman B. 1990. Mapping of herpes simplex virus-1 neurovirulence to gamma 134.5, a gene nonessential for growth in culture. Science **250**: 1262–1266.

133. Papanastassiou V., Rampling R., Fraser M., Petty R., Hadley D., Nicoll J., Harland J., Mabbs R., and Brown M. 2002. The potential for efficacy of the modified (ICP 34.5(-)) herpes simplex virus HSV1716 following intratumoural injection into human malignant glioma: a proof of principle study. Gene Ther. **9**: 398–406.

134. Harrow S., Papanastassiou V., Harland J., Mabbs R., Petty R., Fraser M., Hadley D., Patterson J., Brown S.M., and Rampling R. 2004. HSV1716 injection into the brain adjacent to tumour following surgical resection of high-grade glioma: safety data and long-term survival. Gene Ther. **11**: 1648–1658.

135. Hu J.C., Coffin R.S., Davis C.J., Graham N.J., Groves N., Guest P.J., Harrington K.J., James N.D., Love C.A., McNeish I., Medley L.C., Michael A., Nutting C.M., Pandha H.S., Shorrock C.A., Simpson J., Steiner J., Steven N.M., Wright D., and Coombes R.C. 2006. A phase I study of OncoVEXGM-CSF, a second-generation oncolytic herpes simplex virus expressing granulocyte macrophage colony-stimulating factor. Clin. Cancer Res. **12**: 6737–6747.

136. MacKie R.M., Stewart B., and Brown S.M. 2001. Intralesional injection of herpes simplex virus 1716 in metastatic melanoma. Lancet **357**: 525–526.

137. Nakao A., Takeda S., Shimoyama S., Kasuya H., Kimata H., Teshigahara O., Sawaki M., Kikumori T., Kodera Y., Nagasaka T., Goshima F., Nishiyama Y., and Imai T. 2007. Clinical experiment of mutant herpes simplex virus HF10 therapy for cancer. Curr. Cancer Drug Targets **7**: 169–174.

138. Fujimoto Y., Mizuno T., Sugiura S., Goshima F., Kohno S., Nakashima T., and Nishiyama Y. 2006. Intratumoral injection of herpes simplex virus HF10 in recurrent head and neck squamous cell carcinoma. Acta Otolaryngol. **126**: 1115–1117.

139. Kemeny N., Brown K., Covey A., Kim T., Bhargava A., Brody L., Guilfoyle B., Haag N.P., Karrasch M., Glasschroeder B., Knoll A., Getrajdman G., Kowal K.J., Jarnagin W.R., and Fong Y. 2006. Phase I, open-label, dose-escalating study of a genetically engineered herpes simplex virus, NV1020, in subjects with metastatic colorectal carcinoma to the liver. Hum. Gene Ther. **17**: 1214–1224.

140. Nakao A., Kimata H., Imai T., Kikumori T., Teshigahara O., Nagasaka T., Goshima F., and Nishiyama Y. 2004. Intratumoral injection of herpes simplex virus HF10 in recurrent breast cancer. Ann. Oncol. **15**: 988–989.

141. Kimata H., Imai T., Kikumori T., Teshigahara O., Nagasaka T., Goshima F., Nishiyama Y., and Nakao A. 2006. Pilot study of oncolytic viral therapy using mutant herpes simplex virus (HF10) against recurrent metastatic breast cancer. Ann. Surg. Oncol. **13**: 1078–1084.

142. Ushijima Y., Luo C., Goshima F., Yamauchi Y., Kimura H., and Nishiyama Y. 2007. Determination and analysis of the DNA sequence of highly attenuated herpes simplex virus type 1 mutant HF10, a potential oncolytic virus. Microbes Infect. **9**: 142–149.

143. Liu B.L., Robinson M., Han Z.Q., Branston R.H., English C., Reay P., McGrath Y., Thomas S.K., Thornton M., Bullock P., Love C.A., and Coffin R.S. 2003. ICP34.5 deleted herpes simplex virus with enhanced oncolytic, immune stimulating, and anti-tumour properties. Gene Ther. **10**: 292–303.

144. Fulci G., and Chiocca E.A. 2007. The status of gene therapy for brain tumors. Expert Opin. Biol. Ther. **7**: 197–208.

145. Cole C., Qiao J., Kottke T., Diaz R.M., Ahmed A., Sanchez-Perez L., Brunn G., Thompson J., Chester J., and Vile R.G. 2005. Tumor-targeted, systemic delivery of therapeutic viral vectors using hitchhiking on antigen-specific T cells. Nat. Med. **11**: 1073–1081.

146. Komarova S., Kawakami Y., Stoff-Khalili M.A., Curiel D.T., and Pereboeva L. 2006. Mesenchymal progenitor cells as cellular vehicles for delivery of oncolytic adenoviruses. Mol. Cancer Ther. **5**: 755–766.

147. Power A.T., and Bell J.C. 2007. Cell-based delivery of oncolytic viruses: a new strategic alliance for a biological strike against cancer. Mol. Ther. **15**: 660–665.

148. Stoff-Khalili M.A., Rivera A.A., Mathis J.M., Banerjee N.S., Moon A.S., Hess A., Rocconi R.P., Numnum T.M., Everts M., Chow L.T., Douglas J.T., Siegal G.P., Zhu Z.B., Bender H.G., Dall P., Stoff A., Pereboeva L., and Curiel D.T. 2007. Mesenchymal stem cells as a vehicle for targeted delivery of CRAds to lung metastases of breast carcinoma. Breast Cancer Res. Treat. **105**: 157–167.

149. Thorne S.H., Negrin R.S, and Contag C.H., 2006. Synergistic antitumor effects of immune cell-viral biotherapy. Science. **311**: 1780–1784.

150. Nilaver G., Muldoon L.L., Kroll R.A., Pagel M.A., Breakefield X.O., Davidson B.L., and Neuwelt E.A. 1995. Delivery of herpes virus and adenovirus to nude rat intracerebral tumors after osmotic blood-brain barrier disruption. Proc. Natl. Acad. Sci. U.S.A. **92**: 9829–9833.

151. Schellingerhout D., Rainov N.G., Breakefield X.O., and Weissleder R. 2000. Quantitation of HSV mass distribution in a rodent brain tumor model. Gene Ther. **7**: 1648–1655.

152. Mok W., Stylianopoulos T., Boucher Y., and Jain R.K. 2009. Mathematical modeling of herpes simplex virus distribution in solid tumors: implications for cancer gene therapy. Clin. Cancer Res. **15**: 2352–2360.

153. Kuriyama N., Kuriyama H., Julin C.M., Lamborn K.R., and Israel M.A. 2001. Protease pretreatment increases the efficacy of adenovirus-mediated gene therapy for the treatment of an experimental glioblastoma model. Cancer Res. **61**: 1805–1809.

154. McKee T.D., Grandi P., Mok W., Alexandrakis G., Insin N., Zimmer J.P., Bawendi M.G., Boucher Y., Breakefield X.O., and Jain R.K. 2006. Degradation of fibrillar collagen in a human melanoma xenograft improves the efficacy of an oncolytic herpes simplex virus vector. Cancer Res. **66**: 2509–2513.

155. Mok W., Boucher Y., and Jain R.K. 2007. Matrix metalloproteinases-1 and -8 improve the distribution and efficacy of an oncolytic virus. Cancer Res. **67**: 10664–10668.

156. Zhou G., and Roizman B. 2005. Characterization of a recombinant herpes simplex virus 1 designed to enter cells via the IL13Ralpha2 receptor of malignant glioma cells. J. Virol. **79**: 5272–5277.

157. Immonen A., Vapalahti M., Tyynelä K., Hurskainen H., Sandmair A., Vanninen R., Langford G., Murray N., and Ylä-Herttuala S. 2004. AdvHSV-tk gene therapy with intravenous ganciclovir improves survival in human malignant glioma: a randomised, controlled study. Mol. Ther. **10**: 967–972.

158. Chen M.Y., Hoffer A., Morrison P.F., Hamilton J.F., Hughes J., Schlageter K.S., Lee J., Kelly B.R., and Oldfield E.H. 2005. Surface properties, more than size, limiting convective distribution of virus-sized particles and viruses in the central nervous system. J. Neurosurg. **103**: 311–319.

159. Saito R., Krauze M.T., Bringas J.R., Noble C., McKnight T.R., Jackson P., Wendland M.F., Mamot C., Drummond D.C., Kirpotin D.B., Hong K., Berger M.S., Park J.W., and Bankiewicz K.S. 2005. Gadolinium-loaded liposomes allow for real-time magnetic resonance imaging of convection-enhanced delivery in the primate brain. Exp. Neurol. **196**: 381–389.

160. Wodarz D. 2001. Viruses as antitumor weapons: defining conditions for tumor remission. Cancer Res. **61**: 3501–3507.

161. Friedman A., Tian J.P., Fulci G., Chiocca E.A., and Wang J. 2006. Glioma virotherapy: effects of innate immune suppression and increased viral replication capacity. Cancer Res. **66**: 2314–2319.

162. Paiva L.R., Binny C., Ferreira S.C., Jr., and Martins M.L. 2009. A multiscale mathematical model for oncolytic virotherapy. Cancer Res, **69**: 1205–1211.

163. Troiani T., Schettino C., Martinelli E., Morgillo F., Tortora G., and Ciardiello F. 2008. The use of xenograft models for the selection of cancer treatments with the EGFR as an example. Crit. Rev. Oncol. Hematol. **65**: 200–211.

164. Varghese S., Rabkin S.D., Nielsen P.G., Wang W., and Martuza R.L. 2006. Systemic oncolytic herpes virus therapy of poorly immunogenic prostate cancer metastatic to lung. Clin. Cancer Res. **12**: 2919–2927.

165. Lampson L.A. 2001. New animal models to probe brain tumor biology, therapy, and immunotherapy: advantages and remaining concerns. J. Neurooncol. **53**: 275–287.

166. Wakimoto H., Ikeda K., Abe T., Ichikawa T., Hochberg F.H., Ezekowitz R.A., Pasternack M.S., and Chiocca E.A. 2002. The complement response against an oncolytic virus is species-specific in its activation pathways. Mol. Ther. **5**: 275–282.

167. Brinster R.L., Chen H.Y., Messing A., van Dyke T., Levine A.J., and Palmiter R.D. 1984. Transgenic mice harboring SV40 T-antigen genes develop characteristic brain tumors. Cell. **37**: 367–379.

168. Corcoran R.B., and Scott M.P. 2001. A mouse model for medulloblastoma and basal cell nevus syndrome. J. Neurooncol. **53**: 307–318.

169. Voges J., Reszka R., Gossmann A., Dittmar C., Richter R., Garlip G., Kracht L., Coenen H.H., Sturm V., Wienhard K., Heiss W.D., and Jacobs A.H. 2003. Imaging-guided convection-enhanced delivery and gene therapy of glioblastoma. Ann. Neurol. **54**: 479–487.

170. Rehemtulla A., Hall D.E., Stegman L.D., Prasad U., Chen G., Bhojani M.S., Chenevert T.L., and Ross B.D. 2002. Molecular imaging of gene expression and

efficacy following adenoviral-mediated brain tumor gene therapy. Mol. Imaging **1**: 43–55.

171. Gilad A.A., McMahon M.T., Walczak P., Winnard P.T., Jr., Raman V., van Laarhoven H.W., Skoglund C.M., Bulte J.W., and van Zijl P.C. 2007. Artificial reporter gene providing MRI contrast based on proton exchange. Nat. Biotechnol. **25**: 217–219.
172. Kirn D. 2007. Armed interference: oncolytic viruses engineered to carry antitumor shRNAs. Mol. Ther. **15**: 227–228.
173. Tong A.W., Zhang Y.A., and Nemunaitis J. 2005. Small interfering RNA for experimental cancer therapy. Curr. Opin. Mol. Ther. **7**: 114–124.
174. Huang C., Li M., Chen C., and Yao Q. 2008. Small interfering RNA therapy in cancer: mechanism, potential targets, and clinical applications. Expert Opin. Ther. Targets **12**: 637–645.
175. Pfeffer S., Zavolan M., Grässer F.A., Chien M., Russo J.J., Ju J., John B., Enright A.J., Marks D., Sander C., and Tuschl T. 2004. Identification of virus-encoded microRNAs. Science **304**: 734–736.
176. Scaria V., Hariharan M., Maiti S., Pillai B., and Brahmachari S.K. 2006. Host-virus interaction: a new role for microRNAs. Retrovirology **3**: 68.
177. Stadler B.M., and Ruohola-Baker H. 2008. Small RNAs: keeping stem cells in line. Cell **132**: 563–566.
178. Weidhaas J.B., Babar I., Nallur S.M., Trang P., Roush S., Boehm M., Gillespie E., and Slack F.J. 2007. MicroRNAs as potential agents to alter resistance to cytotoxic anticancer therapy. Cancer Res. **67**: 11111–11116.
179. Edge R.E., Falls T.J., Brown C.W., Lichty B.D., Atkins H., and Bell J.C. 2008. A let-7 MicroRNA-sensitive vesicular stomatitis virus demonstrates tumor-specific replication. Mol. Ther. **16**: 1437–1443.
180. Kelly E.J., Hadac E.M., Greiner S., and Russell S.J. 2008. Engineering microRNA responsiveness to decrease virus pathogenicity. Nat. Med. **14**: 1278–1283.
181. Ylösmäki E., Hakkarainen T., Hemminki A., Visakorpi T., Andino R., and Saksela K. 2008. Generation of a conditionally replicating adenovirus based on targeted destruction of E1A mRNA by a cell type-specific MicroRNA. J. Virol. **82**: 11009–11015.
182. Adusumilli P.S., Stiles B.M., Chan M.K., Mullerad M., Eisenberg D.P., Ben-Porat L., Huq R., Rusch V.W., and Fong Y. 2006. Imaging and therapy of malignant pleural mesothelioma using replication-competent herpes simplex viruses. J. Gene Med. **8**: 603–615.
183. Coen D.M., Kosz-Vnenchak M., Jacobson J.G., Leib D.A., Bogard C.L., Schaffer P.A., Tyler K.L., and Knipe D.M. 1989. Thymidine kinase-negative herpes simplex virus mutants establish latency in mouse trigeminal ganglia but do not reactivate. Proc. Natl. Acad. Sci. U.S.A. **86**: 4736–4740.
184. Goldstein D.J., and Weller S.K. 1988. Herpes simplex virus type 1-induced ribonucleotide reductase activity is dispensable for virus growth and DNA synthesis: isolation and characterization of an ICP6 lacZ insertion mutant. J. Virol. **62**: 196–205.
185. MacLean A.R., ul-Fareed M., Robertson L., Harland J., and Brown SM. 1991. Herpes simplex virus type 1 deletion variants 1714 and 1716 pinpoint neurovirulence-related sequences in Glasgow strain 17+ between immediate early gene 1 and the 'a' sequence. J. Gen. Virol. **72**: 631–639.

186. Meignier B., Longnecker R., and Roizman B. 1988. In vivo behavior of genetically engineered herpes simplex viruses R7017 and R7020: construction and evaluation in rodents. J. Infect. Dis. **158**: 602–614.

187. Thompson R.L., and Stevens J.G. 1983. Biological characterization of a herpes simplex virus intertypic recombinant which is completely and specifically non-neurovirulent. Virology **131**: 171–179.

188. Andreansky S., He B., van Cott J., McGhee J., Markert J.M., Gillespie G.Y., Roizman B., and Whitley R.J. 1998. Treatment of intracranial gliomas in immunocompetent mice using herpes simplex viruses that express murine interleukins. Gene Ther. **5**: 121–130.

189. Wong R.J., Patel S.G., Kim S., DeMatteo R.P., Malhotra S., Bennett J.J., St-Louis M., Shah J.P., Johnson P.A., and Fong Y. 2001. Cytokine gene transfer enhances herpes oncolytic therapy in murine squamous cell carcinoma. Hum. Gene Ther. **12**: 253–265.

190. Parker J.N., Gillespie G.Y., Love C.E., Randall S., Whitley R.J., and Markert J.M. 2000. Engineered herpes simplex virus expressing IL-12 in the treatment of experimental murine brain tumors. Proc. Natl. Acad. Sci. U.S.A. **97**: 2208–2213.

191. Han Z.Q., Assenberg M., Liu B.L., Wang Y.B., Simpson G., Thomas S., and Coffin R.S. 2007. Development of a second-generation oncolytic Herpes simplex virus expressing TNF alpha for cancer therapy. J. Gene Med. **9**: 99–106.

192. Pawlik T.M., Nakamura H., Mullen J.T., Kasuya H., Yoon S.S., Chandrasekhar S., Chiocca E.A., and Tanabe K.K. 2002. Prodrug bioactivation and oncolysis of diffuse liver metastases by a herpes simplex virus 1 mutant that expresses the CYP2B1 transgene. Cancer **95**: 1171–1181.

193. Mahller Y.Y., Vaikunth S.S., Ripberger M.C., Baird W.H., Saeki Y., Cancelas J.A., Crombleholme T.M., and Cripe T.P. 2008. Tissue inhibitor of metalloproteinase-3 via oncolytic herpes virus inhibits tumor growth and vascular progenitors. Cancer Res. **68**: 1170–1179.

194. Mace A.T., Harrow S.J., Ganly I., and Brown S.M. 2007. Cytotoxic effects of the oncolytic herpes simplex virus HSV1716 alone and in combination with cisplatin in head and neck squamous cell carcinoma. Acta Otolaryngol. **127**: 880–887.

195. Toyoizumi T., Mick R., Abbas A.E., Kang E.H., Kaiser L.R., and Molnar-Kimber K.L. 1999. Combined therapy with chemotherapeutic agents and herpes simplex virus type 1 ICP34.5 mutant (HSV-1716) in human non-small cell lung cancer. Hum. Gene Ther. **10**: 3013–3029.

196. Chahlavi A., Todo T., Martuza R.L., and Rabkin S.D. 1999. Replication-competent herpes simplex virus vector G207 and cisplatin combination therapy for head and neck squamous cell carcinoma. Neoplasia **1**: 162–169.

197. Nakano K., Todo T., Zhao G., Yamaguchi K., Kuroki S., Cohen J.B., Glorioso J.C., and Tanaka M. 2005. Enhanced efficacy of conditionally replicating herpes simplex virus (G207) combined with 5-fluorouracil and surgical resection in peritoneal cancer dissemination models. J. Gene Med. **7**: 638–648.

198. Petrowsky H., Roberts G.D., Kooby D.A., Burt B.M., Bennett J.J., Delman K.A., Stanziale S.F., Delohery T.M., Tong W.P., Federoff H.J., and Fong Y. 2001. Functional interaction between fluorodeoxyuridine-induced cellular alterations and replication of a ribonucleotide reductase-negative herpes simplex virus. J. Virol. **75**: 7050–7058.

199. Carroll N.M., Chase M., Chiocca E.A., and Tanabe K.K. 1997. The effect of ganciclovir on herpes simplex virus-mediated oncolysis. J. Surg. Res. **69**: 413–417.

200. Watanabe I., Kasuya H., Nomura N., Shikano T., Shirota T., Kanazumi N., Takeda S., Nomoto S., Sugimoto H., and Nakao A. 2008. Effects of tumor selective replication-competent herpes viruses in combination with gemcitabine on pancreatic cancer. Cancer Chemother. Pharmacol. **61**: 875–882.

201. Naito S., Obayashi S., Sumi T., Iwai S., Nakazawa M., Ikuta K., and Yura Y. 2006. Enhancement of antitumor activity of herpes simplex virus gamma(1)34.5-deficient mutant for oral squamous cell carcinoma cells by hexamethylene bisacetamide. Cancer Gene Ther. **13**: 780–791.

202. Katsura T., Iwai S., Ota Y., Shimizu H., Ikuta K., and Yura Y. 2009. The effects of trichostatin A on the oncolytic ability of herpes simplex virus for oral squamous cell carcinoma cells. Cancer Gene Ther. **16**: 237–245.

203. Eisenberg D.P., Adusumilli P.S., Hendershott K.J., Yu Z., Mullerad M., Chan M.K., Chou T.C., and Fong Y. 2005. 5-Fluorouracil and gemcitabine potentiate the efficacy of oncolytic herpes viral gene therapy in the treatment of pancreatic cancer. J. Gastrointest. Surg. **9**: 1068–1077.

PART II

BACTERIAL PRODUCTS AS ANTICANCER AGENTS

8

PROMISCUOUS ANTICANCER DRUGS FROM PATHOGENIC BACTERIA: RATIONAL VERSUS INTELLIGENT DRUG DESIGN

Arsénio M. Fialho[1] and Ananda M. Chakrabarty[2]

[1]*Institute for Biotechnology and Bioengineering (IBB), Center for Biological and Chemical Engineering, Instituto Superior Tecnico, Lisbon, Portugal*

[2]*Department of Microbiology and Immunology, University of Illinois College of Medicine, Chicago, IL*

INTRODUCTION

Early anticancer drugs were mostly small molecular weight anti-metabolites with varied range of activities, such as cisplatin as DNA damaging agent, doxorubicin as an inhibitor of topoisomerase II, paclitaxel as mitosis-inhibiting stabilizer of microtubule, and methotrexate as inhibitor of DNA/RNA synthesis (1). Since cancer cells grow much faster than normal cells, the inhibitory effect on cancer cells of these drugs is much more pronounced, although they do affect normal cell growth and metabolism, producing certain toxicity symptoms. More recent, newer and better targeting drugs have been characterized from the massive screening efforts by the U.S. National Cancer Institute, assaying the growth inhibitory activity of over 100,000 chemicals when incubated for 3 days at different concentrations with 60 different human cancer cell lines. Such random search for anticancer agents provides many candidate drugs whose modes of action and potential toxicity are then further evaluated.

Emerging Cancer Therapy: Microbial Approaches and Biotechnological Tools, Edited by Arsénio M. Fialho and Ananda M. Chakrabarty

A more recent effort in anticancer drug development does not involve searching for anticancer activity but to design rationally potential anticancer agents that inhibit specific steps in the disease progression pathways, either cancer or other diseases such as HIV/AIDS. Such drugs usually target a key single step so that the disease progression is halted or at least considerably slowed down. An example, in the case of HIV/AIDS, is the development of inhibitors for specific steps in HIV-1 viral replication/maturation machinery, such as protease, integrase, and nucleoside/non-nucleoside reverse transcriptase, or even in entry, such as enfuvirtide (2, 3). In the case of cancer, as mentioned above, the target could be DNA replication, cell division, cell signaling for growth, new blood vessel formation, or even inhibition of apoptosis of cancer cells. Such inhibition could be mediated either by small molecule compounds or by proteins including monoclonal antibodies (mAbs) that target key antigens in cancer cells (4, 5). mAbs can also be conjugated with drugs/prodrugs or toxins (6), as well as a source of radiation (7), for better efficacy, and several such conjugated mAbs have been approved for therapy by the Food and Drug Administration (FDA). A problem with single-targeting drugs is rapid resistance development by the disease agents, as exemplified by HIV-1 virus becoming resistant to single-target drugs (3). Thus, to overcome such drug resistance, a combination of drugs, known as highly active antiretroviral therapy (HAART), is used quite successfully to treat HIV/AIDS. In the case of cancer, combination therapy using mAb and small molecule inhibitors provides better protection than single drugs (8). Single targeting is not only less efficient for a drug to demonstrate efficacy (9, 10) but is also amenable to the presence of genetic mutations in non-small cell lung cancer (NSCLC) patients treated with erlotinib (11). Cancer cells can also mutate to nullify the growth inhibitory effect, as demonstrated by resistance to small molecule epidermal growth factor receptor (EGFR) inhibitors by Akt activation (12). Combinations of cyclin-dependent kinase inhibitors and other agents have shown promise (13). Since the number of anticancer drugs is large and varied, defining what combinations are most effective is the most challenging decision (14). Such combinatorial treatment, in turn, not only adds to the cost of the therapy but also poses potential problems of multidrug resistance.

An alternative approach that is gaining ground in drug design and discovery is to develop multi-targeting promiscuous drugs with low affinity for multiple targets in the disease progression pathway but are often more efficacious with less side effects (1, 15–17). Examples of such promiscuous drugs, often cited, are aspirin, memantine, or the antipsychotic clozapine, which target multiple sites as pain killer, blood thinner, anticancer agent, or as N-methyl d-aspartate receptor (NMDA) receptor antagonist in Alzheimer's disease or as modulator on a group of G protein-coupled receptors (16, 18, 19). While genome and proteome-based high-throughput screening methods are often used in single-target rational drug design, there is no rational way to develop small molecule compounds that can target multiple sites in a disease progression pathway or even in multiple pathways involving multiple diseases. Computer-generated model network approaches, often termed "multi-target drug design games" (20), are being developed to address the need for developing promiscuous drugs. More recently, selectivity filters have been proposed to guide rational drug design, involving changing the wrapping patterns of target proteins, particularly highly cross-reactive proteins (21, 22). The strategy is to modify the drug to turn it into a wrapper of unique dehydrons while removing the potential sources of cross-reactivity (21, 22). An example of such a technique is the redesigning of imatinib, which is therapeutically effective against both chronic myeloid leukemia (CML) and gastrointestinal stromal tumors (GIST) because of the inhibition of both Bcr-Abl and C-KIT. The cardiotoxicity of imatinib is largely due to the specificity exhibited by Bcr-Abl. Thus, Fernández et al. (23) reengineered imatinib to reduce its binding with Bcr-Abl and consequent reduction in cardiotoxicity, but retaining the binding specificity for C-KIT, thereby making the modified imatinib effective for treatment of GIST (23). The modification of imatinib involved introduction of a methyl group that promoted dehydration of the C673–G676 region within the imatinib-C-KIT interface, thereby stabilizing this complex for C-KIT inhibition but not promoting any associative inhibition of Bcr-Abl and consequent efficacy against CML. Unlike imatinib, this methyl derivative, termed WBZ_4, did not affect the levels of brain natriuretic peptide, a marker of myocardial hypertrophy, in a mouse model (23, 24), implying a relative lack of toxicity. Similar wrapping concept for reduction of toxicity for other small molecule tyrosine kinase inhibitors (Table 8.1) such as sunitinib, where methylation in sunitinib at particular locations for turning the drug into a wrapper of dehydron at position T83-Q99, has also been proposed (25).

Promiscuous anticancer drugs have so far been limited primarily to the inhibitors of protein kinases involved in cellular signaling, both serine/threonine and tyrosine kinases. Such rationally designed inhibitors targeting ATP binding site or similar targets of a particular tyrosine kinase may also show cross-reactivity, having inhibitory activity against other cell signaling tyrosine kinases as well, as mentioned above. While this promiscuity allows the drug to attack other forms of cancer, side effects such as cardiotoxicity have become problematic (26–28).

The side effects and cardiotoxicity problems signify the importance of selective inhibition of protein kinases, including tyrosine kinases, associated with many forms of cancer, without affecting other kinases important for normal cellular physiological functions. Targeting the ATP binding site of a kinase, while quite effective in inhibiting the kinase activity by an inhibitor that competes with cellular ATP, creates the toxicity problems mentioned above, since such inhibitors often bind the ATP binding sites of other physiologically important kinases as well. An alternative approach that takes advantage of an inhibitor as an allosteric inhibitor, such as imatinib that stabilizes the inactive conformation of a kinase such as Bcr-Abl, acts by inducing conformational change in the active site motif Asp-Phe-Gly (DFG), preventing ATP-catalyzed phosphorylation (29). Sorafenib, when bound to the kinase domain of B-Raf,

TABLE 8.1. Tyrosine Kinase Inhibitory Drugs and Their Toxicities

Drug	Target	Cancer	Side Effects	Reference
Imatinib (Novartis, Washington, DC)	Bcr-Abl, C-KIT, PDGFR α/β	CML, GIST, CMML	Cardiotoxicity, edema, myelosuppression	(67)
Sunitinib (Pfizer [New York, NY]/SUGE)	VEGFR1,2,3, PDGFR α/β, KIT, FLT3, CSF1R, RET	RCC, GIST	Neutropenia, thrombocytopenia, anemia, hypertension, GI toxicity	(28, 68)
Sorafinib (Bayer [Leverkusen, Germany]/Onyx [Emeryville, CA])	VEGFR2,3, PDGFR β, FLT3, Raf1	RCC, Melanoma, HCC	Skin rash, hypertension, acute coronary syndromes	(26)
Trastuzumab (Genentech, San Francisco, CA)	ERBB2 (HER2)	$HER2^+$—breast cancer	Cardiotoxicity, neutropenia	(69)
Bevacizumab (Genentech/ Roche [Basel, Switzerland])	VEGFA	Colorectal cancer	Hypertension, proteinuria cardiotoxicity	(70)
Cetuximab* (Merck [Whitehouse Station, NJ]/ Imclone [Branchburg, NJ])	EGFR	Colorectal, squamous cell carcinoma of head/ neck	Skin rash, interstitial lung diseases, hypomagnesemia	(28)
Erlotinib* (Genentech/ Roche)	EGFR	NSCLC, pancreatic cancer	Skin rash, diarrhea, nausea, interstitial lung disease	(28)
Gefitinib* (Astrazeneca [London]/Teva [Petah Tikva, Israel])	EGFR	NSCLC	Skin rash, diarrhea, nausea, interstitial lung disease	(28)

*Low cardiotoxicity.

shows similar conformational change in the DFG motif, implying its inhibitory effect on protein kinase conformation.

Given the importance of protein kinases in cellular signaling, and the recent trend to develop rationally designed inhibitors against tyrosine kinases that are often hyperexpressed in specific cancer cells, an important consideration is to define whether an inhibitor truly acts against the particular kinase that it was screened on or whether it would recognize other kinases as well and inhibit their activity, at least to some extent. An important trend in this respect is to assay the inhibitory activity of the most promising drugs or potential drugs against a large number of kinases since more than 500 human genes encode protein kinases. Thus, the profiles of an entire group of compounds can be evaluated against a panel of cellular activities (30) to provide important information on the range of activities of various drugs on a large number of cellular enzymes including kinases. In a typical example, interaction maps of 38 kinase inhibitors, including some of the important anticancer drugs already in the market or undergoing late-phase clinical trials, were constructed against a panel of 317 kinases representing greater than 50% of the predicted human protein kinome (31). Each such compound was assayed at a single concentration of 10 μM against the kinases, and the equilibrium dissociation constant (K_d) values were determined. Such assays demonstrated that most of the compounds bound only a relatively small number of kinases with high affinity, an exception being sunitinib and dasatinib that bound greater than 15% of the kinases with high affinity, while others such as gefitinib, erlotinib, imatinib, and sorafenib had binding affinity about one-third lower (31). Such extended inhibitory activity against other tyrosine kinases has proved to be valuable in the evaluation of a drug against other cancers for which its efficacy was unknown. For example, the inhibitory activity of imatinib, normally targeting Bcr-Abl for CML, toward C-KIT led to the evaluation of its efficacy against GIST as well (32, 33). A downside, of course, is the fact that such cross-reactivity also leads to toxicity by affecting the kinase activities of cellular kinases that are important for normal cell function. The lack of selectivity of tyrosine kinase inhibitors has recently led to the development of fragment-based drug discovery, where small molecule compounds that bind weakly to a target are initially selected. High-affinity compounds are then made through optimization in combination with other fragments. For example, co-crystallization of various fragments with their targets can produce molecules that bind strongly to a particular target kinase but very weakly to other kinases, thus ameliorating toxicity problems.

Another approach has been to develop KINOMEscan methods (Ambit Biosciences, San Diego, CA) to assay for 402 kinases of the 518 protein kinases of the human kinome. In this assay, a potential inhibitory compound is combined with DNA-tagged human kinases and an immobilized ligand on a solid surface. After equilibration, any inhibitory compound that binds the kinase will compete with ligand binding so that the extent of kinase binding to the ligand, and therefore the extent of inhibition, can be measured by quantifying the amount of tagged kinase bound to the solid support by quantitative PCR. The potency of an inhibitor at various concentrations can then be evaluated in a statistically significant number of experiments against the 402 kinase panel to ensure the efficacy of the inhibitor against the target protein kinase but with insignificant activity against other cellular kinases. An ideal promiscuous drug, thus, would be one that inhibits a target kinase with a great deal of specificity. More ideally, such a promiscuous drug should not only inhibit a specific kinase to inhibit cell signaling, but should also inhibit other cellular proteins/enzymes that are important for rapid cancer growth-promoting angiogenesis or inhibiting apoptosis in cancer cells. Even more ideally, a promiscuous drug should not only inhibit cancer growth, but should also prevent its recurrence and attack other agents of diverse diseases such as HIV/AIDS or malaria.

PROMISCUOUS DRUGS TARGETING MULTIPLE STEPS IN CANCER GROWTH PROGRESSION

As discussed above, while receptor tyrosine kinases have been a major focus as validated targets in anticancer drug development, there are other targets that are also important since they modulate cancer growth either by preventing cancer apoptotic cell death or growth, as mediated by the tumor suppressor protein p53, or enhancing angiogenesis to provide nutrients, including oxygen, to rapidly growing cancer cells. Since angiogenesis also involves growth factors and their receptors, including receptor tyrosine kinases, some of the receptor tyrosine kinase inhibitors such as gefitinib can inhibit both cell signaling and angiogenesis. Thus, rational drug design has also involved inhibitor development for these diverse processes (Table 8.2). However, no rationally designed drugs can target all the three major processes that contribute to cancer growth, namely, tumor suppressor proteins such as p53-inducing apoptosis (enhancing its stabilization), angiogenesis, and tyrosine kinase-mediated cell signaling, as shown in Table 8.2 (34–38).

There is an exception to the above statement, which refers to a potential candidate drug. This candidate drug is, however, not a rationally designed drug

TABLE 8.2. Rationally Designed Anticancer Drugs

Target	Drug	Reference
Tumor suppressors (p53) (in p53-positive cancers)	Inhibitors of MDM2, nutlin 3	(34)
Angiogenesis	Avastin (bevacizumab; Genentech) inhibiting VEGF	(35)
	p53 status positively modulates anti-angiogenic activity, but not necessarily in clinical setting	(36, 71)
	Low Mw oral TK inhibitor SU11248 (sunitinib) inhibiting VEGFR1,2,3, PDGFR α, β, FLT3, KIT	(68)
Tyrosine kinase (cell signaling)	Sunitinib, Gleevec (imatinib; Novartis) inhibiting Bcr-Abl fusion protein and activated C-KIT	(37)
	Low Mw Tk inhibitor gefitinib targeting EGFR	(38)

as contemplated by the pharmaceutical industry but an intelligently designed potential candidate drug produced by pathogenic bacteria such as *Pseudomonas aeruginosa* or gonococci/meningococci (39, 40). This candidate drug is a water soluble, low molecular weight (14 kDa), copper-containing protein called azurin (41, 42) or an azurin-like protein called Laz (43). Both azurin and Laz not only have anticancer activity (41–43) but also have strong growth inhibitory activity for viruses, such as HIV-1, and parasites, such as the malarial parasite *Plasmodium falciparum* (44) and toxoplasmosis-causing parasite *Toxoplasma gondii* (45). Thus, a single protein such as azurin or Laz has the potential to be used as a drug against a multitude of diverse diseases such as cancers, HIV/AIDS, malaria, or toxoplasmosis, if proven nontoxic and efficacious in humans. What confers in azurin/Laz such unique characteristics? It has been proposed (39, 40) that azurin or similar proteins are produced by pathogenic bacteria that cause chronic infections in humans through long-term residence and biofilm formation to protect their turf from other invading disease agents such as cancers, viruses, or parasites that can either dislodge the bacteria through rapid growth or inhibitory compound production, or simply by killing the host to deprive the bacteria of their sanctuary. Because the bacteria have 3 billion years of evolutionary wisdom, they have designed multipurpose weapons such as azurin or Laz that has various domains that target the major surface antigens important for invasion or entry of parasites and viruses such as *P. falciparum* and HIV-1. Azurin can also target surface-exposed or intracellular proteins in cancer cells that promote cancer growth, not only to kill cancer cells but also to keep in check their rapid growth.

An example in this regard is the mode of action of azurin toward cancer cells. As previously mentioned, there are at least three critical steps in cancer progression pathway, namely, inhibition of cell death and promotion of cell division/growth through inhibition of the tumor suppressor protein p53, often through mutations in the p53 gene or epigenetic silencing. New blood vessel formation through angiogenesis is critical to supply nutrients to fast-growing cancer cells. Cell signaling and growth promotion through growth factors and receptor tyrosine kinases are important, which is the reason why many such kinases are targets for anticancer drug development (Tables 8.1 and 8.2). Azurin is known to have an impact on all three steps. Azurin enters preferentially to cancer cells (46, 47) and has been shown to stabilize tumor suppressor p53, leading to increased expression of pro-apoptotic Bax and Bax-dependent apoptosis in cancer cells (Figure 8.1) (41, 42). Several physical studies including isothermal calorimetry, docking study, free energy simulation, and single-molecule force spectroscopy have defined the nature of the complex between azurin and p53 (48–51). The specificity of azurin to enter preferentially in cancer cells, but not in normal cells, has been shown to be due to a domain of azurin spanning the amino acids 50–77. A chemically synthesized peptide comprising these 28 amino acids, termed P28, has not only such entry specificity in cancer cells (47, 52) but, similar to azurin, also demonstrates strong anti-angiogenic activity, leading to inhibition of cancer cell growth (52). While

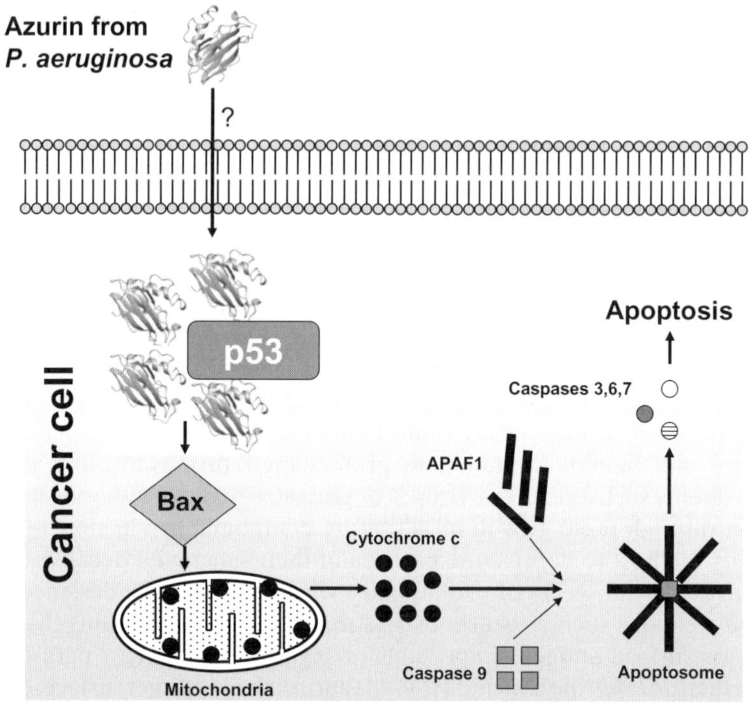

Figure 8.1 Azurin enters preferentially to cancer cells (46, 47), forms a complex with tumor suppressor protein p53 (48–51), thereby stabilizing p53, and induces apoptotic cell death, allowing cancer regression (41, 42).

azurin has been shown to have no toxicity in nude mice when given at 0.5–1.0 mg dosages for 22–28 days (41, 42), the synthetic P28 peptide has been shown not to have any toxicity in animals, including nonhuman primates, and is undergoing Phase I human clinical trials (http://clinicaltrials.gov/; FDA Investigational New Drug [IND] 77, 754). In addition to stabilization of tumor suppressor p53 through complex formation as well as having anti-angiogenic activity, leading to cancer growth inhibition, azurin also inhibits cancer cell growth through interference in the phosphorylation of a receptor tyrosine kinase EphB2 that is often hyperexpressed in cancer cells (Figure 8.2) (53). EphB2 mediates cell signaling, leading to cancer cell growth through its interaction with the ligand ephrinB2. Azurin has structural similarity to ephrinB2. Unlike rationally designed tyrosine kinase inhibitors, such as imatinib, sunitinib, and so on, that act through binding the ATP binding pockets of the kinases (21–26, 29), azurin inhibits EphB2 tyrosine kinase by competitively inhibiting ligand ephrinB2 binding with its receptor EphB2 because of azurin's binding with EphB2 with a higher affinity (Figure 8.2). A domain of azurin separate from P28, azurin 88–113 (26-amino acid peptide P26) was also shown to bind EphB2 with a high affinity. Both azurin and P26 bound EphB2 with Kd values of 6 and 12 nM, while the ligand ephrinB2 bound EphB2 with a Kd

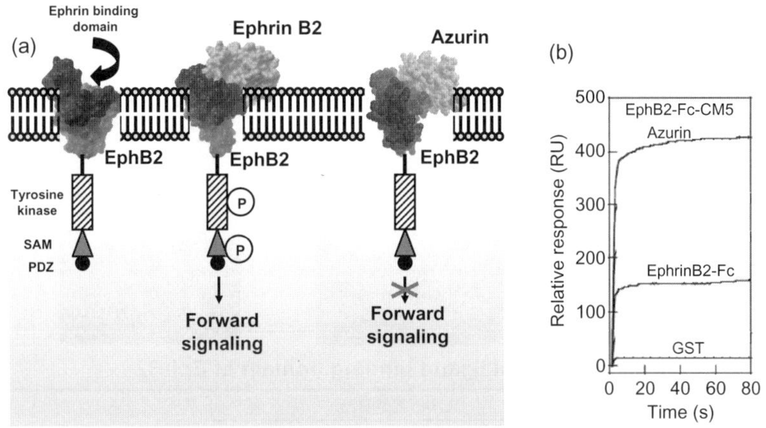

Figure 8.2 (A) Receptor tyrosine kinase EphB2 initiates cell signaling when bound with its ligand ephrinB2. Such ligand binding allows phosphorylation in the tyrosine residues, initiating forward cell signaling and promoting cancer growth. Azurin binds EphB2 with a higher affinity than ephrinB2, as measured (B) by surface plasmon resonance (53), leading to inhibition of forward cell signaling (A). GST, glutathione S-transferase.

value of 30 nM, demonstrating the ability of azurin and P26 to interfere in the ligand binding of EphB2, preventing EphB2's phosphorylation and inhibiting cancer cell growth (53). The ability of azurin or P26 to bind EphB2 has been shown to be due to the presence of a region 96–113 of azurin, which is structurally similar and corresponds to the G–H loop region of ephrinB2 (Figure 8.3). The G–H loop region of ephrinB2 is the region that mediates EphB2 receptor binding. This structural similarity or binding specificity of azurin or P26 to the G–H loop region of ephrinB2 is quite specific for ephrinB2 and not for other ligands of other receptor tyrosine kinases that are often involved in normal physiological processes. It is thus likely, though not proven, that azurin or P26 will not demonstrate cardiotoxicity or other forms of toxicity commonly exhibited by rationally designed drugs as mentioned earlier.

One of the most interesting effects of azurin or P28 is not only their ability to allow cancer regression, but also to prevent precancerous lesion formation. For example, in mouse mammary gland organ culture (MMOC) assays, cells from mammary glands from young, virgin animals are incubated for 6 days in the presence of insulin, prolactin, and aldosterone for alveolar or estrogen/progesterone for ductal cell growth. Such cells can then differentiate into fully grown glands. When exposed to the carcinogen dimethyl-benz-anthracene (DMBA), both the alveolar and ductal cells form preneoplastic lesions. Inhibition of such lesion formation is considered indicative of the efficacy of chemopreventive agents to prevent cancer. Interestingly, both azurin and P28 demonstrate significant inhibition of precancerous lesion formation in alveolar and ductal cells in the MMOC assay (54), suggesting that azurin or its truncated derivative P28 can not only be used for cancer therapy because of their

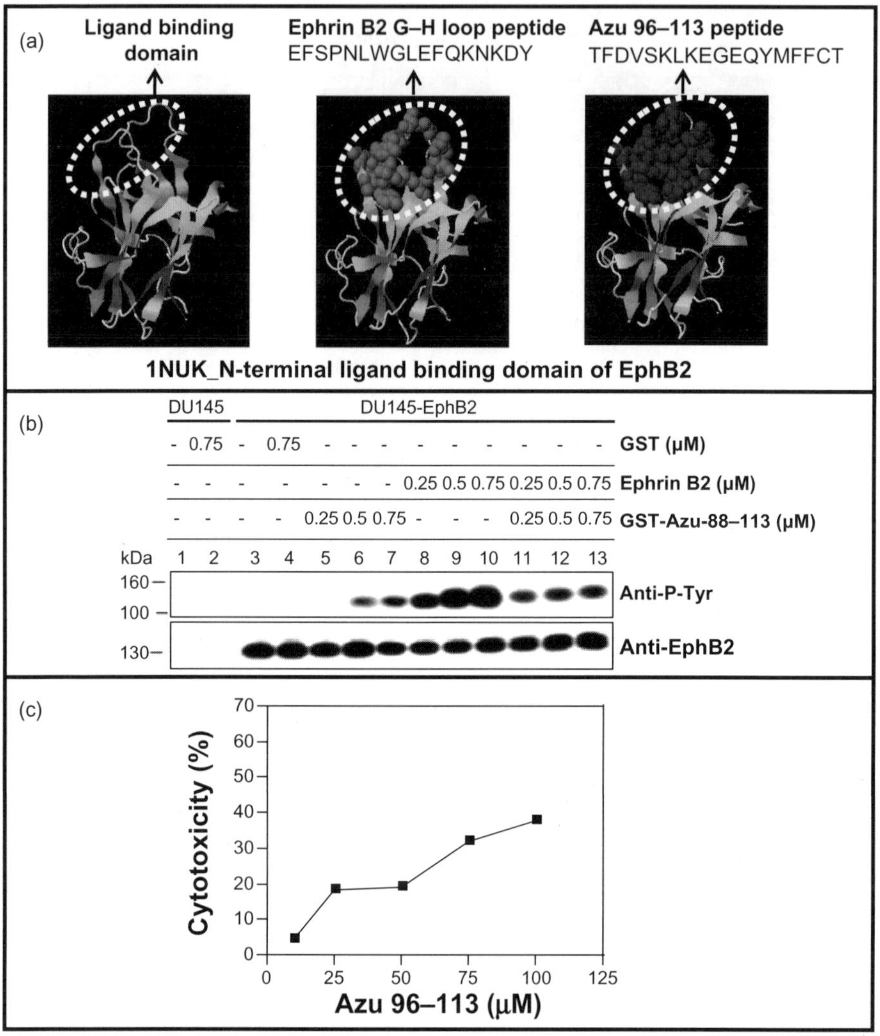

Figure 8.3 The EphB2 receptor-binding domain of azurin, having structural similarity to the G–H loop region of ephrinB2 (A), comprises residues 96–113 (53). A glutathione S-transferase (GST) fusion derivative of azurin 88–113 domain, termed GST-Azu-88–113, competitively inhibits ephrinB2-induced phosphorylation of EphB2 tyrosine residues (B), leading to dose-dependent inhibition (53) of cancer cells (C).

ability to interfere in multiple steps in the cancer progression pathways, but may also be used to prevent cancer emergence or recurrence, provided one or both of them prove to be nontoxic and efficacious in humans. Azurin not only interferes in multiple steps in cancer progression and may thus be less susceptible for drug resistance in cancer cells, but it also binds multiple agents

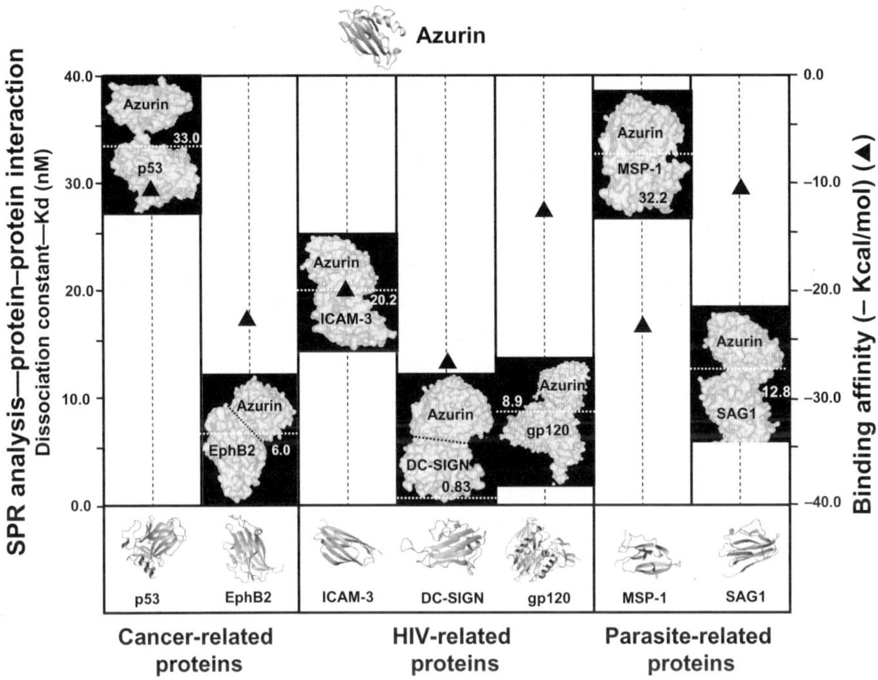

Figure 8.4 Protein–protein interaction studies by surface plasmon resonance (SPR), as well as other techniques (49–51), demonstrate complex formation of azurin with mammalian cancer-related cellular proteins such as tumor suppressor p53 or receptor tyrosine kinase EphB2. Similar complex formation with host cell surface proteins or HIV-1 envelope proteins (44) or parasite cell surface proteins (44, 45), such as *P. falciparum* MSP-1 or *T. gondii* SAG1, has been shown to interfere in viral and parasite growth by azurin, making it a highly promiscuous potential candidate drug (39, 40).

responsible for HIV-1 entry, including host proteins such as CD4, ICAM-3, and DC-SIGN that are important for HIV-1 entry in T cells (Figure 8.4). Blocking host functions, in addition to viral gp120 binding (44), is likely to block HIV-1 to mutate to become drug resistant. It is also important to note that P26, the peptide encompassing azurin amino acids 88–113, not only has anticancer activity (53), but also binds avidly to the dendritic cell surface protein DC-SIGN with a Kd value of 5.98 nM, while the Kd value of azurin binding to DC-SIGN is 0.83 nM (44). DC-SIGN is considered important for binding HIV-1 gp120 to transport the virus from the mucosal surface to the lymphatic T cells. Inhibition of transport of HIV-1 by P26 because of its binding to DC-SIGN could be important for preventing HIV-1 infections. Thus, azurin or its various domains demonstrate a degree of promiscuity that is extremely rare in rationally designed drugs of the pharmaceutical industry.

ANTICANCER ACTIVITY OF Pa-CASPASE RECRUITMENT DOMAIN AND Pa-ARGININE DEIMINASE

Azurin and Laz are not the only bacterial proteins with anticancer activity. An enzyme from *Mycoplasma arginini*, arginine deiminase (ADI), has been known for a long time to have anticancer activity against such cancers as melanoma, leukemia, hepatocellular carcinoma, and renal cell carcinoma (55–58). Such anticancer activity of ADI is believed to be due to its ability to degrade intracellular arginine in cancer cells, thus depleting arginine from such cells. Arginine is a nonessential amino acid for normal human cells, since ADI-promoted conversion of arginine to citrulline can be reversed by enzymes such as argininosuccinate synthetase (ASS) that convert citrulline back to arginine. However, certain cancers such as hepatocellular carcinoma or renal cell carcinoma are deficient in such enzymes as ASS, thereby making them uniquely vulnerable to arginine deprivation by ADI. *M. arginini* ADI, particularly its polyethylene glycol (PEG)ylated form to enhance its stability in blood, is currently undergoing human clinical trials and is discussed in detail in this volume by Feun et al. (Chapter 9). Most interestingly, ADI not only has anticancer activity, and many patents have been issued on its potential application in cancer therapy (59), but also demonstrates promiscuity in having antiviral activity as well against such viruses as HIV-1 and hepatitis C (60, 61).

An interesting aspect of ADI is its low, but discernable, structural similarity in its N-terminal domain with mammalian caspase recruitment domain (CARD) proteins (62). Such CARD proteins have a domain, called CARD, which is a protein–protein interaction motif. Thus, two or more mammalian CARD-carrying proteins can interact among themselves through the CARD-CARD domains, allowing activation or inhibition of cell signaling mediated by NF-κB or cellular processes controlling cell death (63, 64). The anticancer activity of *M. arginini* ADI is thought to be mediated by its enzymatic action that allows degradation of arginine to citrulline, thus depleting arginine in ASS-deficient cancer cells, leading to their growth inhibition (57). However, ADI has also been reported to inhibit angiogenesis in cancer cells and also inhibit the proliferation of cultured human lymphatic leukemia cells by inducing growth arrest in the G1/S phase, leading to apoptosis (55). Such activities, including antiviral activity, are not readily accounted for through arginine depletion by ADI enzymatic activity.

The presence of a CARD-like domain in the N-terminal of *M. arginini* ADI (62) and the known involvement of CARD proteins in promoting cancer growth, thus making CARD-carrying proteins a good target for anticancer drug development (64), raised an interesting question if the CARD-like domain of ADI might account, at least in part, to the anticancer activity of *M. arginini* ADI. Since nothing was known about the anticancer activity of ADI from other bacteria and since *P. aeruginosa* is known to produce ADI (65), Kwan et al. (66) cloned the sequence corresponding to the 5′-end of the *P. aeruginosa* ADI gene harboring the putative CARD domain and expressed it as a 17-kDa polypeptide termed Pa-CARD. The entire ADI protein, designated Pa-ADI, was also purified by Kwan et al. (66). As expected, Pa-ADI had the size of full-length ADI (46 kDa) with ADI enzymatic activity, while the truncated 17-kDa Pa-CARD had no ADI enzymatic activity, since the critical amino acids Cys-406 and His-278 in the catalytic triad of Pa-ADI were missing in Pa-CARD. In spite of the lack of ADI enzymatic activity, Pa-CARD was shown to have higher anticancer activity at identical concentrations than Pa-ADI against fibrosarcoma, breast cancer, and leukemia cells (66). Kwan et al. (66) also demonstrated that Pa-CARD, as well as Laz, inhibited cell cycle progression in the CML K562 cells at the G2/M phase. Both Pa-CARD and Laz enhanced Wee 1 protein level in the nuclear fraction of K562 cells but did not demonstrate such an activity in the acute myeloid leukemia (AML) HL60 cells. In the nuclear fractions of HL60 cells, and in the cytoplasmic fractions of K562 and HL60 cells, the levels of AKT, phosphorylated in the serine 473 residue (AKT-P-S473), were significantly reduced by Pa-CARD (66). Pa-CARD has also been shown by Kundu et al. (72) to induce apoptotic cell death in ovarian and other cancers through modulation of NF-κB-regulated gene expression.

The cell cycle arrest at the G2/M phase of leukemia cells and the inhibition of growth of other cancers, such as ovarian and breast cancer, by Pa-CARD (66, 72) raises many interesting questions about the mode of action of Pa-CARD. Does Pa-CARD exert its anticancer activity through CARD-CARD interactions with other CARD-carrying proteins in cancer cells that promote cancer growth? Since the CARD domains are found in eukaryotic cellular proteins, but not reported in bacterial proteins other than in ADI, are there other CARD domains in other proteins in *P. aeruginosa*, *M. arginini*, or other bacteria? If present, do such proteins interact with one another to mediate cell signaling in bacteria? Can Pa-CARD be further truncated to a size such as the 28-amino acid P28 from azurin with entry specificity in cancer cells and inhibiting cancer growth (47) that can be chemically synthesized for use as a drug? Does Pa-CARD have antiviral or antiparasitic activity similar to azurin? Further experiments in this regard will throw considerable light on the role of various bacterial proteins as anticancer agents.

REFERENCES

1. Shoshan M.C., and Linder, S. 2004. Promiscuous and specific anti-cancer drugs: combating biological complexity with complex therapy. Cancer Ther. **2**: 297–304.
2. Briz V., Poveda E., and Soriano, V. 2006. HIV entry inhibitors: mechanisms of action and resistance pathways. J. Antimicrob. Chemother. **57**: 619–627.
3. Simon V., Vanderhoeven J., Hurley A., Ramratnam B., Louie M., Dawson K., Parkin N., Boden D., and Markowitz M. 2002. Evolving patterns of HIV-1 resistance to antiretroviral agents in newly infected individuals. AIDS. **16**: 1511–1519.
4. Baselga, J., Carbonell, X., Castañeda-Soto, N.J., Clemens, M., Green, M., Harvey, V., Morales, S., Barton, C., and Ghahramani, P. 2005. Phase II study of efficacy, safety

and pharmacokinetics of trastuzumab monotherapy administered on a 3-weekly schedule. J. Clin. Oncol. **23**: 2162–2171.

5. Jonker D.J., O'Callaghan C.J., Karapetis C.S., Zalcberg J.R., Tu D., Au H.J., Berry S.R., Krahn M., Price T., Simes R.J., Tebbutt N.C., van Hazel G., Wierzbicki R., Langer C., and Moore M.J. 2007. Cetuximab for the treatment of colorectal cancer. N. Engl. J. Med. **357**: 2040–2048.
6. Schrama, D., Reisfeld, R.A., and Becker, J.C. 2006. Antibody targeted drugs as cancer therapeutics. Nat. Rev. Drug Discov. **5**: 147–159.
7. Milenic D.E., Brady E.D., and Brechbiel M.W. 2004. Antibody-targeted radiation cancer therapy. Nat. Rev. Drug Discov. **3**: 488–498.
8. Matar P., Rojo F., Cassia R., Moreno-Bueno G., Di Cosimo S., Tabernero J., Guzmán M., Rodriguez S., Arribas J., Palacios J., and Baselga J. 2004. Combined epidermal growth factor receptor targeting with the tyrosine kinase inhibitor gefitinib (2D1839) and the monoclonal antibody cetuximab (IMC-C225): superiority over single-agent receptor targeting. Clin. Cancer Res. **10**: 6487–6501.
9. Kris M.G., Natale R.B., Herbst R.S., Lynch T.J., Jr., Prager D., Belani C.P., Schiller J.H., Kelly K., Spiridonidis H., Sandler A., Albain K.S., Cella D., Wolf M.K., Averbuch S.D., Ochs J.J., and Kay A.C. 2003. Efficacy of gefitinib, an inhibitor of the epidermal growth factor receptor tyrosine kinase, in symptomatic patients with non-small cell lung cancer: a randomized trial. JAMA. **290**: 2149–2158.
10. Cohen E.E., Rosen F., Stadler W.M., Recant W., Stenson K., Huo D., and Vokes E.E. 2003. Phase II trial of ZD 1839 in recurrent or metastatic squamous cell carcinoma of the head and neck. J. Clin. Oncol. **21**: 1980–1987.
11. Pao W., Miller V., Zakowski M., Doherty J., Politi K., Sarkaria I., Singh B., Heelan R., Rusch V., Fulton L., Mardis E., Kupfer D., Wilson R., Kris M., and Varmus H. 2004. EGF receptor gene mutations are common in lung cancers from "never smokers" and are associated with sensitivity of tumors to gefitinib and erlotinib. Proc. Natl. Acad. Sci. U S A. **101**: 13306–13311.
12. Li B., Chang C.M., Yuan M., McKenna W.G., and Shu H.K. 2003. Resistance to small molecule inhibitors of epidermal growth factor receptor in malignant gliomas. Cancer Res. **63**: 7443–7450.
13. Dai Y., and Grant S. 2004. Small molecule inhibitors targeting cyclin-dependent kinases as anticancer agents. Curr. Oncol. Rep. **6**: 123–130.
14. Roberts T.G., Jr., Lynch T.J., Jr., and Chabner B.A. 2003. The phase III trial in the era of targeted therapy: unraveling the "go or no go" decision. J. Clin. Oncol. **21**: 3683–3695.
15. Fialho A.M., Das Gupta T.K., and Chakrabarty A.M. 2007. Designing promiscuous drugs? Look at what nature made! Lett. Drug Des. Discov. **4**: 40–43.
16. Hopkins A.L., Mason J.S., and Overington J.P. 2006. Can we rationally design promiscuous drugs? Curr. Opin. Struct. Biol. **16**: 127–136.
17. Sams-Dodd F. 2005. Target-based drug discovery: is something wrong? Drug Discov. Today. **10**: 139–147.
18. Lipton S.A. 2004. Turning down but not off. Neuroprotection requires a paradigm shift in drug development. Nature. **428**: 473.
19. Roth B.L., Sheffler D.J., and Kroeze W.K. 2004. Magic shotguns versus magic bullets: selectively non-selective drugs for mood disorders and schizophrenia. Nat. Rev. Drug Discov. **3**: 353–359.

20. Csermely P., Agoston V., and Pongor S. 2005. The efficiency of multi-target drugs: the network approach might help drug design. Trends Pharmacol. Sci. **26**: 178–182.
21. Fernandez A. 2007. Rational drug design to overcome drug resistance in cancer therapy: imatinib moving target. Cancer Res. **67**: 4028–4033.
22. Zhang X., Crespo A., and Fernandez A. 2008. Turning promiscuous kinase inhibitors into safer drugs. Trends Biotechnol. **26**: 295–301.
23. Fernández A., Sanguino A., Peng Z., Ozturk E., Chen J., Crespo A., Wulf S., Shavrin A., Qin C., Ma J., Trent J., Lin Y., Han H.D., Mangala L.S., Bankson J.A., Gelovani J., Samarel A., Bornmann W., Sood A.K., and Lopez-Berestein G. 2007. An anticancer C-Kit kinase inhibitor is reengineered to make it more active and less cardiotoxic. J. Clin. Invest. **117**: 4044–4054.
24. Crunkhorn S. 2008. Redesigning kinase inhibitors. Nat. Rev. Drug Discov. **7**: 120–121.
25. Crespo A., Zhang X., and Fernández A. 2008. Redesigning kinase inhibitors to enhance specificity. J. Med. Chem. **51**: 4890–4898.
26. Force T., Krause D.S., and Van Etten R.A. 2007. Molecular mechanisms of cardiotoxicity of tyrosine kinase inhibition. Nat. Rev. Cancer. **7**: 332–344.
27. Chu T.F., Rupnick M.A., Kerkela R., Dallabrida S.M., Zurakowski D., Nguyen L., Woulfe K., Pravda E., Cassiola F., Desai J., George S., Morgan J.A., Harris D.M., Ismail N.S., Chen J.H., Schoen F.J., Van den Abbeele A.D., Demetri G.D., Force T., and Chen M.H. 2007. Cardiotoxicity associated with tyrosine kinase inhibitor sunitinib. Lancet. **370**: 2011–2019.
28. Joensuu H. 2007. Cardiac toxicity of sunitinib. Lancet. **370**: 1978–1980.
29. Bogoyevitch M.A., and Fairlie D.P. 2007. A new paradigm for protein kinase inhibition: blocking phosphorylation without directly targeting ATP binding. Drug Discov. Today. **12**: 622–633.
30. Melnick J.S., Janes J., Kim S., Chang J.Y., Sipes D.G., Gunderson D., Jarnes L., Matzen J.T., Garcia M.E., Hood T.L., Beigi R., Xia G., Harig R.A., Asatryan H., Yan S.F., Zhou Y., Gu X.J., Saadat A., Zhou V., King F.J., Shaw C.M., Su A.I., Downs R., Gray N.S., Schultz P.G., Warmuth M., and Caldwell J.S. 2006. An efficient rapid system for profiling the cellular activities of molecular libraries. Proc. Natl. Acad. Sci. U S A. **103**: 3153–3158.
31. Karaman M.W., Herrgard S., Treiber D.K., Gallant P., Atteridge C.E., Campbell B.T., Chan K.W., Ciceri P., Davis M.I., Edeen P.T., Faraoni R., Floyd M., Hunt J.P., Lockhart D.J., Milanov Z.V., Morrison M.J., Pallares G., Patel H.K., Pritchard S., Wodicka L.M., and Zarrinkar P.P. 2008. A quantitative analysis of kinase inhibitor selectivity. Nat. Biotechnol. **26**: 127–132.
32. Heinrich M.C., Griffith D.J., Druker B.J., Wait C.L., Ott K.A., and Zigler A.J. 2000. Inhibition of c-kit receptor tyrosine kinase activity by STI571, a selective tyrosine kinase inhibitor. Blood. **96**: 925–932.
33. Joensuu H., Roberts P.J., Sarlomo-Rikala M., Andersson L.C., Tervahartiala P., Tuveson D., Silberman S., Capdeville R., Dimitrijevic S., Druker B., and Demetri G.D. 2001. Effect of the tyrosine kinase inhibitor STI571 in a patient with a metastatic gastrointestinal stromal tumor. N. Engl. J. Med. **344**: 1052–1056.
34. Vassilev L.T., Vu B.T., Graves B., Carvajal D., Podlaski F., Filipovic Z., Kong N., Kammlott U., Lukacs C., Klein C., Fotouhi N., and Liu E.A. 2004. In vivo activation

of the p53 pathway by small-molecule antagonists of MDM2. Science. **303**: 844–848.

35. Jubb A.M., Oates A.J., Holden S., and Koeppen H. 2006. Predicting benefit from anti-angiogenic agents in malignancy. Nat. Rev. Cancer. **6**: 626–635.
36. Ince W.L., Jubb A.M., Holden S.N., Holmgren E.B., Tobin P., Sridhar M., Hurwitz H.I., Kabbinavar F., Novotny W.F., Hillan K.J., and Koeppen H. 2005. Association of k-ras, b-raf, and p53 status with the treatment effect of bevacizumab. J. Natl. Cancer Inst. **97**: 981–989.
37. Sawyers C.L. 2003. Opportunities and challenges in the development of kinase inhibitor therapy for cancer. Genes Dev. **17**: 2998–3010.
38. Herbst R.S., Fukuoka M., and Baselga J. 2004. Gefitinib—novel targeted approach to treating cancer. Nat. Rev. Cancer. **4**: 956–965.
39. Fialho A.M., Stevens F.J., Das Gupta T.K., and Chakrabarty A.M. 2007. Beyond host-pathogen interactions: microbial defense strategy in the host environment. Curr. Opin. Biotechnol. **18**: 279–286.
40. Fialho A.M., Das Gupta T.K., and Chakrabarty A.M. 2008. Promiscuous drugs from pathogenic bacteria in the post-antibiotics era. In: Sleator R., and Hill C., eds. Patho-Biotechnology. Austin, TX: Landes Bioscience, pp. 145–162.
41. Yamada T., Goto M., Punj V., Zaborina O., Chen M.L., Kimbara K., Majumdar D., Cunningham E., Das Gupta T.K., and Chakrabarty, A.M. 2002. Bacterial redox protein azurin, tumor suppressor protein p53 and regression of cancer. Proc. Natl. Acad. Sci. U S A. **99**: 14098–14103.
42. Punj V., Bhattacharyya S., Saint-Dic D., Vasu C., Cunningham E.A., Graves J., Yamada T., Constantinou A.I., Christov K., White B., Li G., Majumdar D., Chakrabarty A.M., and Das Gupta T.K. 2004. Bacterial cupredoxin azurin as an inducer of apoptosis and regression in human breast cancer. Oncogene. **23**: 2367–2378.
43. Hong C.S., Yamada T., Hashimoto W., Fialho A.M., Das Gupta T.K., and Chakrabarty A.M. 2006. Disrupting the entry barrier and attacking brain tumors: the role *of the Nei*sseria H.8 epitope and the Laz protein. Cell Cycle. **5**: 1633–1641.
44. Chaudhari A., Fialho A.M., Ratner D., Gupta P., Hong C.S., Kahali S., Yamada T., Haldar K., Murphy S., Cho W., Chauhan V.S, Das Gupta T.K., and Chakrabarty A.M. 2006. Azurin, *Plasmodium* falciparum malaria and HIV/AIDS: inhibition of parasitic and viral growth by azurin. Cell Cycle. **5**: 1642–1648.
45. Naguleswaran A., Fialho A.M., Chaudhari A., Hong C.S., Chakrabarty A.M., and Sullivan W.J., Jr. 2008. Azurin-like protein blocks invasion of *Toxoplasma gondii* through potential interactions with parasite surface antigen SAG1. Antimicrob. Agents Chemother. **52**: 402–408.
46. Yamada T., Fialho A.M., Punj V., Bratescu L., Gupta T.K., and Chakrabarty A.M. 2005. Internalization of bacterial redox protein azurin in mammalian cells: entry domain and specificity. Cell. Microbiol. **7**: 1418–1431.
47. Taylor B.N., Mehta R.R., Yamada T., Lekmine F., Christov K., Chakrabarty A.M., Green A., Bratescu L., Shilkaitis A., Beattie C.W., and Das Gupta T.K. 2009. Noncationic peptides obtained from azurin preferentially enter cancer cells. Cancer Res. **69**: 537–546.
48. Yamada T., Hiraoka Y., Ikehata M., Kimbara K., Avner B.S., Das Gupta T.K., and Chakrabarty A.M. 2004. Apoptosis or growth arrest: modulation of tumor suppres-

sor p53's specificity by bacterial redox protein azurin. Proc. Natl. Acad. Sci. U S A. **101**: 4770–4775.

49. Apiyo D., and Wittung-Stafshede P. 2005. Unique complex between bacterial azurin and tumor-suppressor protein p53. Biochem. Biophys. Res. Commun. **332**: 965–968.
50. De Grandis V., Bizzarri A.R., and Cannistraro S. 2007. Docking study and free energy simulation of the complex between p53 DNA-binding domain and azurin. J. Mol. Recognit. **20**: 215–226.
51. Taranta M., Bizzarri A.R., and Cannistraro S. 2008. Probing the interaction between p53 and the bacterial protein azurin by single molecule force spectroscopy. J. Mol. Recognit. **21**: 63–70.
52. Mehta R.R., Taylor B.N., Yamada T., Beattie C.W., Das Gupta T.K., and Chakrabarty A.M. 2009. Compositions and methods to control angiogenesis with cupredoxins. U.S. Patent 7556810, issued July 7, 2009.
53. Chaudhari A., Mahfouz M., Fialho A.M., Yamada T., Granja A.T., Zhu Y., Hashimoto W., Schlarb-Ridley B., Cho W., Das Gupta T.K., and Chakrabarty A.M. 2007. Cupredoxin-cancer interrelationship: azurin binding with EphB2, interference in EphB2 tyrosine phosphorylation and inhibition of cancer growth. Biochemistry. **46**: 1799–1810.
54. Das Gupta T.K., and Chakrabarty A.M. 2009. Compositions and methods to prevent cancer with cupredoxins. U.S. Patent 7618939, issued November 17, 2009.
55. Gong H., Zolzer F., Von Recklinghausen G., Havers W., and Schweigerer L. 2000. Arginine deiminase inhibits proliferation of human leukemia cells more potently than asparaginase by inducing cell cycle arrest and apoptosis. Leukemia. **14**: 826–829.
56. Ensor C.M., Holtsberg F.W., Bomalaski J.S., and Clark M.A. 2002. Pegylated arginine deiminase (ADI-SS PEG 20,000 mw) inhibits human melanomas and hepatocellular carcinomas in vitro and in vivo. Cancer Res. **62**: 5443–5450.
57. Yoon C.Y., Shim Y.J., Kim E.H., Lee J.H., Won N.H., Kim J.H., Park I.S., Yoon D.K., and Min B.H. 2007. Renal cell carcinoma does not express argininosuccinate synthetase and is highly sensitive to arginine deprivation via arginine deiminase. Int. J. Cancer. **120**: 897–905.
58. Ni Y., Schwaneberg U., and Sun Z.H. 2008. Arginine deiminase, a potential antitumor drug. Cancer Lett. **261**: 1–11.
59. Fialho A.M., and Chakrabarty A.M. 2007. Recent patents on bacterial proteins as potential anticancer agents. Recent Pat. Anticancer Drug Discov. **2**: 224–234.
60. Kubo M., Nishitsuji H., Kurihara K., Hayashi T., Masuda T., and Kannagi M. 2006. Suppression of human immunodeficiency virus type 1 replication by arginine deiminase of *Mycoplasma arginini*. J. Gen. Virol. **87**: 1589–1593.
61. Izzo F., Montella M., Orlando A.P., Nasti G., Beneduce G., Castello G., Cremona F., Ensor C.M., Holtzberg F.W., Bomalaski J.S., Clark M.A., Curley S.A., Orlando R., Scordino F., and Korba B.E. 2007. Pegylated arginine deiminase lowers hepatitis c viral titers and inhibits nitric oxide synthesis. J. Gastroenterol. Hepatol. **22**: 86–91.
62. Das K., Butler G.H., Kwiatkowski V., Clark A.D., Jr., Yadav P., and Arnold E. 2004. Crystal structures of arginine deiminase with covalent reaction intermediates: implications for catalytic mechanism. Structure. **12**: 657–667.

63. Bouchier-Hayes L., and Martin S. 2002. CARD games in apoptosis and immunity. EMBO Rep. **3**: 616–621.
64. Damiano J.S., and Reed J.C. 2004. CARD proteins as therapeutic targets in cancer. Curr. Drug Targets. **5**: 367–374.
65. Lu X., Li L., Wu R., Feng X., Li Z., Yang H., Wang C., Guo H., Galkin A., Herzberg O., Mariano P.S., Martin B.M., and Dunaway-Mariano D. 2006. Kinetic analysis of *Pseudomonas aeruginosa* arginine deiminase mutants and alternate substrates provides insight into structural determinants of function. Biochemistry. **45**: 1162–1172.
66. Kwan J.M., Fialho A.M., Kundu M., Thomas J., Hong C.S., Das Gupta T.K., and Chakrabarty A.M. 2009. Bacterial proteins as potential drugs in the treatment of leukemia. Leuk. Res. **33**: 1392–1399.
67. Kerkelä R., Grazette L., Yacobi R., Iliescu C., Patten R., Beahm C., Walters B., Shevtsov S., Pesant S., Clubb F.J., Rosenzweig A., Salomon R.N., Van Etten R.A., Alroy J., Durand J.B., and Force T. 2006. Cardiotoxicity of the cancer therapeutic agent imatinib mesylate. Nat. Med. **12**: 908–916.
68. Faivre S., Demetri G., Sargent W., and Raymond E. 2007. Molecular basis for sunitinib efficacy and future clinical development. Nat. Rev. Drug Discov. **6**: 734–745.
69. Seidman A., Hudis C., Pierri M.K., Shak S., Paton V., Ashby M., Murphy M., Stewart S.J., and Keefe D. 2002. Cardiac dysfunction in the trastuzumab clinical trials experience. J. Clin. Oncol. **20**: 1215–1221.
70. Blowers E., and Hall K. 2009. Managing adverse events in the use of bevacizumab and chemotherapy. Br. J. Nurs. **18**: 351–358.
71. Yu J.L., Rak J.W., Coomber B.L., Hicklin D.J., and Kerbel R.S. 2002. Effect of p53 status on tumor response to antiangiogenic therapy. Science. **295**: 1526–1528.
72. Kundu M., Thomas J., Fialho A.M., Kwan J., Moreira L., Mahfouz M., Das Gupta T.K., and Chakrabarty A.M. 2009. The anticancer activity of the N-terminal CARD-like domain of arginine deiminase (ADI) from *Pseudomonas aeruginosa*. Lett. Drug Des. Discov. **6**: 403–412, in press.

9

ARGININE DEIMINASE AND CANCER THERAPY

LYNN FEUN,[1] M. TIEN KUO,[2] MING YOU,[1] CHUNG JING WU,[1] MEDHI WANGPAICHITR,[1] AND NIRAMOL SAVARAJ[1]

[1]*Hematology/Oncology, Sylvester Comprehensive Cancer Center, University of Miami School of Medicine, Miami, FL*
[2]*Molecular Pathology, M.D. Anderson Cancer Center, Houston, TX*

INTRODUCTION

Mycoplasma, the smallest of prokaryotic organisms, may alter host cell metabolism *in vitro*. This alteration in host cell metabolism may be utilized as a possible therapeutic method to treat human malignances. Historically, Copperman and Morton first noted reversible inhibition of mitosis in lymphocyte cultures by mycoplasma (1). Subsequently, Barile and Leventhal discovered that the depletion of the amino acid arginine was the mechanism for mycoplasma inhibition of lymphocyte transformation induced by phytohemagglutinin (2). Arginine has long been considered as an alternate energy source for nonglycolytic arginine-utilizing mycoplasma via arginine deiminase (3). Arginine deiminase produced by mycoplasma may appear in multiple forms (4). Cloning and sequence analysis of arginine deiminase gene from *Mycoplasma arginini* has been reported (5). The purified enzyme had a molecular weight of 46,000 Da and a specific activity of 20 units/mg protein. The nucleotide sequence of a 2189 bp fragment encoding the gene was determined. The open reading frame encoded arginine deiminase, a 385 amino acid polypeptide with a molecular weight of 43,900 Da (5).

Emerging Cancer Therapy: Microbial Approaches and Biotechnological Tools, Edited by Arsénio M. Fialho and Ananda M. Chakrabarty

Antitumor Activity

Arginine deiminase was identified by Sugimura et al. as the lymphocyte blastogenesis factor which originated from *M. arginini* (6). This led to the idea that arginine depletion via arginine deiminase may be a potential antitumor agent. In the same year, Miyazaki et al. identified arginine deiminase in culture of a Rous sarcoma virus–transformed liver cell line, RSV-BRL (7). When added to various cultures of human cancer cell lines, arginine deiminase inhibited their growth by depleting L-arginine in the culture media. This active dose was 1000 times lower than that observed using bovine liver arginase.

In a follow-up study, the same group demonstrated the *in vivo* antitumor activity of arginine deiminase purified from *M. arginini* (8). The purified enzyme was found to be composed of two identical subunits with a molecular mass of 45,000 Da. Importantly, the enzyme had its maximal activity at pH 6.0–7.5 and at 50°C and was stable at neutral pH. In tissue culture, arginine deiminase inhibited six different mouse tumor cell lines by depleting L-arginine. When arginine deiminase was injected intravenously into mice, the half-life of the enzyme was found to be 4 h. When injected intraperitoneally into mice, arginine deiminase prolonged the survival of the mice with all kinds of tumor cell lines, but particularly against the hepatoma cell line.

Hematologic Malignancies

In addition to solid tumors, arginine deiminase has demonstrated antitumor activity against a variety of lymphoma and leukemia cell lines. Table 9.1 shows a select list of lymphomas and leukemias in which arginine depletion by arginine deiminase has been shown to produce cell kill or inhibition. Based on these preclinical data, it appears that clinical trials in human lymphomas and leukemias with arginine depleting agents such as arginine deiminase may be indicated.

Arginine Deprivation and Cell Death

Thus, as noted by Wheatley, arginine deprivation or depletion such as caused by arginine deiminase can cause tumor cells to die because they cannot convert urea cycle intermediates into arginine (18–22). The administration of arginine deiminase represents a novel and targeted approach to cancer therapy. How does arginine depletion by enzymes such as arginine deiminase produce tumor cell death?

It is well-known that amino acid depletion is one method for treating certain cancers. The most notable example is asparaginase, an enzyme that lowers blood levels of the amino acid asparagine. Asparagine, like arginine, is considered to be a nonessential amino acid in humans. Asparaginase is especially active in acute lymphoblastic leukemia, which is the most common form of leukemia in children and young adults (23, 24). Most human body cells do not require asparagine, whereas lymphoblastic leukemia cells require this

TABLE 9.1. Inhibition of Leukemia and Lymphoma by ADI in Tissue Culture

	Cancer Cell Type	Reference
Murine	Lymphoblastic leukemia L5178Y	9
		10
		11
		12
		13
	Lymphoblastic leukemia L1210	13
		8
Human	T-lymphoma Jurkat	6
	T-cell leukemia TL-MOR	6
	T-cell leukemia MT-2	6
	B-cell lymphoma Raji	6
	B-cell lymphoma Manaca	6
	Histiocytic lymphoma U937	6
	T-lymphoblastoid	14
	Lymphoblastic leukemia	15
	Myeloid leukemia	15
	T-leukemia	16
	T-lymphoblastic leukemia CCRF-CEM	17

amino acid for their growth and survival. Hence, leukemic cells will die due to starvation, but most human body tissues will tolerate the drug well. Therefore, there is a historical precedence for effective amino acid depletion therapy in cancers which are auxotrophic for a particular amino acid.

Like asparagine, arginine is considered a nonessential amino acid in human adults. Arginine is involved in multiple pathways which are important in critical cellular functions such as nitric oxide production, creatine production, and synthesis of polyamines (25, 26). In humans, arginine is derived mainly from the diet, turnover of body proteins, and de novo synthesis via the intestinal–renal axis (26). Arginine is considered a semi-essential amino acid in adult humans since endogenous production is not adequate when cells are under stress or need to proliferate (27, 28). Thus, in cancer cells, arginine can influence their growth and proliferation, and diet restriction of arginine can inhibit tumor cell growth (20, 27, 29, 30). Arginine is synthesized from citrulline by two enzymes, argininosuccinate synthetase (ASS) and argininosuccinate lyase (ASL). Citrulline is converted to argininosuccinate by ASS. Argininosuccinate is then converted to arginine by ASL. Cancer cells that do not express ASS will be sensitive to arginine depletion by enzymes such as arginine deiminase. Conversely, cells that express ASS are not likely to be affected by arginine depleting enzymes.

Arginine deiminase converts L-arginine to L-citrulline and ammonia. As noted above, in human adults, arginine is not considered to be an essential amino acid. An arginine-deficient diet will not result in hyperammonemia or orotic aciduria, which is a theoretical concern (31–34). Interestingly, while mice appear to tolerate an arginine-free diet, other species such as dogs, rats, and cats require arginine in their diets (35).

Since arginine deiminase is produced by mycoplasma and does not exist normally in humans, it is recognized as a foreign protein by the body. Hence, the drug is immunogenic and has a very short half-life. Multiple and frequent injections are necessary to produce a prolonged depletion of arginine.

Pegylated Arginine Deiminase (ADI-PEG20)

One method of reducing immunogenicity and increasing the half-life of a drug is to link the drug to polyethylene glycol (PEG). A number of proteins and anticancer drugs have been pegylated (36, 37). These include PEG-interferon and asparaginase. Since asparaginase, like arginine deiminase, is produced by microbes, it has been formulated with PEG. This has reduced the frequency and number of injections needed for treatment for leukemia. The half-life of the drug is thus prolonged, the dosage needed is reduced, and consequently, the risk of anaphylactic reaction is less.

A number of chemical linkers for formulation of PEG with arginine deiminase have been tested (38). These PEGs can have varying effects on the immunogenicity, antigenicity, and pharmacokinetics of a protein. The chemical linker used for formulation of arginine deiminase with PEG is succinimidyl succinate, which is the linker also used for formulation of PEG-asparaginase and PEG-adenosine deaminase. Arginine deiminase has been cloned from *Mycoplasma hominis* and produced in *Escherichia coli* and then linked covalently to PEG 20,000 (termed ADI-PEG20 available from Polaris, Inc). This formulation of PEG to arginine deiminase results in approximately 50% of its specific enzyme activity, reduced immunogenicity, and increased circulating half-life in experimental animals (39). ADI-PEG20 is currently undergoing clinical trial in humans with various malignancies.

To determine which human cancers would best respond to arginine dep letion with ADI-PEG20, a number of tumor specimens were analyzed for ASS expression (40). Melanoma, hepatocellular carcinoma, and prostate carcinoma were the most frequent tumors deficient in ASS. These findings are consistent with the observation that melanoma and hepatocellular carcinoma are auxotrophic for arginine (8, 38, 41–44). Ensor et al. demonstrated that ADI-PEG20 inhibited human melanoma and hepatocellular carcinoma both *in vitro* and *in vivo* (45). These sensitive cell lines did not express ASS. However, when these cell lines were transfected with the expression plasmid containing ASS cDNA, they were more resistant to treatment with ADI-PEG20 than the parental cells *in vitro* and *in vivo*. While melanomas and hepatocellular carcinomas are often deficient in ASS, some cancers were almost invariably positive for ASS staining such as lung cancer and colon cancer (40). Some other cancers may be ASS deficient such as breast cancer, sarcomas, and renal cell cancer (40).

Solid Tumors

More recent studies suggest that other cancers may be sensitive to arginine depletion. Thus, Yoon et al. demonstrated that renal cell carcinoma does not express ASS and is highly sensitive to arginine depletion via arginine deiminase (46). While ASS was highly expressed in the epithelium of normal proximal tubules, it was not seen in tumor cells, and arginine deiminase showed remarkable growth retardation. Tumor angiogenesis and vascular endothelial growth factor (VEGF) expression were significantly reduced by arginine deiminase. Pancreatic cancer cell lines that were deficient in ASS expression also were highly sensitive to arginine deprivation (47) as well as prostate cancer cell lines (48). Retinoblastoma is a malignancy which represents a challenge when it is high-stage intraocular or has metastasized. A recent study indicated that arginine deiminase inhibited cell growth and induced cell death in retinoblastoma cells in a dose-dependent manner even with a high expression of ASS (49). Another cancer which may be difficult to treat is malignant mesothelioma. Patients with mesothelioma have a poor prognosis in general with a median survival of less than 1 year. Szlosarek et al. demonstrated downregulation of ASS mRNA in three of seven mesothelioma cell lines and absence of ASS protein in four of seven cell lines (50). The 9q34 locus, the site of the ASS gene, was found to be intact. Consistent with the cell line data, these investigators showed that 70% of patients (57 of 82) had tumors that did not express ASS protein. Tumor cell viability decreased markedly when arginine was removed from the media in ASS negative cell lines but not ASS positive cell lines. Highly significant Bax activation and mitochondrial membrane depolarization were found, indicating that arginine depletion triggered apoptosis of the ASS negative malignant mesothelioma cells. A clinical trial of ADI-PEG20 is planned in patients with ASS negative malignant mesothelioma.

CLINICAL STUDIES

A case report was published in which a patient with unresectable hepatocellular carcinoma was treated with ADI-PEG20 (51). The patient received escalating doses of ADI-PEG20 including successive treatment cycles at the optimal biologic dose of 160 IU/m^2. The tumor appeared to reduce in size and the serum alpha fetoprotein levels decreased. No apparent toxicity occurred in this single patient. As a result of the clinical response observed in this single patient, a cohort dose-escalation Phase I/II trial of ADI-PEG20 was conducted in patients with unresectable hepatocellular carcinoma (52). The study was performed at the outpatient clinic at the Pascale National Cancer Institute in Naples, Italy. In this trial, patients with advanced or metastatic inoperable hepatocellular carcinoma were eligible for treatment. Patients were enrolled onto one of four different cohorts. The first three cohorts were composed of three patients each at initial dose of ADI-PEG20 at 20, 40, or 80 IU/m^2 injected intramuscularly. Subsequently, patients were treated onto cohort 4 at an initial dose of 160 IU/m^2. The 160 IU/m^2 was determined to be the optimum biologic dose that lowered the plasma arginine level to undetectable amounts for 1 week. Patients were treated for three cycles at the optimum biologic dose as long as toxicity was acceptable to assess tumor response to ADI-PEG20. The initial cycle of therapy consisted of three treatments on days 1, 15, and 22. Subsequent cycles consisted of four treatments on days 1, 8, 15, and 22. Subsequent cycles of treatment started on day 36 of the preceding cycle. A total of 19 patients were enrolled onto the study. Of these 19 patients, 15 patients (79%) completed all cycles of therapy.

In terms of toxicity, no patients reported any serious adverse effects after treatment. The most common side effect was pain at the injection site which was mild and did not last beyond 3 days. The most common laboratory abnormalities were elevation of the serum uric acid and increases in fibrinogen. No episodes of gout or coagulation abnormalities attributed to the dose were found. It is of interest that of all the patients who had elevation of the serum uric acid, all but one had radiological evidence of tumor necrosis. The hyperuricemia was treated with intravenous urate oxidase. There were occasional elevation of the lipase and amylase, but no clinical effects such as pancreatitis were noted. Immunogenecity studies were performed using enzyme-linked immunosorbent assay (ELISA). No neutralizing antibody production was noted. There was a slight increase in anti-ADI-PEG20 titers which plateaued on day 20 but did not increase in the following weeks. No allergic reactions, either local or systemic, due to the ADI-PEG20 were noted.

Pharmacodynamics of a single intramuscular injection of ADI-PEG20 at a dose of 160 IU/m^2 showed that no patient had measurable arginine (detection limit less than 2 μM) in the blood circulation for the 7 days after treatment. In terms of antitumor response, two patients (10.5%) had a complete response by CAT scan, seven (36.8%) had a partial response, seven (36.8%) had stable disease, and three patients (15.9%) had progression of disease. The duration of response was noted to be >400 days (range, 37 to >680 days). In summary, this pilot study demonstrated that arginine depletion by ADI-PEG20 was feasible, appeared to be well tolerated, and had promising antitumor activity in patients with advanced hepatocellular carcinoma.

Another Phase I/II trial of arginine depletion therapy using ADI-PEG20 in unresectable hepatocellular carcinoma was performed by Delman et al. at MD Anderson Cancer Institute (53). Similar to the Italian study, the Phase I part involved escalating doses of ADI-PEG20 up to the biologic optimum dose of 160 IU/m^2 given intramuscularly weekly. The Phase II part of the study consisted of a starting dose of 160 IU/m^2 with a dose escalation permitted up to 240 IU/m^2 weekly. Thirty-five patients were entered into the study. Since this was a Phase I/II trial, antitumor response was not the end point of the study. In terms of responses observed, one patient who was considered unresectable prior to entering the study became resectable after therapy. Sixteen patients had stable disease, four patients did not finish the study due to either allergic reaction or intercurrent disease, and twenty-eight patients progressed. The mean time to tumor progression was 3.4 months (range 1–13 months). In terms of pharmacodynamics, all of the patients had undetectable plasma arginine levels (<2 μM) after dosing with ADI-PEG20. For the patient who became resectable after treatment with ADI-PEG20, death occurred after surgery due to portal vein thrombosis and subsequent liver failure. Autopsy showed minimal viable tumor left. The toxicity in this Phase I/II trial was greater than that reported for the Italian study. In the study at MD Anderson, 12 patients had grade 3 toxicity due mainly to liver dysfunction or electrolyte abnormalities, and 3 patients had grade 4 toxicity, with liver function abnormalities in two patients and one patient with elevation of serum lipase.

As part of the Italian study, on a compassionate basis, we treated nine patients with unresectable hepatocellular carcinoma with ADI-PEG20 (54). In terms of toxicity, most patients tolerated the treatment without major side effects except for one patient who appeared to have a grade 4 allergic reaction. This patient had been treated with ADI-PEG20 for over 2 years with stable disease. He developed grade 4 hypotension within 30 min after a treatment, which reversed quickly with intravenous fluids and steroids. Altogether, three patients (33%) had prolonged time to tumor progression of 4+, 17, and 28+ months. We studied ASS expression in tumor specimens as part of our research with hepatocellular carcinoma. Real-time reverse transcription polymerase chain reaction (RT-PCR) was performed using a set of ASS primer and glyceraldehyde 3-phosphate dehydrogenase (GADPH) primer which have similar efficiency of amplification. The ratio of GADPH and ASS was calculated. Normal human fibroblast (BJ-1) which expresses ASS and is resistant to ADI-PEG20 was used as a positive control, and a melanoma cell line which lack ASS expression was used as a negative control. Five tumor samples were obtained for ASS study. The GADPH/ASS ratio ranged from 0.31 to 31, while BJ-1 which is resistant to ADI-PEG20 has a ratio of 6. ADI-PEG20 induced autophagy in cell lines which possess low levels of ASS expression.

Altogether, the data suggest that arginine depletion using ADI-PEG20 has modest antitumor activity in patients with hepatocellular carcinoma with tolerable toxicity. Selection factors, performance status of the patient, tumor bulk, prior treatment, predisposing factors such as hepatitis B or C, or heavy alcoholism may all have played a role in the different response rates reported. It is important to note that none of these studies reported on the ASS status of the primary liver tumor. Since arginine deiminase would be expected to be effective in only tumors which lack ASS expression (55), it is important to try to select out which patients are most likely to benefit from this therapy.

Arginine depletion therapy with ADI-PEG20 has also been reported in malignant melanoma. A Phase I/II trial of this agent was performed in patients with advanced or metastatic melanoma (56). In this trial, two cohort dose-escalation studies were performed. A Phase I trial in the United States enrolled 15 patients. A Phase I/II trial was performed in Italy and enrolled 24 patients. The patients in Italy also received two subsequent cycles of therapy consisting of four once a week intramuscular injections at 160 IU/m^2. Pharmacodynamic studies demonstrated that the dose of 160 IU/m^2 was the biologic optimum dose which lowered the plasma arginine to nondetectable levels (<2 μM) for at least 7 days. Nitric oxide levels were also lowered. In terms of toxicity, no grade 3 or 4 toxicities were attributed to the drug. In the U.S. Phase I trial, no antitumor responses were observed. In the Italian Phase I/II trial, 6 of 24 patients had a response to treatment. Five of these responses were partial, and one was a complete response, for a total response rate of 25%. Fourteen patients had stable disease for at least one cycle of therapy, and 6 patients had stable disease for at least 3 months of the study. The median survival was 15 months with a mean survival time of 19.6 months with stage IV disease.

We have reported the preliminary results of a Phase II clinical and pharmacologic trial of ADI-PEG20 in patients with advanced or metastatic melanoma (57). Patients had weekly intramuscular injections of ADI-PEG20 at a starting dose of 160 IU/m^2. Serum arginine levels were determined every 2 weeks. The dose of ADI-PEG20 was escalated up to 320 IU/m^2 weekly if the arginine levels were still detectable. Tumor samples were assayed for ASS expression using immunohistochemical staining and/or PCR prior to start of therapy and at time of relapse. In the preliminary report, 17 patients were treated and evaluated. Eleven patients had M1c disease and 11 had prior chemotherapy and/or immunotherapy for advanced disease. Responses included partial response four patients (24%), minor response two (12%), and mixed response 2 (12%). Duration of partial response from start of therapy was 4(+), 6(+), 12, and 16 months. Sites of response included soft tissue/nodal and lung. Tumor tissue was available for ASS determination in 12 patients. ASS expression was associated with treatment failure and relapse. Interestingly, two patients had tumors which were ASS negative at the start of therapy, and they responded to therapy. At the time of relapse, tumor samples were obtained again, and the tumors stained ASS positive. The data suggest that arginine depletion using ADI-PEG20 has antitumor activity in malignant melanoma, and ASS expression may correlate with response. This is the first study to our knowledge which demonstrated that ASS expression in tumors may correlate with clinical response in patients with cancer to arginine deiminase. This study is continuing to accrue patients.

MECHANISMS OF ACTION

One mechanism of action of arginine deiminase (e.g., ADI-PEG20) is selective starvation in tumor cells auxotrophic for arginine. However, other mechanism may be as or more important. It is known that arginine is the only endogenous nitrogen-containing substrate of nitric oxide synthase and thus governs the production of nitric oxide. Arginine deiminase can inhibit nitric oxide production which may be the basis for other mechanisms of action. For example, nitric oxide can promote tumor growth and metastasis by stimulating tumor cell migration, invasiveness, and angiogenesis (58–64). The inhibition of nitric oxide production has multiple effects such as protection of mice from the lethal effects of tumor necrosis factor-α and endotoxin (65). The reduction in nitric oxide can also lower hepatitis C viral titers (66) which may be important in the treatment of hepatocellular carcinoma. Since nitric oxide appears to be important for the capillary leak syndrome induced by interleukin (IL)-2 (67), depletion of arginine by arginine deiminase may reduce the hypotensive effects associated with high-dose IL-2 administration and thereby augment this form of immunotherapy (68, 69).

Arginine deiminase may have direct effect on tumor cells but may also have indirect effects as well. Beloussow et al. showed that arginine deiminase had anti-angiogenic effects on cultured human umbilical vein endothelial cells (HUVEC) (70). Serum VEGF levels in animals treated with arginine deiminase are significantly reduced compared to control animals (62). Arginine deiminase can also potentiate the effects of radiation therapy on certain tumors such as neuroblastoma (62). Park et al. demonstrated that recombinant arginine deiminase could inhibit angiogenesis in a dose-dependent fashion which was reversible when arginine was added back to the culture media (60). Arginine deiminase was shown to effectively inhibit tumor growth in Chinese hamster ovary (CHO) and HeLa cancer cell lines (60).

Arginine deiminase can induce apoptosis in certain cell lines such as melanoma. It has been known that arginine depletion can inhibit protein synthesis, but the exact mechanism is not clear. Thus, Gong et al. reported apoptosis induced by arginine deiminase in human lymphoblastic cell lines (15, 71). We have also reported that arginine deiminase induced apoptosis in melanoma cell lines (72). We demonstrated that melanoma cells can undergo apoptosis when arginine is depleted by adding ADI-PEG20 to the media for 3 days. However, the cells may survive when ADI-PEG20 is removed and new media (containing arginine) is added. It appears that some melanoma cells can undergo autophagy when arginine is depleted, but prolonging arginine depletion eventually results in apoptosis. Indeed, our data showed increased LC3 II at 72 h upon treatment which can last for 5 days in certain melanoma cell lines, but apoptosis does occur eventually (72). In this regard, Szlosarek et al. demonstrated that apoptosis induced by arginine depletion in mesothelioma cell lines is related to activation of Bax (50) while Kim et al. has shown that apoptosis in prostate cancer cells upon arginine deprivation is caspase independent (48). Hence, the mechanism of apoptosis may be different among different tumor cells.

Nutrient deprivation such as amino acid depletion can inhibit mammalian target of rapamycin (mTOR) signaling (73–75). Also, it has been demonstrated that inhibition of mTOR can decrease cellular proliferation and lead to autophagy (76). One potential mechanism of arginine deiminase is the negative impact on mTOR signaling. Arginine depletion most likely produces a decrease in ATP. This subsequently leads to activation of 5′ adenosine monophosphate-activated protein kinase (AMPK) which negatively affects mTOR activity (77). Whether the tuberous sclerosis complex (TSC)1/2 complex participates is not clear (78). Amino acid depletion also positively influences the association of mTOR with raptor. When mTOR and raptor are tightly coupled, this will lead to decrease in mTOR activity (79). Our laboratory has shown that after exposure of ASS negative melanoma cell lines to arginine depletion using ADI-PEG20, the phosphorylation of 4E-BP is reduced strikingly in the majority of the cell lines (80). In addition, the amount of phosphorylated p70S6 kinase is also reduced at 72h. On the other hand, phospho-AMPK is highly elevated in certain melanoma cell lines, which negatively influences mTOR activity. These changes in 4E-BP phosphorylation and p70S6 kinase phosphorylation are not observed in cell lines such as non-small cell lung carcinoma which express ASS. Together, our data show that drugs such as ADI-PEG20 promote autophagy in ASS negative tumor cells via inhibition of mTOR. Consistent with our findings, Kim et al. (48) examined the effects of ADI-PEG20 on autophagy in prostate cell lines. They showed that LC3 translocation and cleavage occurred within hours of arginine depletion with ADI-PEG20 treatment. Apoptosis occurred at 96h after ADI-PEG20 exposure and was caspase independent. This suggests that autophagy is an early response to amino acid depletion. They also observed the activation of extracellular signal-related kinase (ERK)1/2 by ADI-PEG20 which has been demonstrated to regulate autophagy when cells are exposed to different stimuli (81, 82).

While autophagy can be a survival mechanism for tumor cells under metabolic stress such as amino acid deprivation, excessive autophagy can lead to cell death. On the other hand, inhibition of autophagy can also lead to apoptosis. Thus, drugs which can be added together to produce autophagy and/or inhibit autophagy may lead to increased tumor cell killing. Inhibition of autophagy by chloroquine and beclin1 siRNA knockdown enhanced ADI-PEG20-induced cell death in prostrate cancer cells (48). We have studied the addition of chloroquine with ADI-PEG20 to inhibit autophagy in melanoma cell lines (83) (Savaraj, unpublished data). Chloroquine was able to enhance cell killing in our melanoma cell lines, but the dose needed to achieve this exceeded the normal clinically achievable dose recommended. Thus, other agents to inhibit autophagy should be explored.

ASS EXPRESSION

The ASS gene is located on chromosome 9q34.1. At least 14 pseudogenes have been reported to reside among other chromosomes (84). The absence of ASS expression could be explained by deletion of this region in the chromosome. Our data in melanoma cell lines indicate that ASS DNA is present as determined by Southern blot analysis, while the ASS expression is low or absent. We (72) and others (45, 84) had shown that the lack of ASS protein corresponded with low levels of mRNA expression, which was undetectable by Northern blot analysis but is able to be detected by real-time RT-PCR.

Levels of ASS expression vary in different tissues. Moreover, expression levels of ASS can be modulated by various extracellular factors either in positive or negative manner. Several factors which may increase ASS expression include glucagon, glucocorticoids, and cyclic adenosine monophosphate (AMP) whereas fatty acids may negatively influence ASS expression (85–92). Interestingly, some agents such as insulin and growth hormone affect ASS in liver tissue but not in other tissues (93–95). Other agents which can affect ASS expression include cytokines, IL-1β, interferon, transforming growth factor-beta (TGF-β), glutamine, and arginine itself (96–99). Our laboratory has shown that arginine in the media can influence ASS expression in melanoma cells (72), while fatty acids and glutamine glucocorticoid have no effect.

Multiple mechanisms have been reported in the regulation of ASS expression by extracellular influences. A prior study indicated that ASS regulation occurs at the posttranslational level (26), but epigenetic and transcriptional regulation mechanisms have also been reported. The promoter sequences in humans and mice have been identified. Multiple transcription initiation sites for ASS gene expression have been identified, and it appears that different initiation sites are used in the expression of ASS in different tissues. Szlosarek et al. found aberrant promoter CpG methylation in mesothelioma cells which express low levels of ASS (84). They theorized that epigenetic regulation of ASS transcription plays a role in controlling ASS expression in their mesothelioma cell lines.

The 5′ flanking sequences of human ASS gene show multiple putative transcription factor binding sites, including Sp1, NF-κB, AP2, E-box, and Egr (100, 101). It has been shown that glutamine stimulates ASS expression in rat hepatocytes and Caco-2 cells through *O*-glycosylation of Sp1 sites (98). Stimulation of ASS expression in Caco-2 cells by IL-1β is mediated by the transcription factor NF-kB (102). Both Sp1 and NF-kB are suggested to be involved in the suppressive effects on the expression of ASS by tumor necrosis factor-alpha (TNF)-α (103). More recently, a cAMP responsive element located about 10kb upstream from the transcription start site of ASS gene has been identified to be responsible for liver-specific enhanced expression (104).

Expression of ASS not only affects tumor sensitivity to ADI-PEG20 treatment but also metabolic consequences. Arginine is used by nitric oxide synthase (NOS) to produce nitric oxide. Nitric oxide is involved in a wide spectrum of regulatory signaling in the cardiovascular (endothelial) system, whereas citrulline is recycled back to arginine in the liver where ASS is highly expressed. Previous studies on ASS regulation have been mainly focused on endothelial and hepatic cells. Research on regulation mechanisms in tumor cells is lacking. It is not known why different tumor cell lines express different levels of ASS. In particular, it is unknown why a majority of melanoma and hepatocellular carcinoma cells do not express ASS. Whether this is part of disease mechanisms associated with these malignancies remains to be investigated. Moreover, the mechanisms on the induction of ASS expression after arginine deiminase (ADI) treatment occurring in some melanoma cell lines but not in others need to be elucidated. Since reexpression of ASS in melanoma cells appears to be associated with the development of resistance to arginine depletion, understanding the underlying mechanisms is important for improving treatment efficacy using ADI-PEG20.

FUTURE DIRECTIONS

Arginine deiminase produced by mycoplasma has shown promise as an antitumor agent for malignancies such as malignant melanoma and hepatocellular carcinoma, and clinical trials are continuing in patients with these types of cancers. In addition, based on *in vitro* data, arginine deiminase may be active in patients with other cancers such as mesothelioma, prostate cancer, some renal cell cancers, and sarcomas, suggesting that arginine depleting therapy may have a broad application in cancer therapy. It is important to select the most appropriate patients for these trials (e.g., tumors which lack or have very low level of ASS expression are more likely to respond to treatment). Arginine deiminase may also potentially reduce nitric oxide-induced hypotension caused by high-dose IL-2, and clinical trials in combination with IL-2 seem appropriate. Recent laboratory investigations have demonstrated that combination therapy consisting of arginine deiminase which induces nutritional stress in tumor cells with other agents such as chemotherapy or drugs which inhibit autophagy may greatly potentiate cell killing. Thus, the future of arginine deiminase therapy may lie in combination drug treatment. Finally, understanding how ASS expression is regulated may be the key to preventing drug resistance in patients responding to arginine deiminase.

REFERENCES

1. Copperman R., and Morton H.E. 1966. Reversible inhibition of mitosis in lymphocyte cultures by non-viable Mycoplasma. Proc. Soc. Exp. Biol. Med. **123**: 790–795.
2. Barile M.F., and Leventhal B.G. 1968. Possible mechanism for Mycoplasma inhibition of lymphocyte transformation induced by phytohaemagglutinin. Nature. **219**: 750–752.
3. Fenske J.D., and Kenny G.E. 1976. Role of arginine deiminase in growth of *Mycoplasma hominis*. J. Bacteriol. **126**: 501–510.
4. Weickmann J.L., and Fahrney D.E. 1977. Arginine deiminase from *Mycoplasma arthritidis*. Evidence for multiple forms. J. Biol. Chem. **252**: 2615–2620.
5. Kondo K., Sone H., Yoshida H., Toida T., Kanatani K., Hong Y.M., Nishino N., and Tanaka J. 1990. Cloning and sequence analysis of the arginine deiminase gene from *Mycoplasma arginini*. Mol. Gen. Genet. **221**: 81–86.
6. Sugimura K., Fukuda S., Wada Y., Taniai M., Suzuki M., Kimura T., Ohno T., Yamamoto K., and Azuma I. 1990. Identification and purification of arginine deiminase that originated from *Mycoplasma arginini*. Infect. Immun. **58**: 2510–2515.

7. Miyazaki K., Takaku H., Umeda M., Fujita T., Huang W.D., Kimura T., Yamashita J., and Horio T. 1990. Potent growth inhibition of human tumor cells in culture by arginine deiminase purified from a culture medium of a Mycoplasma-infected cell line. Cancer Res. **50**: 4522–4527.
8. Takaku H., Takase M., Abe S., Hayashi H., Miyazaki K. 1992. In vivo anti-tumor activity of arginine deiminase purified from *Mycoplasma arginini*. Int. J. Cancer. **51**: 244–249.
9. Kenny G.E., and Pollock M.E. 1963. Mammalian cell cultures contaminated with pleuropneumonia-like organisms. I. Effect of pleuropneumonia-like organisms on growth of established cell strains. J. Infect. Dis. **112**: 7–16.
10. Kraemer P.M., Defendi V., Hayflick L., and Manson L.A. 1963. Mycoplasma (PPLO) strains with lytic activity for murine lymphoma cells in vitro. Proc. Soc. Exp. Biol. Med. **112**: 381–387.
11. Kraemer P.M. 1964. Interaction of Mycoplasma (PPLO) and murine lymphoma cell cultures: prevention of cell lysis by arginine. Proc. Soc. Exp. Biol. Med. **115**: 206–212.
12. Gill P., and Pan J. 1970. Inhibition of cell division in L5178Y cells by arginine-degrading mycoplasmas: the role of arginine deiminase. Can. J. Microbiol. **16**: 415–419.
13. Jones J. 1981. The Effect of Arginine Deiminase on Murine Leukemia Lymphoblasts. Oklahoma City, OK: University of Oklahoma.
14. Komada Y., Zhang X.L., Zhou Y.W., Ido M., and Azuma E. 1997. Apoptotic cell death of human T lymphoblastoid cells induced by arginine deiminase. Int. J. Hematol. **65**: 129–141.
15. Gong H., Zölzer F., von Recklinghausen G., Havers W., and Schweigerer L. 2000. Arginine deiminase inhibits proliferation of human leukemia cells more potently than asparaginase by inducing cell cycle arrest and apoptosis. Leukemia. **14**: 826–829.
16. Di Marzio L., Russo F.P., D'Alò S., Biordi L., Ulisse S., Amicosante G., De Simone C., and Cifone M.G. 2001. Apoptotic effects of selected strains of lactic acid bacteria on a human T leukemia cell line are associated with bacterial arginine deiminase and/or sphingomyelinase activities. Nutr. Cancer. **40**: 185–196.
17. Noh E.J., Kang S.W., Shin Y.J., Choi S.H., Kim C.G., Park I.S., Wheatley D.N., and Min B.H. 2004. Arginine deiminase enhances dexamethasone-induced cytotoxicity in human T-lymphoblastic leukemia CCRF-CEM cells. Int. J. Cancer. **112**: 502–508.
18. Wheatley D.N. 2005. Arginine deprivation and metabolomics: important aspects of intermediary metabolism in relation to the differential sensitivity of normal and tumour cells. Semin. Cancer Biol. **15**: 247–253.
19. Wheatley D.N., and Campbell E. 2002. Arginine catabolism, liver extracts and cancer. Pathol. Oncol. Res. **8**: 18–25.
20. Wheatley D.N., and Campbell E. 2003. Arginine deprivation, growth inhibition and tumour cell death: 3. Deficient utilisation of citrulline by malignant cells. Br. J. Cancer. **89**: 573–576.
21. Wheatley D.N. 2004. Controlling cancer by restricting arginine availability—arginine-catabolizing enzymes as anticancer agents. Anticancer Drugs. **15**: 825–833.

22. Wheatley D.N., Campbell E., Lai P.B.S., and Cheng P.N.M. 2005. A rational approach to the systemic treatment of cancer involving medium-term depletion of arginine. Gene Ther. Mol. Biol. **9**: 33–40.
23. Asselin B.L. 1999. The three asparaginases. Comparative pharmacology and optimal use in childhood leukemia. Adv. Exp. Med. Biol. **457**: 621–629.
24. Muller H.J. and Boos J. 1998. Use of L-asparaginase in childhood ALL. Crit. Rev. Oncol. Hematol. **28**: 97–113.
25. Wu G., and Morris S.M., Jr. 1998. Arginine metabolism: nitric oxide and beyond. Biochem. J. **336**: 1–17.
26. Husson A., Brasse-Lagnel C., Fairand A., Renouf S., and Lavoinne A. 2003. Argininosuccinate synthetase from the urea cycle to the citrulline-NO cycle. Eur. J. Biochem. **270**: 1887–1899.
27. Lind D.S. 2004. Arginine and cancer. J. Nutr. **134**: 2837S–2841S.
28. Morris S.M., Jr. 2006. Arginine: beyond protein. Am. J. Clin. Nutr. **83**: 508S–512S.
29. Caso G., McNurlan M.A., McMillan N.D., Eremin O., and Garlick P.J. 2004. Tumour cell growth in culture: dependence on arginine. Clin. Sci. (Lond). **107**: 371–379.
30. Gonzalez G.G., and Byus C.V. 1991. Effect of dietary arginine restriction upon ornithine and polyamine metabolism during two-stage epidermal carcinogenesis in the mouse. Cancer Res. **51**: 2932–2939.
31. Rose W. 1949. Amino acid requirements of man. Fed. Proc. 546–552.
32. Snyderman S.E., Boyer A., and Holt L.E., Jr. 1959. The arginine requirement of the infant. AMA J. Dis. Child. **97**: 192–195.
33. Barbul A. 1986. Arginine: biochemistry, physiology, and therapeutic implications. J. Parenter. Enteral. Nutr. **10**: 227–238.
34. Carey G.P., Kime Z., Rogers Q.R., Morris J.G., Hargrove D., Buffington C.A., and Brusilow S.W. 1987. An arginine-deficient diet in humans does not evoke hyperammonemia or orotic aciduria. J. Nutr. **117**: 1734–1739.
35. Rogers Q. 1994. Species variation in arginine requirements. In Proceeding from a Symposium Honoring Willard J. Visek—from Ammonia to Cancer and Gene Expression. Special Publication. **86**: 9–21.
36. Mehvar R. 2000. Modulation of the pharmacokinetics and pharmacodynamics of proteins by polyethylene glycol conjugation. J. Pharm. Pharm. Sci. **3**: 125–136.
37. Zalipsky S. and Lee C. 1992. Use of functionalized poly(ethylene glycol)s for modification of polypeptides. In: Harris J.M., ed. Poly(Ethylene Glycol) Chemistry: Biotechnical and Biomedical Applications. New York: Plenum Press, pp. 347–361.
38. Takaku H., Misawa S., Hayashi H., and Miyazaki K. 1993. Chemical modification by polyethylene glycol of the anti-tumor enzyme arginine deiminase from *Mycoplasma arginini*. Jpn. J. Cancer Res. **84**: 1195–1200.
39. Holtsberg F.W., Ensor C.M., Steiner M.R., Bomalaski J.S., and Clark M.A. 2002. Poly(ethylene glycol) (PEG) conjugated arginine deiminase: effects of PEG formulations on its pharmacological properties. J. Control Release. **80**: 259–271.
40. Dillon B.J., Prieto V.G., Curley S.A., Ensor C.M., Holtsberg F.W., Bomalaski J.S., and Clark M.A. 2004. Incidence and distribution of argininosuccinate synthetase deficiency in human cancers: a method for identifying cancers sensitive to arginine deprivation. Cancer. **100**: 826–833.

41. Scott L., Lamb J., Smith S., and Wheatley D.N. 2000. Single amino acid (arginine) deprivation: rapid and selective death of cultured transformed and malignant cells. Br. J. Cancer. **83**: 800–810.

42. Takaku H., Matsumoto M., Misawa S., and Miyazaki K. 1995. Anti-tumor activity of arginine deiminase from Mycoplas*ma arginini and its* growth-inhibitory mechanism. Jpn. J. Cancer Res. **86**: 840–846.

43. Sugimura K., Ohno T., Fukuda S., Wada Y., Kimura T., and Azuma I. 1990. Tumor growth inhibitory activity of a lymphocyte blastogenesis inhibitory factor. Cancer Res. **50**: 345–349.

44. Sugimura K., Ohno T., Kusuyama T., and Azuma I. 1992. High sensitivity of human melanoma cell lines to the growth inhibitory activity of mycoplasmal arginine deiminase in vitro. Melanoma Res. **2**: 191–196.

45. Ensor C.M., Holtsberg F.W., Bomalaski J.S., and Clark MA. 2002. Pegylated arginine deiminase (ADI-SS PEG20,000 mw) inhibits human melanomas and hepatocellular carcinomas *in vitro* and *in vivo*. Cancer Res. **62**: 5443–5450.

46. Yoon C.Y., Shim Y.J., Kim E.H., Lee J.H., Won N.H., Kim J.H., Park I.S., Yoon D.K., and Min B.H. 2007. Renal cell carcinoma does not express argininosuccinate synthetase and is highly sensitive to arginine deprivation via arginine deiminase. Int. J. Cancer. **120**: 897–905.

47. Bowles T.L., Kim R., Galante J., Parsons C.M., Virudachalam S., Kung H.J., and Bold R.J. 2008. Pancreatic cancer cell lines deficient in argininosuccinate synthetase are sensitive to arginine deprivation by arginine deiminase. Int. J. Cancer. **123**: 1950–1955.

48. Kim R.H., Coates J.M., Bowles T.L., McNerney G.P., Sutcliffe J., Jung J.U., Gandour-Edwards R., Chuang F.Y., Bold R.J., and Kung H.J. 2009. Arginine deiminase as a novel therapy for prostate cancer induces autophagy and caspase-independent apoptosis. Cancer Res. **69**: 700–708.

49. Kim J.H., Kim J.H., Yu Y.S., Kim D.H., Min B.H., and Kim K.W. 2007. Anti-tumor activity of arginine deiminase via arginine deprivation in retinoblastoma. Oncol. Rep. **18**: 1373–1377.

50. Szlosarek P.W., Klabatsa A., Pallaska A., Sheaff M., Smith P., Crook T., Grimshaw M.J., Steele J.P., Rudd R.M., Balkwill F.R., and Fennell D.A. 2006. Arginine depletion upregulates Bax and triggers apoptosis of malignant mesothelioma cells deficient in argininosuccinate synthetase. In: Proceedings of 97th Annual Meeting of the American Association for Cancer Research, April 1–5, Washington, DC. Philadelphia: AACR.

51. Curley S.A., Bomalaski J.S., Ensor C.M., Holtsberg F.W., and Clark M.A. 2003. Regression of hepatocellular cancer in a patient treated with arginine deiminase. Hepatogastroenterology. **50**: 1214–1216.

52. Izzo F., Marra P., Beneduce G., Castello G., Vallone P., De Rosa V., Cremona F., Ensor C.M., Holtsberg F.W., Bomalaski J.S., Clark M.A., Ng C., and Curley S.A. 2004. Pegylated arginine deiminase treatment of patients with unresectable hepatocellular carcinoma: results from phase I/II studies. J. Clin. Oncol. **22**: 1815–1822.

53. Delman K., Brown T., and Thomas M. 2005. PhaseI/II trial of pegylated arginine deiminase (ADI-PEG20) in unresectable hepatocellular carcinoma. In: Proceedings of 41st Annual Meeting of the American Society of Clinical Oncology, May 13–17, Orlando, FL. Alexandria, VA: ASCO.

54. Feun L., Savaraj N., Wu C., You M., Wangpaichitr M., Marini A., Levi D., Bomalaski J., and Kuo M.T. 2008. Clinical and pharmacologic study of ADI-PEG 20 in hepatocellular carcinoma. In: Proceedings of 44th Annual Meeting of the American Society of Clinical Oncology, May 30–June 3, Chicago, IL. Alexandria, VA: ASCO.
55. Shen L.J., Lin W.C., Beloussow K., and Shen W.C. 2003. Resistance to the antiproliferative activity of recombinant arginine deiminase in cell culture correlates with the endogenous enzyme, argininosuccinate synthetase. Cancer Lett. **191**: 165–170.
56. Ascierto P.A., Scala S., Castello G., Daponte A., Simeone E., Ottaiano A., Beneduce G., De Rosa V., Izzo F., Melucci M.T., Ensor C.M., Prestayko A.W., Holtsberg F.W., Bomalaski J.S., Clark M.A., Savaraj N., Feun L.G., and Logan T.F. 2005. Pegylated arginine deiminase treatment of patients with metastatic melanoma: results from phase I and II studies. J. Clin. Oncol. **23**: 7660–7668.
57. Feun L.G., Savaraj N., Marini A., Wu C., Robles C., Herrera C., Spector S., Luedemann K., Moffat F., and Bomalaski J. 2005. Phase 2 study of pegylated arginine deiminase (ADI-PEG20), a novel targeted therapy for melanoma. Proc. Am. Soc. Clin. Oncol. **23**: 236s.
58. Jadeski L.C., Hum K.O., Chakraborty C., and Lala P.K. 2000. Nitric oxide promotes murine mammary tumour growth and metastasis by stimulating tumour cell migration, invasiveness and angiogenesis. Int. J. Cancer. **86**: 30–39.
59. Ziche M., Morbidelli L., Masini E., Amerini S., Granger H.J., Maggi C.A., Geppetti P., and Ledda F. 1994. Nitric oxide mediates angiogenesis in vivo and endothelial cell growth and migration in vitro promoted by substance P. J. Clin. Invest. **94**: 2036–2044.
60. Park I.S., Kang S.W., Shin Y.J., Chae K.Y., Park M.O., Kim M.Y., Wheatley D.N., and Min B.H. 2003. Arginine deiminase: a potential inhibitor of angiogenesis and tumour growth. Br. J. Cancer. **89**: 907–914.
61. Shen L.J., Lin W.C., Beloussow K., Hosoya K., Terasaki T., Ann D.K., and Shen W.C. 2003. Recombinant arginine deiminase as a differential modulator of inducible (iNOS) and endothelial (eNOS) nitric oxide synthetase activity in cultured endothelial cells. Biochem. Pharmacol. **66**: 1945–1952.
62. Gong H., Pöttgen C., Stüben G., Havers W., Stuschke M., and Schweigerer L. 2003. Arginine deiminase and other antiangiogenic agents inhibit unfavorable neuroblastoma growth: potentiation by irradiation. Int. J. Cancer. **106**: 723–728.
63. Noh E.J., Kang S.W., Shin Y.J., Kim D.C., Park I.S., Kim M.Y., Chun B.G., and Min B.H. 2002. Characterization of mycoplasma arginine deiminase expressed in E. coli *and its* inhibitory regulation of nitric oxide synthesis. Mol. Cells. **13**: 137–143.
64. Lee J., Ryu H., Ferrante R.J., Morris S.M., Jr., and Ratan R.R. 2003. Translational control of inducible nitric oxide synthase expression by arginine can explain the arginine paradox. Proc. Natl. Acad. Sci. U S A. **100**: 4843–4848.
65. Thomas J.B., Holtsberg F.W., Ensor C.M., Bomalaski J.S., and Clark M.A. 2002. Enzymatic degradation of plasma arginine using arginine deiminase inhibits nitric oxide production and protects mice from the lethal effects of tumour necrosis factor alpha and endotoxin. Biochem. J. **363**: 581–587.
66. Izzo F., Montella M., Orlando A.P., Nasti G., Beneduce G., Castello G., sCremona F., Ensor C.M., Holtzberg F.W., Bomalaski J.S., Clark M.A., Curley S.A., Orlando R., Scordino F., and Korba B.E. 2007. Pegylated arginine deiminase

lowers hepatitis C viral titers and inhibits nitric oxide synthesis. J. Gastroenterol. Hepatol. **22**: 86–91.

67. Orucevic A., and Lala P.K. 1998. Role of nitric oxide in IL-2 therapy-induced capillary leak syndrome. Cancer Metastasis Rev. **17**: 127–142.
68. Kilbourn R.G., Fonseca G.A., Trissel L.A., and Griffith O.W. 2000. Strategies to reduce side effects of interleukin-2: evaluation of the antihypotensive agent NG-monomethyl-L-arginine. Cancer J. Sci. Am. **6**: S21–S30.
69. Kilbourn R.G., Fonseca G.A., Griffith O.W., Ewer M., Price K., Striegel A., Jones E., and Logothetis C.J. 1995. NG-methyl-L-arginine, an inhibitor of nitric oxide synthase, reverses interleukin-2-induced hypotension. Crit. Care Med. **23**: 1018–1024.
70. Beloussow K., Wang L., Wu J., Ann D., and Shen W.C. 2002. Recombinant arginine deiminase as a potential anti-angiogenic agent. Cancer Lett. **183**: 155–162.
71. Gong H., Zölzer F., von Recklinghausen G., Rössler J., Breit S., Havers W., Fotsis T., and Schweigerer L. 1999. Arginine deiminase inhibits cell proliferation by arresting cell cycle and inducing apoptosis. Biochem. Biophys. Res. Commun. **261**: 10–14.
72. Savaraj N., Wu C., Kuo M.T., You M., Wangpaichitr M., Robles C., Spector S., and Feun L. 2007. The relationship of arginine deprivation, argininosuccinate synthetase and cell death in melanoma. Drug Target Insights. **2**: 119–128.
73. Sarbassov D.D., Ali S.M., and Sabatini D.M. 2005. Growing roles for the mTOR pathway. Curr. Opin. Cell Biol. **17**: 596–603.
74. Tokunaga C., Yoshino K., and Yonezawa K. 2004. mTOR integrates amino acid- and energy-sensing pathways. Biochem. Biophys. Res. Commun. **313**: 443–446.
75. Guertin D.A., and Sabatini D.M. 2007. Defining the role of mTOR in cancer. Cancer Cell. **12**: 9–22.
76. Pattingre S., Espert L., Biard-Piechaczyk M., and Codogno P. 2007. Regulation of macroautophagy by mTOR and Beclin 1 complexes. Biochimie. **90**: 313–323.
77. Kimura N., Tokunaga C., Dalal S., Richardson C., Yoshino K., Hara K., Kemp B.E., Witters L.A., Mimura O., and Yonezawa K. 2003. A possible linkage between AMP-activated protein kinase (AMPK) and mammalian target of rapamycin (mTOR) signalling pathway. Genes Cells. **8**: 65–79.
78. Smith E.M., Finn S.G., Tee A.R., Browne G.J., and Proud C.G. 2005. The tuberous sclerosis protein TSC2 is not required for the regulation of the mammalian target of rapamycin by amino acids and certain cellular stresses. J. Biol. Chem. **280**: 18717–18727.
79. Nobukuni T., Joaquin M., Roccio M., Dann S.G., Kim S.Y., Gulati P., Byfield M.P., Backer J.M., Natt F., Bos J.L., Zwartkruis F.J., and Thomas G. 2005. Amino acids mediate mTOR/raptor signaling through activation of class 3 phosphatidylinositol 3OH-kinase. Proc. Natl. Acad. Sci. U S A. **102**: 14238–14243.
80. Feun L., You M., Wu C.J., Kuo M.T., Wangpaichitr M., Spector S., and Savaraj N. 2008. Arginine deprivation as a targeted therapy for cancer. Curr. Pharm. Des. **14**: 1049–1057.
81. Shinojima N., Yokoyama T., Kondo Y., and Kondo S. 2007. Roles of the Akt/mTOR/p70S6K and ERK1/2 signaling pathways in curcumin-induced autophagy. Autophagy. **3**: 635–637.

82. Pattingre S., Bauvy C., and Codogno P. 2003. Amino acids interfere with the ERK1/2-dependent control of macroautophagy by controlling the activation of Raf-1 in human colon cancer HT-29 cells. J. Biol. Chem. **278**: 16667–16674.
83. Carritt B., and Povey S. 1979. Regional asssignments of the loci AK3, ACONS, and ASS on human chromosome 9. Cytogenet. Cell Genet. **23**: 171–181.
84. Szlosarek P.W., Klabatsa A., Pallaska A., Sheaff M., Smith P., Crook T., Grimshaw M.J., Steele J.P., Rudd R.M., Balkwill F.R., and Fennell D.A. 2006. In vivo loss of expression of argininosuccinate synthetase in malignant pleural mesothelioma is a biomarker for susceptibility to arginine depletion. Clin. Cancer Res. **12**: 7126–7131.
85. Haggerty D.F., Spector E.B., Lynch M., Kern R., Frank L.B., and Cederbaum S.D. 1982. Regulation of glucocorticoids of arginase and argininosuccinate synthetase in cultured rat hepatoma cells. J. Biol. Chem. **257**: 2246–2253.
86. Gebhardt R., and Mecke D. 1979. Permissive effect of dexamethasone on glucagon induction of urea-cycle enzymes in perifused primary monolayer cultures of rat hepatocytes. Eur. J. Biochem. **97**: 29–35.
87. Lin R.C., Snodgrass P.J., and Rabier D. 1982. Induction of urea cycle enzymes by glucagon and dexamethasone in monolayer cultures of adult rat hepatocytes. J. Biol. Chem. **257**: 5061–5067.
88. Ulbright C., and Snodgrass P.J. 1993. Coordinate induction of the urea cycle enzymes by glucagon and dexamethasone is accomplished by three different mechanisms. Arch. Biochem. Biophys. **301**: 237–243.
89. Morris S.M., Jr., Moncman C.L., Rand K.D., Dizikes G.J., Cederbaum S.D., and O'Brien W.E. 1987. Regulation of mRNA levels for five urea cycle enzymes in rat liver by diet, cyclic AMP, and glucocorticoids. Arch. Biochem. Biophys. **256**: 343–353.
90. Nebes V.L., and Morris S.M., Jr. 1988. Regulation of messenger ribonucleic acid levels for five urea cycle enzymes in cultured rat hepatocytes. Requirements for cyclic adenosine monophosphate, glucocorticoids, and ongoing protein synthesis. Mol. Endocrinol. **2**: 444–451.
91. Husson A., Bouazza M., Buquet C., and Vaillant R. 1983. Hormonal regulation of two urea-cycle enzymes in cultured foetal hepatocytes. Biochem. J. **216**: 281–285.
92. Tomomura M., Tomomura A., Dewan M.A., and Saheki T. 1996. Long-chain fatty acids suppress the induction of urea cycle enzyme genes by glucocorticoid action. FEBS Lett. **399**: 310–312.
93. Grøfte T., Wolthers T., Jensen S.A., Møller N., Jørgensen J.O., Tygstrup N., Orskov H., and Vilstrup H. 1997. Effects of growth hormone and insulin-like growth factor-I singly and in combination on in vivo capacity of urea synthesis, gene expression of urea cycle enzymes, and organ nitrogen contents in rats. Hepatology. **25**: 964–969.
94. McLean P., and Novello F. 1965. Influence of pancreatic hormones on enzymes concerned with urea synthesis in rat liver. Biochem. J. **94**: 410–422.
95. Grøfte T., Jensen D.S., Grønbaek H., Wolthers T., Jensen S.A., Tygstrup N., and Vilstrup H. 1998. Effects of growth hormone on steroid-induced increase in ability of urea synthesis and urea enzyme mRNA levels. Am. J. Physiol. **275**: E79–E86.

96. Hattori Y., Shimoda S., and Gross S.S. 1995. Effect of lipopolysaccharide treatment in vivo on tissue expression of argininosuccinate synthetase and argininosuccinate lyase mRNAs: relationship to nitric oxide synthase. Biochem. Biophys. Res. Commun. **215**: 148–153.
97. Zhang W.Y., Gotoh T., Oyadomari S., and Mori M. 2000. Coinduction of inducible nitric oxide synthase and arginine recycling enzymes in cytokine-stimulated PC12 cells and high output production of nitric oxide. Brain Res. Mol. Brain Res. **83**: 1–8.
98. Brasse-Lagnel C., Fairand A., Lavoinne A., and Husson A. 2003. Glutamine stimulates argininosuccinate synthetase gene expression through cytosolic O-glycosylation of Sp1 in Caco-2 cells. J. Biol. Chem. **278**: 52504–52510.
99. Jackson M.J., Allen S.J., Beaudet A.L., and O'Brien W.E. 1988. Metabolite regulation of argininosuccinate synthetase in cultured human cells. J. Biol. Chem. **263**: 16388–16394.
100. Jinno Y., Matuo S., Nomiyama H., Shimada K., and Matsuda I. 1985. Novel structure of the 5′ end region of the human argininosuccinate synthetase gene. J. Biochem. (Tokyo). **98**: 1395–1403.
101. Anderson G.M., and Freytag S.O. 1991. Synergistic activation of a human promoter in vivo by transcription factor Sp1. Mol. Cell Biol. **11**: 1935–1943.
102. Brasse-Lagnel C., Lavoinne A., Fairand A., Vavasseur K., and Husson A. 2005. IL-1beta stimulates argininosuccinate synthetase gene expression through NF-kappaB in Caco-2 cells. Biochimie. **87**: 403–409.
103. Goodwin B.L., Pendleton L.C., Levy M.M., Solomonson L.P., and Eichler D.C. 2007. Tumor necrosis factor-alpha reduces argininosuccinate synthase expression and nitric oxide production in aortic endothelial cells. Am. J. Physiol. Heart Circ. Physiol. **293**: H1115–H1121.
104. Guei T.R., Liu M.C., Yang C.P., and Su T.S. 2008. Identification of a liver-specific cAMP response element in the human argininosuccinate synthetase gene. Biochem. Biophys. Res. Commun. **377**: 257–261.

10

CYTOSINE DEAMINASE/ 5-FLUOROCYTOSINE MOLECULAR CANCER CHEMOTHERAPY

Sergey A. Kaliberov and Donald J. Buchsbaum
University of Alabama at Birmingham, Birmingham, AL

Chemotherapy is widely used alone and in combination with surgery and radiotherapy for the treatment of cancer. The antimetabolite 5-fluorouracil (5-FU), first described more than five decades ago, is one of most commonly prescribed chemotherapy agents (1). 5-FU is a pyrimidine antagonist in which the hydrogen atom at the C-5 position of uracil is substituted by fluorine. Since the structure of 5-FU is similar to that of uracil, 5-FU is converted into biologically active metabolites, 5-fluoro-2′-deoxyuridine-5′-monophosphate (FdUMP) or 5-fluorouracil triphosphate (FUTP) as well as being metabolized to an inactive form, 5-fluoro-β-alanine, ammonia, and CO_2, by the same anabolic and catabolic pathways as that of uracil. 5-FU and its derivates remain an essential component of chemotherapy for a number of solid tumors, particularly gastrointestinal (GI) and head and neck malignancies, with modest success. For instance, the single-agent response rate of 5-FU in GI tumors is 10–20% (2). Data from seven randomized clinical trials of 5-FU-based adjuvant therapy for stage II and stage III colon cancer have shown a response rate of 30–40% and a toxicity of 20–30% (3). 5-FU has also been actively investigated during the last 40 years for many non-GI tumors. Examples of such non-GI tumors include head and neck cancer (4), recurrent and/or hormone-refractory localized prostate cancer (5), and malignant gliomas (6). 5-FU is able to diffuse across the cell membrane into adjacent cells without going through the gap

Emerging Cancer Therapy: Microbial Approaches and Biotechnological Tools, Edited by Arsénio M. Fialho and Ananda M. Chakrabarty

junctions, resulting in a significant bystander effect. In addition, a number of studies have demonstrated that a distant bystander effect by host immune responses mediated by natural killer (NK) cells and T cells contributes to the success of 5-FU-mediated therapy (7, 8). However, the role of systemic 5-FU in the therapy of these tumors has been limited by the fact that dose-limiting GI and hematological toxicities (myelosuppression and stomatitis) are usually reached before evidence of antitumor response. Potential explanations for this low therapeutic index include pharmacokinetic factors limiting accessibility of systemic 5-FU to tumor cells and inherent insensitivity of tumor cells to 5-FU. This high systemic toxicity of 5-FU may be overcome by using a gene-directed enzyme/prodrug therapy using the *codA* gene encoding the cytosine deaminase (CD) enzyme.

GENE-DIRECTED ENZYME/PRODRUG CANCER THERAPY

The selectivity of most drugs for malignant cells remains elusive. Unfortunately, an insufficient therapeutic index, a lack of specificity, and the emergence of drug-resistant cell subpopulations often hamper the efficacy of drug therapies. A major problem for cancer treatment is the presence of toxic side effects associated with chemotherapeutic agents that limit their efficacy. The heterogeneity of the tumor cell population is another major drawback. There is a need for the development of new alternative therapeutic strategies. Among these approaches, gene-directed enzyme/prodrug therapy or suicide gene therapy has been developed. The approach for the treatment of hepatocellular carcinoma involving retroviral-mediated enzyme/prodrug therapy using a chimeric varicella–zoster virus thymidine kinase was first described in 1991 (9). Generally, gene-directed enzyme/prodrug therapy involves the delivery of a specific enzyme that can produce cell death through the conversion of an inactive nontoxic prodrug into a cytotoxic drug metabolite. Specifically targeted expression of the prodrug-activating enzyme avoids systemic toxicity and results in a high drug concentration in the tumor mass and an improved therapeutic index compared to systemic drug administration. The key element of a gene-directed enzyme prodrug therapy system is a gene that encodes an enzyme, which converts a prodrug to an active cytotoxic drug. Importantly, prodrug-activating enzymes are normally absent or are poorly expressed in mammalian cells. Ideally, this means tumor targeting of gene therapy restricts enzyme expression to (1) the transduced tumor cells using specific delivery vehicles and (2) adjacent surrounding tumor cells through the diffusion of the drug metabolite to generate a bystander effect.

CD-MEDIATED GENE-DIRECTED ENZYME/PRODRUG CANCER THERAPY

One of the most widely studied gene-directed enzyme/prodrug systems is CD/5-fluorocytosine (5-FC), which has been investigated intensely during the last decade. CD (EC 3.5.4.1, cytosine aminohydrolase) belongs to the cytidine and deoxycytidylate family that is part of the superfamily of metallo-dependent hydrolases (also called amidohydrolase superfamily) that catalyzes the hydrolytic deamination (the removal of an amine group) from carbon number 4 of cytosine to uracil with the release of ammonia in the reaction. Also, CD converts 5-methylcytosine to thymidine. In *Escherichia coli*, CD (bacterial CD [bCD]) is assembled into a hexamer that contains a catalytic iron and most closely resembles human adenosine deaminase. Importantly, CD is present in prokaryotes and fungi but not in higher eukaryotes, where it is an important member of the pyrimidine salvage pathway. 5-FC has been extensively used in the past as an antifungal agent for therapy of *Cryptococcus* and *Candida* infection and requires activation by the fungal CD. Therefore, its toxicological properties in humans are well established. As a part of a gene-directed enzyme/prodrug system, CD can convert the prodrug 5-FC into the chemotherapy agent 5-FU, which is further processed by cellular enzymes into either FUTP or FdUMP. 5-Fluorouracil triphosphate is incorporated into RNA and interferes with RNA processing, while FdUMP with the coenzyme N5,N10-methylene tetrahydrofolate (m-THF) forms a covalent ternary complex with thymidylate synthase (TS), an essential DNA de novo synthetic enzyme that catalyzes the methylation of deoxyuridine monophosphate (dUMP) to deoxythymidine monophosphate (dTMP). This irreversibly inhibits the conversion of dUMP to dTMP and hence the disruption of nucleotide pools of dTMP, a thymidine precursor in DNA synthesis (Figure 10.1).

The *codA* gene encoding bCD has been cloned from *E. coli* into eukaryotic expression vectors and was first used for gene-directed enzyme/prodrug therapy of mouse fibroblasts *in vitro* and in murine tumor models almost two decades ago (8). It was shown that CD-transfected tumor cells demonstrated significantly enhanced sensitivity to 5-FC *in vitro* and *in vivo*. In murine studies with CD-conjugated monoclonal (10), encapsulated CD protein (11), or tumor cells engineered to express the CD gene (12), it was shown that high intratumoral concentrations of 5-FU, with minimal diffusion back into the circulation, may be attained upon conversion of systemically administered 5-FC. Earlier experimental therapy studies revealed the antitumor activity of CD gene therapy in combination with 5-FC administration in several animal models, including mouse fibrosarcoma and adenocarcinoma (8), human colorectal cancer (12–14); hepatocellular carcinoma (15, 16), glioma (17, 18), metastatic murine adenocarcinomas (7), and hepatic metastasis of colon carcinoma (19).

GENE DELIVERY SYSTEM

An appropriate method of gene delivery is important for the efficacy and safety of gene-directed enzyme/prodrug therapy of cancer. The tumor-specific expression of CD and conversion 5-FC to 5-FU should decrease the systemic toxicity of molecular cancer chemotherapy. Gene therapy strategies are based on the

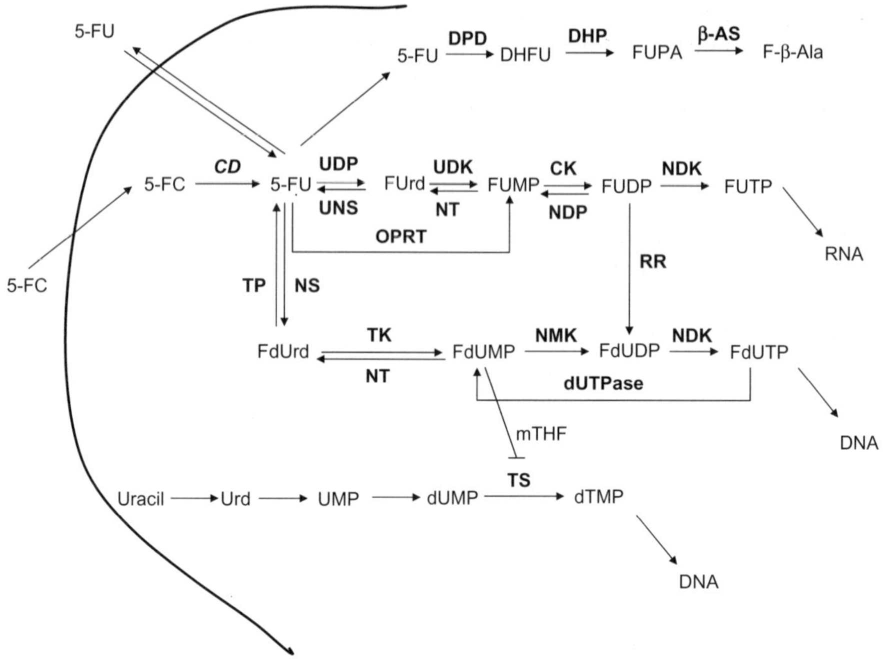

Figure 10.1 Schematic overview of the molecular mechanisms of CD/5-FC-mediated gene-directed enzyme/prodrug therapy and metabolism of 5-FU. CD converts the prodrug 5-FC into the chemotherapy agent 5-FU, which is further processed by cellular enzymes into either 5-fluorouracil triphosphate (FUTP) or FdUMP. These active metabolites produce cytotoxicity by several mechanisms including incorporation into DNA that leads to DNA damage, inhibition of thymidylate synthase (TS) and subsequently DNA de novo synthesis, and incorporation into RNA and interference of RNA processing. 5-FC, 5-fluorocytosine; 5-FU, 5-fluorouracil; CD, cytosine deaminase; CK, cytidylate kinase; DHP, dihydropyrimidinase; DPD, dihydropyrimidine dehydrogenase; dTMP, deoxythymidine monophosphate; dUMP, deoxyuridine monophosphate; dUTPase, deoxyuridine triphosphate nucleotidohydrolase; FdUMP, 5-fluoro-2′-deoxyuridine 5′-monophosphate; FUPA, α-fluoro-β-ureidopropionoic acid; FUTP, 5-fluorouracil triphosphate; F-β-Ala, 5-fluoro-β-alanine; m-THF, N5,N10-methylene tetrahydrofolate; NDK, nucleoside diphosphate kinase; NMK, nucleoside monophosphate kinase; NS, nucleosidase; NT, 5′-nucleotidase; NDP, nucleoside diphosphatase; OPRT, orotate phosphoribosyltrasferase; RR, ribonucleotide diphosphate reductase; TK, thymidine kinase; TP, thymidine phosphorylase; TS, thymidylate synthase; UDK, uridine kinase; UDP, uridine phosphorylase; UMP, uridine monophosphate; UNS, uridine nucleosidase; Urd, uridine; β-AS, β-ureidopropionase.

employment of genetically engineered vectors that promote the transfer of the transgene expression unit into the cancer cells and produce transient or stable expression of the therapeutic gene. Most commonly used in cancer gene therapy, recombinant viral vectors are biological systems derived from naturally evolved pathogens that, as part of the infection process, transfer their genomes into target tumor cells. At the present time, adenoviruses as well as retroviruses are the most commonly used cancer gene therapy vectors in CD/5-FC molecular chemotherapy with varying results. To date, a majority of gene therapy studies utilize only local intratumoral gene delivery such that gene expression is restricted to the area surrounding the injection site. This method is characterized by relatively low toxic effects as well as low therapeutic efficacy due to the limited delivery of the cytotoxic gene to each cancer cell. However, production of a bystander effect following CD/5-FC gene-directed enzyme/prodrug results in increased efficacy of cancer treatment. In an attempt to overcome low levels of transduction of vectors used in gene-directed enzyme/prodrug therapy, more advanced vector systems are being developed to improve CD gene delivery *in vivo*. Replication-competent retroviruses harboring the yeast CD (yCD) gene, conditionally replicating oncolytic adenoviruses as well as oncolytic adenoviral vectors encoding a CD/herpes simplex virus thymidine kinase (HSV-TK) fusion gene, were constructed for intratumoral application with high-risk tumors like glioblastoma and prostate carcinoma (20–22).

The high efficiency of yCD/uracil phosphoribosyltransferase (UPRT) fusion gene therapy for suppressing the growth of colon, breast, pancreatic, and hepatocellular carcinoma cells *in vitro* and LoVo human colon tumor xenografts was shown in yCD/UPRT/5-FC suicide gene therapy using the modified vaccinia virus Ankara in comparison with an E1/E3-deleted adenoviral vector encoding an identical expressing cassette (23). The recombinant vesicular stomatitis virus encoding the CD/UPRT fusion gene (rVSV-C:U) exhibited normal growth properties and generated high levels of biologically active CD/UPRT that could catalyze the conversion of 5-FC into 5-FU. Intratumoral inoculation of rVSV-C:U in the presence of the systemically administered prodrug 5-FC produced statistically significant reductions in the growth of syngeneic lymphoma (A20) or mammary carcinoma (TS/A) in BALB/c mice in comparison with injection of control rVSV vector without the therapeutic gene or 5-FU alone. Notably, prolonged and therapeutic levels of 5-FU in rVSV-C:U treated animals harboring TSA tumors were associated with activation of interferon (IFN)-gamma-secreting cytotoxic T cells (24). Recently, anaerobic bacteria *Bifidobacterium longum* was genetically modified to produce large amounts of bCD. It was shown that after systemic administration, *B. longum* selectively localized and proliferated within hypoxic regions of tumor. Injected i.v. or i.t. *B. longum* with bCD gene produced antitumor activity in rats bearing mammary tumors (25). An attenuated strain of *Salmonella typhimurium* expressing the *E. coli* CD gene was tested in refractory cancer patients without adverse effects and with evidence of intratumoral bacterial colonization and conversion of 5-FC to 5-FU (26). Calcium phosphate nanoparticles have been shown to successfully deliver therapeutic DNA into gastric cancer cells (27). In this study, the cytomegalovirus (CMV) enhancer and carcinoembryonic antigen (CEA) promoter was fused to a chimeric suicide gene yCDglyTK. The expression of yCDglyTK in CEA-positive cancer cells induced cell death following the addition of 5-FC and also produced a bystander effect. Intratumoral injection of nanoparticles containing the yCDglyTK complex followed by administration of 5-FC produced significant regression in gastric tumor xenografts.

In the past few years, the finding that human stem cells can localize to tumor tissues under physiological conditions as a default response to the tumor stroma formation opened the possibility to employ mesenchymal stem cells as an attractive cell-based therapeutic vehicle for cancer gene therapy. Genetically engineered mesenchymal stem cells derived from the bone marrow, from adipose tissues, as well as from neural stem cells were found to be an effective vehicle for drug delivery to tumors and for site-specific gene-directed enzyme/prodrug therapy. It was shown that embryonic endothelial progenitor cells carrying a CD/UPRT fusion gene targeted hypoxic lung metastases after intravenous delivery. In survival experiments, male C57/BL/6 mice were injected via the tail vein with 3LL-D122 Lewis lung carcinoma cells and were treated using a combination of i.v. injection of embryonic endothelial progenitor cells expressing CD/UPRT and i.p. administration of 5-FC. The embryonic endothelial progenitor cell-mediated CD/UPRT fusion gene therapy plus 5-FC significantly prolonged the survival of mice with multiple established lung metastases (28). Nude mice bearing intracranial Daoy medulloblastoma tumor cells were injected in the tumor site with HB1.F3-CD (immortalized clonal human neural stem cells transduced with bCD using a retroviral vector) followed by systemic 5-FC treatment, which resulted in a 76% reduction of tumor volume in the treated animals (29). A similar observation was shown with melanoma brain metastases using CD expressing neural stem/progenitor cells. Neural stem/progenitor cells transduced with a retroviral vector expressing CD gene targeted multiple sites of brain metastases in a syngeneic experimental melanoma model. Mice with established B16/F10 murine melanoma brain metastasis were injected intracranially with CD expressing neural stem/progenitor cells followed by systemic 5-FC administration, which resulted in a significant reduction in tumor growth (30).

BYSTANDER ACTIVITY OF CD/5-FC THERAPY

One of the major advantages of the CD/5-FC-mediated molecular chemotherapy system is a strong bystander effect that does not require gap junctions in contrast to other suicide gene therapy systems such as HSV-TK/ganciclovir (GCV), since 5-FU is a small, uncharged molecule capable of non-facilitated diffusion through cellular membranes (31–35). The bystander effect can be defined as an enlargement of cancer cell cytotoxicity in which non-transfected neighboring cells are killed through the transfer of metabolites of 5-FU from CD-expressing cells in close proximity. This means that even if only a small number of cancer cells express the CD gene, tumor eradication may still be achieved by using CD/5-FC molecular chemotherapy. Thus, the bystander activity is important for successful gene-directed enzyme/prodrug therapy due to the low transfection efficiency of gene delivery vectors currently employed for cancer gene therapy. In a colorectal carcinoma model, it was shown that 4% CD-expressing cells are sufficient to produce a 60% cure rate, while for the same level of regression to be achieved with the HSV-TK/GCV system, at least 50% of the cell population needed to be HSV-TK positive (31).

Generally, molecular mechanisms of therapeutic agent-mediated cytotoxicity can be divided into two groups that involve development of a local immune-mediated effect and distant bystander activity. The bystander effect may be produced via the diffusion of a soluble chemotherapy drug such as 5-FU to adjacent surrounding cells (12, 33, 36, 37), as well as via transfer of toxic products through gap junctions (38–41), or via apoptotic vesicles (42). It was shown that gap junctions produce restricted transfer due to the requirement of cell-to-cell contact, and the levels of intercellular gap junction communication are often decreased between tumor cells (39, 41, 43). Also, cancer cytotoxicity can be induced by cell-to-cell contact due to the transfer of apoptosis-inducing factors (42, 44). A significant antitumor effect was shown on human colorectal tumor cells after 5-FC treatment of a mixed cell culture of cancer cells transduced with the CD gene and untransfected cells (12). Significant amounts of 5-FU were detected in the culture medium of 5-FC-treated CD-positive cells. Bystander killing was also observed in cells negative for gap junctions (33).

Preclinical and clinical evidence points to the important role of cytotoxicity-induced immune response in the development of bystander action *in vivo*. The systemic immune stimulation does not only produce regression of local tumor but also induces a distant metastatic tumor growth inhibition. Evidence of strong inflammatory activity was demonstrated in the CD/5-FC molecular chemotherapy of murine mammary adenocarcinoma using immunocompetent BALB/c mice (7). On the other hand, it was shown that the bystander effect was significantly reduced in athymic mice (45–47). Significant bystander effect has been observed in other experimental therapy studies in animal models. It was shown that 5-FC treatment of human colorectal tumor xenografts composed of CD-expressing and CD-negative cancer cells in nude mice produced tumor regression even when only a small number of the inoculated cells were CD positive (12). As is the case with HSV-TK/GCV, it was shown that CD/5-FC molecular chemotherapy produces an immune-mediated distant bystander effect, associated with NK cell infiltration within the tumor mass. There was a significant immunity response against tumor rechallenge in rats treated with CD/5-FC. Animals bearing experimental liver metastases subsequently vaccinated with CD-expressing tumor cells s.c. or beneath the liver capsule followed by 5-FC treatment showed a 70% reduction in volume of the original liver tumor (48). On the other hand, the CD protein showed immunogenic properties; the employment of CD/5-FC molecular chemotherapy produced significant resistance of immunocompetent mice to rechallenge with wild-type tumor cells (7, 8, 34, 46).

THE MOLECULAR MECHANISMS OF RESISTANCE TO 5-FU

A major obstacle in chemotherapy to treat cancer is resistance to the therapeutic agents. There are generally two major forms of resistance encountered in clinical practice. One is intrinsic resistance, which is an innate property of the cancer cells. The other is acquired resistance, which develops subsequent to chemotherapy. It has been generally accepted that the efficacy of CD/5-FU molecular chemotherapy is limited clinically, in part, by the resistance of tumor cells to 5-FU. The fluoropyrimidines, including 5-FU, have been used in the clinic for the therapy of cancer with modest success due to acquired and intrinsic resistance to the therapeutic agents. Preclinical and clinical studies carried out over many years have revealed a number of possible mechanisms of resistance to 5-FU.

The contribution of the different pathways of 5-FU metabolism is still unclear. However, among the various enzymatic steps depicted in Figure 10.1, it was identified that reactions catalyzed by enzymes dihydropyrimidine dehydrogenase (DPD) (uracil reductase), TS, orotate phosphoribosyltrasferase (OPRT), uridine phosphorylase (UDP), thymidine phosphorylase (TP), and cytidylate kinase (CK) (uridine monophosphate [UMP]/cytidine monophosphate [CMP] kinase) play a significant role in fluoropyrimidine-mediated therapy. This is based on clinical data, which demonstrate that the levels of DPD, TS, OPRT, UDP, TP, and CK expression correlate with therapy response to 5-FU administration. Genetic factors at least partly explain variation in antitumor efficacy and toxicity of 5-FU. Individual variation in the activity of metabolizing enzymes can affect the extent of prodrug activation and, as a result, the efficacy of chemotherapy treatment.

The importance of DPD and TS activity for the antitumor effect of 5-FU has been generally accepted. DPD, a rate-limiting enzyme, catalyzes the major catabolic step in pyrimidine metabolism (Figure 10.1). It was shown that the catabolism of 5-FU is regulated by DPD, which rapidly converts 5-FU to the inactive metabolite dihydrofluorouracil (DHFU). When 5-FU is given to patients with genetic DPD deficiency or to those taking drugs that inhibit DPD activity, blood concentrations of the drug are markedly elevated, resulting in serious adverse effects (49). On the other hand, decreased DPD activity in combination with low TS expression in colorectal tumors was associated with an improved antitumor effect of 5-FU in terms of both response and survival after 5-FU-based chemotherapy (50). The levels of TS activity, the crucial enzyme of the cytotoxic activity of 5-FU, may predict response to 5-FU therapy. Intrinsic resistance to 5-FU has mostly been attributed to high levels of TS expression in tumors. Several studies have demonstrated that high levels of TS mRNA and protein expression generally predict for lack of response, while lower levels are correlated with response of colorectal cancer to chemotherapy (51–53). It has been shown that TS levels differ among metastatic sites. Colon cancer lung metastases were less responsive to fluoropyrimidine-based therapies and demonstrated higher levels of TS gene expression in comparison with hepatic metastases. Also, abdominal metastasis demonstrated a higher level of

TS gene expression in comparison with hepatic metastasis that correlated with therapeutic response (54, 55). In a few cases, the increase in TS activity was attributable to multiplication of copy number or as a result of increased levels of expression of the E2F-1 transcription factor (54). Polymorphisms in the TS promoter have also been reported to correlate with response. The data from a TS polymorphism study of patients with disseminated colorectal cancer suggest that genotyping for TS may have the potential to identify patients with a resistance to 5-FU phenotype, although the role of these mechanisms in human cancers is yet to be established (56). In earlier studies using mutagenesis analysis, several mutations in the TS gene were identified as being responsible for fluoropyrimidine and 5-FU drug resistance (57, 58).

Phosphorylation is necessary to convert 5-FU into active metabolites and involves several pathways. OPRT is a key enzyme in de novo UMP biosynthesis. It catalyzes the reaction between orotic acid and 5-phosphoribosyl-1-pyrophosphate (PRPP) to yield orotidine monophosphate (OMP), which is transformed to UMP by decarboxylation. OPRT phosphorylates 5-FU directly to 5-fluorouridine monophosphate (FUMP) and is the first limiting enzyme involved in the anabolism of 5-FU. Also, small quantities of 5-FU may be converted directly into 5-FUMP. The preferential activation of the OPRT pathway was associated with a higher sensitivity to 5-FU by cancer cells *in vitro* and human xenografts in animal therapy models. Activation of 5-FU to FUMP can be achieved in a sequence of reactions with conversion of 5-FU to 5-fluorouridine (FUrd) catalyzed by UDP in the presence of ribose-1-phosphate (Rib-1-P). UDP is an enzyme that can convert uracil to uridine. UDP gene expression appears featured with developmental regu lation and is strictly controlled at the promoter level by oncogenes, tumor suppressor genes, and cytokines. UDP activity is frequently elevated in various tumor tissues, and this induction appears to confer 5-FU therapeutic advantage to cancer patients. UDP is the most important phosphorylase identified in the regulation of uridine homeostasis and 5-FU activation, although TP and Rib-1-P can also utilize to a certain extent 5-FU as a substrate. It is still controversial as to whether OPRT or UDP expression is a key step to activate 5-FU and to predict the antitumor effect of pyrimidine therapy in patients.

CK (uridine-cytidine kinase) is a pyrimidine ribonucleoside kinase that catalyzes the phosphorylation of uridine and cytidine to UMP and CMP. This enzyme is important in de novo and salvage synthesis of pyrimidine nucleotides, and no other enzyme with the same substrate specificity as CK has been identified (59, 60). The enzyme also catalyzes the transfer of the phosphate group to FUMP using ATP as a cofactor and in the presence of magnesium (Figure 10.1). High frequency of low and undetectable levels of CK expression was shown in colon cancer patients with hepatic metastasis previously treated with 5-FU. The decreased levels of CK mRNA expression were correlated with the metastasis potential stage but not with primary resistance to 5-FU therapy (61). Analysis of CK expression showed decreased levels of mRNA expression in 5-FU-resistant cells (62).

COMBINATION OF CD AND UPRT INCREASES EFFECT OF GENE-DIRECTED ENZYME/PRODRUG CANCER THERAPY

The prodrug activation system formed by the *codA* gene encoding CD and 5-FC developed for selective cancer chemotherapy suffers from a sensitivity limitation in many tumor cells. In an attempt to improve the CD/5-FC killing efficiency, a combination of the *upp* gene encoding UPRT with the *codA* gene was used to create the situation prevailing in *E. coli*, a bacterium very efficient in metabolizing 5-FC. *E. coli* UPRT is a pyrimidine salvage enzyme that catalyzes the synthesis of uridine 5′-monophosphate (UMP), a precursor for pyrimidine nucleotide, from uracil and 5-phosphoribosyl α-diphosphate (PRPP). As 5-FU is directly converted into 5-FUMP by UPRT in microorganisms, transfer of the *upp* gene encoding UPRT into tumor cells reportedly induces a more efficient conversion of 5-FU into 5-FUMP, which results in restoring cell sensitivity to 5-FU and enhanced cytotoxicity of chemotherapy (Figure 10.2). In mammalian cells, which appear to lack UPRT, 5-FU is converted into 5-FdUMP via a two-step procedure that is activated only when 5-FU is present at high intracellular concentrations. Furthermore, a combination of UPRT with the CD/5-FC system has been reported to increase 5-FC killing of cells (63, 64). The results of our studies demonstrated that the Ad-CD:UPRT-mediated expression of *codA* and *upp* fusion genes in human pancreatic cancer and glioma cells generated a cooperative effect resulting in enhanced 5-FC sensitivity of cells compared to the expression of *codA* alone (65).

MUTATION OF CD SIGNIFICANTLY ENHANCES MOLECULAR CHEMOTHERAPY

A major problem associated with this gene-directed enzyme/prodrug therapy approach is the low affinity displayed by the CD gene product toward 5-FC in comparison with cytosine. Thus, high doses of this prodrug must be administered in order to achieve cell killing. The plasma levels of 5-FC required to obtain a significant amount of active metabolites may lead to adverse effects. This is observed with 5-FC, whereas deamination by CD of bacterial intestinal microflora into 5-FU is responsible for side effects. Recent studies have demonstrated that substitution of an alanine (A) for the aspartic acid (D) at position 314 of bCD (bCD-D314A) decreased efficiency for endogenous cytosine, which can compete with the prodrug for the active enzyme site in combination with increased efficiency of the mutant bCD-D314A enzyme for 5-FC that resulted in a 19-fold relative substrate preference for 5-FC in comparison with wild-type bCD (66, 67). Recent studies conducted in our laboratory showed that infection of human glioma and pancreatic cancer cells using an adenoviral vector (AdbCD-D314A) encoding this mutant bCD resulted in increased 5-FC-mediated cancer cell killing that correlated with significantly enhanced

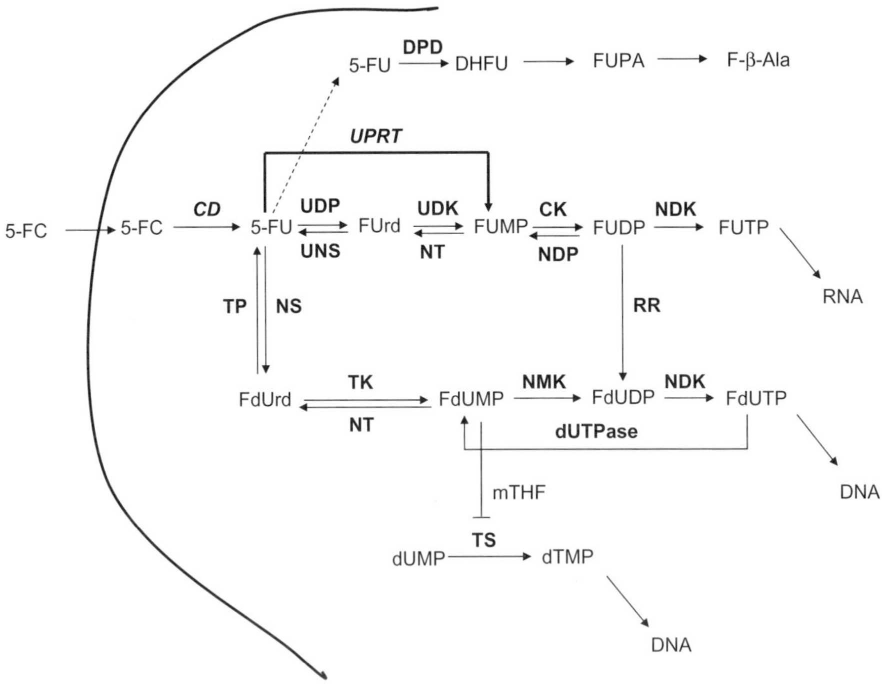

Figure 10.2 Molecular mechanisms of increased cytotoxicity with a combination of CD and UPRT cancer therapy. Uracil phosphoribosyltransferase (UPRT) converts 5-FU directly to 5-FUMP, which is further processed to its active metabolites 5-FdUMP and 5-FUTP. In part, this reduces degradation of 5-FU to DHFU through competition with DPD, which is involved in uracil and 5-FU catabolism. 5-FC, 5-fluorocytosine; 5-FU, 5-fluorouracil; CD, cytosine deaminase; CK, cytidylate kinase; DPD, dihydropyrimidine dehydrogenase; dTMP, deoxythymidine monophosphate; dUMP, deoxyuridine monophosphate; dUTPase, deoxyuridine triphosphate nucleotidohydrolase; FdUMP, 5-fluoro-2′-deoxyuridine 5′-monophosphate; FUPA, α-fluoro-β-ureidopropionoic acid; FUTP, 5-fluorouracil triphosphate; F-β-Ala, 5-fluoro-β-alanine; m-THF, N5,N10-methylene tetrahydrofolate; NDK, nucleoside diphosphate kinase; NMK, nucleoside monophosphate kinase; NS, nucleosidase; NT, 5′-nucleotidase; NDP, nucleoside diphosphatase; RR, ribonucleotide diphosphate reductase; TK, thymidine kinase; TP, thymidine phosphorylase; TS, thymidylate synthase; UDK, uridine kinase; UDP, uridine phosphorylase; UNS, uridine nucleosidase; UPRT, uracil phosphoribosyltransferase.

CD enzyme activity detected by measuring conversion of ^{3}H-5-FC to ^{3}H-5-FU, compared with AdbCDwt encoding wild-type bCD. Animal studies showed significant inhibition of growth of human glioma and pancreatic tumors treated with AdbCD-D314A/5-FC in comparison with AdbCDwt/5-FC. These results also were confirmed using the D54MG intracranial model for survival studies (68, 69).

yCD

In another approach of CD/5-FC gene-directed enzyme/prodrug therapy, yCD gene therapy has been developed. yCD belongs to the amidohydrolase protein fold family and shares homology with bacterial and eukaryotic cytidine deaminases. The enzyme is assembled into a homodimer that contains a catalytic zinc ion. Transient kinetic studies have shown that the deamination reaction is a rate-limiting step in the activation of the prodrug 5-FC by yCD (70). yCD has been observed to display superior kinetic properties toward 5-FC (corresponding to a 22-fold lower K_m for the prodrug) and a slightly improved efficacy for treating tumors in mice than the bCD derived from *E. coli* (71). The enhanced catalytic activity of yCD compared to wild-type bCD led to increased sensitivity to 5-FC *in vitro* in several tumor cell lines. In addition, this enhanced catalytic activity resulted in both increased tumor growth delay and more sustained response when yCD was compared to wild-type bCD in preclinical models. Although wild-type yCD is more efficient at converting 5-FC into the cytotoxic drug 5-FU than wild-type bCD, this enzyme loses all activity by 96 h at 37°C in contrast to bCD, which retains 100% of its activity at 168 h. This loss of activity may limit its performance in therapeutic applications (71). Recently, a number of mutant yCD variants that display elevated unfolding temperatures in denaturation experiments and increased half-lives of CD conversion activity at elevated temperatures were developed using computational protein engineering. These yCD mutants display wild-type yCD catalytic efficiencies and 5- and 30-fold increases in the half-lives of those catalytic activities. Furthermore, yCD double and triple mutants display apparent melting temperatures 6 and 10°C greater, respectively, than wild-type yCD. Also, the yCD triple mutant displayed increased catalytic activity at elevated temperatures in bacteria (72, 73).

COMBINATION OF CD/5-FC AND HSV-TK/GCV THERAPY

Tumor-targeted suicide gene expression and prodrug activation offer the unique possibility of a combination of gene-directed enzyme/prodrug therapy approaches to enhance the efficacy of anticancer therapy, without increasing systemic toxicity. Thus, it was reasoned that the coexpression of CD and HSV-TK, in conjunction with dual prodrug treatment, would be more effective in killing tumor cells than the employment of each enzyme/prodrug system independently. Several studies have integrated the CD/5-FC and HSV-TK/GCV strategies that, when combined together, have been more effective compared with the use of either strategy alone. It was shown that delivery of the HSV-TK gene to target cells resulted in the expression of viral thymidine kinase. The HSV-TK/GCV system is characterized by the effective phosphorylation of prodrug GCV into a toxic compound, GCV monophosphate. Monophosphorylated GCV is further phosphorylated by endogenous cellular kinases into an active triphosphate purine analogue and is incorporated into cellular DNA during its replication. This incorporation into guanine sites in newly synthesized DNA chains causes termination of synthesis and the selective killing of dividing cells by activation of apoptosis pathways. The new-generation bifunctional gene therapy system consisting of CD/5-FC and HSV-TK/GCV-mediated therapy has resulted in enhanced antitumor activity in cultured cancer cells *in vitro* and in several experimental therapy animal models. Delivery of CD:HSV-TK fusion genes followed by GCV and 5-FC treatment conferred upon glioma, mammary carcinoma, and prostate tumor cells prodrug sensitivity (74, 75) and radiosensitization (75, 76), equivalent to or better than that observed for each system independently. However, relatively little is known about the molecular mechanisms of interaction between CD/5-FC and HSV-TK/GCV-mediated pathways, but it appears to be related to the enhancement of GCV phosphorylation by HSV-TK after 5-FU treatment (77). Recently, it was shown that CD/5-FC-mediated reduction of 2′-deoxythymidine 5′-triphosphate (dTTP) was associated with decreased levels of 2′-deoxyguanosine-5′-triphosphate (dGTP) due to allosteric regulation of ribonucleotide reductase. dGTP is the endogenous competitor of GCV triphosphate. Thus, a reduced dGTP pool at the time of GCV addition into the cell can result in increased GCV incorporation into DNA, termination of DNA synthesis, and cell replication. Also, including 2′-deoxyguanosine during the 5-FC treatment reversed the depletion of endogenous dGTP, reduced the amount of GCV incorporated into DNA, and decreased the level of HSV-TK/GCV-mediated cytotoxicity in combination with CD/5-FC molecular chemotherapy (78). Importantly, animal therapy studies showed high efficacy of the combination of CD/5-FC and HSV-1 TK/GCV suicide gene therapy systems, which were further sensitized by radiation therapy (21, 74, 79–81). When administered with prodrugs, the Ad5-CD/TK*rep*, an adenoviral vector that expressed both CD and HSV-TK genes, converted these nontoxic agents into toxic metabolites, thereby providing a local chemotherapeutic effect. The safety of this gene-directed enzyme/prodrug therapy approach has been evaluated in three Phase I/II trials of prostate cancer, alone and in combination with conformal radiotherapy. The investigational therapy was associated with low toxicity, and signs of efficacy have emerged (81–84).

COMBINATION OF CD/5-FC-MEDIATED MOLECULAR CHEMOTHERAPY AND RADIATION THERAPY

It is unlikely that gene therapy, particularly CD gene-directed enzyme/prodrug therapy, would ultimately be used as a single treatment modality in patients. Another advantage of the CD/5FC combination is based on the extensive preclinical and clinical experience with 5-FU as both a chemotherapeutic agent and a strong radiation sensitizer. Several groups have reported *in vivo* results with CD to have a significant bystander effect at clinically relevant 5-FC doses and radiation regimens (79, 85–88). Intrinsic differences in sensitivity to 5-FU and ionizing radiation may contribute to the overall efficacy of molecular chemotherapy in combination with radiation therapy. The precise mechanisms underlying the interaction of radiation and 5-FU are not clear. However, there are a number of potential pathways by which 5-FU could increase radiation sensitivity at the cellular level. Although the disruption of either RNA or DNA synthesis can produce cytotoxicity, a substantial body of evidence suggests that radiosensitization is a result of TS inhibition (89, 90). It was shown that 5-FU produces tumor cell killing particularly during the S phase, when cells are relatively radioresistant (91). Also, the combination of 5-FU and radiation sensitization probably results in cells that are progressing inappropriately through the S phase due to misrepair of DNA damage imposed by ionizing radiation. This loss of the S-phase checkpoint in cancer cells may provide the molecular basis for the selective killing of tumors compared to normal tissues (92, 93). Ionizing radiation can generate reactive oxygen species through radiolysis of water that results in a multitude of oxidized base lesions, DNA strand breaks, and base loss, largely caused by hydroxyl radicals. The hydrolytic deamination of cytosine that gives a miscoding uracil residue is a frequent DNA lesion induced by oxidative damage, through exposure to reactive oxygen species-producing agents such as ionizing radiation. Cytostatic agents interfering with pyrimidine metabolism may enhance uracil incorporation by increasing the dUTP:dTTP ratio. Thus, cytostatic drugs inhibiting TS, for example, 5-FU, may enhance uracil incorporation and thus may contribute to the cytotoxicity of combined CD/5-FC gene-directed enzyme/prodrug therapy with radiation therapy. An analysis of the linear quadratic or single-hit-multi-target parameters of the survival curves indicated a significant reduction in cell survival at both low and high doses of radiation therapy (85, 86). The combination of CD/5-FC and ionizing radiation produced a measurable antitumor effect using 2 or 5 Gy per fraction over 1 week, which is a clinically relevant dose regimen of radiation therapy (86). It was shown that a significantly greater inhibition of growth of human pancreatic tumor xenografts was produced by the combination of AdbCD-D314A/5-FC with radiation as compared to either agent alone (69). Also, animal studies showed significant inhibition of subcutaneous or intracranial tumor growth of human glioma xenografts by the combination of AdbCD-D314A/5-FC with ionizing radiation, as compared to either agent alone, and with AdbCDwt/5-FC plus radiation. The results suggest that the combination of CD/5-FC with radiation produces markedly increased cytotoxic effects in cancer cells *in vitro* and *in vivo* (68).

CD/5-FC MOLECULAR CHEMOTHERAPY IN CLINICAL TRIALS

Currently, a number of clinical trials are being conducted using CD-mediated enzyme/prodrug therapy, including fusion genes with HSV-TK. A Phase I clinical trial involving local injection of a plasmid encoding the CD gene, regulated by the tumor-specific erbB-2 promoter, and systemic administration of 5-FC at 200 mg/kg/day demonstrated the safety of this gene-directed enzyme/prodrug therapy approach for the treatment of breast cancer. Using CD immunohistochemistry and mRNA *in situ* hybridization showed tumor-selective gene expression in 11 of 12 patients. In four patients, there was evidence of tumor regression, even though two of them did not receive the prodrug, which may be due to the immunogenicity of CD (94). Two Phase I clinical trials have reported the safety of the CD/5-FC combination. In one, the adenoviral vector expressing CD was administered to 18 patients diagnosed with metastatic liver disease associated with colorectal carcinoma, followed by oral administration of 5-FC. The other Phase I trial used a combination of oncolytic virotherapy, HSV-TK, and CD genes for the treatment high-risk prostate cancer. In an extensive series of preclinical studies and Phase I clinical trials, the replication-competent Ad5-CD/TK*rep* encoding the *CD:HSV-TK* fusion gene demonstrated potent tumor cell radiosensitization activity. Both trials showed signs of biological activity, and the prostate trial had two of 16 patients clear of carcinoma at 1-year follow-up (81, 95). Recently, patients with clinically localized prostate cancer were administered an intraprostatic injection of a replication-competent adenovirus, Ad5-yCD/*mut*TK_{SR39}*rep*-hNIS, encoding yCD and mutant HSV-TK suicide genes and armed with the sodium iodide symporter (NIS) gene. NIS gene expression was imaged noninvasively by uptake of $Na^{99m}TcO_4$ in infected cells using single photon emission computed tomography (SPECT). The investigational therapy was safe; NIS gene expression was detected in the prostate in seven of nine (78%) patients; levels of reporter gene expression intensity peaked 1–2 days after the adenovirus injection and were detectable in the prostate up to 7 days (96). A pilot trial has also been conducted using an attenuated strain of *S. typhimurium* for cellular delivery of CD rather than attempting *in vivo* modification of the tumor cells. Data showed evidence of bacterial colonization and increased levels of 5-FU without significant adverse events (26, 97).

CONCLUSION

The poor prognosis for patients with aggressive or metastatic tumors and systemic toxic side effects of currently available treatments including 5-FU require the development of more effective tumor-selective therapies. CD/5-FC-mediated enzyme/prodrug therapy, which concentrates the cytotoxic effect in the tumor site, may be one alternative. Targeted cancer gene therapy offers the possibility of a specific treatment that destroys tumor cells and metastases, but not normal tissues. Cancer-specific expression of the CD/5-FC prodrug-activating enzyme reduces systemic toxicity and results in high drug concentration in the tumor mass and in an improved therapeutic index compared to systemic drug administration. However, CD/5-FC molecular chemotherapy is still at the early stages of development, and some major problems remain to be solved before this approach becomes routinely adopted in the clinic.

Therapeutic efficacy is limited generally by insufficient tumor-specific expression of the therapeutic gene. Substantial efforts have been made to develop gene therapy vectors that are capable of targeting cancer cells. A wide variety of gene carriers, including viral and synthetic vectors, have been explored as CD gene transfer systems. Modifications of gene delivery systems, such as using tumor-specific promoters for selective expression of the CD gene and development of new oncolytic virotherapy vectors that are designed to selectively replicate in tumor cells, should significantly increase the efficacy and safety of cancer therapy. For instance, the evolution of several generations of vector systems from replication-deficient adenoviruses toward conditionally replicating vectors, which combined transcriptional or/and transductional targeting with therapeutic gene expression, significantly increases the efficacy of cancer gene therapy in experimental models. However, development of new vectors will continue to be critical for CD/5-FC gene-directed enzyme prodrug cancer therapy. In recent years, stem/progenitor cells that display inherent tumor-tropic properties have been successfully applied for targeted delivery of suicide genes to invasive and metastatic tumors.

Despite the promise of such therapy, the development of resistance to 5-FU could decrease the efficacy of CD/5-FC-mediated molecular chemotherapy. A number of studies elucidated the mechanisms for the development of resistance to 5-FU treatment. These include rapid degradation of 5-FU to nontoxic derivatives through the action of DPD (98, 99), different expression levels of enzymes involved in 5-FU phosphorylation (100), and amplification of the gene encoding TS (101, 102) or reduction in the synthesis of folylpolyglutamine (103). Also, poor efficiency of conversion of 5-FU into its toxic metabolites plays an important role in resistance to 5-FU cancer therapy. Thus, the low levels of DPD and/or TS enzyme activity, as well as a high level of OPRT and CK expression, are associated with enhancement of the antitumor effect of 5-FU.

Promising data have been generated in studies where CD/5-FC cancer gene therapy was employed in combination with chemotherapy or radiotherapy. Thus, it appears that a combination of CD/5-FC gene therapy approaches with conventional cancer treatments will increase the efficacy of treatment.

REFERENCES

1. Heidelberg C., Chaudhuri N.K., Danneberg P., Mooren D., Griesbach L., Duschinsky R., Schnitzer R.J., Pleven E., and Scheiner J. 1957. Fluorinated pyrimidines, a new class of tumour-inhibitory compounds. Nature. **179**: 663–666.
2. Sotos G.A., Grogan L., and Allegra C.J. 1994. Preclinical and clinical aspects of biomodulation of 5-fluorouracil. Cancer Treat. Rev. **20**: 11–49.
3. Gill S., Loprinzi C.L., Sargent D.J., Thomé S.D., Alberts S.R., Haller D.G., Benedetti J., Francini G., Shepherd L.E., Francois Seitz J., Labianca R., Chen W., Cha S.S., Heldebrant M.P., and Goldberg R.M. 2004. Pooled analysis of fluorouracil-based

adjuvant therapy for stage II and III colon cancer: who benefits and by how much? J. Clin. Oncol. **22**: 1797–1806.

4. Khuri F.R., Shin D.M., Glisson B.S., Lippman S.M., and Hong W.K. 2000. Treatment of patients with recurrent or metastatic squamous cell carcinoma of the head and neck: current status and future directions. Semin. Oncol. **27**: 25–33.
5. Russell P.J., and Khatri A. 2006. Novel gene-directed enzyme prodrug therapies against prostate cancer. Expert Opin. Investig. Drugs. **15**: 947–961.
6. Chambers R., Gillespie G.Y., Soroceanu L., Andreansky S., Chatterjee S., Chou J., Roizman B., and Whitley R.J. 1995. Comparison of genetically engineered herpes simplex viruses for the treatment of brain tumors in a scid mouse model of human malignant glioma. Up-regulation of urokinase and urokinase receptor genes in malignant astrocytoma. Proc. Natl. Acad. Sci. U.S.A. **92**: 1411–1415.
7. Consalvo M., Mullen C.A., Modesti A., Musiani P., Allione A., Cavallo F., Giovarelli M., and Forni G. 1995. 5-Fluorocytosine-induced eradication of murine adenocarcinomas engineered to express the cytosine deaminase suicide gene requires host immune competence and leaves an efficient memory. J. Immunol. **154**: 5302–5312.
8. Mullen C.A., Coale M., Lowe R., and Blaese R.M. 1994. Tumors expressing the cytosine deaminase suicide gene can be eliminated in vivo with 5-fluorocytosine and induce protective immunity to wild type tumor. Cancer Res. **54**: 1503–1506.
9. Huber B.E., Richards C.A., and Krenitsky T.A. 1991. Retroviral-mediated gene therapy for the treatment of hepatocellular carcinoma: an innovative approach for cancer therapy. Proc. Natl. Acad. Sci. U.S.A. **88**: 8039–8043.
10. Wallace P.M., MacMaster J.F., Smith V.F., Kerr D.E., Senter P.D., and Cosand W.L. 1994. Intratumoral generation of 5-fluorouracil mediated by an antibody-cytosine deaminase conjugate in combination with 5-fluorocytosine. Cancer Res. **54**: 2719–2723.
11. Nishiyama T., Kawamura Y., Kawamoto K., Matsumura H., Yamamoto N., Ito T., Ohyama A., Katsuragi T., and Sakai T. 1985. Antineoplastic effects in rats of 5-fluorocytosine in combination with cytosine deaminase capsules. Cancer Res. **45**: 1753–1761.
12. Huber B.E., Austin E.A., Richards C.A., Davis S.T., and Good S.S. 1994. Metabolism of 5-fluorocytosine to 5-fluorouracil in human colorectal tumor cells transduced with cytosine deaminase gene: significant antitumor effects when only a small percentage of tumor cells express cytosine deaminase. Proc. Natl. Acad. Sci. U.S.A. **91**: 8302–8306.
13. Huber B.E., Austin E.A., Good S.S., Knick V.C., Tibbels S., and Richards C.A. 1993. *In vivo* antitumor activity of 5-fluorocytosine on human colorectal carcinoma cells genetically modified to express cytosine deaminase. Cancer Res. **53**: 4619–4626.
14. Ohwada A., Hirschowitz E.A., and Crystal R.G. 1996. Regional delivery of an adenovirus vector containing the *Escherichia coli* cytosine deaminase gene to provide local activation of 5-fluorocytosine to suppress the growth of colon carcinoma metastatic to liver. Hum. Gene Ther. **7**: 1567–1576.
15. Bentires-Alj M., Hellin A.C., Lechanteur C., Princen F., Lopez M., Fillet G., Gielen J., Merville M.P., and Bours V. 2000. Cytosine deaminase suicide gene therapy for peritoneal carcinomatosis. Cancer Gene Ther. **7**: 20–26.

16. Kanai F., Lan K.H., Shiratori Y., Tanaka T., Ohashi M., Okudaira T., Yoshida Y., Wakimoto H., Hamada H., Nakabayashi H., Tamaoki T., and Omata M. 1997. *In vivo* gene therapy for a-fetoprotein-producing hepatocellular carcinoma by adenovirus-mediated transfer of cytosine deaminase gene. Cancer Res. **57**: 461–465.
17. Ichikawa T., Tamiya T., Adachi Y., Ono Y., Matsumoto K., Furuta T., Yoshida Y., Hamada H., and Ohmoto T. 2000. In vivo efficacy and toxicity of 5-fluorocytosine/cytosine deaminase gene therapy for malignant gliomas mediated by adenovirus. Cancer Gene Ther. **7**: 74–82.
18. Miller C.R., Williams C.R., Buchsbaum D.J., and Gillespie G.Y. 2002. Intratumoral 5-fluorouracil produced by cytosine deaminase/5-fluorocytosine gene therapy is effective for experimental human glioblastomas. Cancer Res. **62**: 773–780.
19. Topf N., Worgall S., Hackett N.R., and Crystal R.G. 1998. Regional "pro-drug" gene therapy: intravenous administration of an adenoviral vector expressing the *E. coli* cytosine deaminase gene and systemic administration of 5-fluorocytosine suppresses growth of hepatic metastasis of colon carcinoma. Gene Ther. **5**: 507–513.
20. Conrad C., Miller C.R., Ji Y., Gomez-Manzano C., Bharara S., McMurray J.S., Lang F.F., Wong F., Sawaya R., Yung W.K., and Fueyo J. 2005. D24-hyCD adenovirus suppresses glioma growth *in vivo* by combining oncolysis and chemosensitization. Cancer Gene Ther. **12**: 284–294.
21. Barton K.N., Paielli D., Zhang Y., Koul S., Brown S.L., Lu M., Seely J., Kim J.H., and Freytag S.O. 2006. Second-generation replication-competent oncolytic adenovirus armed with improved suicide genes and ADP gene demonstrates greater efficacy without increased toxicity. Mol. Ther. **13**: 347–356.
22. Tai C.K., Wang W.J., Chen T.C., and Kasahara N. 2005. Single-shot, multicycle suicide gene therapy by replication-competent retrovirus vectors achieves long-term survival benefit in experimental glioma. Mol. Ther. **12**: 842–851.
23. Erbs P., Findeli A., Kintz J., Cordier P., Hoffmann C., Geist M., and Balloul J.M. 2008. Modified vaccinia virus Ankara as a vector for suicide gene therapy. Cancer Gene Ther. **15**: 18–28.
24. Porosnicu M., Mian A., and Barber G.N. 2003. The oncolytic effect of recombinant vesicular stomatitis virus is enhanced by expression of the fusion cytosine deaminase/uracil phosphoribosyltransferase suicide gene. Cancer Res. **63**: 8366–8376.
25. Sasaki T., Fujimori M., Hamaji Y., Hama Y., Ito K., Amano J., and Taniguchi S. 2006. Genetically engineered *Bifidobacterium longum* for tumor-targeting enzyme-prodrug therapy of autochthonous mammary tumors in rats. Cancer Sci. **97**: 649–657.
26. Nemunaitis J., Cunningham C., Senzer N., Kuhn J., Cramm J., Litz C., Cavagnolo R., Cahill A., Clairmont C., and Sznol M. 2003. Pilot trial of genetically modified, attenuated *Salmonella* expressing the *E. coli* cytosine deaminase gene in refractory cancer patients. Cancer Gene Ther. **10**: 737–744.
27. Liu T., Zhang G., Chen Y.H., Chen Y., Liu X., Peng J., Xu M.H., and Yuan J.W. 2006. Tissue specific expression of suicide genes delivered by nanoparticles inhibits gastric carcinoma growth. Cancer Biol. Ther. **5**: 1683–1690.
28. Wei J., Blum S., Unger M., Jarmy G., Lamparter M., Geishauser A., Vlastos G.A., Chan G., Fischer K.D., Rattat D., Debatin K.M., Hatzopoulos A.K., and Beltinger

C. 2004. Embryonic endothelial progenitor cells armed with a suicide gene target hypoxic lung metastases after intravenous delivery. Cancer Cell. **5**: 477–488.

29. Kim S.K., Kim S.U., Park I.H., Bang J.H., Aboody K.S., Wang K.C., Cho B.K., Kim M., Menon L.G., Black P.M., and Carroll R.S. 2006. Human neural stem cells target experimental intracranial medulloblastoma and deliver a therapeutic gene leading to tumor regression. Clin. Cancer Res. **12**: 5550–5556.
30. Aboody K.S., Najbauer J., Schmidt N.O., Yang W., Wu J.K., Zhuge Y., Przylecki W., Carroll R., Black P.M., and Perides G. 2006. Targeting of melanoma brain metastases using engineered neural stem/progenitor cells. Neuro-oncology. **8**: 119–126.
31. Trinh Q.T., Austin E.A., Murray D.M., Knick V.C., and Huber B.E. 1995. Enzyme/prodrug gene therapy: comparison of cytosine deaminase/5-fluorocytosine versus thymidine kinase/ganciclovir enzyme/prodrug systems in a human colorectal carcinoma cell line. Cancer Res. **155**: 4808–4812.
32. Hoganson D.K., Batra R.K., Olsen J.C., and Boucher R.C. 1996. Comparison of the effects of three different toxin genes and their levels of expression on cell growth and bystander effect in lung adenocarcinoma. Cancer Res. **56**: 1315–1323.
33. Lawrence T.S., Rehemtulla A., Ng E.Y., Wilson M., Trosko J.E., and Stetson P.L. 1998. Preferential cytotoxicity of cells transduced with cytosine deaminase compared to bystander cells after treatment with 5-flucytosine. Cancer Res. **58**: 2588–2593.
34. Kuriyama S., Masui K., Sakamoto T., Nakatani T., Kikukawa M., Tsujinoue H., Mitoro A., Yamazaki M., Yoshiji H., Fukui H., Ikenaka K., Mullen C.A., and Tsujii T. 1998. Bystander effect caused by cytosine deaminase gene and 5-fluorocytosine in vitro is substantially mediated by generated 5-fluorouracil. Anticancer Res. **18**: 3399–3406.
35. Nishihara E., Nagayama Y., Narimatsu M., Namba H., Watanabe M., Niwa M., and Yamashita S. 1998. Treatment of thyroid carcinoma cells with four different suicide gene/prodrug combinations in vitro. Anticancer Res. **18**: 1521–1525.
36. Greco O., Folkes L.K., Wardman P., Tozer G.M., and Dachs G.U. 2000. Development of a novel enzyme/prodrug combination for gene therapy of cancer: horseradish peroxidase/indole-3-acetic acid. Cancer Gene Ther. **7**: 1414–1420.
37. Stribbling S.M., Friedlos F., Martin J., Davies L., Spooner R.A., Marais R., and Springer C.J. 2000. Regressions of established breast carcinoma xenografts by carboxypeptidase G2 suicide gene therapy and the prodrug CMDA are due to a bystander effect. Hum Gene Ther. **11**: 285–292.
38. Elshami A.A., Saavedra A., Zhang H., Kucharczuk J.C., Spray D.C., Fishman G.I., Amin K.M., Kaiser L.R., and Albelda S.M. 1996. Gap junctions play a role in the "bystander effect" of the herpes simplex virus thymidine kinase/ganciclovir system *in vitro*. Gene Ther. **3**: 85–92.
39. Mesnil M., Piccoli C., Tiraby G., Willecke K., and Yamasaki H. 1996. Bystander killing of cancer cells by herpes simplex virus thymidine kinase gene is mediated by connexins. Proc. Natl. Acad. Sci. U.S.A. **93**: 1831–1835.
40. Dilber M.S., Abedi M.R., Christensson B., Björkstrand B., Kidder G.M., Naus C.C., Gahrton G., and Smith C.I. 1997. Gap junctions promote the bystander effect of herpes simplex virus thymidine kinase in vivo. Cancer Res. **57**: 1523–1528.

41. Touraine R.L., Ishii-Morita H., Ramsey W.J., and Blaese R.M. 1998. The bystander effect in the HSVtk/ganciclovir system and its relationship to gap junctional communication. Gene Ther. **5**: 1705–1711.
42. Freeman S.M., Abboud C.N., Whartenby K.A., Packman C.H., Koeplin D.S., Moolten F.L., and Abraham G.N. 1993. The "bystander effect": tumor regression when a fraction of the tumor mass is genetically modified. Cancer Res. **53**: 5274–5283.
43. Holder J.W., Elmore E., and Barrett J.C. 1993. Gap junction function and cancer. Cancer Res. **53**: 3475–3485.
44. Frank D.K., Frederick M.J., Liu T.J., and Clayman G.L. 1998. Bystander effect in the adenovirus-mediated wild-type p53 gene therapy model of human squamous cell carcinoma of the head and neck. Clin. Cancer Res. **14**: 2521–2528.
45. Gagandeep S., Brew R., Green B., Christmas S.E., Klatzmann D., Poston G.J., and Kinsella A.R. 1996. Prodrug-activated gene therapy: involvement of an immunological component in the "bystander effect." Cancer Gene Ther. **3**: 83–88.
46. Kuriyama S., Kikukawa M., Masui K., Okuda H., Nakatani T., Sakamoto T., Yoshiji H., Fukui H., Ikenaka K., Mullen C.A., and Tsujii T. 1999. Cytosine deaminase/5-fluorocytosine gene therapy can induce efficient anti-tumor effects and protective immunity in immunocompetent mice but not in athymic nude mice. Int. J. Cancer. **81**: 592–597.
47. Ramesh R., Marrogi A.J., Munshi A., Abboud C.N., and Freeman S.M. 1996. In vivo analysis of the "bystander effect": a cytokine cascade. Exp. Hematol. **24**: 829–838.
48. Pierrefite-Carle V., Baque P., Gavelli A., Benchimol D., Bourgeon A., Milano G., Saint-Paul M.C., and Rossi B. 2000. Regression of experimental liver tumor after distant intra-hepatic injection of cytosine deaminase-expressing tumor cells and 5-fluorocytosine treatment. Int. J. Mol. Med. **5**: 275–278.
49. Okuda H., Ogura K., Kato A., Takubo H., and Watabe T. 1998. A possible mechanism of eighteen patient deaths caused by interactions of sorivudine, a new antiviral drug, with oral 5-fluorouracil prodrugs. J. Pharmacol. Exp. Ther. **287**: 791–799.
50. Diasio R.B., and Lu Z. 1994. Dihydropyrimidine dehydrogenase activity and fluorouracil chemotherapy. J. Clin. Oncol. **12**: 2239–2242.
51. Aschele C., Debernardis D., Casazza S., Antonelli G., Tunesi G., Baldo C., Lionetto R., Maley F., and Sobrero A. 1999. Immunohistochemical quantitation of thymidylate synthase expression in colorectal cancer metastases predicts for clinical outcome to fluorouracil-based chemotherapy. J. Clin. Oncol. **17**: 1760–1770.
52. Leichman C.G., Lenz H.J., Leichman L., Danenberg K., Baranda J., Groshen S., Boswell W., Metzger R., Tan M., and Danenberg P.V. 1997. Quantitation of intratumoral thymidylate synthase expression predicts for disseminated colorectal cancer response and resistance to protracted-infusion fluorouracil and weekly leucovorin. J. Clin. Oncol. **15**: 3223–3229.
53. Lenz H.J., Hayashi K., Salonga D., Danenberg K.D., Danenberg P.V., Metzger R., Banerjee D., Bertino J.R., Groshen S., Leichman L.P., and Leichman C.G. 1998. p53 point mutations and thymidylate synthase messenger RNA levels in disseminated colorectal cancer: an analysis of response and survival. Clin. Cancer Res. **4**: 1243–1250.

54. Banerjee D., Gorlick R., Liefshitz A., Danenberg K., Danenberg P.C., Danenberg P.V., Klimstra D., Jhanwar S., Cordon-Cardo C., Fong Y., Kemeny N., and Bertino J.R. 2000. Levels of E2F-1 expression are higher in lung metastasis of colon cancer as compared with hepatic metastasis and correlate with levels of thymidylate synthase. Cancer Res. **60**: 2365–2367.

55. Cascinu S., Aschele C., Barni S., Debernardis D., Baldo C., Tunesi G., Catalano V., Staccioli M.P., Brenna A., Muretto P., and Catalano G. 1999. Thymidylate synthase protein expression in advanced colon cancer: correlation with the site of metastasis and the clinical response to leucovorin-modulated bolus 5-fluorouracil. Clin. Cancer Res. **5**: 1996–1999.

56. Pullarkat S.T., Stoehlmacher J., Ghaderi V., Xiong Y.P., Ingles S.A., Sherrod A., Warren R., Tsao-Wei D., Groshen S., and Lenz H.J. 2001. Thymidylate synthase gene polymorphism determines response and toxicity of 5-FU chemotherapy. Pharmacogenomics J. **1**: 65–70.

57. Kawate H., Landis D.M., and Loeb L.A. 2002. Distribution of mutations in human thymidylate synthase yielding resistance to 5-fluorodeoxyuridine. J. Biol. Chem. **277**: 36304–36311.

58. Kitchens M.E., Forsthoefel A.M., Barbour K.W., Spencer H.T., and Berger F.G. 1999. Mechanisms of acquired resistance to thymidylate synthase inhibitors: the role of enzyme stability. Mol. Pharmacol. **56**: 1063–1070.

59. Van Rompay A.R., Johansson M., and Karlsson A. 1999. Phosphorylation of deoxycytidine analog monophosphates by UMP-CMP kinase: molecular characterization of the human enzyme. Mol. Pharmacol. **56**: 562–569.

60. Hsu C.H., Liou J.Y., Dutschman G.E., and Cheng Y.C. 2005. Phosphorylation of cytidine, deoxycytidine, and their analog monophosphates by human UMP/CMP kinase is differentially regulated by ATP and magnesium. Mol. Pharmacol. **67**: 806–814.

61. Schmidt W.M., Kalipciyan M., Dornstauder E., Rizovski B., Steger G.G., Sedivy R., Mueller M.W., and Mader R.M. 2004. Dissecting progressive stages of 5-fluorouracil resistance in vitro using RNA expression profiling. Int. J. Cancer. **112**: 200–212.

62. Wang W., Cassidy J., O'Brien V., Ryan K.M., and Collie-Duguid E. 2004. Mechanistic and predictive profiling of 5-fluorouracil resistance in human cancer cells. Cancer Res. **64**: 8167–8176.

63. Kawamura K., Tasaki K., Hamada H., Takenaga K., Sakiyama S., and Tagawa M. 2000. Expression of *Escherichia coli* uracil phosphoribosyltransferase gene in murine colon carcinoma cells augments the antitumoral effect of 5-fluorouracil and induces protective immunity. Cancer Gene Ther. **7**: 637–643.

64. Koyama F., Sawada H., Hirao T., Fujii H., Hamada H., and Nakano H. 2000. Combined suicide gene therapy for human colon cancer cells using adenovirus-mediated transfer of *Escherichia coli* cytosine deaminase gene and *Escherichia coli* uracil phosphoribosyltransferase gene with 5-fluorocytosine. Cancer Gene Ther. **7**: 1015–1022.

65. Kaliberov S.A., Chiz S., Kaliberova L.N., Krendelchtchikova V., Della Manna D., Zhou T., and Buchsbaum D.J. 2006. Combination of cytosine deaminse suicide gene expression with DR5 antibody treatment increases cancer cell cytotoxicity. Cancer Gene Ther. **13**: 203–214.

66. Mahan S.D., Ireton G.C., Knoeber C., Stoddard B.L., and Black M.E. 2004. Random mutagenesis and selection of *Escherichia coli* cytosine deaminase for cancer gene therapy. Protein Eng. Des. Sel. **17**: 625–633.
67. Mahan S.D., Ireton G.C., Stoddard B.L., and Black M.E. 2004. Alanine-scanning mutagenesis reveals a cytosine deaminase mutant with altered substrate preference. Biochemistry. **43**: 8957–8964.
68. Kaliberova L.N., Krendelchtchikova V., Harmon D.K., Stockard C.R., Petersen A.S., Markert J.M., Gillespie G.Y., Grizzle W.E., Buchsbaum D.J., and Kaliberov S.A. 2007. Mutation of *Escherichia coli* cytosine deaminase significantly enhances molecular chemotherapy of human glioma. Gene Ther. **14**: 1111–1119.
69. Kaliberova L.N., Della Manna D.L., Krendelchtchikova V., Black M.E., Buchsbaum D.J., and Kaliberov S.A. 2008. Molecular chemotherapy of pancreatic cancer using novel mutant bacterial cytosine deaminase gene. Mol. Cancer Ther. **7**: 2845–2854.
70. Yao L., Li Y., Wu Y., Liu A., and Yan H. 2005. Product release is rate-limiting in the activation of the prodrug 5-fluorocytosine by yeast cytosine deaminase. Biochemistry. **44**: 5940–5947.
71. Kievit E., Bershad E., Ng E., Sethna P., Dev I., Lawrence T.S., and Rehemtulla A. 1999. Superiority of yeast over bacterial cytosine deaminase for enzyme/prodrug gene therapy in colon cancer xenografts. Cancer Res. **59**: 1417–1421.
72. Korkegian A., Black M.E., Baker D., and Stoddard B.L. 2005. Computational thermostabilization of an enzyme. Science. **308**: 857–860.
73. Stolworthy T.S., Korkegian A.M., Willmon C.L., Ardiani A., Cundiff J., Stoddard B.L., and Black M.E. 2008. Yeast cytosine deaminase mutants with increased thermostability impart sensitivity to 5-fluorocytosine. J. Mol. Biol. **377**: 854–869.
74. Uckert W., Kammertöns T., Haack K., Qin Z., Gebert J., Schendel D.J., and Blankenstein T. 1998. Double suicide gene (cytosine deaminase and herpes simplex virus thymidine kinase) but not single gene transfer allows reliable elimination of tumor cells *in vivo*. Hum. Gene Ther. **9**: 855–865.
75. Blackburn R.V., Galoforo S.S., Corry P.M., and Lee Y.J. 1998. Adenoviral-mediated transfer of a heat-inducible double suicide gene into prostate carcinoma cells. Cancer Res. **58**: 1358–1362.
76. Rogulski K.R., Kim J.H., Kim S.H., and Freytag S.O. 1997. Glioma cells transduced with an *Escherichia coli* CD/HSV-1 TK fusion gene exhibit enhanced metabolic suicide and radiosensitivity. Hum. Gene Ther. **8**: 73–85.
77. Aghi M., Kramm C.M., Chou T-C., Breakefield X.O., and Chiocca E.A. 1998. Synergistic anticancer effects of ganciclovir/thymidine kinase and 5-fluorocytosine/cytosine deaminase gene therapies. J. Natl. Cancer Inst. **90**: 370–380.
78. Boucher P.D., Im M.M., Freytag S.O., and Shewach D.S. 2006. A novel mechanism of synergistic cytotoxicity with 5-fluorocytosine and ganciclovir in double suicide gene therapy. Cancer Res. **66**: 3230–3237.
79. Kim J.H., Kolozsvary A., Rogulski K., Khil M.S., Brown S.L., and Freytag S.O. 1998. Selective radiosensitization of 9L glioma in the brain transduced with double suicide fusion gene. Cancer J. Sci. Am. **4**: 364–369.
80. Rogulski K.R., Wing M.S., Paielli D.L., Gilbert J.D., Kim J.H., and Freytag S.O. 2000. Double suicide gene therapy augments the antitumor activity of a

replication-competent lytic adenovirus through enhanced cytotoxicity and radiosensitization. Hum. Gene Ther. **11**: 67–76.

81. Freytag S.O., Paielli D., Wing M., Rogulski K., Brown S., Kolozsvary A., Seely J., Barton K., Dragovic A., and Kim J.H. 2002. Efficacy and toxicity of replication-competent adenovirus-mediated double suicide gene therapy in combination with radiation therapy in an orthotopic mouse prostate cancer model. Int. J. Radiat. Oncol. Biol. Phys. **54**: 873–885.
82. Freytag S.O., Movsas B., Aref I., Stricker H., Peabody J., Pegg J., Zhang Y., Barton K.N., Brown S.L., Lu M., Savera A., and Kim J.H. 2007. Phase I trial of replication-competent adenovirus-mediated suicide gene therapy combined with IMRT for prostate cancer. Mol. Ther. **15**: 1016–1023.
83. Freytag S.O., Stricker H., Peabody J., Pegg J., Paielli D., Movsas B., Barton K.N., Brown S.L., Lu M., and Kim J.H. 2007. Five-year follow-up of trial of replication-competent adenovirus-mediated suicide gene therapy for treatment of prostate cancer. Mol. Ther. **15**: 636–642.
84. Freytag S.O., Stricker H., Pegg J., Paielli D., Pradhan D.G., Peabody J., DePeralta-Venturina M., Xia X., Brown S., Lu M., and Kim J.H. 2003. Phase I study of replication-competent adenovirus-mediated double-suicide gene therapy in combination with conventional-dose three-dimensional conformal radiation therapy for the treatment of newly diagnosed, intermediate- to high-risk prostate cancer. Cancer Res. **63**: 7497–7506.
85. Pederson L.C., Buchsbaum D.J., Vickers S.M., Kancharla S.R., Mayo M.S., Curiel D.T., and Stackhouse M.A. 1997. Molecular chemotherapy combined with radiation therapy enhances killing of cholangiocarcinoma cells *in vitro* and *in vivo*. Cancer Res. **57**: 4325–4332.
86. Stackhouse M.A., Pederson L.C., Grizzle W.E., Curiel D.T., Gebert J., Haack K., Vickers S.M., Mayo M.S., and Buchsbaum D.J. 2000. Fractionated radiation therapy in combination with adenoviral delivery of the cytosine deaminase gene and 5-fluorocytosine enhances cytotoxic and antitumor effects in human colorectal and cholangiocarcinoma models. Gene Ther. **7**: 1019–1026.
87. Kievit E., Nyati M.K., Ng E., Stegman L.D., Parsels J., Ross B.D., Rehemtulla A., and Lawrence T.S. 2000. Yeast cytosine deaminase improves radiosensitization and bystander effect by 5-fluorocytosine of human colorectal cancer xenografts. Cancer Res. **60**: 6649–6655.
88. Kambara H., Tamiya T., Ono Y., Ohtsuka S., Terada K., Adachi Y., Ichikawa T., Hamada H., and Ohmoto T. 2002. Combined radiation and gene therapy for brain tumors with adenovirus-mediated transfer of cytosine deaminase and uracil phosphoribosyltransferase genes. Cancer Gene Ther. **9**: 840–845.
89. Lawrence T.S., Tepper J.E., and Blackstock A.W. 1997. Fluoropyrimidine-radiation interactions in cells and tumors. Semin. Radiat. Oncol. **7**: 260–266.
90. Miller E.M., and Kinsella T.J. 1992. Radiosensitization by fluorodeoxyuridine: effects of thymidylate synthase inhibition and cell synchronization. Cancer Res. **52**: 1687–1694.
91. McGinn C.J., and Lawrence T.S. 2001. Recent advances in the use of radiosensitizing nucleosides. Semin. Radiat. Oncol. **11**: 270–280.
92. Davis M.A., Tang H.Y., Maybaum J., and Lawrence T.S. 1995. Dependence of fluorodeoxyuridine-mediated radiosensitization on S phase progression. Int. J. Radiat. Biol. **67**: 509–517.

93. Lawrence T.S., Davis M.A., and Loney T.L. 1996. Fluoropyrimidine-mediated radiosensitization depends on cyclin E-dependent kinase activation. Cancer Res. **56**: 3203–3206.
94. Pandha H.S., Martin L.A., Rigg A., Hurst H.C., Stamp G.W., Sikora K., and Lemoine N.R. 1999. Genetic prodrug activation therapy for breast cancer: a phase I clinical trial of *erb*B-2-directed suicide gene expression. J Clin Oncol. **17**: 2180–2189.
95. Crystal R.G., Hirschowitz E., Lieberman M., Daly J., Kazam E., Henschke C., Yankelevitz D., Kemeny N., Silverstein R., Ohwada A., Russi T., Mastrangeli A., Sanders A., Cooke J., and Harvey B.G. Phase I study of direct administration of a replication deficient adenovirus vector containing the *E. coli* cytosine deaminase gene to metastatic colon carcinoma of the liver in association with the oral administration of the pro-drug 5-fluorocytosine. Hum. Gene Ther. **8**: 985–1001.
96. Barton K.N., Stricker H., Brown S.L., Elshaikh M., Aref I., Lu M., Pegg J., Zhang Y., Karvelis K.C., Siddiqui F., Kim J.H., Freytag S.O., and Movsas B. 2008. Phase I study of noninvasive imaging of adenovirus-mediated gene expression in the human prostate. Mol. Ther. **16**: 1761–1769.
97. Cunningham C., and Nemunaitis J. 2001. A phase I trial of genetically modified *Salmonella typhimurium* expressing cytosine deaminase (TAPET-CD, VNP20029) administered by intratumoral injection in combination with 5-fluorocytosine for patients with advanced or metastatic cancer. Protocol no: CL-017. Version: April 9, 2001. Hum. Gene Ther. **12**: 1594–1596.
98. Etienne M.C., Chéradame S., Fischel J.L., Formento P., Dassonville O., Renée N., Schneider M., Thyss A., Demard F., and Milano G. 1995. Response to fluorouracil therapy in cancer patients: the role of tumoral dihydropyrimidine dehydrogenase activity. J Clin. Oncol. **13**: 1663–1670.
99. Harris B.E., Song R., Soong S.J., and Diasio R.B. 1990. Relationship between dihydropyrimidine dehydrogenase activity and plasma 5-fluorouracil levels with evidence for circadian variation of enzyme activity and plasma drug levels in cancer patients receiving 5-fluorouracil by protracted continuous infusion. Cancer Res. **50**: 197–201.
100. Salonga D., Danenberg K.D., Johnson M., Metzger R., Groshen S., Tsao-Wei D.D., Lenz H.J., Leichman C.G., Leichman L., Diasio R.B., and Danenberg P.V. 2000. Colorectal tumors responding to 5-fluorouracil have low gene expression levels of dihydropyrimidine dehydrogenase, thymidylate synthase, and thymidine phosphorylase. Clin. Cancer Res. **6**: 1322–1327.
101. Inaba M., Naoe Y., and Mitsuhashi J. 1998. Mechanisms for 5-fluorouracil resistance in human colon cancer DLD-1 cells. Biol. Pharm. Bull. **21**: 569–573.
102. Lenz H.J., Leichman C.G., Danenberg K.D., Danenberg P.V., Groshen S., Cohen H., Laine L., Crookes P., Silberman H., Baranda J., Garcia Y., Li J., and Leichman L. 1996. Thymidylate synthase mRNA level in adenocarcinoma of the stomach: a predictor for primary tumor response and overall survival. J. Clin. Oncol. **14**: 176–182.
103. Wang F.S., Aschele C., Sobrero A., Chang Y.M., and Bertino J.R. 1993. Decreased folylpolyglutamate synthetase expression: a novel mechanism of fluorouracil resistance. Cancer Res. **53**: 3677–3680.

11

BACTERIAL PROTEINS AGAINST METASTASIS

ANNA MARIA ELISABETH WALENKAMP

Department of Medical Oncology, University Medical Center Groningen, University of Groningen, Groningen, The Netherlands

INTRODUCTION

Metastases are the cause of 90% of human cancer deaths. The current treatment of cancer with chemotherapy and/or radiotherapy is based on cell death by DNA damage, neglecting the fact that cancer cell invasion into surrounding tissues and metastasizing are fundamental features of these cells and the major reason for treatment failure. The control of metastasis is critical for the control of cancer progression. In addition to cytotoxic and targeted therapies, drugs that target receptors on malignant cells responsible for their metastasizing capacity are of great value for treatment of most cancers.

In recent years, striking similarities between leukocyte trafficking and tumor cell migration revealed that they are both critically regulated by chemokines and their receptors (1). This was also found for receptors involved in leukocyte and tumor cell adhesion. Pathogen-derived peptides, chemokines, and activated complement fragments guide the recruitment of leukocytes to sites of infections. Bacteria are natural producers of chemokine/adhesion receptor inhibitors that prevent leukocyte migration toward the site of infection. These evolutionary tailored bacterial proteins can be explored for their capacity to antagonize receptors that play a role in malignant cell behavior as well.

Emerging Cancer Therapy: Microbial Approaches and Biotechnological Tools, Edited by Arsénio M. Fialho and Ananda M. Chakrabarty

For more than a century, bacterial products have been used for the treatment of cancer (2). Starting from the practical observation of tumor regression in individuals with concomitant bacterial infections, the field has evolved, and there are now some standard clinical practices, such as the use of Bacillus Calmette–Guerin (BCG) for the treatment of superficial bladder cancer. New applications have started to emerge that may profoundly change the perspective of the field. Bacteria such as *Salmonella* and *Listeria* can be attenuated by genetically defined mutations and can provide effective vehicles for DNA vaccines encoding tumor-associated antigens. *Salmonella* and nonpathogenic strains of *Clostridium* can selectively accumulate in tumors *in vivo*, providing attractive delivery systems to target immunomodulatory molecules and therapeutic agents to the tumor site (3). Staphylococcal superantigen-like (SSL) 10 was found to bind CXCR4 expressed on human T-ALL, lymphoma, and cervical carcinoma cell lines (4). SSL5 was recently described to bind platelets selectin (P-selectin) glycoprotein ligand-1 (PSGL-1) on leukemia cells and to inhibit rolling these malignant cells on a P-selectin surface (5).

Since the first description in 1878 by Eisenlohr, about 100 cases of spontaneous remission of acute leukemia after recovery from sepsis have been described. The mechanisms inducing spontaneous remission are thought to be related to an overwhelming immune response, leading to raised levels of various cytokines and resulting in an increased activation of immune cells. Bacterial proteins directly target chemokine and adhesion receptors primarily to prevent clearance by innate immune cells. But as these same receptors are also expressed in cancer cells and play a role in malignant cell behavior, including migration, this phenomenon would be an additional explanation for the observed remissions.

CHEMOKINES AND THEIR RECEPTORS

Chemokines are defined by their ability to induce directional migration of cells toward a gradient of the chemokine (chemotaxis) through binding to a subset of G protein-coupled (chemokine) receptors (GPCRs). Directional migration plays a role during (embryonic) development, immunity, and cancer cell migration. During these processes, cells expressing GPCRs on their surface migrate toward a gradient of chemokines/chemoattractants (Figure 11.1). For example, during embryonic development, progenitor cells expressing GPCRs on their surface migrate from their place of origin toward a gradient of chemokines to their final destination, where they will differentiate into organs and tissues.

The innate immune system can clear invading pathogens by phagocytes mainly by neutrophils and macrophages. These cells are attracted to the site of infection through sensing of chemotactic factor gradients. Chemoattractants are secreted by activated host cells and form a gradient in the tissue, thereby directing leukocyte migration to the site of infection and inflammation.

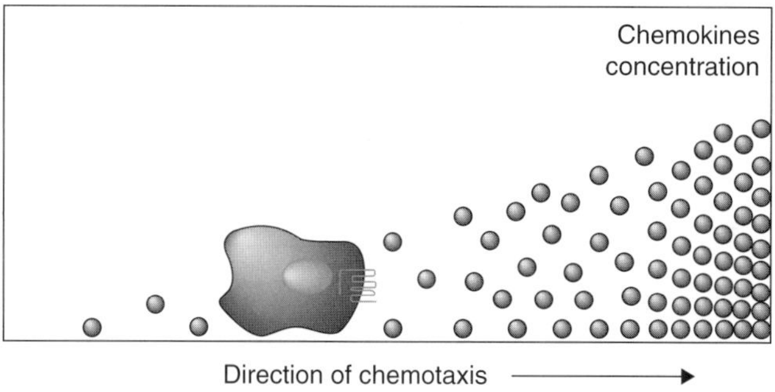

Figure 11.1 Chemokine-induced chemotaxis. Cells expressing a chemokine receptor migrate along a chemokine gradient.

Chemokines play a critical role in malignancy, both in transformation from premalignant cells to growth and metastasis of established tumors. A complex network of chemokines and their receptors influences the development of primary tumors and metastases by several important mechanisms: activation of a host tumor-specific immunological response, initiation of primary tumor growth and survival, regulation of site-specific spread and proliferation of cancer cells, and influencing tumor-induced angiogenesis.

GPCR STRUCTURE AND LIGAND BINDING

GPCRs form a large family of receptors that can bind a number of ligands, such as chemokines, hormones, and neurotransmitters, and vary in size from small molecules to peptides to large proteins. GPCRs can be grouped into six classes based on sequence homology and functional similarity (6). Examples are GPCRs sensitive to chemokines that are called chemokine receptors and GPCRs sensitive to "classical" chemoattractants like formylated peptides (fMLP), complement 5a (C5a), C3a, platelet-activating factor (PAF), and leukotriene B_4 (7).

All GPCRs are known to have a common motif of seven transmembrane structures with seven helical regions connected by six extramembrane loops (8). The N-terminus and three loops face outside the cell, while the C-terminus and three remaining loops extend inside the cell.

GPCR ACTIVATION BY CHEMOKINES

Chemokines are a large family of small proteins of 8–12 kDa that attract a variety of effector cells. Most chemokines have four characteristic cysteines, and depending on the motif displayed by the first two cysteines, they have been classified into CXC or alpha, CC or beta, C or gamma, and CX3C or delta chemokine classes (9, 10) (Figure 11.2 and Table 11.1).

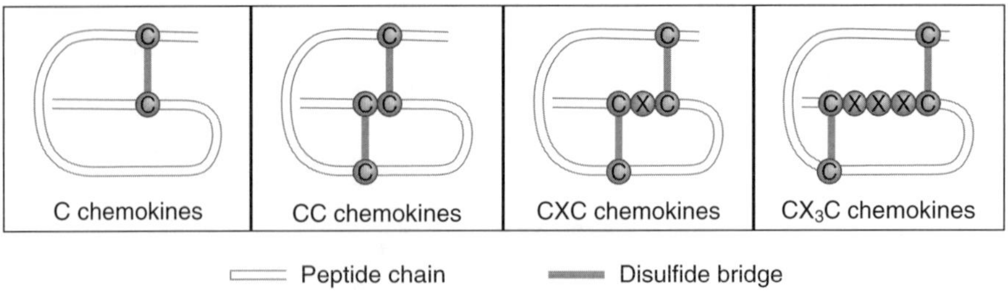

Figure 11.2 Structure of chemokine classes. A schematic overview of the four subfamilies of chemokines.

TABLE 11.1. Chemokine Receptors/Chemokines

Family	Structure	Receptor	Ligand
CXC (α)	NH2–C-X-C– 2 disulfide bridges	CXCR1	CXCL6–8
		CXCR2	CXCL1–3, CXCL5–8
		CXCR3	CXCL9–11
		CXCR4	CXCL12
		CXCR5	CXCL13
		CXCR6	CXCL16
		CXCR7	CXCL12
CC (β)	NH2–C-C– 2 disulfide bridges	CCR1	CCL3, CCL3L1, CCL5, CCL9–10, CCL14–16
		CCR2	CCL2, CCL7–8, CCL12–13
		CCR3	CCL5–6, CCL11, CCL15–16, CCL23–24, CCL26
		CCR4	CCL17, CCL22
		CCR5	CCL3–4, CCL3L1, CCL5–6, CCL8, CCL12
		CCR6	CCL20
		CCR7	CCL19, CCL21
		CCR8	CCL1
		CCR9	CCL25
		CCR10	CCL27–28
C/XC (γ)	NH2–C– 1 disulfide bridge	XCR1	XCL1–2
CX3C (δ)	NH2–C-X-X-X– 2 disulfide bridges Mucin-like stalk	CX3CR1	CX3CL1

The CXC chemokines are further subdivided into two subfamilies, based on the presence or absence of a Glu-Leu-Arg (ELR) sequence preceding the first cysteine residue. ELR-CXC chemokines are potent angiogenic factors, able to stimulate endothelial cell chemotaxis, while most non-ELR-CXC chemokines are strong angiostatic factors, which inhibit the endothelial cell

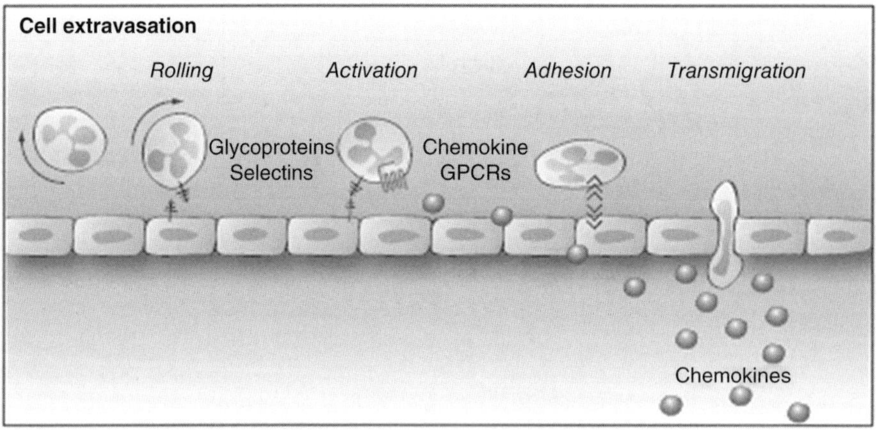

Figure 11.3 Cell migration and extravasation. Cell extravasation and migration from blood into tissue require multiple sequential activation steps.

chemotaxis induced by ELR-CXC chemokines (11). ELR-CXC chemokines preferentially attract neutrophils, whereas non-ELR-CXC chemokines act on lymphocytes. The CC chemokines are involved in the attraction of various cell types, for example, monocytes, eosinophils, and lymphocytes. Chemokines have a low sequence homology, but all show conserved structural homology.

Interaction between chemokines and the extracellular domain of the chemokine receptors leads to a cascade of intracellular events mediated, in part, by G protein-coupled signal transduction and, thus, the activation of effector cells.

Chemokine binding of target GPCRs varies from highly specific to highly promiscuous. Based on the chemokine class they bind, the receptors have been named CXCR1–7 (bind CXC chemokines), CCR1 through CCR11 (bind CC chemokines), CX_3CR1 (bind CX_3CL), and XCR1 (bind CXCL1).

Chemokines create a chemotactic gradient that directs cell transmigration of the endothelial layer or movement through the tissue, respectively (Figure 11.3).

GPCR SIGNALING

After ligand binding, GPCRs are activated, and the signal is transduced to enable effector functions (12). The receptors are coupled to heterotrimeric G proteins that are associated with their intracellular loops (Figure 11.4). Upon activation, GPCRs expose intracellular sites involved in the interaction with the G protein heterotrimer, which contains α, β, and γ subunits (13, 14). This catalyzes the dissociation of guanosine diphosphate (GDP) bound to the Gα subunit and its replacement with GTP, and leads to dissociation of Gα from G$\beta\gamma$ subunits. Both αG protein subunit and G$\beta\gamma$ subunit complexes then

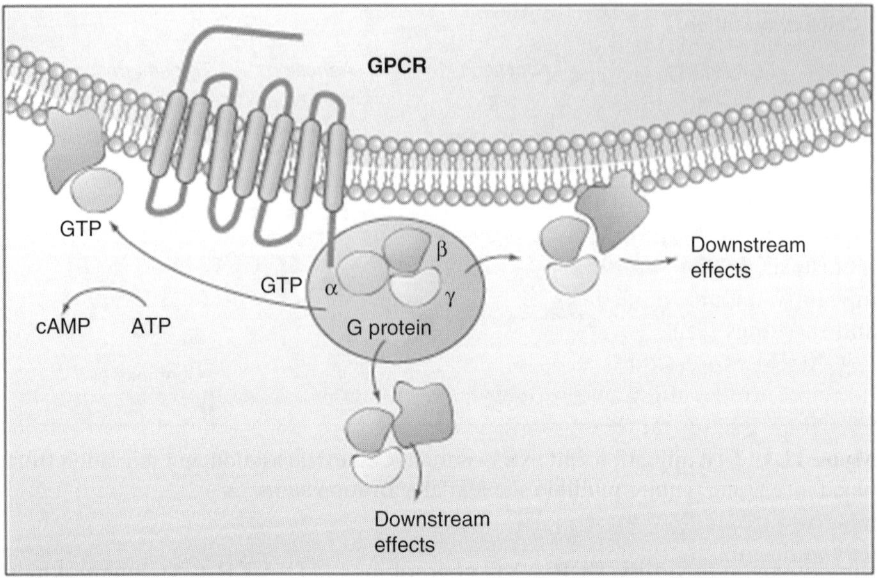

Figure 11.4 GPCR signaling. Upon binding of the ligand to GPCRs, an intracellular signaling cascade is initiated.

stimulate several downstream effectors. Ultimately, the G protein coupling specificity of each receptor determines the nature of its downstream signaling targets.

GPCRS AND CANCER

Chemokines modulate tumor behavior, such as the control of tumor cell adhesion, invasion, and migration; regulation of tumor-associated angiogenesis; activation of host tumor-specific immunological responses; and direct stimulation of tumor cell proliferation in an autocrine fashion. In this way, they play an important role in the development and progression of different cancers, and a growing number of therapeutic approaches are under investigation. Targets for antitumor therapy include individual chemokines and chemokine receptors that are known to be associated with specific cancers and modulation of attracted cytotoxic leukocytes to the site of the tumor. Several examples of cancer-related chemokines and their receptors are discussed below.

CHEMOKINE RECEPTOR 4 (CXCR4)

Tumor cells express functional chemokine receptors to sustain proliferation, angiogenesis, and survival, and to promote organ-specific localization of distant metastases (15, 16). Increasing evidence suggests the pivotal role of the chemokine stromal cell-derived factor-1 (CXCL12/SDF-1α) and its CXCR4 in the regulation of growth of both primary and metastatic cancers (1, 17, 18). CXCR4 is involved in the dissemination of breast cancer, prostate cancer to the bone marrow (19), colon cancer to the liver (20), and undifferentiated thyroid cancer (21). CXCR4 is highly expressed in human breast cancer cells and metastases. The specific ligand CXCL12 exhibits peak levels of expression in organs representing the first destination of breast cancer metastasis. *In vivo*, neutralizing the interactions of CXCL12 and CXCR4 significantly impairs metastasis of breast cancer cells to regional lymph nodes and the lungs (22).

CXCR4 expression is associated with cervical adenocarcinoma cell migration and proliferation, and primary cervical adenocarcinoma cells expressing CXCR4 are significantly more likely to metastasize to pelvic lymph nodes (23). Myeloid and lymphoid leukemia cells express high levels of CXCR4, which plays a critical role in leukemia cell chemotaxis and migration into bone marrow stroma (24–26). *In vitro*, CXCR4 antagonists strongly inhibit migratory and signaling responses to CXCL12 and partially antagonize the protective effects of marrow stromal cells to spontaneous or drug-induced apoptosis of chronic lymphocytic leukemia (CLL), acute lymphocytic leukemia (ALL), and acute myeloid leukemia (AML) cells (27–29). CXCR4 expression levels have a major prognostic impact in AML (30). CXCR4 is expressed by primary T-ALL cells of patients with childhood T-ALL (31), and a high CXCR4 expression predicts extramedullary organ infiltration in childhood (32). Promising results in preclinical tumor models indicate that CXCR4 antagonists may have additional value to conventional cytotoxic therapy in patients with various malignancies and immune diseases. Recent findings indicate that CXCL12-1alpha/CXCR4 interactions contribute to the resistance of leukemic cells to signal transduction inhibitor and chemotherapy-induced apoptosis in systems mimicking the physiological microenvironment (33). Disruption of these interactions with CXCR4 inhibitors represents a novel strategy of sensitizing leukemic cells by targeting their protective bone marrow microenvironment. The potential of this strategy was illustrated in a pilot experiment, which demonstrated six complete remissions out of eight highly chemo-resistant AML patients after combined treatment with CXCR4 inhibitor, plerixafor, and chemotherapy (34). In addition, this study demonstrated that the combined therapy of plerixafor with conventional chemotherapy is safe and does not affect the hematological recovery.

The acquired ability of a localized tumor to metastasize is a multistep process involving many pathways, including those involved in angiogenesis, focal adhesion, invasion, and, eventually, colonization of a distant site (35). CXCR4 inhibitors may have, like other inhibitors targeting malignant cell migration (36), a role in advanced disease, but even if no activity is observed in this setting, their role in invasion and metastasis might still enable a potential role in the adjuvant setting to reduce the risk of recurrence after definitive therapy. Further studies are necessary to identify potential combinations that will be of benefit, including combinations with cytotoxic chemotherapy agents in the frontline setting and after the development of resistance.

CHEMOKINE RECEPTOR 3 (CXCR3)

The chemokine receptor CXCR3 has three ligands: CXCL9, CXCL10, and CXCL11. It is expressed in various hematological (37) and solid cancers (38, 39), including prostate and renal cell carcinoma (RCC) (40). In RCC patients, presence of high CXCR3 expression in their tumors was an independent predictor of improved disease-free survival (DFS) following nephrectomy for localized disease (40). CXCR3 expression in human colon carcinoma cell lines has been associated with tumor cell migration, and CXCL10 was found in the preferred metastatic sites of colorectal cancer (CRC) (41). Several CXCR3 ligands are chemoattractants for malignant B cells (42), melanoma cells (43), and mucosa-associated lymphoid tissue (MALT) cells (44).

Upregulated CXCR3 ligands in tissue from RCC patients are reported to be produced by tumor vessels and may interact with CXCR3 expressed in tumor cells, with migratory activities leading to RCC progression. CXCR3 is expressed by tumor cells and infiltrating leukocytes. Furthermore, CXCR3 ligands' expressions are much higher both at the mRNA and the protein level than in normal kidney tissue. CXCL11 is present in pericytes and vascular smooth muscle cells in tumor vessels; CXCL9 is present in tumor endothelial cells and in infiltrating leukocytes. In CRC patients, CXCL10 expression was found in the primary tumor and in the preferred metastatic sites of CRC. CXCL10 upregulates the secretion of matrix metalloproteinase-9 (MMP9) by CRC cells and promotes the adhesion of (metastatic) CRC cells to laminin *in vitro*. Migration in response to CXCL10 could, *in vitro*, only be shown for the cell lines derived from the metastases. Therefore, it is suggested that CXCL10 has an interstitial motility and adherence activity in CRC cells that have reached their metastatic loci (41).

In a murine metastatic mammary cancer model, systemic administration of AMG487, a potent and selective orally bioavailable CXCR3 antagonist, inhibited development of lung metastases, while local tumor growth was not affected (39). A total of nine healthy Caucasian male volunteers were enrolled into a randomized, double-blind, multiple-dose, two-period crossover study to evaluate the pharmacokinetics after multiple oral dosing of AMG487 (45). AMG487 exhibited linear pharmacokinetics on both days 1 and 7 at the 25-mg dose with dose- and time-dependent kinetics at the two higher doses.

CHEMOKINE RECEPTOR 5 (CCR5)

CCR5 is expressed both in hematologic malignancies (46–49) and in solid tumors, and has been associated with cancer metastasis (50–53) and poor prognosis in patients with gastric cancer. Following gastrectomy, patients with CCR5 expression in their tumor had a 5-year survival rate of 54%, compared to 19% in patients without CCR5 expression (54). Furthermore, a 10-year DFS of 75% was observed in primary breast cancer patients with CCR5 expression in their tumor as compared to 55% in those without functional CCR5 expression (55).

Several CCR5 ligands interact with the tumor microenvironment. In a patient with CRC, an accumulation of CCR5-expressing T cells was detected along the invasive margin of the primary tumor (51). Its ligand CCL5 was localized within the infiltrating $CD8^+$ T cells. Earlier studies showed that the number of intraepithelial $CD8^+$ T cells in CRC was an independent favorable prognostic factor for patient survival. The T cells infiltrating the tumor may be involved in antitumor immunity and, thus, lead to survival advantage. The CCL5, released by the infiltrating T cells, is possibly directly involved in antitumor immunity by enhancing the levels of antigen-specific Th-1 and cytotoxic T lymphocyte responses, by upregulating the lyses of target cells, and costimulatory functions of antigen-presenting cells (56, 57). CCR5 is the coreceptor for the most commonly transmitted HIV-1 strains (58). Various drugs have been developed to target the CCR5 receptor, including maraviroc (59), E-913 (60), SCH-C (61), TAK-779 (62), and vicriviroc (63).

CHEMOKINE RECEPTOR 7 (CCR7)

CCR7 has as ligands CCL19 and CCL21. Expression of CCR7 is an independent prognostic factor for metastatic non-small cell lung cancer (64) and gastric carcinoma (65), and associated with shorter overall survival (OS) in cervical cancer (66).

Cancer cells may well exploit similar mechanisms to access the lymphatics as compared to T cells. CCR7 is known for its role in homing of T cells to the lymph nodes (67). One of the CCR7 ligands, CCL21, is expressed in the high endothelial venules of lymph nodes and is necessary for homing of CCR7-positive T cells into the lymph nodes (68). Indeed, the incidence of lymph node metastases correlated with the presence of CCR7 on tissue sections of human cancers, including breast cancer, melanoma, CRC, head and neck, prostate, non-small cell lung, esophageal squamous cell, and gastric cancers. In lymph node samples of breast cancer patients, both CCR7 ligands CCL19 and CCL21 show abundant expression (22). CCL21 exhibits peak levels of expression in organs representing the first destinations of breast cancer metastases like liver, lung, and bone tissue. In breast cancer cells, signaling through CCR7 mediated actin polymerization and pseudopodia formation, which leads to chemotactic and invasive responses of the CCR7-expressing cells toward ligands-expressing organs.

Human CCR7 monoclonal antibodies (mAbs) are used *in vitro* to block migration of human CLL cells in response to CCL19. Moreover, the CCR7 antibodies mediate a potent, complement-dependent cytotoxicity against CLL cells while sparing normal T lymphocytes from the same patient (69). In a metastatic CCR7 cDNA-transduced melanoma mouse model, the expression of CCR7 promotes the metastasis of melanoma cells to regional lymph nodes. This process is completely blocked by neutralizing anti-CCL21 antibodies but is not affected by control immunoglobulin (Ig)G (70). These results suggest that blocking CCR7 plays an important role in prevention of metastasis.

CHEMOKINE RECEPTOR 1/2 (CXCR1/2)

CXCR1 and 2 are expressed in glioblastoma multiforme (GBM) (71), in prostate cancers (72), and in RCC (CXCR2) (73). CXCL8 (one of the ligands) expression is directly correlated with poor survival in non-small cell lung cancer (74–78) and ovarium cancer (79). Moreover, elevated CXCL8 expression level is positively associated with cancer progression in prostate cancer (80, 81), bladder cancer (82), ovarian cancer (83), and melanoma patients (84, 85). The aggressiveness of malignant melanoma is attributed, in part, to the expression of CXCL8 and its receptors CXCR1 and CXCR2. Analysis of CXCR2 and CXCL8 in human melanomas shows that CXCR2 is expressed predominantly by higher grade melanomas and melanoma metastases (86). Enhanced CXCR2 expression is also correlated with tumor progression in GBM (71). In experimental models of human non-small cell lung cancer tumorigenesis, CXCL5 (one of the ligands) expression is directly correlated with tumor growth, tumor-derived angiogenesis, and metastatic potential (87).

Furthermore, CXCL8 correlates with the angiogenesis, tumorigenicity, and metastatic potential of many solid cancers in xenograft and orthotopic *in vivo* models (81, 88). CXCL8 secreted by tumor cells induces the differentiation and activation of osteoclasts, supporting the finding of the characteristic osteolytic metastasis of breast cancer cells that have disseminated to the bone (89). Interestingly, CXCL8 expression and secretion can be induced by chemotherapeutics such as 5-fluorouracil, doxorubicin, dacarbazine, and paclitaxel in various cancer cell lines (90–92). Moreover, chemotherapeutic agents also induce transcriptional regulation of CXCR1/2 genes and increase the level of CXCL8 signaling by the prostate cancer cell lines (93). By using liposome-encapsulated small interfering RNA (siRNA) as a treatment within ovarian tumor xenografts to suppress CXCL8 expression, Merritt et al. showed growth retardation, reduced microvessel density, and an increased response to docetaxel (94, 95).

Several successful treatment interventions have been described. Many small molecule inhibitors of CXCR1/2 signaling with appropriate pharmacokinetic properties, and therefore suitable for preclinical animal models, are now emerging (96–101).

BACTERIAL GPCR INHIBITORS

For successful infection of the host, invading pathogens first need to cross the mucosal surfaces and the skin. Upon subsequent entrance of host tissue by the pathogen, the acute inflammatory response of the innate immune system is initiated. Neutrophils and macrophages are the main initial effector cells of the innate immune system that can clear the invading pathogen through phagocytosis. These cells are attracted to the site of infection through sensing of chemotactic factor gradients. Bacterial-derived products also serve as effective chemoattractants. All these chemoattractants activate phagocytes by binding to GPCRs. Activation through GPCRs not only directs phagocyte chemotaxis but also primes and activates the cells for effector functions such as phagocytosis.

Chemoattractants and GPCRs thus play an essential role in directing the innate immune defense against invading pathogens. It is therefore not surprising that pathogens have evolved a large variety of strategies to evade activation of phagocytes by chemoattractants. A number of these bacterial proteins were discovered: chemotaxis inhibitory protein of *Staphylococcus aureus* (CHIPS) (102), staphylococcal complement inhibitor (SCIN) and its homologues (103, 104), FPRL1 inhibitory protein (FLIPr) (105), SSL5 (106, 107), and SSL10 (4).

CHIPS is an excreted virulence factor of *S. aureus* and inhibits fMLP- and complement factor C5a-induced responses in neutrophils through direct binding to the formyl peptide receptor and C5a receptor (C5aR), respectively. Thereby, CHIPS inhibits the initial activation and migration of leukocytes to the site of infection, and thus, it hampers clearance of *S. aureus* by innate immune cells. Recently, the structure of CHIPS consisting of residues 31–121 (CHIPS31–121) was resolved (108). CHIPS31–121 is composed of an α-helix packed onto a four-stranded antiparallel β-sheet, a domain also present in the C-terminal domain of superantigens. This protein also revealed to be homologous to the C-terminal domain of SSL5 and SSL7. SSLs are a family of secreted proteins identified through sequence homology to staphylococcal and streptococcal superantigens. Eleven different SSLs exist that are encoded on staphylococcal pathogenicity island 2 in a conserved order. Staphylococci contain 7–11 different SSLs, and their homology varies between 36% and 67%. Allelic variants show 85–100% homology. Determination of the crystal structures of SSL5 and SSL7 also revealed their high structural homology to superantigens; the N-terminal oligonucleotide/oligosaccharide-binding fold and the C-terminal β-grasp domain characteristic for superantigens are also observed in SSLs. However, residues important for major histocompatibility complex (MHC) Class II and T cell receptor (TCR) binding of superantigens are not conserved in SSLs, which may explain their inability to display superantigenic activities. Three proteins of the SSL family are thus far functionally described: SSL7 binds IgA and complement C5, and inhibits IgA-Fc alpha RI binding and serum killing of bacteria (109). SSL5 inhibits P-selectin-mediated neutrophil rolling by binding PSGL-1 (107). SSL5 binding to PSGL-1 was found to be dependent on the presence of sugar moieties. SSL11 was also shown to interact with sialyl Lewis x (sLex). SSL10 inhibits CXCL12-induced human CXCR4-positive tumor cell migration.

A number of other examples of bacterial proteins targeting GPCRs exist, like cholera and pertussis toxin that covalently modify the α subunits of numerous G proteins by ADP-ribosylating-specific amino acid residues (110). Staphylococcal superantigens' stimulation of human peripheral blood monocytes results in a rapid, dose-dependent, and specific downregulation of CCR1, CCR2, and CCR5, which correlates with a concomitant hyporesponsiveness of human monocytes to these CC chemokine ligands (111). Lipopolysaccharide (LPS) causes a drastic and rapid downregulation of the expression of CCR2 (112) and is able to downmodulate chemokine receptor 4 (CXCR4) in neutrophils and monocytes (113).

BACTERIAL INHIBITOR SSL10 AGAINST CANCER-RELATED GPCR CXCR4

SSL10 was found to bind CXCR4 expressed on human T-ALL, lymphoma, and cervical carcinoma cell lines. SSL10 inhibited CXCL12-induced responses at different levels of the various involved signal transduction pathways, as shown by inhibition of CXCL12-induced calcium mobilization, Akt phosphorylation, and migration.

For expression of recombinant SSL10 (protein identity YP_498982.1), the SSL10 gene (genomic locus tag SAOUHSC_00395) of the *S. aureus* strain NCTC8325 (minus signal sequence coding for the first 30 amino acids) was cloned into the expression vector pRSETB. The pRSET/SSL10 expression vector was transformed in BL21(DE3) *Escherichia coli*. Then, expression of histidine-tagged SSL10 (HIS-SSL10) was induced. HIS-tagged SSL10 was isolated under denaturing conditions, and the protein was renatured on the column by gradual exchanging denaturing buffer. After dialysis, the HIS-tag was removed from SSL10 by cleavage with enterokinase (EK).

SSL10 competes with the binding of an antibody directed against CXCR4 on T cells and HeLa cells, a cervical carcinoma cell line. The hallmark of chemokine receptors is a rapid and transient increase in the free intracellular calcium level upon ligand binding. This signaling pathway was used to show that SSL10 inhibits the CXCL12-induced calcium mobilization in Jurkat cells. CXCL12/CXCR induced a rapid increase in phosphorylation of Akt, a pathway associated with cell survival. Akt phosphorylation was inhibited by AMD3100, a bicyclam antagonist of CXCR4, and, to a lesser extent, by SSL10. The combination of both CXCR4-inhibiting agents showed a synergistic effect of this inhibiting capacity. SSL10 clearly inhibits the chemotactic response of both HeLa and Jurkat cells toward CXCL12.

The above-mentioned results proved that bacterial proteins like SSL10 targeting chemokine receptors on leukocytes can be used to inhibit malignant cell migration.

PSGL-1

PSGL-1 is a glycoprotein found on white blood cell, endothelial cells, leukemic, and prostate cancer cells (114) that binds to its ligand P-selectin (Figure 11.3). It is one of a family of selectins that includes E-selectin (endothelial selectin) and L-selectin (leukocyte selectin). Selectins are part of the broader family of cell adhesion molecules. Initial leukocyte adhesion is mediated by selectins that interact with sialylated, fucosylated lactosaminoglycans such as sLex on the cell surface or, predominantly, when displayed by mucin-like glycoproteins.

Successful metastasis of tumor cells often involves hematogenous dissemination. Thereby, tumor cells need to pass through the blood stream and extravasate from the blood vessels into peripheral organ tissues. While traveling through the bloodstream, interactions with normal host cells like leukocytes, platelets, or endothelial cells are of crucial importance. The ability of tumor cells to bind platelets in the blood stream may be of particular importance as stabilized platelet-enriched thrombi can physically protect tumor cells from destruction, or facilitate tissue adherence and formation of metastatic foci (115, 116). Adhesion to the endothelial lining is also required and involves specific adhesion receptors present on the tumor cells and various adhesive ligands on the endothelium.

Receptors important for leukocyte adhesion were shown to play an important role in malignancies as well (117, 118). In this respect, P-selectin was shown to mediate rolling of human HL-60 leukemia cells on activated endothelial cells through PSGL-1 (119). Immunohistochemical analysis of PSGL-1 in normal prostate tissue and in localized and metastatic prostate tumors revealed that PSGL-1 was detected on the surfaces of bone-metastatic prostate tumor cells. These findings implicated a functional role of PSGL-1 in the bone tropism of prostate tumor cells (114). The understanding that receptors important for leukocyte adhesion also play an important role in malignancies and are thus important candidates to target led to an increasing literature on cell adhesion molecules and cancer metastasis. Several cancer cell lines are described to bind P-selectin (120), and positive staining for P-selectin on intratumoral vessels was found to be negatively associated with OS in melanoma patients (121). Studies with P-selectin-deficient mice demonstrated significantly slower growth of subcutaneously implanted human colon carcinoma cells and generated fewer lung metastases from intravenously injected cells (122). In addition to binding P-selectin on endothelial cells, platelets provide a significant source of P-selectin for carcinoma binding (123). Platelets promote survival of tumor cells as they directly protect them from lysis by natural killer

(NK) cells *in vitro* as well as *in vivo* (124). In a mouse model of experimental metastasis, tumor seeding of three different tumor cell lines in the target organs was reduced when the host was platelet depleted, but only if the tumor cells were NK sensitive. Additionally, P-selectin on platelets was shown to allow the formation of platelet–tumor aggregates as P-selectin-deficient mice show less coating of platelets on tumor cells (122).

In addition, interactions among tumor cells, platelets, and endothelial cells are not only important for dissemination and metastasis formation, but also play an important role in tumor-induced angiogenesis. Inhibition of these interactions inhibits tumor-induced angiogenesis (125, 126).

Clinical relevance of PSGL-1 in metastasis was supported in an invasive and metastatic T cell lymphoma model. Raes et al. (127) demonstrated that, although overexpression of PSGL-1 may not be sufficient for successful dissemination, PSGL-1 played a crucial role in the hematogenous, disseminative properties in organ colonization of the lymphoid cancer cells investigated. Malignant cell transformation is commonly associated with increased and altered expression of glycans at the cell surface, and the aberrant glycosylation is a prominent feature of carcinoma progression (128). Among the glycans associated with cancer cells, sLex expression on cancer cells is known to be correlated with a poor prognosis and an increased rate of metastasis (129). Many anticancer therapies have therefore focused on targeting sLex epitopes. Compounds mimicking sLex, compounds, and antisense DNA strategies directed against enzymes responsible for sLex synthesis (130) and antibodies against sLex (131) have all proven to be beneficial in mouse models.

BACTERIAL PSGL-1 INHIBITOR: SSL5

SSL5 was recently described to bind PSGL-1 on leukocytes and to inhibit neutrophil rolling on a P-selectin surface (106). Thereby, SSL5 inhibits neutrophil rolling on endothelial cells by disrupting the interaction of PSGL-1 with its natural ligand P-selectin. Sugar moieties were shown to be crucial in this effect. Recently, the crystal structure of SSL5 in complex with sLex was determined, and it was shown that SSL5 binds sLex through its C-terminal domain (132). SSL5 thus inhibits the initial activation and migration of neutrophils to the site of infection.

The SSL5 gene of the NCTC8325 *S. aureus* strain was cloned into the pRSETB expression vector directly downstream of an EK cleavage site and a polyhistidine tag. The vector was transformed in BL21(DE3) *E. coli* for protein expression. The histidine-tagged protein was purified using nickel affinity chromatography and cleaved with EK.

As leukocytes and tumor cells share many characteristics in migration and dissemination, the potential of SSL5 as an antagonist of malignant cell behavior was explored. Previously, it was demonstrated that rolling of human HL-60 leukemia cells on activated endothelial cells was mediated by P-selectin. SSL5 targets HL-60 cells. Binding of SSL5 was rapid and without observed toxicity. Competition of SSL5 with the binding of three anti-PSGL-1 antibodies and P-selectin to HL-60 cells identified PSGL-1 as the ligand on HL-60 cells. Presence of sLex epitopes on PSGL-1 was crucial for its interaction with SSL5. Importantly, SSL5 inhibited the interaction of HL-60 cells not only with activated endothelial cells but also with platelets, which both play an important role in growth and metastasis of cancers. These data support the concept that a bacterial protein like SSL5 could be a lead in the search for novel strategies against hematological malignancies.

In conclusion, SSL5 can alter malignant cell behavior by targeting PSGL-1 and interfere with its binding to P-selectin. SSL5 blocks adhesion of HL-60 malignant leukemia cells not only to endothelial cells but also to platelets. Not unimportantly, SSL5 binding to HL-60 cells and PSGL-1 is glycan dependent. Therefore, SSL5 is a promising, potential bacterial-derived lead for novel cancer therapies as it inhibits the adhesive properties of leukemia cells.

REFERENCES

1. Zlotnik A. 2006. Chemokines and cancer. Int. J. Cancer **119**: 2026–2029.
2. Sinha G. 2003. Bacterial battalions join war against cancer. Nat. Med. **9**: 1229.
3. Chabalgoity J.A., Dougan G., Mastroeni P., and Aspinall R.J. 2002. Live bacteria as the basis for immunotherapies against cancer. Expert Rev. Vaccines. **1**: 495–505.
4. Walenkamp A.M., Boer I.G., Bestebroer J., Rozeveld D., Timmer-Bosscha H., Hemrika W., van Strijp J.A., and de Haas C.J. 2009. Staphylococcal superantigen-like 10 inhibits CXCL12-induced human tumor cell migration. Neoplasia. **11**: 333–344.
5. Walenkamp A.M., Bestebroer J., Boer I.G., Kruizinga R., Verheul H.M., van Strijp J.A., and de Haas C.J. 2010. Staphylococcal SSL5 binding to human leukemia cells inhibits cell adhesion to endothelial cells and platelets. Cell Oncol. **32**: 1–10.
6. Kristiansen K. 2004. Molecular mechanisms of ligand binding, signaling, and regulation within the superfamily of G-protein-coupled receptors: molecular modeling and mutagenesis approaches to receptor structure and function. Pharmacol. Ther. **103**: 21–80.
7. Rollins B.J. 1997. Chemokines. Blood. **90**: 909–928.
8. Palczewski K. 2006. G protein-coupled receptor rhodopsin. Annu. Rev. Biochem. **75**: 743–767.
9. Allen S.J., Crown S.E., and Handel T.M. 2007. Chemokine: receptor structure, interactions, and antagonism. Annu. Rev. Immunol. **25**: 787–820.
10. Fernandez E.J., and Lolis E. 2002. Structure, function, and inhibition of chemokines. Annu. Rev. Pharmacol. Toxicol. **42**: 469–499.
11. Strieter R.M., Polverini P.J., Kunkel S.L., Arenberg D.A., Burdick M.D., Kasper J., Dzuiba J., Van Damme J., Walz A., and Marriott D. 1995. The functional role of the ELR motif in CXC chemokine-mediated angiogenesis. J. Biol. Chem. **270**: 27348–27357.

12. Hamm H.E. 1998. The many faces of G protein signaling. J. Biol. Chem. **273**: 669–672.
13. Dorsam R.T., and Gutkind J.S. 2007. G-protein-coupled receptors and cancer. Nat. Rev. Cancer. **7**: 79–94.
14. Neves S.R., Ram P.T., and Iyengar R. 2002. G protein pathways. Science. **296**: 1636–1639.
15. Ruffini P.A., Morandi P., Cabioglu N., Altundag K., and Cristofanilli M. 2007. Manipulating the chemokine-chemokine receptor network to treat cancer. Cancer. **109**: 2392–2404.
16. Balkwill F. 2004. Cancer and the chemokine network. Nat. Rev. Cancer. **4**: 540–550.
17. Takeuchi H., Kitago M., and Hoon D.S. 2007. Effects of chemokines on tumor metastasis. Cancer Treat. Res. **135**: 177–184.
18. Smith M.C., Luker K.E., Garbow J.R., Prior J.L., Jackson E., Piwnica-Worms D., and Luker G.D. 2004. CXCR4 regulates growth of both primary and metastatic breast cancer. Cancer Res. **64**: 8604–8612.
19. Taichman R.S., Cooper C., Keller E.T., Pienta K.J., Taichman N.S., and McCauley L.K. 2002. Use of the stromal cell-derived factor-1/CXCR4 pathway in prostate cancer metastasis to bone. Cancer Res. **62**: 1832–1837.
20. Zeelenberg I.S., Ruuls-Van S.L., and Roos E. 2003. The chemokine receptor CXCR4 is required for outgrowth of colon carcinoma micrometastases. Cancer Res. **63**: 3833–3839.
21. De F.V., Guarino V., Avilla E., Castellone M.D., Salerno P., Salvatore G., Faviana P., Basolo F., Santoro M., Melillo R.M. 2007. Biological role and potential therapeutic targeting of the chemokine receptor CXCR4 in undifferentiated thyroid cancer. Cancer Res. **67**: 11821–11829.
22. Muller A., Homey B., Soto H., Ge N., Catron D., Buchanan M.E., McClanahan T., Murphy E., Yuan W., Wagner S.N., Barrera J.L., Mohar A., Verastegui E., and Zlotnik A. 2001. Involvement of chemokine receptors in breast cancer metastasis. Nature. **410**: 50–56.
23. Zhang J.P., Lu W.G., Ye F., Chen H.Z., Zhou C.Y., and Xie X. 2007. Study on CXCR4/SDF-1alpha axis in lymph node metastasis of cervical squamous cell carcinoma. Int. J. Gynecol. Cancer. **17**: 478–483.
24. Burger J.A., and Burkle A. 2007. The CXCR4 chemokine receptor in acute and chronic leukaemia: a marrow homing receptor and potential therapeutic target. Br. J. Haematol. **137**: 288–296.
25. Burger J.A., and Kipps T.J. 2006. CXCR4: a key receptor in the crosstalk between tumor cells and their microenvironment. Blood. **107**: 1761–1767.
26. Redondo-Munoz J., Escobar-Diaz E., Samaniego R., Terol M.J., Garcia-Marco J.A., and Garcia-Pardo A. 2006. MMP-9 in B-cell chronic lymphocytic leukemia is up-regulated by alpha4beta1 integrin or CXCR4 engagement via distinct signaling pathways, localizes to podosomes, and is involved in cell invasion and migration. Blood. **108**: 3143–3151.
27. Juarez J., Bradstock K.F., Gottlieb D.J., and Bendall L.J. 2003. Effects of inhibitors of the chemokine receptor CXCR4 on acute lymphoblastic leukemia cells in vitro. Leukemia. **17**: 1294–1300.
28. Burger M., Hartmann T., Krome M., Rawluk J., Tamamura H., Fujii N., Kipps T.J., and Burger J.A. 2005. Small peptide inhibitors of the CXCR4 chemokine receptor

(CD184) antagonize the activation, migration, and antiapoptotic responses of CXCL12 in chronic lymphocytic leukemia B cells. Blood. **106**: 1824–1830.

29. Zeng Z., Samudio I.J., Munsell M., An J., Huang Z., Estey E., Andreeff M., and Konopleva M. 2006. Inhibition of CXCR4 with the novel RCP168 peptide overcomes stroma-mediated chemoresistance in chronic and acute leukemias. Mol. Cancer Ther. **5**: 3113–3121.
30. Rombouts E.J., Pavic B., Lowenberg B., and Ploemacher R.E. 2004. Relation between CXCR-4 expression, Flt3 mutations, and unfavorable prognosis of adult acute myeloid leukemia. Blood. **104**: 550–557.
31. Dialynas D.P., Shao L., Billman G.F., and Yu J. 2001. Engraftment of human T-cell acute lymphoblastic leukemia in immunodeficient NOD/SCID mice which have been preconditioned by injection of human cord blood. Stem Cells. **19**: 443–452.
32. Crazzolara R., Kreczy A., Mann G., Heitger A., Eibl G., Fink F.M., Mohle R., and Meister B. 2001. High expression of the chemokine receptor CXCR4 predicts extramedullary organ infiltration in childhood acute lymphoblastic leukaemia. Br. J. Haematol. **115**: 545–553.
33. Zeng Z., Shi Y.X., Samudio I.J., Wang R.Y., Ling X., Frolova O., Levis M., Rubin J.B., Negrin R.R., Estey E.H., Konoplev S., Andreeff M., and Konopleva M. 2008. Targeting the leukemia microenvironment by CXCR4 inhibition overcomes resistance to kinase inhibitors and chemotherapy in AML. Blood. **113**: 6215–6224.
34. Uy G.L., Retting M., McFarland K., et al. 2008. Mobilization and chemo sensitization of AML with the CXCR4 antagonist Plerixafor [AMD3100]: a phase I/II study of AMD3100 +MEC in patients with relapsed or refractory disease. Abstract, American Society of Hematology.
35. Chambers A.F., Groom A.C., and MacDonald I.C. 2002. Dissemination and growth of cancer cells in metastatic sites. Nat. Rev. Cancer. **2**: 563–572.
36. Finn R.S. 2008. Targeting Src in breast cancer. Ann. Oncol. **19**: 1379–1386.
37. Jones D., Benjamin R.J., Shahsafaei A., and Dorfman D.M. 2000. The chemokine receptor CXCR3 is expressed in a subset of B-cell lymphomas and is a marker of B-cell chronic lymphocytic leukemia. Blood. **95**: 627–632.
38. Ohtani H., Jin Z., Takegawa S., Nakayama T., and Yoshie O. 2009. Abundant expression of CXCL9 (MIG) by stromal cells that include dendritic cells and accumulation of CXCR3+ T cells in lymphocyte-rich gastric carcinoma. J. Pathol. **217**: 21–31.
39. Walser T.C., Rifat S., Ma X., Kundu N., Ward C., Goloubeva O., Johnson M.G., Medina J.C., Collins T.L., and Fulton A.M. 2006. Antagonism of CXCR3 inhibits lung metastasis in a murine model of metastatic breast cancer. Cancer Res. **66**: 7701–7707.
40 Klatte T., Seligson D.B., Leppert J.T., Riggs S.B., Yu H., Zomorodian N., Kabbinavar F.F., Strieter R.M., Belldegrun A.S., and Pantuck A.J. 2008. The chemokine receptor CXCR3 is an independent prognostic factor in patients with localized clear cell renal cell carcinoma. J. Urol. **179**: 61–66.
41. Zipin-Roitman A., Meshel T., Sagi-Assif O., Shalmon B., Avivi C., Pfeffer R.M., Witz I.P., and Ben-Baruch A. 2007. CXCL10 promotes invasion-related properties in human colorectal carcinoma cells. Cancer Res. **67**: 3396–3405.
42. Trentin L., Agostini C., Facco M., Piazza F., Perin A., Siviero M., Gurrieri C., Galvan S., Adami F., Zambello R., and Semenzato G. 1999. The chemokine receptor CXCR3 is expressed on malignant B cells and mediates chemotaxis. J. Clin. Invest. **104**: 115–121.

43. Kawada K., Sonoshita M., Sakashita H., Takabayashi A., Yamaoka Y., Manabe T., Inaba K., Minato N., Oshima M., Taketo M.M. 2004. Pivotal role of CXCR3 in melanoma cell metastasis to lymph nodes. Cancer Res. **64**: 4010–4017.

44. Suefuji H., Ohshima K., Karube K., Kawano R., Nabeshima K., Suzumiya J., Hayabuchi N., and Kikuchi M. 2005. CXCR3-positive B cells found at elevated frequency in the peripheral blood of patients with MALT lymphoma are attracted by MIG and belong to the lymphoma clone. Int. J. Cancer. **114**: 896–901.

45. Tonn G.R., Wong S.G., Wong S.C., Johnson M.G., Ma J., Cho R., Floren L.C., Kersey K., Berry K., Marcus A.P., Wang X., Van L.B., Medina J.C., Pearson P.G., and Wong B.K. 2009. An inhibitory metabolite leads to dose- and time-dependent pharmacokinetics of (R)-N-{1-[3-(4-ethoxy-phenyl)-4-oxo-3,4-dihydro-pyrido[2,3-d]pyrimidin-2-yl]-ethyl}-N-pyridin-3-yl-methyl-2-(4-trifluoromethoxy-phenyl) acetamide (AMG 487) in human subjects after multiple dosing. Drug Metab. Dispos. **37**: 502–513.

46. Trentin L., Cabrelle A., Facco M., Carollo D., Miorin M., Tosoni A., Pizzo P., Binotto G., Nicolardi L., Zambello R., Adami F., Agostini C., and Semenzato G. 2004. Homeostatic chemokines drive migration of malignant B cells in patients with non-Hodgkin lymphomas. Blood. **104**: 502–508.

47. Lentzsch S., Gries M., Janz M., Bargou R., Dorken B., and Mapara M.Y. 2003. Macrophage inflammatory protein 1-alpha (MIP-1 alpha) triggers migration and signaling cascades mediating survival and proliferation in multiple myeloma (MM) cells. Blood. **101**: 3568–3573.

48. Maggio E.M., Van Den B.A., Visser L., Diepstra A., Kluiver J., Emmens R., and Poppema S. 2002. Common and differential chemokine expression patterns in rs cells of NLP, EBV positive and negative classical Hodgkin lymphomas. Int. J. Cancer. **99**: 665–672.

49. Makishima H., Ito T., Asano N., Nakazawa H., Shimodaira S., Kamijo Y., Nakazawa Y., Suzuki T., Kobayashi H., Kiyosawa K., and Ishida F. 2005. Significance of chemokine receptor expression in aggressive NK cell leukemia. Leukemia. **19**: 1169–1174.

50. Luboshits G., Shina S., Kaplan O., Engelberg S., Nass D., Lifshitz-Mercer B., Chaitchik S., Keydar I., and Ben-Baruch A. 1999. Elevated expression of the CC chemokine regulated on activation, normal T cell expressed and secreted (RANTES) in advanced breast carcinoma. Cancer Res. **59**: 4681–4687.

51. Musha H., Ohtani H., Mizoi T., Kinouchi M., Nakayama T., Shiiba K., Miyagawa K., Nagura H., Yoshie O., and Sasaki I. 2005. Selective infiltration of CCR5(+) CXCR3(+) T lymphocytes in human colorectal carcinoma. Int. J. Cancer. **116**: 949–956.

52. Vaday G.G., Peehl D.M., Kadam P.A., and Lawrence D.M. 2006. Expression of CCL5 (RANTES) and CCR5 in prostate cancer. Prostate. **66**: 124–134.

53. van Deventer H.W., O'Connor W., Jr., Brickey W.J., Aris R.M., Ting J.P., and Serody J.S. 2005. C-C chemokine receptor 5 on stromal cells promotes pulmonary metastasis. Cancer Res. **65**: 3374–3379.

54. Sugasawa H., Ichikura T., Tsujimoto H., Kinoshita M., Morita D., Ono S., Chochi K., Tsuda H., Seki S., and Mochizuki H. 2008. Prognostic significance of expression of CCL5/RANTES receptors in patients with gastric cancer. J. Surg. Oncol. **97**: 445–450.

55. Manes S., Mira E., Colomer R., Montero S., Real L.M., Gomez-Mouton C., Jimenez-Baranda S., Garzon A., Lacalle R.A., Harshman K., Ruiz A., and Martinez A. 2003. CCR5 expression influences the progression of human breast cancer in a p53-dependent manner. J. Exp. Med. **198**: 1381–1389.

56. Kim J.J., Nottingham L.K., Sin J.I., Tsai A., Morrison L., Oh J., Dang K., Hu Y., Kazahaya K., Bennett M., Dentchev T., Wilson D.M., Chalian A.A., Boyer J.D., Agadjanyan M.G., and Weiner D.B. 1998. CD8 positive T cells influence antigen-specific immune responses through the expression of chemokines. J. Clin. Invest. **102**: 1112–1124.

57. Taub D.D., Ortaldo J.R., Turcovski-Corrales S.M., Key M.L., Longo D.L., and Murphy W.J. 1996. Beta chemokines costimulate lymphocyte cytolysis, proliferation, and lymphokine production. J. Leukoc. Biol. **59**: 81–89.

58. Samson M., Libert F., Doranz B.J., Rucker J., Liesnard C., Farber C.M., Saragosti S., Lapoumeroulie C., Cognaux J., Forceille C., Muyldermans G., Verhofstede C., Burtonboy G., Georges M., Imai T., Rana S., Yi Y., Smyth R.J., Collman R.G., Doms R.W., Vassart G., and Parmentier M. 1996. Resistance to HIV-1 infection in caucasian individuals bearing mutant alleles of the CCR-5 chemokine receptor gene. Nature. **382**: 722–725.

59. Dorr P., Westby M., Dobbs S., Griffin P., Irvine B., Macartney M., Mori J., Rickett G., Smith-Burchnell C., Napier C., Webster R., Armour D., Price D., Stammen B., Wood A., and Perros M. 2005.Maraviroc (UK-427,857), a potent, orally bioavailable, and selective small-molecule inhibitor of chemokine receptor CCR5 with broad-spectrum anti-human immunodeficiency virus type 1 activity. Antimicrob. Agents Chemother. **49**: 4721–4732.

60. Maeda K., Yoshimura K., Shibayama S., Habashita H., Tada H., Sagawa K., Miyakawa T., Aoki M., Fukushima D., and Mitsuya H. 2001. Novel low molecular weight spirodiketopiperazine derivatives potently inhibit R5 HIV-1 infection through their antagonistic effects on CCR5. J. Biol. Chem. **276**: 35194–35200.

61. Strizki J.M., Xu S., Wagner N.E., Wojcik L., Liu J., Hou Y., Endres M., Palani A., Shapiro S., Clader J.W., Greenlee W.J., Tagat J.R., McCombie S., Cox K., Fawzi A.B., Chou C.C., Pugliese-Sivo C., Davies L., Moreno M.E., Ho D.D., Trkola A., Stoddart C.A., Moore J.P., Reyes G.R., and Baroudy B.M. 2001. SCH-C (SCH 351125), an orally bioavailable, small molecule antagonist of the chemokine receptor CCR5, is a potent inhibitor of HIV-1 infection in vitro and in vivo. Proc. Natl. Acad. Sci. U S A. **98**: 12718–12723.

62. Baba M., Nishimura O., Kanzaki N., Okamoto M., Sawada H., Iizawa Y., Shiraishi M., Aramaki Y., Okonogi K., Ogawa Y., Meguro K., and Fujino M. 1999. A small-molecule, nonpeptide CCR5 antagonist with highly potent and selective anti-HIV-1 activity. Proc. Natl. Acad. Sci. U S A. **96**: 5698–5703.

63. Wilkin T.J., Su Z., Kuritzkes D.R., Hughes M., Flexner C., Gross R., Coakley E., Greaves W., Godfrey C., Skolnik P.R., Timpone J., Rodriguez B., and Gulick R.M. 2007. HIV type 1 chemokine coreceptor use among antiretroviral-experienced patients screened for a clinical trial of a CCR5 inhibitor: AIDS Clinical Trial Group A5211. Clin. Infect. Dis. **44**: 591–595.

64. Takanami I. 2003. Overexpression of CCR7 mRNA in nonsmall cell lung cancer: correlation with lymph node metastasis. Int. J. Cancer. **105**: 186–189.

65. Mashino K., Sadanaga N., Yamaguchi H., Tanaka F., Ohta M., Shibuta K., Inoue H., and Mori M. 2002. Expression of chemokine receptor CCR7 is

associated with lymph node metastasis of gastric carcinoma. Cancer Res. **62**: 2937–2941.

66. Kodama J., Hasengaowa, Kusumoto T., Seki N., Matsuo T., Ojima Y., Nakamura K., Hongo A., and Hiramatsu Y. 2007. Association of CXCR4 and CCR7 chemokine receptor expression and lymph node metastasis in human cervical cancer. Ann. Oncol. **18**: 70–76.
67. Hedrick J.A., and Zlotnik A. 1997. Identification and characterization of a novel beta chemokine containing six conserved cysteines. J. Immunol. **159**: 1589–1593.
68. Gunn M.D., Kyuwa S., Tam C., Kakiuchi T., Matsuzawa A., Williams L.T., and Nakano H. 1999. Mice lacking expression of secondary lymphoid organ chemokine have defects in lymphocyte homing and dendritic cell localization. J. Exp. Med. **189**: 451–460.
69. Alfonso-Perez M., Lopez-Giral S., Quintana N.E., Loscertales J., Martin-Jimenez P., and Munoz C. Anti-CCR7 monoclonal antibodies as a novel tool for the treatment of chronic lymphocyte leukemia. J. Leukoc. Biol. **79**: 1157–1165.
70. Wiley H.E., Gonzalez E.B., Maki W., Wu M.T., and Hwang S.T. 2001. Expression of CC chemokine receptor-7 and regional lymph node metastasis of B16 murine melanoma. J. Natl. Cancer Inst. **93**: 1638–1643.
71. Bajetto A., Barbieri F., Dorcaratto A., Barbero S., Daga A., Porcile C., Ravetti J.L., Zona G., Spaziante R., Corte G., Schettini G., and Florio T. 2006. Expression of CXC chemokine receptors 1-5 and their ligands in human glioma tissues: role of CXCR4 and SDF1 in glioma cell proliferation and migration. Neurochem. Int. **49**: 423–432.
72. Murphy C., McGurk M., Pettigrew J., Santinelli A., Mazzucchelli R., Johnston P.G., Montironi R., and Waugh D.J. 2005. Nonapical and cytoplasmic expression of interleukin-8, CXCR1, and CXCR2 correlates with cell proliferation and microvessel density in prostate cancer. Clin. Cancer Res. **11**: 4117–4127.
73. Mestas J., Burdick M.D., Reckamp K., Pantuck A., Figlin R.A., and Strieter R.M. 2005. The role of CXCR2/CXCR2 ligand biological axis in renal cell carcinoma. J. Immunol. **175**: 5351–5357.
74. Yuan A., Yang P.C., Yu C.J., Chen W.J., Lin F.Y., Kuo S.H., and Luh K.T. 2000. Interleukin-8 messenger ribonucleic acid expression correlates with tumor progression, tumor angiogenesis, patient survival, and timing of relapse in non-small-cell lung cancer. Am. J. Respir. Crit. Care Med. **162**: 1957–1963.
75. Masuya D., Huang C., Liu D., Kameyama K., Hayashi E., Yamauchi A., Kobayashi S., Haba R., and Yokomise H. 2001. The intratumoral expression of vascular endothelial growth factor and interleukin-8 associated with angiogenesis in nonsmall cell lung carcinoma patients. Cancer. **92**: 2628–2638.
76. Orditura M., De V.F., Catalano G., Infusino S., Lieto E., Martinelli E., Morgillo F., Castellano P., Pignatelli C., and Galizia G. 2002. Elevated serum levels of interleukin-8 in advanced non-small cell lung cancer patients: relationship with prognosis. J. Interferon Cytokine Res. **22**: 1129–1135.
77. White E.S., Flaherty K.R., Carskadon S., Brant A., Iannettoni M.D., Yee J., Orringer M.B., and Arenberg D.A. 2003. Macrophage migration inhibitory factor and CXC chemokine expression in non-small cell lung cancer: role in angiogenesis and prognosis. Clin. Cancer Res. **9**: 853–860.
78. Chen J.J., Yao P.L., Yuan A., Hong T.M., Shun C.T., Kuo M.L., Lee Y.C., and Yang P.C. 2003. Up-regulation of tumor interleukin-8 expression by infiltrating

macrophages: its correlation with tumor angiogenesis and patient survival in non-small cell lung cancer. Clin. Cancer Res. **9**: 729–737.

79. Yoneda J., Kuniyasu H., Crispens M.A., Price J.E., Bucana C.D., and Fidler I.J. 1998. Expression of angiogenesis-related genes and progression of human ovarian carcinomas in nude mice. J. Natl. Cancer Inst. **90**: 447–454.
80. Fregene T.A., Khanuja P.S., Noto A.C., Gehani S.K., Van Egmont E.M., Luz D.A., and Pienta K.J. 1993. Tumor-associated angiogenesis in prostate cancer. Anticancer Res. **13**: 2377–2381.
81. Inoue K., Slaton J.W., Eve B.Y., Kim S.J., Perrotte P., Balbay M.D., Yano S., Bar-Eli M., Radinsky R., Pettaway C.A., and Dinney C.P. 2000. Interleukin 8 expression regulates tumorigenicity and metastases in androgen-independent prostate cancer. Clin. Cancer Res. **6**: 2104–2119.
82. Inoue K., Slaton J.W., Kim S.J., Perrotte P., Eve B.Y., Bar-Eli M., Radinsky R., and Dinney C.P. 2000. Interleukin 8 expression regulates tumorigenicity and metastasis in human bladder cancer. Cancer Res. **60**: 2290–2299.
83. Merogi A.J., Marrogi A.J., Ramesh R., Robinson W.R., Fermin C.D., and Freeman S.M. 1997. Tumor-host interaction: analysis of cytokines, growth factors, and tumor-infiltrating lymphocytes in ovarian carcinomas. Hum. Pathol. **28**: 321–331.
84. Singh R.K., Gutman M., Radinsky R., Bucana C.D., and Fidler I.J. 1994. Expression of interleukin 8 correlates with the metastatic potential of human melanoma cells in nude mice. Cancer Res. **54**: 3242–3247.
85. Ugurel S., Rappl G., Tilgen W., and Reinhold U. 2001. Increased serum concentration of angiogenic factors in malignant melanoma patients correlates with tumor progression and survival. J. Clin. Oncol. **19**: 577–583.
86. Varney M.L., Johansson S.L., and Singh R.K. 2006. Distinct expression of CXCL8 and its receptors CXCR1 and CXCR2 and their association with vessel density and aggressiveness in malignant melanoma. Am. J. Clin. Pathol. **125**: 209–216.
87. Zhu Y.M., Bagstaff S.M., and Woll P.J. 2006. Production and upregulation of granulocyte chemotactic protein-2/CXCL6 by IL-1beta and hypoxia in small cell lung cancer. Br. J. Cancer. **94**: 1936–1941.
88. Huang S., Mills L., Mian B., Tellez C., McCarty M., Yang X.D., Gudas J.M., and Bar-Eli M. 2002. Fully humanized neutralizing antibodies to interleukin-8 (ABX-IL8) inhibit angiogenesis, tumor growth, and metastasis of human melanoma. Am. J. Pathol. **161**: 125–134.
89. Bendre M.S., Margulies A.G., Walser B., Akel N.S., Bhattacharrya S., Skinner R.A., Swain F., Ramani V., Mohammad K.S., Wessner L.L., Martinez A., Guise T.A., Chirgwin J.M., Gaddy D., and Suva L.J. 2005. Tumor-derived interleukin-8 stimulates osteolysis independent of the receptor activator of nuclear factor-kappaB ligand pathway. Cancer Res. **65**: 11001–11009.
90. Collins T.S., Lee L.F., and Ting J.P. 2000. Paclitaxel up-regulates interleukin-8 synthesis in human lung carcinoma through an NF-kappaB- and AP-1-dependent mechanism. Cancer Immunol. Immunother. **49**: 78–84.
91. De Larco J.E., Wuertz B.R., Manivel J.C., and Furcht L.T. 2001. Progression and enhancement of metastatic potential after exposure of tumor cells to chemotherapeutic agents. Cancer Res. **61**: 2857–2861.
92. Lev D.C., Onn A., Melinkova V.O., Miller C., Stone V., Ruiz M., McGary E.C., Ananthaswamy H.N., Price J.E., and Bar-Eli M. 2004. Exposure of melanoma cells to dacarbazine results in enhanced tumor growth and metastasis in vivo. J. Clin. Oncol. **22**: 2092–2100.

93. Wilson C., Purcell C., Seaton A., Oladipo O., Maxwell P.J., O'Sullivan J.M., Wilson R.H., Johnston P.G., and Waugh D.J. 2008. Chemotherapy-induced CXC-chemokine/CXC-chemokine receptor signaling in metastatic prostate cancer cells confers resistance to oxaliplatin through potentiation of nuclear factor-kappaB transcription and evasion of apoptosis. J. Pharmacol. Exp. Ther. **327**: 746–759.

94. Merritt W.M., Lin Y.G., Spannuth W.A., Fletcher M.S., Kamat A.A., Han L.Y., Landen C.N., Jennings N., De G.K., Langley R.R., Villares G., Sanguino A., Lutgendorf S.K., Lopez-Berestein G., Bar-Eli M.M., and Sood A.K. 2008. Effect of interleukin-8 gene silencing with liposome-encapsulated small interfering RNA on ovarian cancer cell growth. J. Natl. Cancer Inst. **100**: 359–372.

95. Waugh D.J., and Wilson C. 2008. The interleukin-8 pathway in cancer. Clin. Cancer Res. **14**: 6735–6741.

96. Winters M.P., Crysler C., Subasinghe N., Ryan D., Leong L., Zhao S., Donatelli R., Yurkow E., Mazzulla M., Boczon L., Manthey C.L., Molloy C., Raymond H., Murray L., McAlonan L., and Tomczuk B. 2008. Carboxylic acid bioisosteres acylsulfonamides, acylsulfamides, and sulfonylureas as novel antagonists of the CXCR2 receptor. Bioorg. Med. Chem. Lett. **18**: 1926–1930.

97. Lai G., Merritt J.R., He Z., Feng D., Chao J., Czarniecki M.F., Rokosz L.L., Stauffer T.M., Rindgen D., and Taveras A.G. 2008. Synthesis and structure-activity relationships of new disubstituted phenyl-containing 3,4-diamino-3-cyclobutene-1,2-diones as CXCR2 receptor antagonists. Bioorg. Med. Chem. Lett. **18**: 1864–1868.

98. Walters I., Austin C., Austin R., Bonnert R., Cage P., Christie M., Ebden M., Gardiner S., Grahames C., Hill S., Hunt F., Jewell R., Lewis S., Martin I., David N., and David R. 2008. Evaluation of a series of bicyclic CXCR2 antagonists. Bioorg. Med. Chem. Lett. **18**: 798–803.

99. Gonsiorek W., Fan X., Hesk D., Fossetta J., Qiu H., Jakway J., Billah M., Dwyer M., Chao J., Deno G., Taveras A., Lundell D.J., and Hipkin R.W. 2007. Pharmacological characterization of Sch527123, a potent allosteric CXCR1/CXCR2 antagonist. J. Pharmacol. Exp. Ther. **322**: 477–485.

100. McCleland B.W., Davis R.S., Palovich M.R., Widdowson K.L., Werner M.L., Burman M., Foley J.J., Schmidt D.B., Sarau H.M., Rogers M., Salyers K.L., Gorycki P.D., Roethke T.J., Stelman G.J., Azzarano L.M., Ward K.W., and Busch-Petersen J. 2007. Comparison of N,N′-diarylsquaramides and N,N′-diarylureas as antagonists of the CXCR2 chemokine receptor. Bioorg. Med. Chem. Lett. **17**: 1713–1717.

101. Dwyer M.P., Yu Y., Chao J., Aki C., Chao J., Biju P., Girijavallabhan V., Rindgen D., Bond R., Mayer-Ezel R., Jakway J., Hipkin R.W., Fossetta J., Gonsiorek W., Bian H., Fan X., Terminelli C., Fine J., Lundell D., Merritt J.R., Rokosz L.L., Kaiser B., Li G., Wang W., Stauffer T., Ozgur L., Baldwin J., Taveras A.G. 2006. Discovery of 2-hydroxy-N,N-dimethyl-3-{2-[[(R)-1-(5-methylfuran-2-yl)propyl]amino]-3,4 dioxocyclobut-1-enylamino}benzamide (SCH 527123): a potent, orally bioavailable CXCR2/CXCR1 receptor antagonist. J. Med. Chem. **49**: 7603–7606.

102. de Haas C.J., Veldkamp K.E., Peschel A., Weerkamp F., Van Wamel W.J., Heezius E.C., Poppelier M.J., Van Kessel K.P., and van Strijp J.A. 2004. Chemotaxis inhibitory protein of *Staphylococcus aureus*, a bacterial antiinflammatory agent. J. Exp. Med. **199**: 687–695.

103. Rooijakkers S.H., Ruyken M., Roos A., Daha M.R., Presanis J.S., Sim R.B., van Wamel W.J., van Kessel K.P., and van Strijp J.A. 2005. Immune evasion by a staphylococcal complement inhibitor that acts on C3 convertases. Nat. Immunol. **6**: 920–927.

104. Jongerius I., Kohl J., Pandey M.K., Ruyken M., Van Kessel K.P., van Strijp J.A., and Rooijakkers S.H. Staphylococcal complement evasion by various convertase-blocking molecules. J. Exp. Med. **204**: 2461–2471.

105. Prat C., Bestebroer J., de Haas C.J., van Strijp J.A., and Van Kessel K.P. 2006. A new staphylococcal anti-inflammatory protein that antagonizes the formyl peptide receptor-like 1. J. Immunol. **177**: 8017–8026.

106. Bestebroer J., Poppelier M.J., Ulfman L.H., Lenting P.J., Denis C.V., Van Kessel K.P., van Strijp J.A., and de Haas C.J. 2007. Staphylococcal superantigen-like 5 binds PSGL-1 and inhibits P-selectin-mediated neutrophil rolling. Blood. **109**: 2936–2943.

107. Bestebroer J., van Kessel K.P., Azouagh H., Walenkamp A.M., Boer I.G., Romijn R.A., van Strijp J.A., and de Haas C.J. 2008. Staphylococcal SSL5 inhibits leukocyte activation by chemokines and anaphylatoxins. Blood. **113**: 328–337.

108. Postma B., Kleibeuker W., Poppelier M.J., Boonstra M., Van Kessel K.P., van Strijp J.A., and de Haas C.J. 2005. Residues 10-18 within the C5a receptor N terminus compose a binding domain for chemotaxis inhibitory protein of *Staphylococcus aureus*. J. Biol. Chem. **280**: 2020–2027.

109. Langley R., Wines B., Willoughby N., Basu I., Proft T., and Fraser J.D. 2005. The staphylococcal superantigen-like protein 7 binds IgA and complement C5 and inhibits IgA-Fc alpha RI binding and serum killing of bacteria. J. Immunol. **174**: 2926–2933.

110. Ui M. 1990. ADP-Ribosylating Toxins and G Proteins. Washington, DC: American Society for Microbiology, p. 64.

111 Rahimpour R., Mitchell G., Khandaker M.H., Kong C., Singh B., Xu L., Ochi A., Feldman R.D., Pickering J.G., Gill B.M., and Kelvin D.J. 1999. Bacterial superantigens induce down-modulation of CC chemokine responsiveness in human monocytes via an alternative chemokine ligand-independent mechanism. J. Immunol. **162**: 2299–2307.

112. Sica A., Saccani A., Borsatti A., Power C.A., Wells T.N., Luini W., Polentarutti N., Sozzani S., and Mantovani A. 1997. Bacterial lipopolysaccharide rapidly inhibits expression of C-C chemokine receptors in human monocytes. J. Exp. Med. **185**: 969–974.

113. Kim H.K., Kim J.E., Chung J., Han K.S., and Cho H.I. 2007. Surface expression of neutrophil CXCR4 is down-modulated by bacterial endotoxin. Int. J. Hematol. **85**: 390–396.

114. Dimitroff C.J., Descheny L., Trujillo N., Kim R., Nguyen V., Huang W., Pienta K.J., Kutok J.L., and Rubin M.A. 2005. Identification of leukocyte E-selectin ligands, P-selectin glycoprotein ligand-1 and E-selectin ligand-1, on human metastatic prostate tumor cells. Cancer Res. **65**: 5750–5760.

115. Dardik R., Kaufmann Y., Savion N., Rosenberg N., Shenkman B., and Varon D.1997. Platelets mediate tumor cell adhesion to the subendothelium under flow conditions: involvement of platelet GPIIb-IIIa and tumor cell alpha(v) integrins. Int. J. Cancer. **70**: 201–207.

116. Chen M., and Geng J.G. 2006. P-selectin mediates adhesion of leukocytes, platelets, and cancer cells in inflammation, thrombosis, and cancer growth and metastasis. Arch. Immunol. Ther. Exp. (Warsz). **54**: 75–84.
117. Kannagi R. 1997. Carbohydrate-mediated cell adhesion involved in hematogenous metastasis of cancer. Glycoconj. J. **14**: 577–584.
118. Witz I.P. 2006. The involvement of selectins and their ligands in tumor-progression. Immunol. Lett. **104**: 89–93.
119. Aigner S., Ramos C.L., Hafezi-Moghadam A., Lawrence M.B., Friederichs J., Altevogt P., and Ley K. 1998. CD24 mediates rolling of breast carcinoma cells on P-selectin. FASEB J. **12**: 1241–1251.
120. Aruffo A., Dietsch M.T., Wan H., Hellstrom K.E., and Hellstrom I. 1992. Granule membrane protein 140 (GMP140) binds to carcinomas and carcinoma-derived cell lines. Proc. Natl. Acad. Sci. U S A. **89**: 2292–2296.
121. Schadendorf D., Heidel J., Gawlik C., Suter L., and Czarnetzki B.M. 1995. Association with clinical outcome of expression of VLA-4 in primary cutaneous malignant melanoma as well as P-selectin and E-selectin on intratumoral vessels. J. Natl. Cancer Inst. **87**: 366–371.
122. Kim Y.J., Borsig L., Varki N.M., and Varki A. 1998. P-selectin deficiency attenuates tumor growth and metastasis. Proc. Natl. Acad. Sci. U S A. **95**: 9325–9330.
123. Stone J.P., and Wagner D.D. 1993. P-selectin mediates adhesion of platelets to neuroblastoma and small cell lung cancer. J. Clin. Invest. **92**: 804–813.
124. Nieswandt B., Hafner M., Echtenacher B., and Mannel D.N. 1999. Lysis of tumor cells by natural killer cells in mice is impeded by platelets. Cancer Res. **59**: 1295–1300.
125. Verheul H.M., Jorna A.S., Hoekman K., Broxterman H.J., Gebbink M.F., and Pinedo H.M. 2000. Vascular endothelial growth factor-stimulated endothelial cells promote adhesion and activation of platelets. Blood. **96**: 4216–4221.
126. Kisucka J., Butterfield C.E., Duda D.G., Eichenberger S.C., Saffaripour S., Ware J., Ruggeri Z.M., Jain R.K., Folkman J., and Wagner D.D. 2006. Platelets and platelet adhesion support angiogenesis while preventing excessive hemorrhage. Proc. Natl. Acad. Sci. U S A. **103**: 855–860.
127. Raes G., Ghassabeh G.H., Brys L., Mpofu N., Verschueren H., Vanhecke D., and De B.P. 2007. The metastatic T-cell hybridoma antigen/P-selectin glycoprotein ligand 1 is required for hematogenous metastasis of lymphomas. Int. J. Cancer. **121**: 2646–2652.
128. Gorelik E., Galili U., and Raz A. On the role of cell surface carbohydrates and their binding proteins (lectins) in tumor metastasis. Cancer Metastasis Rev. **20**: 245–277.
129. Nakamori S., Kameyama M., Imaoka S., Furukawa H., Ishikawa O., Sasaki Y., Kabuto T., Iwanaga T., Matsushita Y., and Irimura T. 1993. Increased expression of sialyl Lewisx antigen correlates with poor survival in patients with colorectal carcinoma: clinicopathological and immunohistochemical study. Cancer Res. **53**: 3632–3637.
130. Weston B.W., Hiller K.M., Mayben J.P., Manousos G.A., Bendt K.M., Liu R., and Cusack J.C., Jr. 1999. Expression of human alpha(1,3)fucosyltransferase antisense sequences inhibits selectin-mediated adhesion and liver metastasis of colon carcinoma cells. Cancer Res. **59**: 2127–2135.

131. Kishimoto T., Ishikura H., Kimura C., Takahashi T., Kato H., and Yoshiki T. 1996. Phenotypes correlating to metastatic properties of pancreas adenocarcinoma in vivo: the importance of surface sialyl Lewis(a) antigen. Int. J. Cancer. **69**: 290–294.

132. Baker H.M., Basu I., Chung M.C., Caradoc-Davies T., Fraser J.D., and Baker E.N. 2007. Crystal structures of the staphylococcal toxin SSL5 in complex with sialyl Lewis X reveal a conserved binding site that shares common features with viral and bacterial sialic acid binding proteins. J. Mol. Biol. **374**: 1298–1308.

12

PSEUDOMONAS EXOTOXIN A-BASED IMMUNOTOXINS FOR TARGETED CANCER THERAPY

Philipp Wolf and Ursula Elsässer-Beile

Department of Urology, Experimental Urology, University Hospital Freiburg, Freiburg, Germany

INTRODUCTION

Classical therapeutic modalities such as surgery, radiation, and chemotherapy often fail to cure neoplastic diseases. Moreover, their use leads to severe and debilitating side effects. Therefore, tumor-directed immunotherapy has been established as a fourth modality of cancer therapy being specifically lethal for malignant cells and less toxic to normal cells.

One immunotherapeutic strategy is the use of PE-based immunotoxins, whereby the enzymatic active domain of *Pseudomonas aeruginosa* exotoxin A (PE) is specifically directed to tumor-associated antigens.

In this chapter, an overview of the structure and the cytotoxic pathways of PE in host cells is given. Additionally, the construction and clinical outcome of different PE-based immunotoxins are discussed, and further prospects are given on the main research focus in the future.

PSEUDOMONAS AERUGINOSA

P. aeruginosa is a gram-negative, opportunistic, and highly versatile bacterium that ranges among the six most frequently encountered pathogens in hospital-

Emerging Cancer Therapy: Microbial Approaches and Biotechnological Tools, Edited by Arsénio M. Fialho and Ananda M. Chakrabarty

acquired infections (1). Moreover, it is one of the most common bacterial isolate from blood-borne infections and is the most frequent cause of nosocomial pneumonia. Its ubiquitous occurrence is a result of its ability to adapt to various adverse environmental conditions and to utilize many environmental compounds for energy extraction (2).

The most serious infections caused by *P. aeruginosa* in humans range from acute infections as endophtalmitis, endocarditis, meningitis, and septicemia, to chronic lung infections. Immunocompromised people, such as AIDS patients, burn victims, or patients undergoing chemotherapy, are at high risk for *P. aeruginosa* infections. Additionally, a high mortality rate is found in persons with underlying diseases like cystic fibrosis or cancer (3). The pathogenesis of *P. aeruginosa* infections is multifactorial, as manifested by the numerous virulence factors produced. For instance, surface factors, including pili, lipopolysaccharides, and polysaccharide alginates play a decisive role in bacterial adherence and colonization, whereas secreted proteins, like proteases, endotoxins, and exotoxins, contribute to dissemination and tissue damage (4).

PSEUDOMONAS EXOTOXIN A

One virulence factor of *P. aeruginosa* is the exotoxin A (PE). This protein is produced in about 95% of the strains and is classified as the bacterium's most toxic substance on the basis of weight. PE is an enzyme with ADP-ribosyltransferase activity, in detail a NAD^+-diphthamide-ADP-ribosyltransferase (EC 2.4.2.36). Its importance for the toxicity of the bacterium became apparent in a first study by Iglewski and Kabat in 1975 (5). These authors discovered that PE catalyzes the ADP-ribosylation of the eukaryotic elongation factor 2 (eEF-2) in host cells. eEF-2 belongs to a set of proteins that facilitate the events of translational elongation at the ribosomes. Its modification by PE significantly affects the protein synthesis by terminating peptide chain elongation and ultimately leads to cell death.

Structural and Functional Domains

PE conforms to the A-B structure-function model of the mono-ADP-ribosyltransferase family, which, among others, includes cholera toxin from *Vibrio cholerae*, diphtheria toxin from *Corynebacterium diphteriae*, heat-labile enterotoxin from *Escherichia coli*, pertussis toxin from *Bordetella pertussis*, and C2 toxin from *Clostridium botulinum* (6). The B-domain of these molecules is responsible for the interaction with an eukaryotic cell receptor, whereas the A-domain represents the catalytic moiety. All A-B structured toxins share a similar multistep mechanism in which (1) the B-domain binds to a receptor on the membrane surface of a target cell, (2) the catalytic A-domain is translocated into the cytoplasm, and (3) the A-domain modifies its target substrate.

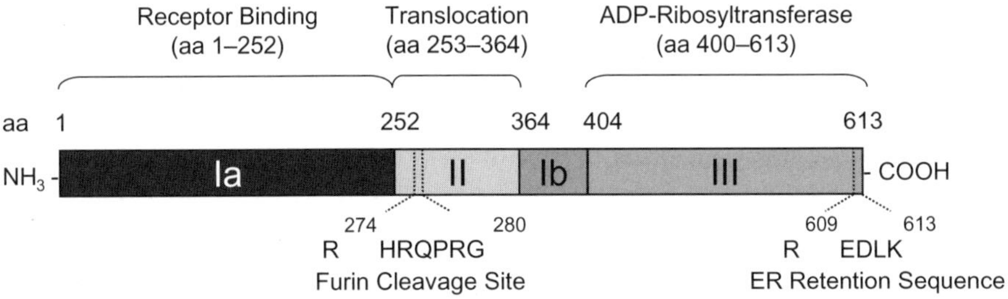

Figure 12.1 Structural and functional domains of *Pseudomonas* exotoxin A (PE). Domain Ia (aa 1–252) represents the receptor binding domain. Domain II (aa 253–364) is required for the translocation of the toxin across cellular membranes. The catalytic subunit of PE with ADP ribosyltransferase activity (aa 400–613) is located inside the structural domains Ib (aa 365–404) and III (aa 405–613). The furin cleavage site (aa 274–280) in domain II and the ER retention sequence (aa 609–613) at the C-terminus represent further essential motifs for the cytotoxicity of PE.

The 66 kDa PE protein is expressed as a 638 amino acid precursor in *P. aeruginosa* from which a highly hydrophobic leader peptide of 25 amino acids (aa) is removed during the secretion process. PE consists of four functional domains: the domain Ia (aa 1–252) at the N-terminus represents the receptor binding domain, whereas domain II (aa 253–364) is required for the translocation of the toxin across cellular membranes (Figure 12.1). The exact function of domain Ib (aa 365–404) is so far unknown and may be required for the secretion of the toxin by the bacterium. Nevertheless, its last four residues (aa 400–404) together with domain III (aa 405–613) form the catalytic subunit of the protein with ADP-ribosyltransferase activity (7).

Two amino acid motifs have been characterized by mutation analyses, which have vital importance with regard to the cytotoxic action of PE. The first motif (aa 274–280, RHRQPRG) is located inside domain II. It is exposed on the surface of the protein and is therewith accessible for the cleavage catalyzed by furin, a subtilisin-like, Ca^{2+}-dependent processing endoprotease (8). The furin cleavage of PE leads to the separation of the receptor binding domain Ia from the catalytic subunit of the protein. This proteolytic activation of PE is common with other intracellularly acting bacterial toxins. In these toxins, the catalytic domain can only enter the target cells if it is physically attached to the receptor-recognition and translocation domains. However, it has to be cleaved intracellularly to fully express its activity. The second essential motif of PE is the C-terminal REDLK sequence (aa 609–613). Deletions and substitutions inside that motif indicated that this sequence is required for full cytotoxic activity (9). Similar sequences were also found at the C-terminus of the *Cholera* toxin A domain or the *E. coli* heat-labile toxin A domain. Today, these sequences are known to bind to the KDEL receptor in the trans-Golgi network (TGN) of the host cell and therefore to act as endoplasmatic reticulum (ER) retention sequences.

Cytotoxic Pathways

Figure 12.2 illustrates the cytotoxic pathways of PE to reach its target protein eEF-2 in the cytosol. The first essential step of PE activity is the cleavage of the lysine residue (aa 613) at the C-terminus in the extracellular space, presumably by plasma carboxypeptidases. This alters the terminal motif from REDLK to REDL (10). Then PE binds via its receptor binding domain Ia to the alpha2-macroglobulin receptor/low-density lipoprotein receptor-related protein (α2MR/LRP) on the surface of the host cells (11). α2MR/LRP, also known as CD91 or LRP1, is a broadly expressed, large dimeric cell surface glycoprotein. It is thought to be responsible for the binding and endocytosis of activated α2-macroglobulin and apoE-enriched and very low-density lipoproteins. Once bound, PE uses at least two pathways to reach the ER of the host cell: the KDEL receptor-mediated pathway and the lipid-dependent sorting pathway (12, 13).

During the KDEL receptor-mediated pathway, the internalization occurs via clathrin-coated pits and PE dissociates from α2MR/LRP in the early endosomes. Then it undergoes a conformational change, and is cleaved by furin between residues R-279 and G-280 within domain II (14). This leads to the separation of PE into an N-terminal fragment of about 28 kDa and a C-terminal fragment of 37 kDa, the latter holding the full ADP-ribosyltransferase activity of the molecule (15). Both fragments are still connected by a disulfide bond, encompassing the furin cleavage site. There is evidence that under the mildly acidic conditions in the endosomes, there is an unfolding event, possibly by the binding of chaperone proteins. This leads to a surface exposure of the disulfide bond with subsequent reduction, perhaps by protein disulfide isomerases, followed by the release of the C-terminal 37 kDa fragment (16). This fragment then travels via late endosomes to the TGN at a route, which is dependent on the GTPase Rab 9 (17). In the TGN, the 37 kDa fragment of PE binds with its C-terminal retention sequence REDL (aa 609–612) to the KDEL receptor, followed by the retrograde transport from the TGN to the ER (18).

During the lipid-dependent sorting pathway, α2MR/LRP-bound PE can associate with detergent-resistant microdomains, and is then transported via caveosomes to the early endosomes under control of Rab5. After furin cleavage, the enzymatic active domain of PE reaches the ER in a pathway, controlled by Rab6 (12).

From the ER, the PE fragment is dislocated into the cytosol. It is suggested that the Sec61p translocon is involved in this step, which is responsible for the dislocation of unfolded or misfolded cellular proteins in eukaryotic cells for proteasomal degradation (19).

Once the enzymatic subunit of PE has reached the cytosol, it catalyzes the ADP-ribosylation of its target protein eEF-2 (20). This ADP-ribosylation mechanism was studied in detail (21): in brief, PE binds to the coenzyme NAD^+ and facilitates the cleavage of the glycosidic bond between nicotinamide and the N-ribose of NAD^+, resulting in a reactive oxacarbenium

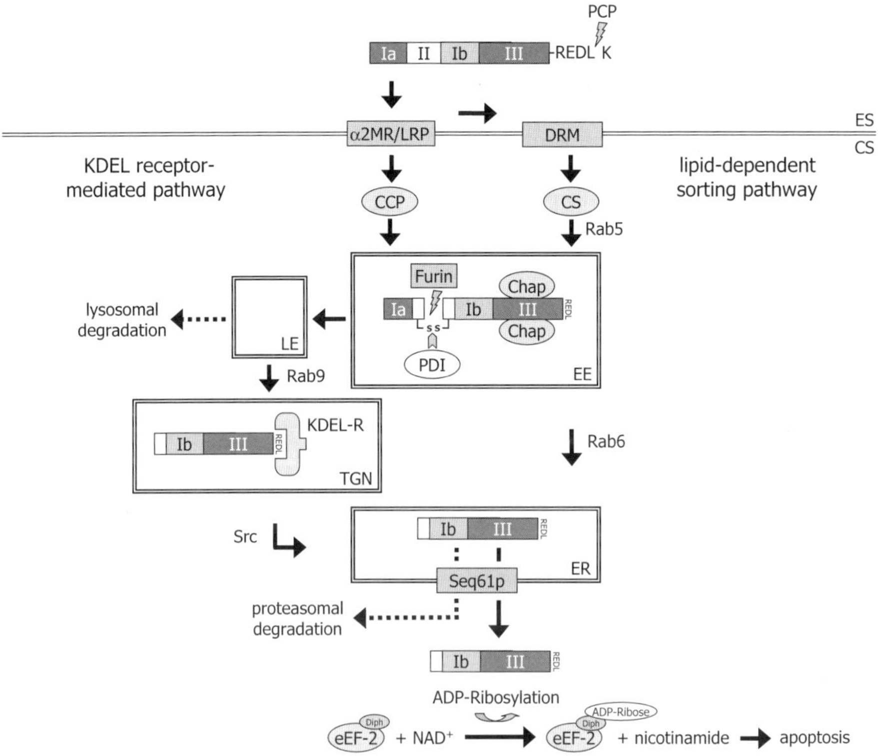

Figure 12.2 Cytotoxic pathways of *Pseudomonas* exotoxin A (PE). After cleavage of the C-terminal lysine (K) by plasma carboxypeptidases (PCP) in the extracellular space (ES), PE binds to the alpha2-macroglobulin receptor/low-density lipoprotein receptor-related protein (α2MR/LRP) on the surface of the host cell. Then, PE can exploit two different pathways to reach the endoplasmatic reticulum (ER): the KDEL receptor-mediated pathway and the lipid-dependent sorting pathway. Using the KDEL receptor-mediated pathway PE is internalized via clathrin-coated pits (CCP) and is cleaved by furin in the early endosomes (EE). After that, there is an unfolding event presumably after chaperone (Chap) binding with subsequent disulfide bond cleavage by protein disulfide isomerases (PDI). Then, the enzymatic active 37 kDa PE fragment travels via late endosomes (LE) in a Rab9-dependent manner to the trans-Golgi network (TGN). There it can bind to the KDEL receptor via its C-terminal ER-retention signal REDL and is then transported to the ER under control of the tyrosine kinase Src. Using the lipid-dependent sorting pathway, PE associates with detergent-resistant microdomains (DRM) after α2MR/LRP binding and is transported via caveosomes (CV) to the EE in a Rab5-dependent manner. After cleavage in the EE, the toxic PE fragment directly travels under the control of Rab6 to the ER. In the ER, the PE fragment is secreted via the translocon Sec61p into the cytosol (CS). There it ADP-ribosylates the eukaryotic elongation factor-2 (eEF-2) by transferring the ADP-ribose group from NAD^+ to the diphthamide (Diph) residue of eEF-2. This leads to the inactivation of eEF-2, to the inhibition of protein biosynthesis on the ribosomes, and finally, to apoptosis.

intermediate. This step is followed by a nucleophilic attack of the eEF-2 protein, based on the nucleophilic residue diphthamide, a post-translationally modified histidine residue (2-[3-carboxyamido-3-(trimethylammonio)propyl] histidine), which was exclusively found in eEF-2 (22). The ADP-ribose group is then transferred to the diphthamide residue, which finally results in the ADP-ribosylated eEF-2 protein. Interestingly, X-ray structure analyses showed evidence that NAD^+-bound PE mimics the normal interaction between eEF-2 and the eukaryotic 80S ribosome during protein translation (21).

The ADP-ribosylation leads to an inactivation of eEF-2, which is followed by peptide chain elongation termination and apoptosis of the target cell. The exact mechanism of PE-induced apoptosis has not yet been clarified. *In vitro* studies showed a decrease in cdc2 and cyclin B protein expression and an increase of the 14-3-3 delta regulatory protein on the one side and an activation of caspase-8 and caspase-3 on the other side (23, 24). This suggests that PE first induces cell cycle arrests and then leads to the activation of the receptor-mediated caspase-8-dependent apoptotic pathway.

Molecular Strategies

PE is a fascinating example of how pathogenic microorganisms developed structure-based, molecular strategies for the effective damage of their host cells, because:

1. it recognizes the randomly distributed α2MR/LRP receptor via its receptor binding domain and therefore reaches many different host cells,
2. it chooses the lipid-dependent sorting pathway to bypass late endosomes with subsequent lysosomal degradation or back transfer to the extracellular space,
3. it is masqueraded, presumably by the unfolding events in the endosomes, as an unfolded/misfolded cellular protein for later secretion into the cytosol,
4. it takes advantage of the recycling machinery of the host cell to reach the ER by binding its modified C-terminus to the KDEL receptor,
5. it uses the translocon Seq61p for the secretion into the cytosol, which usually transports unfolded or misfolded cellular proteins for subsequent proteasomal degradation,
6. it specifically inactivates the eukaryotic protein synthetic machinery by ADP-ribosylating the unique diphthamide residue of eEF-2 and therefore does not affect prokaryotic elongation factors that are present in the toxin producing *P. aeruginosa* bacterium itself, and
7. it evolves its ADP-ribosylation activity by ribosome mimicry before it is degraded by the proteasome.

In the last three decades, the great value of PE as an effective toxin was recognized and many PE-based immunotoxins as therapeutic agents against cancer were developed.

PSEUDOMONAS EXOTOXIN A-BASED IMMUNOTOXINS

Construction

Immunotoxins are generally composed of two main domains. The targeting domain is specific for a tumor surface antigen. The toxin domain consists of a cytotoxic enzyme derived from a plant or bacterial source and exhibits the advantage of non-stoichiometric activity, since one enzyme can convert thousands of substrate molecules within a limited time frame. The toxin domain is linked to the targeting domain and is directed to the tumor cell surface. After this, it enters the cell and induces cell death, predominantly by inactivating protein biosynthesis (25, 26).

For the design of PE-based immunotoxins, the receptor binding domain Ia (aa 1–252) of PE is removed and is substituted by a targeting domain, which specifically binds to a tumor-associated antigen. There are two classes of targeting domains used in the construction of PE-based immunotoxins: one group is composed of monoclonal antibodies or antibody fragments, the other group consists of physiologically important ligands such as growth factors or cytokines.

The truncated form of PE, consisting of domains II, Ib, and III (aa 253–613), represents the toxin domain and is also known as PE40 according to its molecular weight of 40 kDa (Figure 12.3A). Additionally, a part of the domain Ib (aa 365–380), which does not influence the toxicity of PE, can be deleted. This modified form is called PE38.

The first PE-based immunotoxins were produced by chemically coupling the full native toxin to antibodies (Figure 12.3B). These immunotoxins had several drawbacks, including lack of specificity, poor stability, and heterogeneous composition (27). After further knowledge about the structure and function of PE, the cell binding domain Ia was removed, and the resulting toxin fragment that no longer bound to normal cells was coupled to a tumor-specific antibody (Figure 12.3B). This approach increased the amount of immunotoxin that could be safely given in preclinical and clinical studies, but problems of heterogeneity or increased unspecific toxicity due to the binding of the antibody to certain plasma proteins (opsonins) via its Fc domains persisted.

Recombinant PE-based immunotoxins represent the actual state of immunotoxin construction (Figure 12.3C). Genes of the PE38 or PE40 domains are fused to genes of the heavy (V_H) and light (V_L) chains of an antibody. Thereby, the V_H and V_L genes can either be linked by a flexible peptide linker (single-chain antibody fragment [scFv]), or by a disulfide bond (disulfide-stabilized

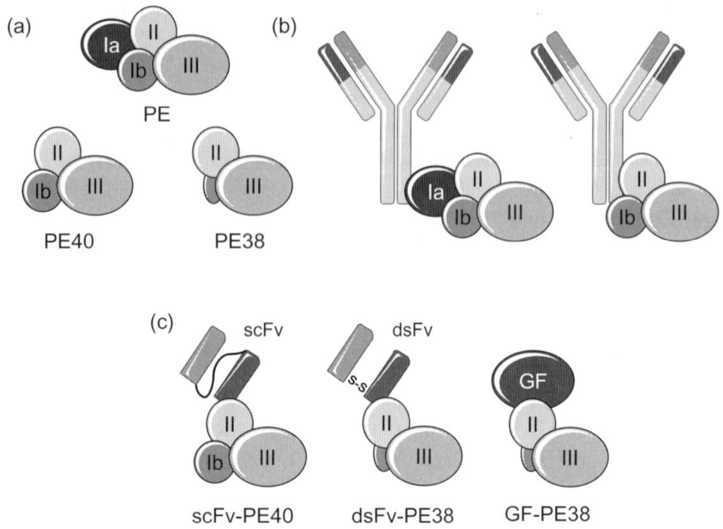

Figure 12.3 Schematic representation of the construction of PE-based immunotoxins: (A) the PE protein molecule consists of the domains Ia, II, Ib, and III and forms a three-dimensional structure. The truncated form of PE, missing the cell binding domain Ia and consisting of domains II, Ib, and III, is called PE40. A part of domain Ib, which does not influence the toxicity of PE, can be additionally deleted and is then called PE38; (B) first-generation immunotoxins were constructed by chemically coupling the whole PE molecule or the PE40 domain to full tumor-specific monoclonal antibodies; (C) recombinant made PE-based immunotoxins represent the actual state of immunotoxin construction and consist of single-chain antibody fragments (scFvs), disulfide stabilized antibody fragments (dsFvs), or growth factors (GF) as binding domains and PE40 or PE38 as toxin domains.

antibody fragment [dsFv]). Alternatively, genes of growth factors or other natural ligands are used as targeting domains. Recombinant immunotoxins combine many advantages to be superior to chemically linked immunotoxins: (1) they can be expressed in *E. coli* in large-scale production, (2) they are expressed as single stranded, homogeneous proteins, (3) they allow the insertion of site-specific mutations, and (4) their small sizes improve tumor penetration.

Preclinical Trials

Many PE-based immunotoxins against various overexpressed antigens in different tumors have been developed and tested in preclinical studies (Table 12.1), and the number of newly synthesized constructs is continually growing. *In vitro* criteria usually include successful bacterial expression and purification of the construct, as well as specific binding, internalization, and cytotoxicity against tumor target cells. Additionally, sufficient serum stability

TABLE 12.1. Preclinical Trials of PE-Based Immunotoxins

Immunotoxin	Construction	Target Antigen	Tumors	Ref.
OVB3-PE	mAb linked via disulfide bond to PE	Ovary	Ovarian	53
B3-Lys-PE38 [LMB-1]	mAb chemical linked to PE38	LeY	Various	54
9.2.27-PE	mAb chemical linked to PE	HMW-MAA	Gliomblastoma	55
CD19-ETA′	scFv fused to PE38KDEL	CD19	Lymphoma, leukemia	56
anti-Tac(Fv)-PE38KDEL [LMB2]	scFv fused to PE38KDEL	CD25	CD25 positive tumors	57
RTF5(scFv)-ETA′	scFv fused to PE40	CD25	Lymphoma	58
RFB(dsFv)-PE38 [BL22]	dsFvfused to PE38	CD22	B cell leukemia	59
G28-5 sFv-PE40	scFv fused to PE40	CD40	Burkitt's lymphoma	60
B1(dsFv)-PE38	dsFv fused to PE38	LeY	LeY positive tumor cells	61
B3(dsFv)-PE38	dsFv fused to PE38	LeY	LeY positive tumor cells	61
BR96sFv-PE40 [SGN-10]	scFv fused to PE40	LeY	LeY positive tumor cells	62–64
IL4(38-37) PE38KDEL [NBI-3001]	IL4mut fused to PE38KDEL	IL4-R	Breast, HNSCC, pancreas, BTC	65, 66
IL13-PE38QQR [Cintredekin Besudotox]	IL13 fused to PE38QQR	IL13-R	HNSCC	67
anti-IL-13Ralpha2(scFv)-PE38	IL13 fused to PE38	IL13-Rα2	Glioma	68
scFv(FRP5)-ETA	scFv fused to PE40	erbB2	Ovarian, prostate	69, 70
AR209 [e23(Fv) PE38KDEL]	scFv fused to PE38KDEL	erbB-2	Lung, prostate	71, 72
Erb-38	dsFv fused to PE38	erbB2	Epidermoid carcinoma, breast	73
TP38	TGFα fused to PE38	EGFR	Glioma	74
TP40	TGFα fused to PE40	EGFR	Glioma, prostate, epidermoid	75, 76
A5-PE40	scFv fused to PE40	PSMA	Prostate	77, 78

TABLE 12.1. *Continued*

Immunotoxin	Construction	Target Antigen	Tumors	Ref.
SS1(dsFv)PE38 [SS1P]	dsFv fused to PE38	Mesothelin	Ovarian, cervical	79
8H9(dsFv)-PE38	dsFv fused to PE38	Cell surface glycoprotein	Breast, osteosarcoma, neuroblastoma	80
4D5MOCB-ETA [VB4-845]	scFv fused to PE40KDEL	Ep-CAM	Lung, colon, SCC	81
HB21(Fv)-PE40	scFv fused to PE40	TfR	Colon	82
TARC-PE38	TARC fused to PE38	CCR4	T cell lymphoma/ leukemia	83
Fab 2F1-PE38KDEL	Fab fused to PE38KDEL	MART-1(26–35)	Melanoma	84

BTC, biliary tract carcinoma; CCR4, Chemokine (C-C motif) receptor 4; dsFv, disulfide-stabilized antibody fragment; EGFR, epidermal growth factor receptor; erbB2, HER2/neu-receptor; HMW-MAA, high-molecular-weight melanoma-associated antigen; HNSCC, head and neck squamous cell carcinoma; IL13-R, interleukin 13-receptor; IL4mut, interleukin 4 mutant with increased binding affinity; IL4-R, interleukin 4-receptor; Le^{Y}, Lewis-related antigen; mAb, monoclonal antibody; MART-1(26–35), melanoma-associated antigen peptide; PE, *Pseudomonas* exotoxin A (aa 1–613); PE38, truncated form of *Pseudomonas* exotoxin A (aa 253–364 and 381–613); PE38KDEL, PE38 with C-terminal added endoplasmatic reticulum retention motif KDEL; PE38QQR, PE38 with replaced lysines 590 and 606 with glutamines and lysine 613 with arginine; PE40, truncated form of *Pseudomonas* exotoxin A (aa 253–613); PE40KDEL, PE40 with C-terminal added endoplasmatic reticulum retention motif KDEL; PSMA, prostate-specific membrane antigen; SCC, squamous cell carcinoma; scFv, single-chain antibody fragment; TARC, thymus and activation-regulated chemokine; TfR, transferrin receptor; TGFα, transforming growth factor alpha.

and thermostability and possible cross-reactivities toward normal tissues are examined. As soon as an immunotoxin has successfully passed the *in vitro* testing, maximal tolerated doses (MTD) and antitumorous effects are examined *in vivo* in animals bearing human tumor xenografts.

Clinical Trials

Several PE-based immunotoxins, which had preclinically shown great promise for further application in cancer patients, were also tested in clinical trials (Table 12.2). Especially in studies against hematologic malignancies, a good clinical outcome was demonstrated. For instance, the immunotoxin BL22, consisting of a dsFv against CD22 as the binding domain and PE38 as the toxin domain, caused 11 complete responses (CR) and 7 partial responses (PR) in 18 patients with hairy-cell leukemia (HCL) on the one side, and 19 CR and 7 PR in 46 lymphoma or leukemia patients on the other side in two Phase I studies (28, 29). The immunotoxin LMB-2 is directed against CD25 and led to 1 CR and 7 PR in 35 patients with hematologic tumors (30, 31).

TABLE 12.2. Clinical Trials of PE-Based Immunotoxins

Name	Phase	Cancer Type	No. of Patients	Application	Outcome	Ref.
BL22 [CAT-3888;	I	HCL	16	i.v. infusion	11 CR, 2 PR, VLS toxicity	28
dsFv(RFB4)-PE38]	I	Lymphoma, leukemia	46	i.v. infusion	19 CR, 7 PR, VLS toxicity	29
LMB-2	I	Lymphoma, leukemia	35	i.v. infusion	1 CR, 7 PR, hepatotoxicity	30, 31
VB4-845	I/II	Bladder	64	Intravesicular catheter	27 CR, 13 PR	32
[4D5MOC-B-ETA]	II	Bladder	29	Intravesicular catheter	9 CR of the first evaluated 18 patients	
	I	HNSCC	24	Intratumoral injection	2 significant and 4 minor tumor regressions	
	I	HNSCC	20	Intratumoral injection	4 CR, 6 PR	
IL4(38-37)PE38KDEL	I	Glioma	9	Intratumoral infusion	1CR, 6 TN, CNS toxicity	33
[NBI-3001]	I	Glioma	31	Intratumoral infusion	22/31 TN, CNS toxicity	34
	I	Solid tumors	14	i.v. infusion	Hepatotoxicity	35
OVB3-PE	I	Ovarian	23	i.v. infusion	CNS toxicity	36
LMB-1 [B3-Lys-PE38]	I	Solid tumors	38	i.v. infusion	1 CR, 1 PR, VLS toxicity	37
SGN-10 [BR96sFv-PE40]	I	Solid tumors	46	i.v. infusion	GI toxicity	38
Erb-38	I	Breast, ovarian	6	i.v. infusion	Hepatotoxicity	39
scFv(FRP5)-ETA	CCR	Breast, colon, melanoma	11	Intratumoral injection	4 complete and 2 partial regressions of nodules	40
	I	Solid tumors	18	i.v. infusion	Hepatotoxicity	41
SS1P [SS1(dsFv)PE38]	I	Mesotheliona, ovarian, pancreas	34	Infusion and bolus	1 PR, 10 patients with indication of antitumor activity, pleuritis	42
Cintredekin Besudotox [IL13-PE38QQR]	I	Glioma	51	Intraparenchymal infusion	CNS toxicity	85
TP38	I	Glioma	20	Intracerebral infusion	1 CR, 1 PR, fatique, CNS toxicity	86
TP40	I	Bladder	43	Transurethral instillation	1 CR, urinary tract irritations	87

CCR, collected case reports; CNS, central nervous system; CR, complete response; GI, gastrointestinal; HCL, hairy-cell leukemia; i.v., intravenous; PR, partial response; TN, tumor necrosis; VLS, vascular leak syndrome.

Another example for a clinically successful acting PE-bound immunotoxin is VB4-845, which targets the epithelial cell adhesion molecule (EpCAM) that is expressed in almost all carcinomas. In a Phase I/II dose-escalation trial, VB4-845 showed a high antitumor efficacy with 27 CR and 13 PR in 64 bladder cancer patients after intravesical application (32). In an ongoing Phase II study, 9 CR in the first 18 evaluated patients bearing bladder tumors, which are unsuitable for Bacillus Calmette-Guerin (BCG) treatment, were registered (32). VB4-845 was also tested in two Phase I trials against head and neck squamous cell cancer (HNSCC) and caused 4 CR and 6 PR in a total of 44 patients after intratumoral injections (32).

Immunotoxin treatment of tumors within the intracerebral compartment poses a great challenge because these tumors are isolated from the systemic circulation by the restrictive blood–brain barrier. Therefore, clinical studies with PE-based immunotoxins against gliomas were performed by intratumoral or intracerebral infusion. For example, the anti-IL4-R immunotoxin IL4(38-37)-PE38KDEL was intratumorally applied to treat high-grade or recurrent gliomas leading to a high number of tumor necroses (33, 34). However, the same immunotoxin showed no clinical responses against solid tumors after intravenous application (35). This is in accordance with few other systemically applied immunotoxins directed against different solid malignancies, which showed only low antitumor activity or failed to cause partial or complete responses (36–42).

With regard to the results of the clinical trials, there is evidence that only PE-based immunotoxins, which are directed against hematologic malignancies or which were intratumorally or tumor adjacent applied, elicited efficient clinical responses.

The reasons for this can be outlined as follows: first, cells from hematologic tumors are more accessible to circulating immunotoxin molecules than cells from solid tumors, because they are located intravascularly or perivascularly in highly perfused lymph nodes. Second, systemically applied immunotoxins against solid tumors have to overcome several hurdles before they can reach the tumor: the generally normal slow rate of protein release into the extracellular space, tight junctions between tumor cells with high interstitial pressure, as well as heterogenous blood supply because of abnormal blood vessels within the tumor masses.

Problems and Further Development

Problems with PE-based immunotoxins in clinical trials often include insufficient cytotoxic activity, lack of tumor specifity, and high immunogenicity resulting in a reduced efficacy or undesirable side effects. Therefore, many approaches focus on the further development of more effective and safer PE-based immunotoxins.

Protein engineering can be used to increase the efficacy of PE-based immunotoxins. For example, BL22 (CAT-3888) was affinity-matured by phage display, and a variant, called HA22 (CAT-8015), with three mutated residues in the complement determining region 3 (CDR3) of the V_H chain, demonstrated an approximately 14-fold increased affinity toward the target antigen CD22 compared to its predecessor (43). In the mutant form of HA22, called HA22-LR, cleavage clusters in the PE38 domain, which are susceptible to lysosomal proteases, were additionally deleted (44). HA22-LR killed CD22 positive cells more potently and uniformly than HA22. Additionally, mice tolerated at least 10-fold higher doses of HA22-LR than lethal doses of HA22.

The lack of specificity of an immunotoxin is often based on a target antigen expression not only on tumorous but also on normal tissues. Nonspecific binding to normal tissues reduces the antitumor activity of the immunotoxin and often leads to severe side effects. For example, hepatotoxicity was dose limiting in the study with the immunotoxin Erb-38 and was traced back to the expression of the target antigen erbB2 on normal hepatocytes (39). In the study with SGN-10, the presence of the Lewis Y antigen on normal stomach cells resulted in gastrointestinal toxicity causing nausea, vomiting, and diarrhea (38). Similarly, SS1P induced pleuritis due to mesothelin expression on pleural mesothelial cells (42).

As shown in Table 12.2, ligand-independent toxicity of PE-based immunotoxins in clinical trials often manifests in hepatotoxicity or the so-called vascular leak syndrome (VLS). There is evidence that PE causes T cell- and Kupffer cell (KC)-dependent hepatotoxicity and that tumor necrosis factor (TNF)-α, interleukin (IL)-18, and perforin are important mediators of liver damage following PE injection (45). VLS usually occurs 4–6 days after immunotoxin treatment. It involves the damage of epithelial cells with increase in vascular permeability, extravasation of fluids and proteins resulting in weight gain and, in most severe form, kidney damage and pulmonary edema (46). There is evidence that PE-based immunotoxins cause VLS in two ways. On the one hand, it was demonstrated that specific amino acid motifs of the PE domain directly bind to vascular endothelial cells, which is followed by endothelial injury (47). Therefore, several strategies are under investigation to reduce VLS by mutation of the vascular endothelial binding residues within the PE domain. On the other hand, VLS can be indirectly induced by the activation of vascular endothelial cells and/or macrophages via elevated circulating cytokines such as TNF-α and interferon (IFN)-γ (48).

In nearly all clinical studies with PE-based immunotoxins, a high number of patients generated anti-immunotoxin antibodies. These antibodies are neutralizing and may form immune complexes with an accelerated clearance from circulation. This results in a reduced antitumor activity of the immunotoxin and limits the treatment to only few applications. To reduce the immunogenicity of the cell binding domains, the antibody fragments are humanized by protein engineering, library technologies, high-throughput screening techniques, or antibody generation in transgenic mice (49). The immunogenicity of the PE domain can be reduced by elimination of immunodominant epitopes. This was achieved by mutation of specific large hydrophilic residues of PE38 in the immunotoxin HA22 (50).

Pharmacologic problems have also been observed in many preclinical and clinical studies. Small immunotoxins show a reduced endothelial toxicity and an increased tumor penetration. However, they are rapidly cleared via renal excretion, leading to a short half-life, a reduced area under the curve, and tubular damage in the kidney. To overcome such problems, the anti-CD25 immunotoxin LMB-2 was modified with 5 or 20 kDa polyethylene glycol (51). Compared to the unmodified construct, both PEGylated immunotoxins showed similar cytotoxicities. However, they both had prolonged serum stabilities and 5–8-fold increased plasma half-lives as well as 3–4-fold increased antitumor activities accompanied by a substantial decrease of animal toxicity. Another interesting approach was to encapsulate the PE38KDEL domain into poly(lactic-co-glycolic acid) (PLGA) antibody modified nanoparticles (NPs), which were conjugated to Fab′ fragments of a humanized anti-HER2 monoclonal antibody (52). Administration of the so-called PE-NP-HER construct showed an increased inhibition of tumor growth as well as a significant reduction of systemic toxicity in mice compared to the PE-HER counterpart.

CONCLUSIONS

Increased knowledge of the cytotoxic pathways of PE, the availability of new antibodies against different tumor antigens, new molecular technologies, and rapid progress in understanding molecular mechanisms of cancer development enable the generation of new PE-based immunotoxins as alternative agents against different cancers. Remarkable clinical outcomes were especially achieved in trials against hematologic malignancies or against tumors that were treated by intratumoral or tumor-directed catheter application. Nevertheless, PE-based immunotoxins also represent valuable, specific acting tools for the future treatment of solid tumors in view of adjuvant application against residual tumor cells after primary therapies or against metastases of low burden.

REFERENCES

1. Vasil M.L. 1986. *Pseudomonas aeruginosa*: biology, mechanisms of virulence, epidemiology. J. Pediatr. **108**: 800–805.
2. Green S.K., Schroth M.N., Cho J.J., Kominos S.K., and Vitanza-Jack V.B. 1974. Agricultural plants and soil as a reservoir for *Pseudomonas aeruginosa*. Appl. Microbiol. **28**: 987–991.
3. Driscoll J.A., Brody S.L., and Kollef M.H. 2007. The epidemiology, pathogenesis and treatment of *Pseudomonas aeruginosa* infections. Drugs. **67**: 351–368.
4. Bitter W. 2003. Secretins of *Pseudomonas aeruginosa*: large holes in the outer membrane. Arch. Microbiol. **179**: 307–314.
5. Iglewski B.H., and Kabat D. 1975. NAD-dependent inhibition of protein synthesis by *Pseudomonas aeruginosa* toxin. Proc. Natl. Acad. Sci. U S A. **72**: 2284–2288.

6. Deng Q., and Barbieri J.T. 2008. Molecular mechanisms of the cytotoxicity of ADP-ribosylating toxins. Annu. Rev. Microbiol. **62**: 271–288.
7. Wedekind J.E., Trame C.B., Dorywalska M., Koehl P., Raschke T.M., McKee M., FitzGerald D., Collier R.J., and McKay D.B. 2001. Refined crystallographic structure of *Pseudomonas aeruginosa* exotoxin A and its implications for the molecular mechanism of toxicity. J. Mol. Biol. **314**: 823–837.
8. Gordon V.M., Klimpel K.R., Arora N., Henderson M.A., and Leppla S.H. 1995. Proteolytic activation of bacterial toxins by eukaryotic cells is performed by furin and by additional cellular proteases. Infect. Immun. **63**: 82–87.
9. Chaudhary V.K., Jinno Y., FitzGerald D., and Pastan I. 1990. *Pseudomonas* exotoxin contains a specific sequence at the carboxyl terminus that is required for cytotoxicity. Proc. Natl. Acad. Sci. U S A. **87**: 308–312.
10. Hessler J.L., and Kreitman R.J. 1997. An early step in *Pseudomonas* exotoxin action is removal of the terminal lysine residue, which allows binding to the KDEL receptor. Biochemistry. **36**: 14577–14582.
11. Kounnas M.Z., Morris R.E., Thompson M.R., FitzGerald D.J., Strickland D.K., and Saelinger C.B. 1992. The alpha 2-macroglobulin receptor/low density lipoprotein receptor-related protein binds and internalizes *Pseudomonas* exotoxin A. J. Biol. Chem. **267**: 12420–12423.
12. Smith D.C., Spooner R.A., Watson P.D., Murray J.L., Hodge T.W., Amessou M., Johannes L., Lord J.M., and Roberts L.M. 2006. Internalized *Pseudomonas* exotoxin A can exploit multiple pathways to reach the endoplasmic reticulum. Traffic. **7**: 379–393.
13. Wolf P., and Elsasser-Beile U. 2009. *Pseudomonas* exotoxin A: from virulence factor to anti-cancer agent. Int. J. Med. Microbiol. **299**: 161–176.
14. Ogata M., Fryling C.M., Pastan I., and FitzGerald D.J. 1992. Cell-mediated cleavage of *Pseudomonas* exotoxin between Arg279 and Gly280 generates the enzymatically active fragment which translocates to the cytosol. J. Biol. Chem. **267**: 25396–25401.
15. Ogata M., Chaudhary V.K., Pastan I., and FitzGerald D.J. 1990. Processing of *Pseudomonas* exotoxin by a cellular protease results in the generation of a 37,000-Da toxin fragment that is translocated to the cytosol. J. Biol. Chem. **265**: 20678–20685.
16. McKee M.L., and FitzGerald D.J. 1999. Reduction of furin-nicked *Pseudomonas* exotoxin A: an unfolding story. Biochemistry. **38**: 16507–16513.
17. Lombardi D., Soldati T., Riederer M.A., Goda Y., Zerial M., and Pfeffer S.R. 1993. Rab9 functions in transport between late endosomes and the trans Golgi network. EMBO J. **12**: 677–682.
18. Jackson M.E., Simpson J.C., Girod A., Pepperkok R., Roberts L.M., and Lord J.M. 1999. The KDEL retrieval system is exploited by *Pseudomonas* exotoxin A, but not by Shiga-like toxin-1, during retrograde transport from the Golgi complex to the endoplasmic reticulum. J. Cell Sci. **112**: 467–475.
19. Koopmann J.O., Albring J., Hüter E., Bulbuc N., Spee P., Neefjes J., Hämmerling G.J., and Momburg F. 2000. Export of antigenic peptides from the endoplasmic reticulum intersects with retrograde protein translocation through the Sec61p channel. Immunity. **13**: 117–127.

20. Iglewski B.H., Liu P.V., and Kabat D. 1977. Mechanism of action of *Pseudomonas aeruginosa* exotoxin A in adenosine diphosphate-ribosylation of mammalian elongation factor 2 in vitro and in vivo. Infect. Immun. **15**: 138–144.
21. Jørgensen R., Merrill A.R., Yates S.P., Marquez V.E., Schwan A.L., Boesen T., and Andersen G.R. 2005. Exotoxin A-eEF2 complex structure indicates ADP ribosylation by ribosome mimicry. Nature. **436**: 979–984.
22. Ortiz P.A., and Kinzy T.G. 2005. Dominant-negative mutant phenotypes and the regulation of translation elongation factor 2 levels in yeast. Nucleic Acids Res. **33**: 5740–5748.
23. Chang J.H., and Kwon H.Y. 2007. Expression of 14-3-3delta, cdc2 and cyclin B proteins related to exotoxin A-induced apoptosis in HeLa S3 cells. Int. Immunopharmacol. **7**: 1185–1191.
24. Jenkins C.E., Swiatoniowski A., Issekutz A.C., Lin T.J. 2004. *Pseudomonas aeruginosa* exotoxin A induces human mast cell apoptosis by a caspase-8 and -3-dependent mechanism. J. Biol. Chem. **279**: 37201–37207.
25. Frankel A.E., Kreitman R.J., and Sausville E.A. 2000. Targeted toxins. Clin. Cancer Res. **6**: 326–334.
26. Michl P., and Gress T.M. 2004. Bacteria and bacterial toxins as therapeutic agents for solid tumors. Curr. Cancer Drug Targets. **4**: 689–702.
27. Pastan I., Hassan R., FitzGerald D.J., and Kreitman R.J. 2007. Immunotoxin treatment of cancer. Annu. Rev. Med. **58**: 221–237.
28. Kreitman R.J., Wilson W.H., Bergeron K., Raggio M., Stetler-Stevenson M., FitzGerald D.J., and Pastan I. 2001. Efficacy of the anti-CD22 recombinant immunotoxin BL22 in chemotherapy-resistant hairy-cell leukemia. N. Engl. J. Med. **345**: 241–247.
29. Kreitman R.J., Squires D.R., Stetler-Stevenson M., Noel P., FitzGerald D.J., Wilson W.H., and Pastan I. 2005. Phase I trial of recombinant immunotoxin RFB4(dsFv)-PE38 (BL22) in patients with B-cell malignancies. J. Clin. Oncol. **23**: 6719–6729.
30. Kreitman R.J., Squires D.R., Stetler-Stevenson M., Noel P., FitzGerald D.J., Wilson W.H., and Pastan I. 2000. Phase I trial of recombinant immunotoxin anti-Tac(Fv)-PE38 (LMB-2) in patients with hematologic malignancies. J. Clin. Oncol. **18**: 1622–1636.
31. Kreitman R.J., Wilson W.H., Robbins D., Margulies I., Stetler-Stevenson M., Waldmann T.A., and Pastan I. 1999. Responses in refractory hairy cell leukemia to a recombinant immunotoxin. Blood. **94**: 3340–3348.
32. Biggers K., and Scheinfeld N. 2008. VB4-845, a conjugated recombinant antibody and immunotoxin for head and neck cancer and bladder cancer. Curr. Opin. Mol. Ther. **10**: 176–186.
33. Rand R.W., Kreitman R.J., Patronas N., Varricchio F., Pastan I., and Puri R.K. 2000. Intratumoral administration of recombinant circularly permuted interleukin-4-*Pseudomonas* exotoxin in patients with high-grade glioma. Clin. Cancer Res. **6**: 2157–2165.
34. Weber F., Asher A., Bucholz R., et al. 2003. Safety, tolerability, and tumor response of IL4-*Pseudomonas* exotoxin (NBI-3001) in patients with recurrent malignant glioma. J. Neurooncol. **64**: 125–137.
35. Garland L., Gitlitz B., Ebbinghaus S., Pan H., de Haan H., Puri R.K., Von Hoff D., and Figlin R. 2005. Phase I trial of intravenous IL-4 pseudomonas exotoxin protein (NBI-3001) in patients with advanced solid tumors that express the IL-4 receptor. J. Immunother. **28**: 376–381.

36. Pai, L.H., Bookman M.A., Ozols R.F., Young R.C., Smith J.W., Longo D.L., Gould B., Frankel A., McClay E.F., Howell S., Reed E., Willingham M.C., Fitzgerald D.J., and Paston I. 1991. Clinical evaluation of intraperitoneal *Pseudomonas* exotoxin immunoconjugate OVB3-PE in patients with ovarian cancer. J. Clin. Oncol. **9**: 2095–2103.
37. Pai L.H., Wittes R., Setser A., Willingham M.C., and Pastan I. 1996. Treatment of advanced solid tumors with immunotoxin LMB-1: an antibody linked to *Pseudomonas* exotoxin. Nat. Med. **2**: 350–353.
38. Posey J.A., Khazaeli M.B., Bookman M.A., et al. 2002. A phase I trial of the single-chain immunotoxin SGN-10 (BR96 sFv-PE40) in patients with advanced solid tumors. Clin. Cancer Res. **8**: 3092–3099.
39. Pai-Scherf L.H., Villa J., Pearson D., Watson T., Liu E., Willingham M.C., and Pastan I. 1999. Hepatotoxicity in cancer patients receiving erb-38, a recombinant immunotoxin that targets the erbB2 receptor. Clin. Cancer Res. **5**: 2311–2315.
40. Azemar M., Djahansouzi S., Jäger E., Solbach C., Schmidt M., Maurer A.B., Mross K., Unger C., von Minckwitz G., Dall P., Groner B., Wels W.S. 2003. Regression of cutaneous tumor lesions in patients intratumorally injected with a recombinant single-chain antibody-toxin targeted to ErbB2/HER2. Breast Cancer Res. Treat. **82**: 155–164.
41. von Minckwitz G., Harder S., Hövelmann S., Jäger E., Al-Batran S.E., Loibl S., Atmaca A., Cimpoiasu C., Neumann A., Abera A., Knuth A., Kaufmann M., Jäger D., Maurer A.B., and Wels W.S. 2005. Phase I clinical study of the recombinant antibody toxin scFv(FRP5)-ETA specific for the ErbB2/HER2 receptor in patients with advanced solid malignomas. Breast Cancer Res. **7**: R617–R626.
42. Hassan R., Bullock S., Premkumar A., Kreitman R.J., Kindler H., Willingham M.C., and Pastan I. 2007. Phase I study of SS1P, a recombinant anti-mesothelin immunotoxin given as a bolus I.V. infusion to patients with mesothelin-expressing mesothelioma, ovarian, and pancreatic cancers. Clin. Cancer Res. **13**: 5144–5149.
43. Alderson R.F., Kreitman R.J., Chen T., Yeung P., Herbst R., Fox J.A., and Pastan I. 2009. CAT-8015: a second-generation pseudomonas exotoxin A-based immunotherapy targeting CD22-expressing hematologic malignancies. Clin. Cancer Res. **15**: 832–839.
44. Weldon J.E., Xiang L., Chertov O., Margulies I., Kreitman R.J., FitzGerald D.J., and Pastan I. 2009. A protease-resistant immunotoxin against CD22 with greatly increased activity against CLL and diminished animal toxicity. Blood. **113**: 3792–3800.
45. Schumann J., Angermuller S., Bang R., Lohoff M., and Tiegs G. 1998. Acute hepatotoxicity of *Pseudomonas aeruginosa* exotoxin A in mice depends on T cells and TNF. J. Immunol. **161**: 5745–5754.
46. Vitetta E.S. 2000. Immunotoxins and vascular leak syndrome. Cancer J. **6**(Suppl. 3): S218–S224.
47. Baluna R., Rizo J., Gordon B.E., Ghetie V., Vitetta E.S. 1999. Evidence for a structural motif in toxins and interleukin-2 that may be responsible for binding to endothelial cells and initiating vascular leak syndrome. Proc. Natl. Acad. Sci. U S A. **96**: 3957–3962.
48. Baluna R., and Vitetta E.S. 1997. Vascular leak syndrome: a side effect of immunotherapy. Immunopharmacology. **37**: 117–132.
49. Almagro J.C., and Fransson J. 2008. Humanization of antibodies. Front. Biosci. **13**: 1619–1633.

50. Onda M., Beers R., Xiang L., Nagata S., Wang Q.C., and Pastan I. 2008. An immunotoxin with greatly reduced immunogenicity by identification and removal of B cell epitopes. Proc. Natl. Acad. Sci. U S A. **105**: 11311–11316.
51. Tsutsumi Y., Onda M., Nagata S., Lee B., Kreitman R.J., and Pastan I. 2000. Site-specific chemical modification with polyethylene glycol of recombinant immunotoxin anti-Tac(Fv)-PE38 (LMB-2) improves antitumor activity and reduces animal toxicity and immunogenicity. Proc. Natl. Acad. Sci. U S A. **97**: 8548–8553.
52. Chen H., Gao J., Lu Y., Kou G., Zhang H., Fan L., Sun Z., Guo Y., and Zhong Y. 2008. Preparation and characterization of PE38KDEL-loaded anti-HER2 nanoparticles for targeted cancer therapy. J. Control Release. **128**: 209–216.
53. Willingham M.C., FitzGerald D.J., and Pastan I. 1987. *Pseudomonas* exotoxin coupled to a monoclonal antibody against ovarian cancer inhibits the growth of human ovarian cancer cells in a mouse model. Proc. Natl. Acad. Sci. U S A. **84**: 2474–2478.
54. Pastan I. 1997. Targeted therapy of cancer with recombinant immunotoxins. Biochim. Biophys. Acta. **1333**: C1–C6.
55. Hjortland G.O., Garman-Vik S.S., Juell S., Olsen O.E., Hirschberg H., Fodstad O., and Engebraaten O. 2004. Immunotoxin treatment targeted to the high-molecular-weight melanoma-associated antigen prolonging the survival of immunodeficient rats with invasive intracranial human glioblastoma multiforme. J. Neurosurg. **100**: 320–327.
56. Schwemmlein M., Stieglmaier J., Kellner C., Peipp M., Saul D., Oduncu F., Emmerich B., Stockmeyer B., Lang P., Beck J.D., and Fey G.H. 2007. A CD19-specific single-chain immunotoxin mediates potent apoptosis of B-lineage leukemic cells. Leukemia. **21**: 1405–1412.
57. Kreitman R.J., Bailon P., Chaudhary V.K., FitzGerald D.J., and Pastan I. 1994. Recombinant immunotoxins containing anti-Tac(Fv) and derivatives of *Pseudomonas* exotoxin produce complete regression in mice of an interleukin-2 receptor-expressing human carcinoma. Blood. **83**: 426–434.
58. Barth S., Huhn M., Wels W., Diehl V., and Engert A. 1998. Construction and in vitro evaluation of RFT5(scFv)-ETA', a new recombinant single-chain immunotoxin with specific cytotoxicity toward CD25+ Hodgkin-derived cell lines. Int. J. Mol. Med. **1**: 249–256.
59. Kreitman R.J., Margulies I., Stetler-Stevenson M., Wang Q.C., FitzGerald D.J., and Pastan I. 2000. Cytotoxic activity of disulfide-stabilized recombinant immunotoxin RFB4(dsFv)-PE38 (BL22) toward fresh malignant cells from patients with B-cell leukemias. Clin. Cancer Res. **6**: 1476–1487.
60. Francisco J.A., Schreiber G.J., Comereski C.R., Mezza L.E., Warner G.L., Davidson T.J., Ledbetter J.A., and Siegall C.B. 1997. In vivo efficacy and toxicity of a single-chain immunotoxin targeted to CD40. Blood. **89**: 4493–4500.
61. Benhar I., Reiter Y., Pai L.H., and Pastan I. 1995. Administration of disulfide-stabilized Fv-immunotoxins B1(dsFv)-PE38 and B3(dsFv)-PE38 by continuous infusion increases their efficacy in curing large tumor xenografts in nude mice. Int. J. Cancer. **62**: 351–355.
62. Friedman P.N., McAndrew S.J., Gawlak S.L., Chace D., Trail P.A., Brown J.P., and Siegall C.B. 1993. BR96 sFv-PE40, a potent single-chain immunotoxin that selectively kills carcinoma cells. Cancer Res. **53**: 334–339.

63. Leland P., Taguchi J., Husain S.R., Kreitman R.J., Pastan I., and Puri R.K. 2000. Human breast carcinoma cells express type II IL-4 receptors and are sensitive to antitumor activity of a chimeric IL-4-*Pseudomonas* exotoxin fusion protein in vitro and in vivo. Mol. Med. **6**: 165–178.
64. Kawakami K., Leland P., and Puri R.K. 2000. Structure, function, and targeting of interleukin 4 receptors on human head and neck cancer cells. Cancer Res. **60**: 2981–2987.
65. Kawakami K., Kawakami M., Husain S.R., and Puri R.K. 2002. Targeting interleukin-4 receptors for effective pancreatic cancer therapy. Cancer Res. **62**: 3575–3580.
66. Ishige K., Shoda J., Kawamoto T., Matsuda S., Ueda T., Hyodo I., Ohkohchi N., Puri R.K., and Kawakami K. 2008. Potent in vitro and in vivo antitumor activity of interleukin-4-conjugated *Pseudomonas* exotoxin against human biliary tract carcinoma. Int. J. Cancer. **123**: 2915–2922.
67. Kawakami K., Kawakami M., Joshi B.H., and Puri R.K. 2001. Interleukin-13 receptor-targeted cancer therapy in an immunodeficient animal model of human head and neck cancer. Cancer Res. **61**: 6194–6200.
68. Kioi M., Seetharam S., and Puri R.K. 2008. Targeting IL-13Ralpha2-positive cancer with a novel recombinant immunotoxin composed of a single-chain antibody and mutated *Pseudomonas* exotoxin. Mol. Cancer Ther. **7**: 1579–1587.
69. Wels W., Harwerth I.M., Mueller M., Groner B., Hynes N.E. 1992. Selective inhibition of tumor cell growth by a recombinant single-chain antibody-toxin specific for the erbB-2 receptor. Cancer Res. **52**: 6310–6317.
70. Wang L., Liu B., Schmidt M., Lu Y., Wels W., and Fan Z. 2001. Antitumor effect of an HER2-specific antibody-toxin fusion protein on human prostate cancer cells. Prostate. **47**: 21–28.
71. Skrepnik N., Araya J.C., Qian Z., Xu H., Hamide J., Mera R., and Hunt J.D. 1996. Effects of anti-erbB-2 (HER-2/neu) recombinant oncotoxin AR209 on human non-small cell lung carcinoma grown orthotopically in athymic nude mice. Clin. Cancer Res. **2**: 1851–1857.
72. Skrepnik N., Zieske A.W., Bravo J.C., Gillespie A.T., and Hunt J.D. 1999. Recombinant oncotoxin AR209 (anti-P185erbB-2) diminishes human prostate carcinoma xenografts. J. Urol. **161**: 984–989.
73. Reiter Y., and Pastan I. 1996. Antibody engineering of recombinant Fv immunotoxins for improved targeting of cancer: disulfide-stabilized Fv immunotoxins. Clin. Cancer Res. **2**: 245–252.
74. Sampson J.H., Akabani G., Archer G.E., Bigner D.D., Berger M.S., Friedman A.H., Friedman H.S., Herndon J.E. II, Kunwar S., Marcus S., McLendon R.E., Paolino A., Penne K., Provenzale J., Quinn J., Reardon D.A., Rich J., Stenzel T., Tourt-Uhlig S., Wikstrand C., Wong T., Williams R., Yuan F., Zalutsky M.R., and Pastan I. 2003. Progress report of a Phase I study of the intracerebral microinfusion of a recombinant chimeric protein composed of transforming growth factor (TGF)-alpha and a mutated form of the *Pseudomonas* exotoxin termed PE-38 (TP-38) for the treatment of malignant brain tumors. J. Neurooncol. **65**: 27–35.
75. Pai L.H., Gallo M.G., FitzGerald D.J., Pastan I. 1991. Antitumor activity of a transforming growth factor alpha-*Pseudomonas* exotoxin fusion protein (TGF-alpha-PE40). Cancer Res. **51**: 2808–2812.

76. Kunwar S., Pai L.H., and Pastan I. 1993. Cytotoxicity and antitumor effects of growth factor-toxin fusion proteins on human glioblastoma multiforme cells. J. Neurosurg. **79**: 569–576.

77. Wolf P., Gierschner D., Buhler P., Wetterauer U., Elsasser-Beile U. 2006. A recombinant PSMA-specific single-chain immunotoxin has potent and selective toxicity against prostate cancer cells. Cancer Immunol. Immunother. **55**: 1367–1373.

78. Wolf P., Alt K., Bühler P., Katzenwadel A., Wetterauer U., Tacke M., and Elsässer-Beile U. 2008. Anti-PSMA immunotoxin as novel treatment for prostate cancer? High and specific antitumor activity on human prostate xenograft tumors in SCID mice. Prostate. **68**: 129–138.

79. Hassan R., Lerner M.R., Benbrook D., Lightfoot S.A., Brackett D.J., Wang Q.C., and Pastan I. 2002. Antitumor activity of SS(dsFv)PE38 and SS1(dsFv)PE38, recombinant antimesothelin immunotoxins against human gynecologic cancers grown in organotypic culture in vitro. Clin. Cancer Res. **8**: 3520–3526.

80. Onda M., Wang Q.C., Guo H.F., Cheung N.K., and Pastan I. 2004. *In vitro* and *in vivo* cytotoxic activities of recombinant immunotoxin 8H9(Fv)-PE38 against breast cancer, osteosarcoma, and neuroblastoma. Cancer Res. **64**: 1419–1424.

81. Di Paolo C., Willuda J., Kubetzko S., Lauffer I., Tschudi D., Waibel R., Plückthun A., Stahel R.A., and Zangemeister-Wittke U. 2003. A recombinant immunotoxin derived from a humanized epithelial cell adhesion molecule-specific single-chain antibody fragment has potent and selective antitumor activity. Clin. Cancer Res. **9**: 2837–2848.

82. Shinohara H., Fan D., Ozawa S., Yano S., Van Arsdell M., Viner J.L., Beers R., Pastan I., and Fidler I.J. 2000. Site-specific expression of transferrin receptor by human colon cancer cells directly correlates with eradication by antitransferrin recombinant immunotoxin. Int. J. Oncol. **17**: 643–651.

83. Baatar D., Olkhanud P., Newton D., Sumitomo K., Biragyn A. 2007. CCR4-expressing T cell tumors can be specifically controlled via delivery of toxins to chemokine receptors. J. Immunol. **179**: 1996–2004.

84. Klechevsky E., Gallegos M., Denkberg G., Palucka K., Banchereau J., Cohen C., and Reiter Y. 2008. Antitumor activity of immunotoxins with T-cell receptor-like specificity against human melanoma xenografts. Cancer Res. **68**: 6360–6367.

85. Kunwar S., Prados M.D., Chang S.M., Berger M.S., Lang F.F., Piepmeier J.M., Sampson J.H., Ram Z., Gutin P.H., Gibbons R.D., Aldape K.D., Croteau D.J., Sherman J.W., and Puri R.K. 2007. Direct intracerebral delivery of cintredekin besudotox (IL13-PE38QQR) in recurrent malignant glioma: a report by the Cintredekin Besudotox Intraparenchymal Study Group. J. Clin. Oncol. **25**: 837–844.

86. Sampson J.H., Akabani G., Archer G.E., Berger M.S., Coleman R.E., Friedman A.H., Friedman H.S., Greer K., Herndon J.E. II, Kunwar S., McLendon R.E., Paolino A., Petry N.A., Provenzale J.M., Reardon D.A., Wong T.Z., Zalutsky M.R., Pastan I., and Bigner D.D. 2008. Intracerebral infusion of an EGFR-targeted toxin in recurrent malignant brain tumors. Neuro. Oncol. **10**: 320–329.

87. Goldberg M.R., Heimbrook D.C., Russo P., Sarosdy M.F., Greenberg R.E., Giantonio B.J., Linehan W.M., Walther M., Fisher H.A.G., Messing E., Crawford E.D., Ollif A.I. and Pastan I.H. 1995. Phase I clinical study of the recombinant oncotoxin TP40 in superficial bladder cancer. Clin. Cancer Res. **1**: 57–61.

13

DENILEUKIN DIFTITOX IN NOVEL CANCER THERAPY

Lin-Chi Chen[1] and Nam H. Dang[2]

[1]*Nevada Cancer Institute, Las Vegas, NV*
[2]*University of Florida, Gainesville, FL*

DENILEUKIN DIFTITOX: STRUCTURE AND BIOLOGY

Denileukin diftitox is a recombinant fusion protein that combines the receptor binding properties of interleukin-2 (IL-2) with the enzymatic amino acid sequences of diphtheria toxin (DT). The ribbon structures of both IL-2 and DT are shown in Figure 13.1. DT is composed of three domains: A, the enzymatic domain; B, the heparin-binding epidermal growth factor (HB-EGF)-like receptor binding domain; and T, the membrane translocation domain. The fusion construct replaces the receptor binding domain of DT with amino acids 1–133 of IL-2. The recombinant protein thus binds to the interleukin-2 receptor (IL-2R) and delivers the enzymatic portion of DT to cells expressing this receptor (Figure 13.2).

The linkage site for DT proved important in the structure and function of the recombinant protein. The initial construct fused IL-2 to the DT at residue Ala486, resulting in DAB486-IL-2 (1, 2). This protein was shown to be active and well tolerated but to have a serum half-life of 5 min at the mean tolerated dose of 0.1 mg/kg/day over 10 doses. A second form of the fusion protein fused the DT fragment at Ala389. This fusion protein lacks 97 amino acids from the native DT receptor binding domain, resulting in a smaller 57-kDa protein (vs. 68 kDa for the DAB486IL-2 construct). The modification resulted in a fivefold

Emerging Cancer Therapy: Microbial Approaches and Biotechnological Tools, Edited by Arsénio M. Fialho and Ananda M. Chakrabarty

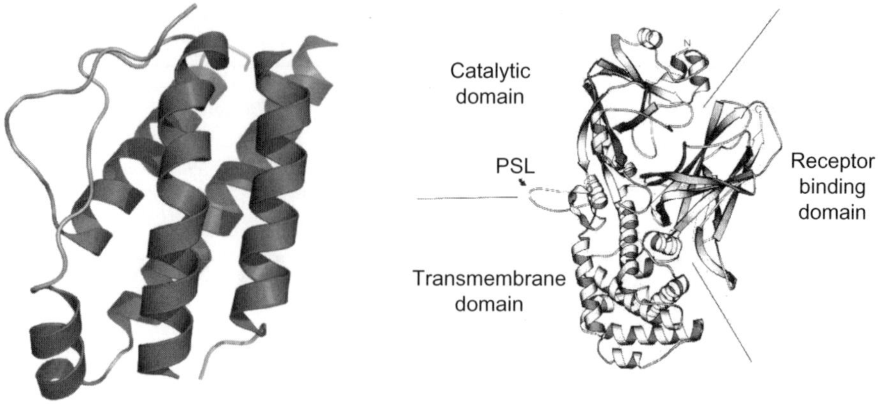

IL-2 from Wikimedia   Diphtheria toxin from http://gsbs.utmb.edu/.

Figure 13.1 Ribbon diagrams of (A) IL-2 (from Wikimedia free-access images). (B) Diphtheria toxin (http://gsbs.utmb.edu/). PSL, protease-sensitive loop.

increased affinity for the IL-2R and an approximate 10-fold increase in potency as well as an increase in half-life to 72 min at the mean tolerated dose of 31 μg/kg/day on days 1–5 every 3–4 weeks (3, 4).

MECHANISM OF ACTION

Denileukin diftitox binds to the IL-2R protein on the cell surface. The IL-2R is found on many hematopoietic cells, including T cells, B cells, natural killer cells, and monocytes (5). IL-2R is composed of combinations of the alpha (CD25), beta (CD122), and gamma (CD132) subunits (6, 7). Different combinations of these subunits result in low-affinity (CD25), intermediate-affinity (CD122/CD132), and high-affinity (CD25/CD122/CD132) receptors (6–8). The high-affinity receptor is usually expressed only on activated T and B cells and macrophages. Certain leukemias and lymphomas, in particular cutaneous T cell lymphoma (CTCL), express IL-2R (9). Expression in solid tumor malignancies is controversial. There are reports of preferential expression of the beta subunit CD122, but not the alpha subunit CD25. While the beta subunit is detected, the alpha subunit appears to be weakly expressed (10).

The receptor binding region of IL-2 was originally thought to bind to the alpha subunit (CD25, originally called the Tac protein) of the IL-2R. However, IL-2 also binds the intermediate-affinity receptor comprising the beta and gamma subunits where CD25 is not present, thus suggesting that the epitope is structural rather than sequence dependent. IL-2 binding to the intermediate- and high-affinity receptors results in internalization of the complex via

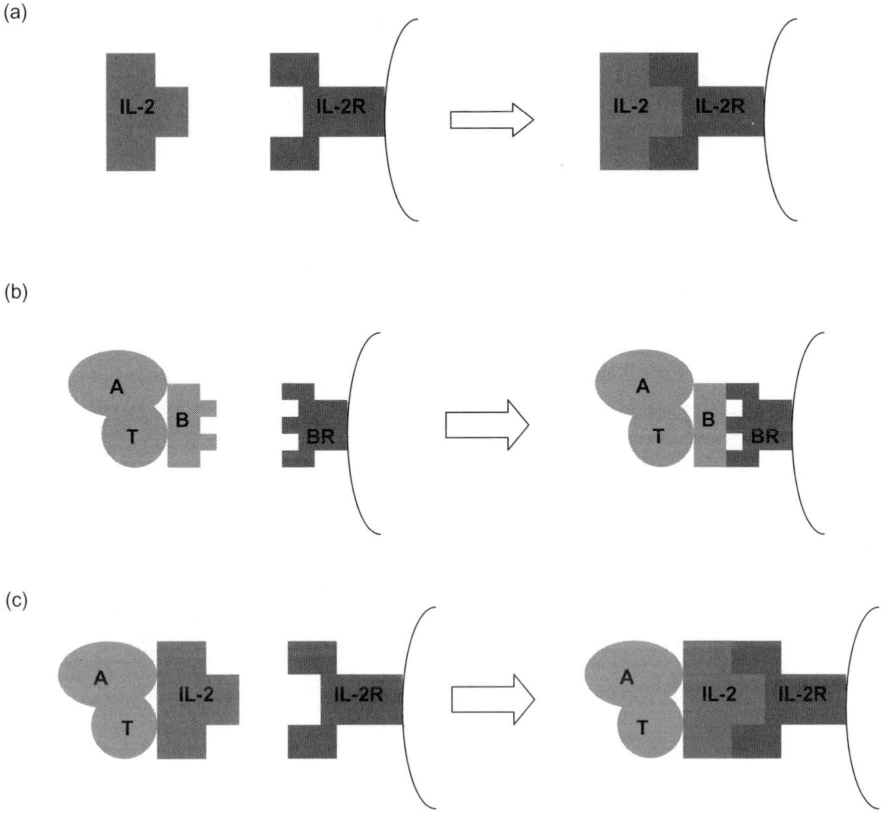

Figure 13.2 Schematic underlying fusion protein for denileukin diftitox. (A) IL-2 binding to native IL-2R. (B) Native diphtheria toxin binding to native receptor. (C) Denileukin diftitox binding to IL-2R-bearing cells. There is preferential binding to the high-affinity receptor containing alpha, beta, and gamma subunits. There is little binding to the low-affinity (alpha subunit only) and intermediate-affinity (beta and gamma subunits) receptors. IL-2, interleukin-2; IL-2R, interleukin-2 receptor; A, enzymatic/catalytic domain of diphtheria toxin (DT); B, native binding domain of DT; T, transmembrane domain of DT; BR, HB-EGF-like receptor.

receptor-mediated endocytosis. The acidic pH of the endocytic compartment results in a conformational change of the DT, enabling translocation of the A chain into the cytosol.

DT is produced by *Corynebacterium diphtheriae*. Infection results in an upper respiratory illness but can progress to myocarditis or polyneuritis and can be fatal. The toxin catalyzes the ADP ribosylation of eukaryotic elongation factor-2 (eEF2), thus inactivating this protein. This, in turn, inhibits protein synthesis (11–13), thus causing cell death (Figure 13.3).

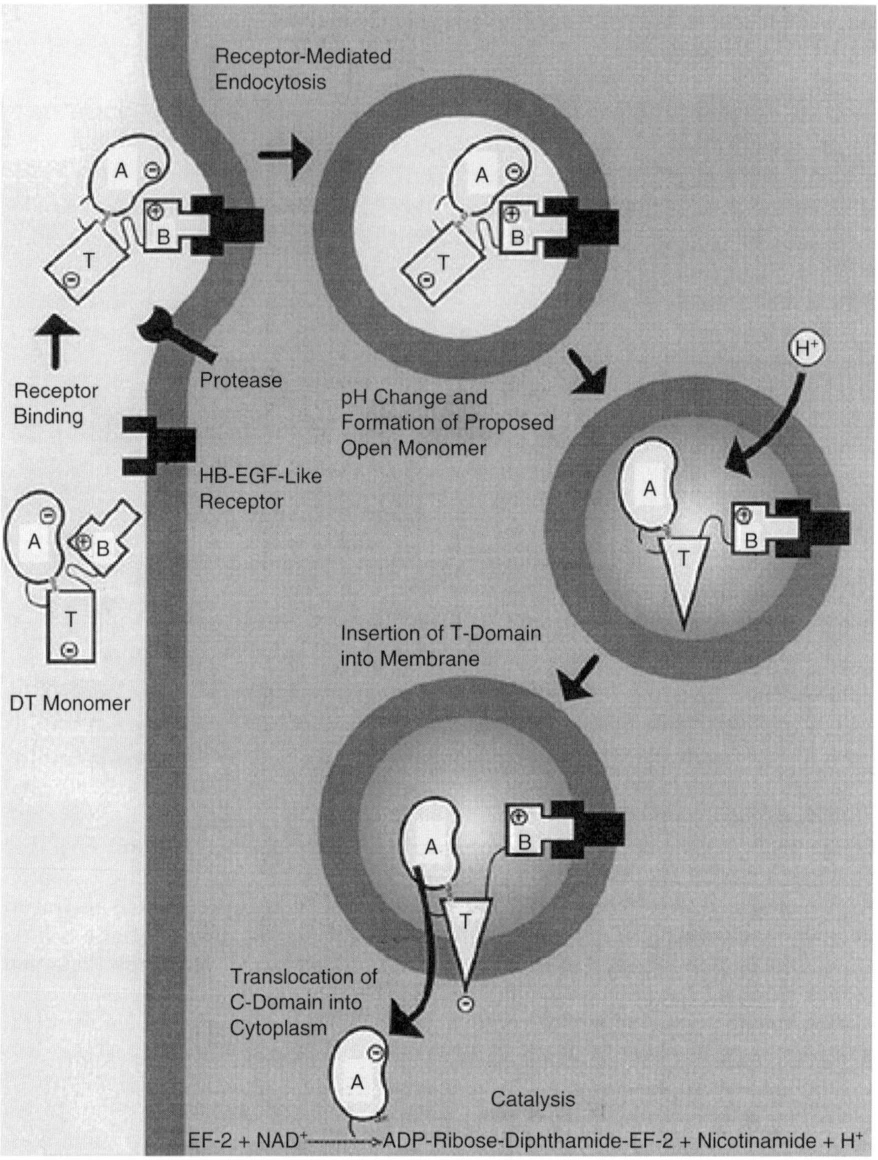

Figure 13.3 Mechanism of action of DT: denileukin diftitox functions in a similar manner to deliver the enzymatic/catalytic domain of DT to IL-2R expressing cells. From www.textbookofbacteriology.net/themicrobialworld/diphtheria.html.

ACTIVITY IN T CELL LYMPHOID MALIGNANCIES AND T CELL-MEDIATED PHENOMENA

Since denileukin diftitox is engineered to bind to IL2-R, which is primarily expressed on T cells, early studies addressing its utility focused on T cell-mediated diseases. Initial studies showed that the fusion protein eliminated IL-2R-positive T cell subsets, thus affecting T cell-mediated immune responses (14, 15). Specific diseases that have been shown to be modulated include myasthenia gravis (14), multiple sclerosis (16), psoriasis and orbital inflammatory syndromes (17, 18), diabetes (19), schistosomiasis (20), as well as human immunodeficiency syndrome (21–23). Data from these studies show that a common theme is activation of the disease phenotype by an IL-2R-mediated immune response. Exposure with denileukin diftitox appears to inhibit the T cell responses. However, response is not maintained without repeated dosing in these experimental models.

Phase I Clinical Trials

Because of these effects seen on T cell subsets, this toxin–protein construct was hypothesized to potentially be active in CD25 expressing malignancies. In three Phase I clinical trials, patients with CD25-expressing hematologic malignancies were treated with different schedules of DAB486-IL-2. In one study, 18 patients with non-Hodgkin's lymphoma (NHL), Hodgkin's disease (HD), CTCL, and chronic lymphocytic leukemia were treated with one complete response and two partial responses (3). A second study utilized an every 8-h dosing every 21 days in 20 patients. Two partial responses (PRs) were seen in mycosis fungoides and NHL (24). One patient with HD in a 15-patient Phase I/II study using daily dosing for 5 days every 28 days showed a complete response (25). A Phase II study in 14 CTCL patients only resulted in 1 PR, 2 minor responses, and 3 patients with stable disease (26). All of these studies demonstrated reasonable tolerability to the drug. Common toxicities included hepatotoxicity with transaminase elevations, flushing, fever, proteinuria, and elevated serum creatinine. The half-life of DAB486-IL-2 was on the order of 5 min, hence necessitating more frequent dosing or prolonged infusion (3, 4, 25, 26).

Another version of the toxin fusion protein, DAB389IL-2 (denileukin diftitox, Ontak), was developed and studied. Denileukin diftitox showed a prolonged serum half-life (72 min vs. 5 min), decreased immunogenicity, and increased affinity for IL-2R, as well as higher potency compared with DAB486-IL-2 (3, 4). Seventy-three patients were enrolled in a Phase I trial of the redesigned molecule: 21 HD, 35 CTCL, and 17 NHL. There was a 37% response rate with 5 complete responses (CRs) and 8 PRs out of 35 patients with CTCL, and 1 CR and 2 PR in 17 NHL patients. No responses were detected in the 21 HD patients. Unlike the precursor molecule, asthenia, fever, and nausea were the major dose-limiting toxicities. Hypotension, edema, hypoalbuminemia, transient rash, and elevated transaminases were also reported, suggestive of a vascular leak syndrome, which was seen only in CTCL patients.

CTCL

CTCLs are a heterogeneous group of lymphomas that have a common underlying feature of infiltration of the skin and other organs by mature T cells. These include mycosis fungoides and Sezary's syndrome. Since IL-2R is expressed on normal activated T cells, denileukin diftitox was studied in this population and was found to have responses in patients with refractory CTCL. An early report of four patients noted a pathologic CR lasting >11 months, a clinical CR lasting >6 months, and two PRs lasting >5 months with tumor IL-2R negative at recurrence (27). Saleh et al. conducted a 35-patient trial in patients with CTCL/mycosis fungoides. Thirty four patients had not responded to conventional treatment available at the time. Thirteen patients demonstrated clinical response, with five CRs (28). The results from these studies, along with the 73-patient Phase I study in NHL, CTCL, and HD patients, led to a confirmatory, pivotal Phase III trial. Seventy-one patients with recurrent/persistent CTCL, mycosis fungoides, or Sezary's syndrome despite prior treatment were randomized to one of two dose levels of denileukin diftitox, 9 or 18 mcg/kg/d daily for five consecutive days every 21 days. The overall response rate was 30% (21/71 patients) with 10% CR or complete CR (cCR). The mean duration of response was 8.2 ± 5.0 months (range 2.7–46.1 months, median 6.9 months). There was a trend suggesting that the higher dose corresponded to higher response, although this was not statistically significant (29). An additional study of 15 patients with CTCL reported a 60% response rate with the addition of steroid premedications (30).

Following conditional Food and Drug Administration (FDA) approval of denileukin diftitox for treatment of CTCL, additional reports have confirmed the efficacy of this drug in this setting. In an analysis of three Phase III trials, the overall response rates were 15.9% in placebo-treated patients ($n = 44$), 38.5% in all denileukin diftitox-treated patients ($n = 263$), 47.5% in the high-dose group ($n = 118$), 30.6% in patients who were CD25 negative (<20% of cells expressed CD25; $n = 36$), and a 27.6% response rate in retreated patients ($n = 29$). Progression-free survival was also significantly improved across all treated groups compared to placebo (794 days vs. 124 days, $p < 0.02$) (31–33). Altogether, 300+ patients with CTCL have been treated on Phase III clinical trials with response rates ranging from 30% to 50%, confirming activity of this drug as a single agent. These subsequent trials highlight that there are responses to placebo (15.9%) and that placebo-controlled studies are needed to adequately analyze treatment options in this disease. Table 13.1 summarizes the results of the Phase III trials in CTCL with overall good concordance.

Activity in Other T Cell Lymphoid Malignancies and Combination Therapy

Denileukin diftitox is active in other T cell lymphoid malignancies, including human T cell lymphotropic virus-1 (HTLV-1) associated adult T cell leukemia

TABLE 13.1. Response Data for Denileukin Diftitox in CTCL

Study	Arm (Dose)	N	Response Rate	Complete Response	Partial Response	Response Duration (Months) (Range)
Olsen (29)	9 mcg/kg/d	35	23% (n = 8)	8.5% (n = 3)	14% (n = 5)	8.7 ± 7.0 (2.7 – 46.1)
	18 mcg/kg/d	36	37% (n = 13)	11.1 (n = 4)	25% (n = 9)	8.0 ± 3.7 (4.0 – 17.5)
	Combined	71	30% (n = 21)	10% (n = 7)	(20%) (n = 14)	8.2 ± 5.0 (2.7 – 46.1)
						PFS in Days (Months)
Negro-Vilar (33)	Placebo	44	15.9%	2.3	13.6	124 (4.1)
	All DD	263	38.0	9.1	28.9	794 (24.5)
	DD (CD25+)	118	47.5	11.0	36.4	870 (29.0)
	DD (CD25–)	36	30.6	8.3	22.2	>487 (16.2)
	DD (R)	29	27.6	6.9	20.7	205 (6.8)

DD, denileukin diftitox; DD (CD25+), CD25-positive tumors treated with denileukin diftitox; DD (CD25–), CD25-negative tumors treated with denileukin diftitox; DD(R), prior denileukin diftitox treatment. PRS, progression free survival.

(adult T-cell leukemia/lymphoma [ATLL]), as well as T cell NHL. A case report of a patient with HTLV-1-associated ATLL who had clinical remission lasting >1 year following therapy with denileukin diftitox and hyper-CVAD (hyperfractionated cyclophosphamide, vincristine, doxorubicin, dexamethasone alternating cycles with methotrexate and cytarabine) and maintenance denileukin diftitox suggests a role in the treatment of a historically difficult-to-treat disease (34). Other T cell lymphomas/leukemias that have been treated with denileukin diftitox include peripheral T cell lymphoma (PTCL), angioimmunoblastic T cell lymphoma (AITL), anaplastic large cell lymphoma (ALCL), and natural killer/T cell lymphoma (NKTCL). In a Phase II study, 27 patients with relapsed, refractory T cell NHL, excluding CTCL, were treated with the IL-2 DT fusion protein. Objective response was reported in 13/27 patients, with 6 CRs and 7 PRs. As in the Phase III trials of CTCL, there were responses in both CD25+ and CD25– patients (35).

Once the single-agent activity of denileukin diftitox was noted, combination therapy in T cell malignancies has naturally been of interest. Current systemic therapies for the CTCL include bexarotene and interferon. *In vitro* studies showed bexarotene upregulated expression of the alpha and beta subunits of IL-2R and enhanced sensitivity of T cell leukemia cells to the effects of denikeulin diftitox. Investigators proceeded with a Phase I study in 14 relapsed CTCL patients. There was a 67% response rate with four CRs and four PRs. One patient with a CR had previously had a suboptimal response to single-agent denileukin diftitox, suggesting that modulation of IL-2R expression could enhance activity. The major adverse events were four patients with grade 2 or 3 leukopenia, six patients with grade 3 lymphopenia, and two patients with grade 4 lymphopenia (36). Response was also seen in a case report of a patient with NKTCL who had a regression of his tumors with the combination of bexarotene and monthly denileukin diftitox (37). Data also suggest that initial therapy with denileukin diftitox may enhance sensitivity to subsequent therapies (38).

Combination therapy is also being investigated with cytotoxic regimens for aggressive T cell lymphomas. In the Phase I trials, there was minimal myelosuppression seen with denileukin diftitox and thus combinations with myelosuppressive regimens such as cyclophosphamide, doxorubicin, vincristine, prednisone (CHOP), or hyper-CVAD. Sequential and maintenance therapy was shown in an HTLV-1-associated ATLL patient who received denileukin diftitox as initial therapy. Following relapse, the patient achieved a clinical remission with hyper-CVAD and maintenance denileukin diftitox (34).

PTCL is one of the more common subtypes of T cell NHL. Patients typically present with advanced-stage disease and generally have higher scores on the International Prognostic Index. Treatment is with aggressive multi-agent chemotherapy similar to regimens used for aggressive B cell malignancies. However, patients have a high rate of relapse (39). In recent Phase II studies, patients with newly diagnosed PTCL were treated with denileukin diftitox at 18 mcg/kg/d on days 1 and 2 followed by CHOP chemotherapy on day 3 every 21 days for up to six cycles. Growth factor support was started on day 4. As of the most recent update, 41 patients were enrolled, of which 31 were evaluable for response. The overall response rate is 90% (28/31 patients), with 71% (22/31) CR/CR unconfirmed (CRu) and 19% (6/31) PR. There was expected grades 3–4 hematologic toxicity in 17/40 patients, and 40% (16/40) of patients experience grades 3–4 non-hematologic toxicity. Median response duration is 13 months (40). Overall, the combination has been reasonably well tolerated and is early evidence that it may be potentially useful in other combination regimens (Table 13.2).

With the high relapse/recurrence seen in T cell lymphomas, there may be a role for denileukin diftitox as a maintenance therapy or as a bridge to other therapies, which has been reported in patients with ATLL, PTCL, and NKTCL. In one case described earlier in this chapter, the patient presented with significant hematologic abnormalities and bone marrow infiltration and fibrosis, which improved after four cycles of denileukin diftitox. Upon relapse, the patient achieved a CR with hyper-CVAD and, at the time of the report, had remained disease free at 1 year on maintenance denileukin diftitox (34). Another patient with anthracycline refractory PTCL achieved remission with salvage ifosfamide, carboplatin, and etoposide. He has been on maintenance denileukin diftitox and remains disease free at greater than 3 years (41; N. Dang, unpublished observations). The previously reported patient with NKTCL treated with bexarotene and denileukin diftitox had quick recurrence following decision to stop therapy but otherwise had a 5-month response

TABLE 13.2. Response in B Cell Lymphoid Malignancies

Tumor Type	Other Concurrent Tx	*N* (Evaluable)	ORR	CR	PR	Minimal Response*	Median TTP† (Range)
B-NHL (50)	Single agent	45	24%	6.7% (*n* = 3)	17.8% (*n* = 8)		7 months
B-NHL (51)	Single agent	29	10.3%	0	10.3% (*n* = 3)		23 months (3–33 months)
B-NHL (52)	Rituxan	38	32%	16% (*n* = 6)	16% (*n* = 6)		8 months (2–36 months)
PTCL (40)	CHOP	31	90%	71% (*n* = 22)	19% (*n* = 6)		13 months
CLL/SLL only	Single agent (47)	12	NA	NA	2	7	9 months (1–19 months)‡
	Single agent (48)	7	NA	NA	2	2	NA
	Single agent (49)	22	27.3%	4.5% (*n* = 1)	22.7% (*n* = 5)	12 (PB CLL)	5 months (2–12 months)§

*Minimal response: decrease in peripheral blood CLL counts.
†Median TTP: includes response duration, time to treatment failure, and progression-free interval.
‡Peripheral blood responses and "minor responses."
§Includes only objective responders.
B-NHL, B-non-Hodgkin's lymphoma; SLL, small lymphocytic lymphoma; Tx, treatment; ORR, overall response rate; NA, not available; PB, peripheral blood; TTP, time to progression.

during which he was treated with monthly denileukin diftitox along with bexarotene (37). A patient with relapsed/recurrent ATLL with central nervous system (CNS) involvement demonstrated a rapid response to denileukin diftitox and was subsequently rendered disease free following a sibling matched allogeneic transplant (42).

Activity in B Cell Lymphoid Malignancies

While denileukin diftitox has demonstrated activity in T cell lymphoid malignancies, the Phase I studies also showed activity in B cell malignancies. While its biological role in these malignancies is not well understood, the CD25 antigen is expressed in most malignant B cell lymphomas: HD, NHL, B cell chronic lymphoblastic leukemia (B-CLL), and acute lymphoblastic leukemia (ALL) (5, 9) and is thought that it may play a role in stimulating proliferation of the malignant cells (43–45).

There are approximately 10,000 new cases of CLL diagnosed per year (46). Some patients will not require therapy for many years. However, the majority of patients develop progressive disease requiring therapy. There are multiple chemotherapy agents used in CLL, including fludarabine, chlorambucil, rituximab, and cyclophosphamide. Most patients respond to therapy, but relapse/progression is common. An initial study was conducted in 18 patients with relapsed/refractory CLL, who were treated with denileukin diftitox. Twelve patients received at least three cycles of therapy and were evaluable for response. Eleven patients showed reductions in peripheral lymphocyte counts; seven patients showed a decrease in lymph node size. Bone marrow response was also seen in 11 patients. There were two PRs and seven minimal responses, with progression-free intervals ranging from 1 to 19+ months (47). A small report of seven patients demonstrated two objective partial responses and two minor responses. One of the partial responders had a 17p deletion, which has typically been associated with poor response and prognosis. In addition, patients who were CD25– demonstrated response (48). Two multi-institutional trials were subsequently implemented to further evaluate the role of denileukin diftitox in CLL. Twenty-eight patients with relapsed CLL following fludarabine therapy were enrolled, of which 22 patients received at least two cycles of denileukin diftitox and were evaluable for response. Twelve patients showed a decrease in peripheral blood counts. Six of 22 patients showed response by a radiographic measurement of the lymph nodes. There was one CR lasting >1 year and five PRs. Responses lasted for 2–12+ months. Toxicity and response did not appear to be related to the number of cycles of therapy nor CD25 expression (49). The molecule is active in CLL, yet we do not have prognostic factors to help identify patients who are more likely to respond to therapy.

Studies in other B cell malignancies showed activity in low- and intermediate-grade NHL. In a single-institution study of 45 assessable patients, 32 patients were refractory to their last chemotherapy and all patients had previously been treated with rituximab. There was an overall response rate of 24.5%

(three CRs, 6.7%, and eight PRs, 17.8%). Nine patients (20%) had stable disease. There was a similar response between CD25+ and CD25– patients. Of note, patients were considered CD25+ if the tissue assay showed ≥10% of cells expressing CD25. Most other studies use a 20% cutoff for determining whether the tumor is considered CD25+ or CD25–; thus, there may actually be a higher percentage of CD25– tumors responding to denileukin diftitox in this study (50). An additional study was intended to enroll 77 patients with indolent lymphomas but was closed after 35 patients. Twenty-nine patients were evaluated, 8 patients with small lymphocytic lymphoma and 21 patients with follicular grade I/II lymphoma. Tumors were considered IL-2R+ if ≥20% of cells expressed the IL-2R. There were no CRs and three partial responses (10%). Of the responders, one was considered IL-2R+ (SLL). The other responders were IL-2R– and included an SLL and a grade 1 follicle center cell lymphoma. Response duration ranged from 3 to 33 months with the IL-2R+ patient at 23 months (51). Tumor response did not correlate with IL-2R (CD25) expression and thus leads to additional questions regarding the mechanism in B cell malignancies.

As with the T cell malignancies, combination therapy in B cell neoplasms is being studied. A Phase II trial of denileukin diftitox and rituximab in relapsed/refractory B cell NHL was reported in 2007. In this study, patients with relapsed/refractory low- or intermediate-grade B cell NHL were eligible. Patients were required to have had prior therapy with a rituximab-containing regimen and a good performance status (Zubrod PS ≤ 2). Thirty-eight evaluable patients were entered with 15 diffuse large cell lymphoma (DLCL), 14 follicular lymphoma (FL), 6 mantle cell lymphoma (MCL), 2 SLL, and 1 marginal zone lymphoma. There were six CRs (16%) and six PRs (16%) for an ORR of 32% with an additional seven patients with stable disease (19%). Thirty patients were rituximab refractory; eight of the responders were previously rituximab refractory. Results suggest a slightly higher response to the combination of rituximab and denileukin diftitox (32% ORR) than to the latter drug alone in the refractory setting (24.5% ORR) (52). In addition, 9 of 14 FL patients demonstrated response, a higher response rate than that seen with the indolent lymphomas in the single-agent study by Kuzel. While this is the first report of denileukin diftitox in combination with other agents for B cell lymphoid malignancies, it heralds hope for improved responses in patients with incurable indolent lymphomas (Table 13.2).

OTHER DISEASES

Denileukin diftitox has the potential to be effective in other tumor types, including solid tumors. IL-2R has been shown to be expressed in melanoma, in renal cell carcinoma, and in head and neck, esophageal, and lung cancers (53–57). Many reports have shown that retinoic acid can induce expression of all three subunits of IL-2R, and hypoxia appears to preferentially upregulate gamma subunit expression (58, 59). In hepatocellular carcinoma cell lines, treatment with retinoic acid and denileukin diftitox suppresses cell growth with effects more prominent in hypoxic conditions (60). Induction of IL-2R expression may enable use of denileukin diftitox in other tumor types.

Due to its engineered design, denileukin diftitox binds IL-2R. However, the mechanism by which denileukin diftitox exerts its effects is more complex than just causing cell death in the IL-2R expressing cells. Besides the direct activity on malignant T-lymphoid cells, data are now accumulating that the toxin fusion protein exerts some of its effects on the immune system, both directly and indirectly. In graft versus host disease (GVHD), donor immune cells are activated. The reconstituted immune system "sees" the host cells as "foreign" or dangerous and stimulates a form of autoimmune response. GVHD can manifest as mild skin toxicity or can blossom into fulminant disease following transplant. Mild forms of GVHD are generally easily managed with mild immune suppressants. However, severe reactions may include compromise of the gastrointestinal tract, liver, or other organs. Severe GVHD may be fatal if immune suppressants are inadequate to control the disease. Since GVHD is the result of the stimulation of the effector arm of the immune system, investigators hypothesized that downregulation of the effector cells could potentially abrogate the response. Thirty patients with steroid refractory GVHD were treated with one of three schedules of denileukin diftitox (18). There were eight CRs (33%) and nine PRs (38%) among the 24 evaluable patients. Four patients with initial PRs achieved CRs after day 29 without additional therapy. Additional studies confirm improvement in patients with steroid refractory GVHD (61). Correlative studies demonstrated that there was a transient decrease in CD3+ peripheral T cells, although CD3+CD25+ cell numbers were not significantly affected, suggesting that depletion of effector T cells could modulate this effect.

There is emerging data that denileukin diftitox may have additional effects on hypersensitivity and other autoimmune syndromes. In a murine model, treatment with denileukin diftitox decreases IL-2R+ T cells, thus diminishing hypersensitivity but upregulating specific IgE antibody production (62). Experimental models in myasthenia gravis also demonstrate downregulation of anti-acetylcholine receptor antibodies following treatment with denileukin diftitox, thus affecting the humoral response arm of the immune system (14). It is unclear from available data whether these observations are due to the direct effect of denileukin diftitox on the immune surveillance or whether there is another arm of the immune system that is affected.

Data are accumulating that denileukin diftitox has effects on immune surveillance through its effects on regulatory T cells which express CD3, CD4, CD25, and FOXP3. Murine models demonstrate enhanced T cell-specific responses to vaccination with exposure to denileukin diftitox prior to vaccination (63, 64), although effects on effector T cells may also be muted. This mechanism is under investigation in animal models and in human clinical trials in various solid tumor malignancies. Melanoma has been the model tumor type for vaccine studies since it expresses tumor-specific antigens, which have been studied for immunogenicity. In murine models, a decrease in CD4+CD25+ T cells is associated with stimulation or enhancement of antigen-specific responses. In patients with metastatic melanoma, treatment with denileukin diftitox caused a transient depletion of Treg cells (CD4+CD25hi+FOXP3+). Five patients experienced responses in the size of their metastases, suggesting that treatment with denileukin diftitox downregulates Treg cells, permitting the development or stimulation of an immune response to tumor-specific (melanoma) antigens (65). Additional studies have shown downregulation of Treg when denileukin diftitox was administered in patients with ovarian cancer (66) treated with high-dose IL-2 for renal cell carcinoma (67) or melanoma (68). Antigen-specific immune responses to vaccines have also been amplified with denileukin diftitox in renal cell carcinoma, carcinoembryonic antigen (CEA)-expressing tumors, as well as melanomas (69–71).

CONCLUSION

Denileukin diftitox is an immunotoxin protein designed to harness the well-studied mechanism of action of DT but with the vision of directing this toxin to specific cell populations. Development of this molecule began prior to 1987 (1, 2), received accelerated FDA approval in 1999, and only received full approval in late 2008. Its development is an excellent example of the research and testing required to develop a biologic compound for therapeutic use.

The fusion protein has shown direct activity in CTCL, ostensibly by binding to CD25+ malignant cells and causing cell death by inhibition of cellular protein synthesis. It was granted accelerated/conditional FDA approval in CTCL based upon a 30% RR with a 10% CR. Additional placebo-controlled trials following initial approval confirmed a 38.5% RR but also showed a 15.9% RR in placebo-treated trials, illustrating that placebo-controlled trials continue to be needed in this disease. The molecule is also active in other T cell malignancies: PTCL, ATLL, and NKTCL. Treatment with combination chemotherapy may yield improved responses or time to progression in these difficult-to-treat malignancies, although trials are ongoing. Phase I trials had also demonstrated activity in B cell malignancies: CLL, NHL, and some activity in HD. Although later studies have not confirmed activity in HD, there has been promising data in CLL/SLL and in low- to intermediate-grade NHL as a single agent, but also in rituximab refractory disease. The mechanism by which denileukin diftitox overcomes rituximab resistance in FL is unknown but warrants further studies.

Denileukin diftitox is the result of rational protein engineering to create a fusion protein with a specific target. In the most simplistic model, interaction of the fusion protein with IL-2R expressing cells causes the selective apoptosis of these cells due to abrogation of protein synthesis by the toxin. In T cell lymphoid malignancies, the result is typically the diminution of this malignant subset of cells. However, there is activity in B cell malignancies, which does not appear to be restricted to CD25+ tumors, thus suggesting alternate mechanisms of action on the immune system. Emerging data in autoimmune diseases including myasthenia gravis, multiple sclerosis, and GVHD suggests effects on the effector arm of the immune system. Current active research is focusing on the role of denileukin diftitox on the regulatory T cells. Data from these studies will likely lead to a better understanding of the immune system and may enable us to harness the yet unrealized potential of the immune system in the treatment or modulation of many diseases. Studies to uncover the mechanisms of action and thus potential applications, both in research and in the clinical setting, are continuing.

REFERENCES

1. Williams D.P., Parker K., Bacha P., Bishai W., Borowski M., Genbauffe F., Strom T.B., and Murphy J.R. 1987. Diphtheria toxin receptor binding domain substitution with interleukin-2: genetic construction and properties of a diphtheria toxin-related interleukin-2 fusion protein. Protein Eng. **1**: 493–498.
2. Williams D.P., Snider C.E., Strom T.B., and Murphy J.R. 1990. Structure/function analysis of interleukin-2-toxin (DAB486-IL-2). Fragment B sequences required for the delivery of fragment A to the cytosol of target cells. J. Biol. Chem. **265**: 11885–11889.
3. LeMaistre C.F., Meneghetti C., Rosenblum M., Reuben J., Parker K., Shaw J., Deisseroth A., Woodworth T., and Parkinson D.R. 1992. Phase I trial of an interleukin-2 (IL-2) fusion toxin (DAB486IL-2) in hematologic malignancies expressing the IL-2 receptor. Blood. **79**: 2547–2554.
4. LeMaistre C.F., Saleh M.N., Kuzel T.M., Foss F., Platanias L.C., Schwartz G., Ratain M., Rook A., Freytes C.O., Craig F., Reuben J., and Nichols J.C. 1998. Phase I trial of a ligand fusion-protein (DAB389IL-2) in lymphomas expressing the receptor for interleukin-2. Blood. **91**: 399–405.
5. Waldmann T.A. 2002. The IL-2/IL-15 receptor systems: targets for immunotherapy. J. Clin. Immunol. **22**: 51–56.
6. Robb R.J., Rusk C.M., Yodoi J., and Greene W.C. 1987. Interleukin 2 binding molecule distinct from the Tac protein: analysis of its role in formation of high-affinity receptors. Proc. Natl. Acad. Sci. U.S.A. **84**: 2002–2006.
7. Waldmann T.A. 1991. The interleukin-2 receptor. J. Biol. Chem. **266**: 2681–2684.
8. Nelson B.H., and Willerford D.M. 1998. Biology of the interleukin-2 receptor. Adv. Immunol. **70**: 1–81.
9. Nakase K., Kita K., Nasu K., Ueda T., Tanaka I., Shirakawa S., and Tsudo M. 1994. Differential expression of interleukin-2 receptors (alpha and beta chain) in mature lymphoid neoplasms. Am. J. Hematol. **46**: 179–183.
10. McMillan D.N., Kernohan N.M., Flett M.E., Heys S.D., Deehan D.J., Sewell H.F., Walker F., and Eremin O. 1995. Interleukin 2 receptor expression and interleukin 2 localisation in human solid tumor cells in situ and in vitro: evidence for a direct role in the regulation of tumour cell proliferation. Int. J. Cancer. **60**: 766–772.

11. Bacha P., Williams D.P., Waters C., Williams J.M., Murphy J.R., and Strom T.B. 1988. Interleukin 2 receptor-targeted cytotoxicity. Interleukin 2 receptor-mediated action of a diphtheria toxin-related interleukin 2 fusion protein. J. Exp. Med. **167**: 612–622.
12. Waters C.A., Schimke P.A., Snider C.E., Itoh K., Smith K.A., Nichols J.C., Strom T.B., and Murphy J.R. 1990. Interleukin 2 receptor-targeted cytotoxicity. Receptor binding requirements for entry of a diphtheria toxin-related interleukin 2 fusion protein into cells. Eur. J. Immunol. **20**: 785–791.
13. vanderSpek J.C., Sutherland J.A., Ratnarathorn M., Howland K., Ciardelli T.L., and Murphy J.R. 1996. DAB389 interleukin-2 receptor binding domain mutations. Cytotoxic probes for studies of ligand-receptor interactions. J. Biol. Chem. **271**: 12145–12149.
14. Balcer L.J., McIntosh K.R., Nichols J.C., and Drachman D.B. 1991. Suppression of immune responses to acetylcholine receptor by interleukin 2-fusion toxin: in vivo and in vitro studies. J. Neuroimmunol. **31**: 115–122.
15. Bastos M.G., Pankewycz O., Rubin-Kelley V.E., Murphy J.R., and Strom T.B. 1990. Concomitant administration of hapten and IL-2-toxin (DAB486-IL-2) results in specific deletion of antigen-activated T cell clones. J. Immunol. **145**: 3535–3539.
16. Phillips S.M., Bhopale M.K., Constantinescu C.S., Ciric B., Hilliard B., Ventura E., Lavi E., and Rostami A. 2007. Effect of DAB(389)IL-2 immunotoxin on the course of experimental autoimmune encephalomyelitis in Lewis rats. J. Neurol. Sci. **263**: 59–69.
17. Gottlieb S.L., Gilleaudeau P., Johnson R., Estes L., Woodworth T.G., Gottlieb A.B., and Krueger J.G. 1995. Response of psoriasis to a lymphocyte-selective toxin (DAB389IL-2) suggests a primary immune, but not keratinocyte, pathogenic basis. Nat. Med. **1**: 442–447.
18. Ho V.H., Chevez-Barrios P., Jorgensen J.L., Silkiss R.Z., and Esmaeli B. 2007. Receptor expression in orbital inflammatory syndromes and implications for targeted therapy. Tissue Antigens. **70**: 105–109.
19. Pacheco-Silva A., Bastos M.G., Muggia R.A., Pankewycz O., Nichols J., Murphy J.R., Strom T.B., and Rubin-Kelley V.E. 1992. Interleukin 2 receptor targeted fusion toxin (DAB486-IL-2) treatment blocks diabetogenic autoimmunity in non-obese diabetic mice. Eur. J. Immunol. **22**: 697–702.
20. Ramadan M.A., Gabr N.S., Bacha P., Gunzler V., and Phillips S.M. 1995. Suppression of immunopathology in schistosomiasis by interleukin-2-targeted fusion toxin, DAB389IL-2. I. Studies of in vitro and in vivo efficacy. Cell Immunol. **166**: 217–226.
21. Finberg R.W., Wahl S.M., Allen J.B., Soman G., Strom T.B., Murphy J.R., and Nichols J.C. 1991. Selective elimination of HIV-1-infected cells with an interleukin-2 receptor-specific cytotoxin. Science. **252**: 1703–1705.
22. Zhang L., Waters C., Nichols J., and Crumpacker C. 1992. Inhibition of HIV-1 RNA production by the diphtheria toxin-related IL-2 fusion proteins DAB486IL-2 and DAB389IL-2. J. Acquir. Immune Defic. Syndr. **5**: 1181–1187.
23. Zhang L.J., Waters C.A., Poisson L.R., Estis L.F., and Crumpacker C.S. 1997. The interleukin-2 fusion protein, DAB389IL-2, inhibits the development of infectious virus in human immunodeficiency virus type 1-infected human peripheral blood mononuclear cells. J. Infect. Dis. **175**: 790–794.

24. Kuzel T.M., Rosen S.T., Gordon L.I., Winter J., Samuelson E., Kaul K., Roenigk H.H., Nylen P., and Woodworth T. 1993. Phase I trial of the diphtheria toxin/interleukin-2 fusion protein DAB486IL-2: efficacy in mycosis fungoides and other non-Hodgkin's lymphomas. Leuk. Lymphoma. **11**: 369–377.
25. Tepler I., Schwartz G., Parker K., Charette J., Kadin M.E., Woodworth T.G., and Schnipper L.E. 1994. Phase I trial of an interleukin-2 fusion toxin (DAB486IL-2) in hematologic malignancies: complete response in a patient with Hodgkin's disease refractory to chemotherapy. Cancer. **73**: 1276–1285.
26. Foss F.M., Borkowski T.A., Gilliom M., Stetler-Stevenson M., Jaffe E.S., Figg W.D., Tompkins A., Bastian A., Nylen P., and Woodworth T. 1994. Chimeric fusion protein toxin DAB486IL-2 in advanced mycosis fungoides and the Sezary syndrome: correlation of activity and interleukin-2 receptor expression in a phase II study. Blood. **84**: 1765–1774.
27. Duvic M., Cather J., Maize J., and Frankel A.E. 1998. DAB389IL2 diphtheria fusion toxin produces clinical responses in tumor stage cutaneous T cell lymphoma. Am. J. Hematol. **58**: 87–90.
28. Saleh M.N., LeMaistre C.F., Kuzel T.M., Foss F., Platanias L.C., Schwartz G., Ratain M., Rook A., Freytes C.O., Craig F., Reuben J., Sams M.W., and Nichols J.C. 1998. Antitumor activity of DAB389IL-2 fusion toxin in mycosis fungoides. J. Am. Acad. Dermatol. **39**: 63–73.
29. Olsen E., Duvic M., Frankel A., Kim Y., Martin A., Vonderheid E., Jegasothy B., Wood G., Gordon M., Heald P., Oseroff A., Pinter-Brown L., Bowen G., Kuzel T., Fivenson D., Foss F., Glode M., Molina A., Knobler E., Stewart S., Cooper K., Stevens S., Craig F., Reuben J., Bacha P., and Nichols J. 2001. Pivotal phase III trial of two dose levels of denileukin diftitox for the treatment of cutaneous T-cell lymphoma. J. Clin. Oncol. **19**: 376–388.
30. Foss F.M., Bacha P., Osann K.E., Demierre M.F., Bell T., and Kuzel T. 2001. Biological correlates of acute hypersensitivity events with DAB(389)IL-2 (denileukin diftitox, ONTAK) in cutaneous T-cell lymphoma: decreased frequency and severity with steroid premedication. Clin. Lymphoma. **1**: 298–302.
31. Negro-Vilar A., Dziewanowska Z., Groves E., Lombardy E., and Stevens V. 2006. Phase III study of denileukin diftitox (ONTAK(R)) to evaluate efficacy and safety in CD25+ and CD25- cutaneous T-cell lymphoma (CTCL) patients. Abstracts. ASH Annu. Meet. **108**: 696.
32. Negro-Vilar A., Dziewanowska Z., and Groves E.S., 2007. Efficacy and safety of denileukin diftitox (Dd) in a phase III, double-blind, placebo-controlled study of CD25+ patients with cutaneous T-cell lymphoma (CTCL). J. Clin. Oncol. **25**: 8026.
33. Negro-Vilar A., Prince H.M., Duvic M., Richardson S., Sun Y., and Acosta M. 2008. Efficacy and safety of denileukin diftitox (Dd) in cutaneous T-cell lymphoma (CTCL) patients: integrated analysis of three large phase III trials. J. Clin. Oncol. **26**: 8551.
34. Di Venuti G., Nawgiri R., and Foss F. 2003. Denileukin diftitox and hyper-CVAD in the treatment of human T-cell lymphotropic virus 1-associated acute T-cell leukemia/lymphoma. Clin. Lymphoma. **4**: 176–178.
35. Dang N.H., Pro B., Hagemeister F.B., Samaniego F., Jones D., Samuels B.I., Rodriguez M.A., Goy A., Romaguera J.E., McLaughlin P., Tong A.T., Turturro F.,

Walker P.L., and Fayad L. 2007. Phase II trial of denileukin diftitox for relapsed/refractory T-cell non-Hodgkin lymphoma. Br. J. Haematol. **136**: 439–447.

36. Foss F., Demierre M.F., and DiVenuti G. 2005. A phase-1 trial of bexarotene and denileukin diftitox in patients with relapsed or refractory cutaneous T-cell lymphoma. Blood. **106**: 454–457.
37. Kerl K., Prins C., Cerroni L., and French L.E. 2006. Regression of extranodal natural killer/T-cell lymphoma, nasal type with denileukin diftitox (Ontak) and bexarotene (Targretin): report of a case. Br. J. Dermatol. **154**: 988–991.
38. Talpur R., and Duvic M. 2006. Treatment of mycosis fungoides with denileukin diftitox and oral bexarotene. Clin. Lymphoma Myeloma. **6**: 488–492.
39. Escalón M.P., Liu N.S., Yang Y., Hess M., Walker P.L., Smith T.L., and Dang N.H. 2005. Prognostic factors and treatment of patients with T-cell non-Hodgkin lymphoma: the M. D. Anderson Cancer Center experience. Cancer. **103**: 2091–2098.
40. Foss F., Sjak-Shie N., and Goy A. 2007. Denileukin diftitox (ONTAK) plus CHOP chemotherapy in patients with peripheral T-cell lymphomas (PTCL), the CONCEPT trial. Abstract. Blood. **110**: 3449.
41. Wong B.Y., Ma Y., Fitzwilson R., and Dang N.H. 2008. De novo maintenance therapy with denileukin diftitox (Ontak) in a patient with peripheral T-cell lymphoma is associated with prolonged remission. Am. J. Hematol. **83**: 596–598.
42. Evens A.M., Ziegler S.L., Gupta R., Augustyniak C., Gordon L.I., and Mehta J. 2007. Sustained hematologic and central nervous system remission with single-agent denileukin diftitox in refractory adult T-cell leukemia/lymphoma. Clin. Lymphoma Myeloma. **7**: 472–474.
43. Nakase K., Kita K., Shirakawa S., Tanaka I., and Tsudo M. 1994. Induction of cell surface interleukin 2 receptor alpha chain expression on non-T lymphoid leukemia cells. Leuk. Res. **18**: 855–859.
44. Tesch H., Günther A., Abts H., Jücker M., Klein S., Krueger G.R., and Diehl V. Expression of interleukin-2R alpha and interleukin-2R beta in Hodgkin's disease. Am. J. Pathol. **142**: 1714–1720.
45. Waldmann T.A. 2007. Anti-Tac (daclizumab, Zenapax) in the treatment of leukemia, autoimmune diseases, and in the prevention of allograft rejection: a 25-year personal odyssey. J. Clin. Immunol. **27**: 1–18.
46. Jemal A., Siegel R., Ward E., Hao Y., Xu J., Murray T., and Thun M.J. 2008. Cancer statistics, 2008. CA Cancer J. Clin. **58**: 71–96.
47. Frankel A.E., Fleming D.R., Hall P.D., Powell B.L., Black J.H., Leftwich C., and Gartenhaus R. 2003. A phase II study of DT fusion protein denileukin diftitox in patients with fludarabine-refractory chronic lymphocytic leukemia. Clin. Cancer Res. **9**: 3555–3561.
48. Morgan S.J., Seymour J.F., Prince H.M., Westerman D.A., and Wolf M.M. 2004. Confirmation of the activity of the interleukin-2 fusion toxin denileukin diftitox against chemorefractory chronic lymphocytic leukemia, including cases with chromosome 17p deletions and without detectable CD25 expression. Clin. Cancer Res. **10**: 3572–3575.
49. Frankel A.E., Surendranathan A., Black J.H., White A., Ganjoo K., and Cripe L.D. 2006. Phase II clinical studies of denileukin diftitox diphtheria toxin fusion protein in patients with previously treated chronic lymphocytic leukemia. Cancer. **106**: 2158–2164.

50. Dang N.H., Hagemeister F.B., Pro B., McLaughlin P., Romaguera J.E., Jones D., Samuels B., Samaniego F., Younes A., Wang M., Goy A., Rodriguez M.A., Walker P.L., Arredondo Y., Tong A.T., and Fayad L. 2004. Phase II study of denileukin diftitox for relapsed/refractory B-cell non-Hodgkin's lymphoma. J. Clin. Oncol. **22**: 4095–4102.
51. Kuzel T.M., Li S., Eklund J., Foss F., Gascoyne R., Abramson N., Schwerkoske J.F., Weller E., and Horning S.J. 2007. Phase II study of denileukin diftitox for previously treated indolent non-Hodgkin lymphoma: final results of E1497. Leuk Lymphoma. **48**: 2397–2402.
52. Dang N.H., Fayad L., McLaughlin P., Romaguara J.E., Hagemeister F., Goy A., Neelapu S., Samaniego F., Walker P.L., Wang M., Rodriguez M.A., Tong A.T., and Pro B. 2007. Phase II trial of the combination of denileukin diftitox and rituximab for relapsed/refractory B-cell non-Hodgkin lymphoma. Br. J. Haematol. **138**: 502–505.
53. Huang A., Quinn H., Glover C., Henderson D.C., and Allen-Mersh T.G. 2002. The presence of interleukin-2 receptor alpha in the serum of colorectal cancer patients is unlikely to result only from T cell up-regulation. Cancer Immunol. Immunother. **51**: 53–57.
54. Rimoldi D., Salvi S., Hartmann F., Schreyer M., Blum S., Zografos L., Plaisance S., Azzarone B., and Carrel S. 1993. Expression of IL-2 receptors in human melanoma cells. Anticancer Res. **13**: 555–564.
55. Tartour E., Mosseri V., Jouffroy T., Deneux L., Jaulerry C., Brunin F., Fridman W.H., and Rodriguez J. 2001. Serum soluble interleukin-2 receptor concentrations as an independent prognostic marker in head and neck cancer. Lancet. **357**: 1263–1264.
56. Wang L.S., Chow K.C., Li W.Y., Liu C.C., Wu Y.C., and Huang M.H. 2000. Clinical significance of serum soluble interleukin 2 receptor-alpha in esophageal squamous cell carcinoma. Clin. Cancer Res. **6**: 1445–1451.
57. Yano T., Fukuyama Y., Yokoyama H., Takai E., Tanaka Y., Asoh H., and Ichinose Y. 1996. Interleukin-2 receptors in pulmonary adenocarcinoma tissue. Lung Cancer. **16**: 13–19.
58. Carswell K.S., Weiss J.W., and Papoutsakis E.T. 2000. Low oxygen tension enhances the stimulation and proliferation of human T lymphocytes in the presence of IL-2. Cytotherapy. **2**: 25–37.
59. Gorgun G., and Foss F. 2002. Immunomodulatory effects of RXR rexinoids: modulation of high-affinity IL-2R expression enhances susceptibility to denileukin diftitox. Blood. **100**: 1399–1403.
60. Kim B.H., Yoon J.H., Myung S.J., Lee J.H., Lee S.H., Lee S.M., and Lee H.S. 2009. Enhanced interleukin-2 diphtheria toxin conjugate-induced growth suppression in retinoic acid-treated hypoxic hepatocellular carcinoma cells. Cancer Lett. **274**: 259–265.
61. Shaughnessy P.J., Bachier C., Grimley M., Freytes C.O., Callander N.S., Essell J.H., Flomenberg N., Selby G., and Lemaistre C.F. 2005. Denileukin diftitox for the treatment of steroid-resistant acute graft-versus-host disease. Biol. Blood Marrow Transplant. **11**: 188–193.
62. Pullerits T., Lundin S., Dahlgren U., Telemo E., Hanson L.A., and Lotvall J. 1999. An IL-2-toxin, DAB389IL-2, inhibits delayed-type hypersensitivity but enhances IgE antibody production. J. Allergy Clin. Immunol. **103**: 843–849.

63. Litzinger M.T., Fernando R., Curiel T.J., Grosenbach D.W., Schlom J., and Palena C. 2007. IL-2 immunotoxin denileukin diftitox reduces regulatory T cells and enhances vaccine-mediated T-cell immunity. Blood. **110**: 3192–3201.

64. Matsushita N., Pilon-Thomas S.A., Martin L.M., and Riker A.I. 2008. Comparative methodologies of regulatory T cell depletion in a murine melanoma model. J. Immunol. Methods. **333**: 167–179.

65. Rasku M.A., Clem A.L., Telang S., Taft B., Gettings K., Gragg H., Cramer D., Lear S.C., McMasters K.M., Miller D.M., and Chesney J. 2008. Transient T cell depletion causes regression of melanoma metastases. J. Transl. Med. **6**: 12.

66. Salazar L.G., Swensen R., and Markle V. 2008. Phase I study of intraperitoneal (IP) denileukin diftitox in patients with advanced ovarian cancer (OC). Abstract. J. Clin. Oncol. **26**: 3036.

67. Gidron A., Eklund J., and Martone B. 2006.Concurrent treatment with denileukin diftitox (DD) to deplete T-regulatory cells enhances rebound lymphocytosis and eosinophilia in patients treated with high-dose IL-2 (HDIL-2) for metastastic renal cell cancer (MRCC). Abstract. Blood. **108**: 1729.

68. Mahnke K., Schönfeld K., Fondel S., Ring S., Karakhanova S., Wiedemeyer K., Bedke T., Johnson T.S., Storn V., Schallenberg S., and Enk A.H. 2007. Depletion of CD4+CD25+ human regulatory T cells in vivo: kinetics of Treg depletion and alterations in immune functions in vivo and in vitro. Int. J. Cancer. **120**: 2723–2733.

69. Chesney J., Rasku M., and Clem A. 2008. Transient T-cell depletion causes regression of melanoma metastases. Abstract. J. Clin. Oncol. **26**: 9030.

70. Dannull J., Su Z., Rizzieri D., Yang B.K., Coleman D., Yancey D., Zhang A., Dahm P., Chao N., Gilboa E., and Vieweg J. 2005. Enhancement of vaccine-mediated antitumor immunity in cancer patients after depletion of regulatory T cells. J. Clin. Invest. **115**: 3623–3633.

71. Morse M.A., Hobeika A.C., Osada T., Serra D., Niedzwiecki D., Lyerly H.K., and Clay T.M. 2008. Depletion of human regulatory T cells specifically enhances antigen-specific immune responses to cancer vaccines. Blood. **112**: 610–618.

14

THE APPLICATION OF CATIONIC ANTIMICROBIAL PEPTIDES IN CANCER TREATMENT: LABORATORY INVESTIGATIONS AND CLINICAL POTENTIAL

ASHLEY L. HILCHIE[1] AND DAVID W. HOSKIN[1,2]

[1]*Department of Microbiology and Immunology, Faculty of Medicine, Dalhousie University, Halifax, Nova Scotia, Canada*
[2]*Department of Pathology, Faculty of Medicine, Dalhousie University, Halifax, Nova Scotia, Canada*

INTRODUCTION

Chemotherapy is the most widely used treatment for metastatic cancer and functions by targeting rapidly dividing cells. Consequently, chemotherapy fails to discriminate normal proliferating cells from cancer cells and is unable to target indolent or dormant cancers (1, 2). The pervasive problem of chemoresistant cancer cells further reduces the therapeutic value of current chemotherapeutic agents (3). Furthermore, certain anticancer drugs, such as tamoxifen, are associated with the development of secondary malignancies (4). For these reasons, significant efforts have been made to generate anticancer agents that are able to target and kill cancer cells while sparing normal healthy cells, regardless of their rate of growth. Such efforts have led to the development of targeted therapies such as trastuzumab (Herceptin®; Genentech,

Emerging Cancer Therapy: Microbial Approaches and Biotechnological Tools, Edited by Arsénio M. Fialho and Ananda M. Chakrabarty

San Francisco, CA), which is a humanized monoclonal antibody against human epidermal growth factor receptor 2 (HER2), that is used to treat breast cancer. Unfortunately, cancer cell resistance to certain targeted therapies has already been reported (5). Therefore, although many advances have been made in the discovery of novel anticancer drugs, the need for agents that can selectively kill cancer cells, including those that are slow growing and multidrug-resistant, remains a focus in the development of new therapeutic approaches.

Cationic antimicrobial peptides (CAPs) with anticancer activity are a promising alternative to conventional chemotherapy. CAPs are small proteins, typically less than 40 amino acids in length, that are predominantly composed of basic (e.g., lysine and arginine) and hydrophobic (e.g., tryptophan) amino acids (6). CAPs are inherently antimicrobial, which makes them important host defense molecules in many species, including insects, fish, and mammals, owing to their ability to kill pathogenic microorganisms at low concentrations. CAPs typically form an amphipathic secondary structure when they interact with biological membranes. The secondary structure often consists of one face that is predominantly cationic and a second face that is hydrophobic. This amphipathic structure is thought to contribute to direct cellular cytotoxicity through a mechanism that involves the destabilization of microbial and cancer cell membranes (6–9).

CAPs are classified as α-helical, β-sheet, loop, or extended peptides on the basis of their secondary structure (Table 14.1) (10). Most CAPs are directly cytotoxic toward cancer cells through a process that is initiated by electrostatic interactions between the positively charged peptide and the negatively charged outer membrane leaflet of the cancer cell (6, 7, 9). However, in addition to their membranolytic activities, certain CAPs also exert anticancer activity by an indirect mechanism that may involve receptor binding, alterations in key signal transduction pathways, and/or other intracellular effects. In this regard, the antibacterial activity of certain CAPs is the result of two mutually independent mechanisms, that is, an indirect effect that occurs at lower peptide concentrations followed by cellular destabilization that takes place at higher peptide concentrations (11). Although this phenomenon has not yet been reported in cancer cells, the possibility of dual action by CAPs against cancer cells would give them a great advantage over conventional chemotherapy. For example, cancer cells that are resistant to peptide killing by an indirect mechanism might still be susceptible to a direct killing mechanism, or vice versa. Furthermore, because the cytotoxic activity of CAPs often involves a membranolytic event rather than an effect on an intracellular target, CAPs are predicted to be effective against multidrug-resistant cancer cells that express high levels of P-glycoprotein and other multidrug resistance proteins (7). Additionally, because CAPs target cancer cells on the basis of electrostatic interactions, these peptides are also expected to kill cancer cells regardless of their proliferative capacity; therefore, CAPs should be able to target slow growing as well as rapidly growing tumors without harming normal cells that are characterized by a net neutral charge.

TABLE 14.1. CAPs with Anticancer Activity

Peptide	Amino Acid Sequence	Source	Number of Residues	Net Charge	Class	Mechanism of Anticancer Activity	References
BMAP-27	G**R**F**KR**F**RKK**F**KK**LF-**KK**LSPVIPLL**H**L	*Bos taurus*	26	+10.5	α-helix	Membranolytic	(75)
BMAP-28	GGL**R**SLG**RK**IL**R**-AW**KK**YGPIIVPII**R**I	*Bos taurus*	27	+7	α-helix	Membranolytic	(75)
LfcinB	F**K**C_1**RR**WQW**R**M**KK**-LGAPSITC_1V**RR**AF	*Bos taurus*	25	+8	β-sheet	Membranolytic, anti-angiogenic, apoptosis	(25–27, 46)
Brevinin-2R	**K**L**K**NFA**K**GVAQSLLN-**K**ASC_1**K**LSGQC_1	*Rana ridibunda*	25	+5	β-sheet	Membranolytic, autophagy	(76)
Cecropin A	**K**W**K**LF**KK**IE**K**VGQNI**R**DGII**K**-AGPAVAVVGQATQIA**K**	*Hyalophora cecropia*	37	+7	α-helix	Membranolytic	(77)
Cecropin B	**K**W**K**VF**KK**IE**K**MG**R**NI**R**NGIV**K**-AGPAIAVLGEA**K**AL	*Hyalophora cecropia*	35	+7	α-helix	Membranolytic	(77)
Epinecidin-1	GFIF**H**IIKGLF**H**AG-**K**MI**H**GLV-**NH_2**	*Epinephelus coioides*	21	+3.5	Unknown	Membranolytic	(78)
Gomesin	ZC**RR**LCY**K**Q**R**-CVTYC**R**G**R**-**NH_2**	*Acanthoscurria gomesiana*	18	+6	Unknown	Membranolytic	(79, 80)
Human α-defensin-1	ACYC**R**IPACIAGE**RR**-YGTCIYQG**R**LWAFCC	*Homo sapiens*	30	+3	β-sheet	Membranolytic, anti-angiogenic	(81, 82)
Human α-defensin-2	CYC**R**IPACIAGE**RR**-YGTCIYQG**R**LWAFCC	*Homo sapiens*	29	+3	β-sheet	Membranolytic	(81)

TABLE 14.1. *Continued*

Peptide	Amino Acid Sequence	Source	Number of Residues	Net Charge	Class	Mechanism of Anticancer Activity	References
Human α-defensin-3	DCYC**R**IPACIAGE**RR**-YGTCIYQG**R**LWAFCC	*Homo sapiens*	30	+2	β-sheet	Membranolytic	(81)
Human α-defensin-4	VCSC**R**LVFC**RR**TEL**R**V-GNCLIGGVSFTYCCT**R**V	*Homo sapiens*	33	+4	β-sheet	Membranolytic	(83)
Human β-defensin-1	D**H**YNCVSSGGQCLYSACPIFT**K**-IQGTCY**R**G**K**A**K**CC**K**	*Homo sapiens*	36	+4.5	β-sheet	Membranolytic	(84)
Human β-defensin-2	GIGDPVTCL**K**SGAIC**H**-PVFCP**RR**Y**K**QIGTCGLPGT**K**	*Homo sapiens*	36	+4.5	β-sheet	Membranolytic	(85)
LL-37	LLGDFF**RK**S**K**E**K**IG**K**EF**KR**-IVQ**RIK**DFL**R**NLVP**R**TES	*Homo sapiens*	37	+6	α-helix	Membranolytic	(86, 87)
Magainin 2	GIG**K**FLHSA**KK**-FG**K**AFVGEIMNS	*Xenopus laevis*	23	+3	α-helix	Membranolytic	(88, 89)
Melittin	GIGAVL**K**VLTTGLPALISWI-**KRKR**QQ	*Apis mellifera*	26	+4	α-helix	Membranolytic, phospholipase A2, and D activator	(90–94)
PR-39	**R**$_3$P**R**P$_2$YLP**R**P**R**P$_3$F$_2$P$_2$-**R**LP$_2$**R**IP$_2$GFP$_2$**R**FP$_2$**R**FP	Porcine sp.	39	+10	Extended	Apoptosis, PI3K inhibitor	(34, 95)
Tachyplesin I	**K**WC$_1$F**R**VC$_2$Y**R**GIC$_2$Y**RR**C$_1$**R**	*Tachypleus tridentatus*	17	+6	β-sheet	Binds hyaluronan and activates complement	(96, 97)

Bold text denotes positively charged amino acids; histidine has a charge of +0.5 at neutral pH.
Z denotes the novel amino acid, pyroglutamic acid.
PI3K, phosphoinositide3-kinase.

MOLECULAR BASIS FOR CANCER CELL TARGETING BY CAPs

All of the models describing CAP-mediated membrane disruption are alike in that they are initiated by peptide binding to the target cell membrane. CAP binding to the cell membrane is initiated by electrostatic interactions between the positively charged amino acids in the peptide and negatively charged cell surface molecules present on the outer membrane leaflet of the target cell (7). CAPs are thought to preferentially bind to cancer cell membranes because, like bacterial cells, cancer cells have an overall negative charge due to the outer membrane leaflet containing three to nine times more of the negatively charged phospholipid phosphatidylserine, in addition to greater levels of negatively charged O-glycosylated mucins, than normal eukaryotic cells (7, 12, 13). In contrast to cancer cells, the outer membrane leaflet of normal eukaryotic cells has a neutral charge owing to the predominance of phosphatidylcholine and other zwitterionic phospholipids (7, 12, 14). As a result, the binding affinity of CAPs for neoplastic cells is reported to be 10-fold higher than for untransformed cells (15). The notion that the net negative charge on the outer membrane leaflet of cancer cells is essential for CAP binding and subsequent membrane destabilization is supported by reports that certain CAPs, including cecropin and magainin, bind to model membranes containing acidic phospholipids to a greater extent than those containing zwitterionic phospholipids (16–18). The lipid composition of a cell is therefore crucial for determining its susceptibility to CAP-induced cytotoxicity. The importance of lipid composition is further supported by evidence that the cholesterol content of the target cell is inversely correlated with cell death (19–21). It has also been suggested that the greater transmembrane potential of neoplastic cells promotes peptide-mediated cytotoxicity by enhancing peptide binding (22). Taken together, these data strongly suggest that the initial electrostatic interaction between the CAP and the target cell is crucial for peptide binding and subsequent membrane destabilization. Furthermore, the development of cancer cell resistance to CAP-mediated killing is predicted to be a rare occurrence because the cancer cell would have to fundamentally change its membrane composition; however, to date, cancer cell resistance to CAPs has not been studied.

Cancer cells possess a greater surface area than normal cells due to an abundance of microvilli on their surface (13). One report suggests that cecropin B killing of cancer cells correlates with microvilli number, as determined by scanning electron microscopy (23); however, the precise role of microvilli in CAP-mediated cytotoxicity has not been extensively studied. It has also been suggested that increased membrane fluidity, which is characteristic of cancer cells, renders neoplastic cells more susceptible to CAP-mediated membrane disruption (12, 24), possibly because hydrophobic amino acid side chains more easily integrate into more fluid cell membranes. However, this hypothesis has not yet been tested in any model membrane system.

Unique membrane composition, a net negative charge on the outer membrane leaflet, negative transmembrane potential, greater surface area, and increased membrane fluidity likely contribute to cancer cell susceptibility to CAP-mediated membrane disruption. Moreover, it is likely that several of these factors combine to promote peptide-mediated cytotoxicity, rather than a single dominant factor dictating whether or not a cancer cell will be killed by CAPs. Furthermore, the importance of each factor is likely to be very much dependent on the amino acid sequence of the CAP and the secondary structure that the peptide adopts when it is in contact with the cancer cell membrane.

MECHANISMS OF CANCER CELL KILLING BY CAPs

To date, most data substantiating the mechanism of CAP-mediated cancer cell death have been obtained using artificial membranes and bacteria as model systems. However, cytotoxic activity in one model system does not always translate into cytotoxicity in another system. Furthermore, cytotoxic activity by a CAP against one cancer cell line does not necessarily indicate that the peptide will be able to kill a different cancer cell line. These discrepancies are likely due to differences among model membranes, bacteria, and cancer cells, at the level of membrane composition, transmembrane potential, and the presence or absence of lipid rafts. Differences in CAP-mediated cytotoxicity may also be due to the presence or absence of a particular receptor required for peptide-induced cell death. The underlying cytotoxic mechanism depends heavily on the amino acid sequence of the CAP, which dictates its secondary structure. These properties explain why different peptides have different antimicrobial and/or anticancer activities. Thus, the amino acid sequence of the CAP and its secondary structure, in addition to the characteristics of the target cell, determine whether or not a given cancer cell will be susceptible to peptide-mediated cytotoxicity. This is an important distinction to make when evaluating the mechanism(s) of anticancer activity by CAPs.

Direct-Acting CAPs

Many CAPs directly kill cancer cells by causing irreparable damage to the cell membrane (6, 7, 24), as depicted in panel B of Figure 14.1. Several models have been proposed to describe the mechanism by which CAPs, such as cecropin B and magainin, bind to and disrupt the cell membrane of cancer cells. These models include the barrel-stave, carpet, toroidal pore, and detergent models (14). The barrel-stave model describes a process by which α-helical or β-sheet CAPs form transmembrane pores in the cell membrane in a fashion akin to staves in a barrel (Figure 14.2A). Pore formation is initiated by peptide binding to the outer membrane leaflet of the target cell, followed by a conformational change that results in the exposure of hydrophobic amino acids that subsequently insert into the hydrophobic core of the cell membrane. When a threshold peptide concentration is reached, peptide monomers aggregate and penetrate deeper into the hydrophobic core of the membrane, leading to the formation of a transmembrane pore that is composed solely of CAPs.

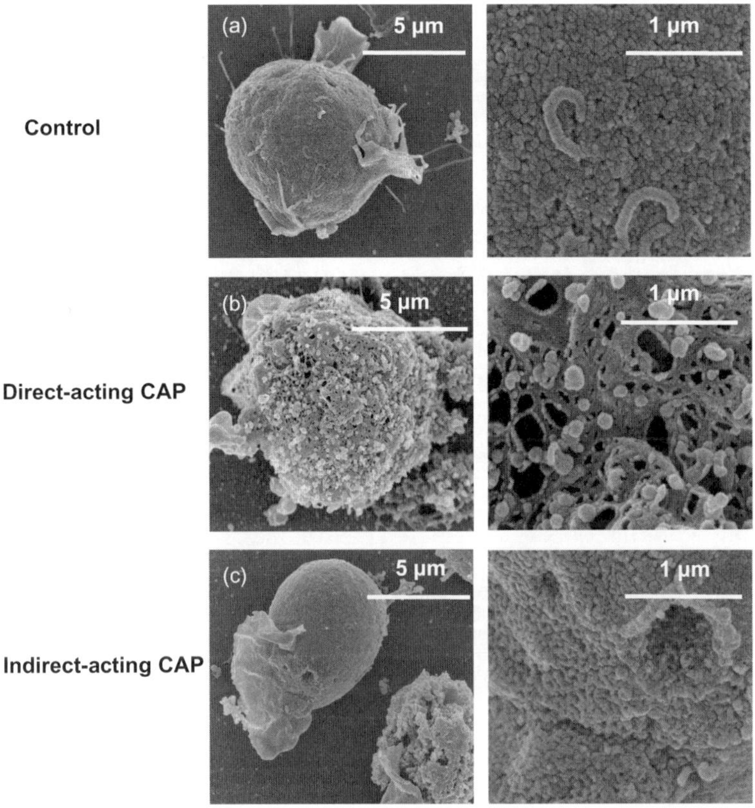

Figure 14.1 Direct and indirect killing of cancer cells by CAPs. The images shown are scanning electron microscopy images depicting (A) control untreated Jurkat T leukemia cells, (B) Jurkat T leukemia cells treated with a direct-acting derivative of LfcinB that causes irreparable damage to the cell membrane, or (C) Jurkat T leukemia cells treated with indirect-acting native LfcinB that induces apoptosis without causing extensive damage to the cell membrane. These images were obtained with the support of the Canada Foundation for Innovation, the Atlantic Innovation Fund, NSERC, and other partners which fund the Facilities for Materials Characterization, managed by the Institute for Research in Materials.

A minimal peptide length of ~22 and ~8 amino acids for α-helical and β-sheet peptides, respectively, is needed to traverse the plasma membrane; thus, shorter CAPs are not able to cause membrane damage by the barrel-stave model.

The carpet model depicts a scenario whereby CAPs bind to the cell membrane and cover the surface of the cell like a carpet (Figure 14.2B). Upon reaching a threshold peptide concentration, membrane destabilization occurs as a result of significant curvature strain imposed by peptide binding and membrane dissolution, or as a result of a reduction in the barrier capacity of the cell due to phospholipid displacement. A reduction in the barrier capacity

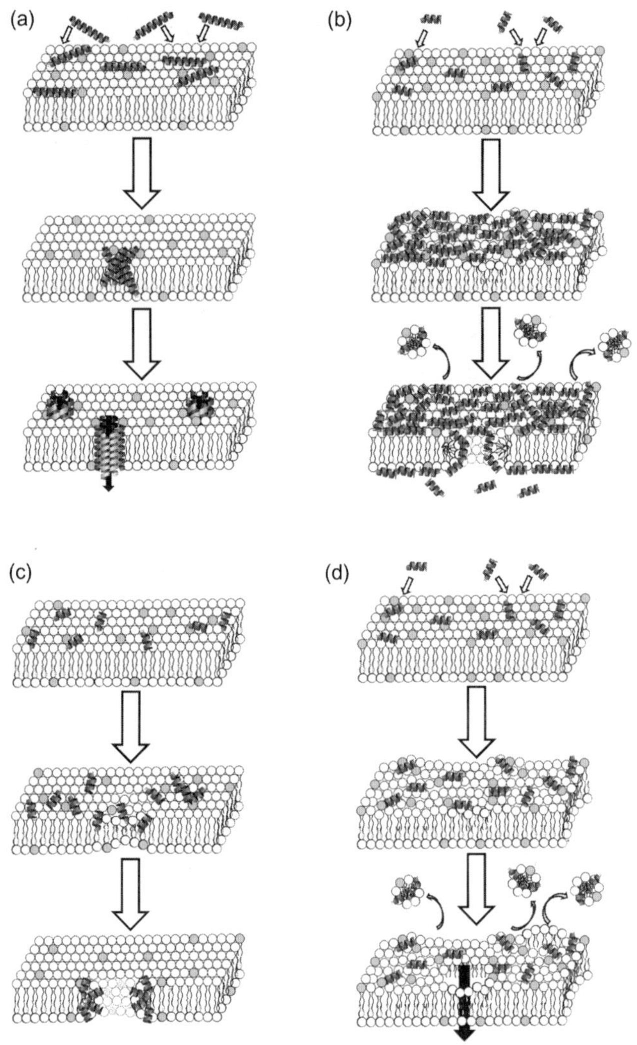

Figure 14.2 Models of CAP-mediated membrane destabilization. (A) The barrel-stave model involves peptide binding to the surface of the target cell, followed by the aggregation of peptide monomers. Subsequently, the peptides insert into the hydrophobic core of the cell membrane, leading to the formation of transmembrane pores that bring about membrane destabilization. (B) In the carpet model, peptide monomers cover the surface of the target cell like a carpet. Peptide binding causes the displacement of phospholipids, membrane curvature, and the formation of transient pores lined with peptides. Cell death may be a consequence of a reduced barrier capacity of the cell or by the indirect action of CAPs that enter the cytoplasmic compartment. (C) The toroidal pore model involves peptide binding to the target cell, integration of peptides into the lipid core of the cell membrane, and the formation of a toroidal pore that consists of peptide and lipid. (D) The detergent model proposes that peptide binding to the target cell causes micelle-like structures to form, resulting in cell membrane destabilization and cell death.

of the cell may also allow peptide translocation into the target cell in the absence of pore formation. Cell membrane destabilization by the carpet model does not involve stable pore formation and does not necessarily involve the insertion of hydrophobic amino acids into the hydrophobic core of the membrane. However, in the carpet model, CAPs may gain access to the cytoplasmic compartment in the absence of membrane destabilization, allowing the peptide to initiate cell death by an indirect mechanism.

The toroidal pore model proposes that both peptides and lipids line the pores that permeate the cell membrane (Figure 14.2C). Like the barrel-stave model, the toroidal pore model is initiated by peptide binding to the outer membrane leaflet of the target cell. Peptide binding to cell surface structures is followed by a conformational change that subsequently leads to the insertion of hydrophobic amino acid side chains into the hydrophobic core of the membrane. CAP insertion destabilizes the cell membrane, which facilitates the formation of torus-like pores composed of both peptide and lipid. The toroidal pore model may explain how highly charged peptides can form stable pores that would not be possible by the barrel-stave model due to excessive electrostatic repulsion between cationic peptides; pores consisting of CAPs in combination with neutral and/or anionic lipids would have reduced electrostatic repulsion.

In the detergent model (Figure 14.2D), CAPs bind at a high density to the surface of the target cell by electrostatic interactions, which results in the formation of micelle-like structures that are composed of membrane lipids and peptides. These micelle-like structures, which can be observed as membrane blebs, are then extruded, leading to rapid cell membrane destabilization. However, it is important to note that there is currently little evidence to suggest that CAPs cause cancer cell death by a detergent-like effect.

Indirect-Acting CAPs

As shown in panel C of Figure 14.1, certain CAPs are able to kill cancer cells by an indirect mechanism that is characterized by an absence of extensive membrane damage. Indirect-acting CAPs may induce apoptosis, inhibit macromolecular synthesis, or alter receptor-mediated signal transduction pathways. Perhaps the best studied example of an indirect-acting CAP is bovine lactoferricin (LfcinB), which induces caspase-dependent apoptosis in carcinoma and hematological cell lines by a mechanism that involves mitochondrial membrane destabilization (25, 26). LfcinB also inhibits basic fibroblast growth factor- and vascular endothelial growth factor (VEGF)-induced angiogenesis, which may also contribute to its anticancer activity (27). Interestingly, scrambled LfcinB fails to inhibit angiogenesis, suggesting that the structure dictated by the amino acid sequence of the peptide is more important for LfcinB-mediated anti-angiogenic activity than the peptide's net positive charge.

Arginine-glycine-aspartic acid (RGD) is the receptor recognition sequence that allows many adhesion molecules, including fibronectin and collagen, to bind integrin receptors (28, 29). Alterations in adhesion molecule interactions may influence tumor angiogenesis and metastasis. Studies have shown that the RGD "homing" sequence in the peptide RGD-4C (CDCRGDCFC) promotes preferential binding of the peptide to tumor cells and tumor-associated endothelium *in vivo* (30). In this regard, the RGD-containing CAP, RGD-tachyplesin, induces death receptor-dependent apoptosis in cancer cells and proliferating endothelial cells *in vitro*, as well as a reduction in tumor volume *in vivo* (31). The ability of RGD-tachyplesin to induce apoptosis is eliminated when its amino acid sequence is scrambled, implying that the structure dictated by the peptide's amino acid sequence is important for the anticancer activity of this CAP.

Certain CAPs cause target cells to die by interfering with normal cellular processes, such as DNA replication and gene expression. PR-39 is a proline- and arginine-rich CAP that binds to bacterial model membranes and bacterial cells (32, 33). PR-39 is believed to kill bacteria by inhibition of DNA and protein synthesis rather than by cell lysis (32). Although a similar effect has not yet been observed in cancer cells, it is conceivable that certain internalized CAPs may also kill cancer cells by this mechanism owing to the negative charge of both RNA and DNA. PR-39 also inhibits phosphoinositide 3-kinase signaling, in addition to interfering with the cellular architecture through interactions with actin, without causing any significant membrane damage (34). Clearly, there are many different mechanisms by which CAPs can kill cancer cells, and certain peptides are likely to kill cancer cells by multiple mechanisms.

MODULATION OF IMMUNE FUNCTION BY CAPs

In addition to direct or indirect cytotoxic activity against cancer cells, certain CAPs are able to modulate immune function, and thereby promote and enhance antitumor immune responses (35). To date, most of this research has been carried out with human α- and β-defensins, and LL-37, as well as their murine analogues.

Human β-defensins are typically produced by epithelial cells and can be expressed constitutively, upregulated in response to pathogenic challenge, or passively released from injured, dying, or necrotic cells (36–39). Human β-defensins might therefore be present at high concentrations in tissues proximal to or within solid tumors that possess a necrotic core. Human and murine β-defensins possess chemotactic activity and have been shown to mobilize memory T cells, immature dendritic cells, mast cells, and activated neutrophils (40–43). Murine β-defensins also activate dendritic cells and stimulate a Th-1-type cytokine response, which promotes cytotoxic T lymphocyte-mediated tumor clearance (44). Interestingly, murine β-defensins have also been found to enhance angiogenesis in the presence of physiologically relevant concentrations of VEGF (45). Tumor angiogenesis is a double-edged sword since the development of new blood vessels may lead to increased tumor growth by transporting nutrients to the tumor or may aid in the transport of soluble cytotoxic mediators and/or immune effector cells to the tumor site.

Although relatively few studies have evaluated the capacity of β-defensins and other CAPs to favorably enhance antitumor immunity, it is at least conceivable that certain endogenous CAPs may establish chemotactic gradients emanating from solid tumors owing to their release from necrotic cells at the center of the tumor. For example, β-defensins released from necrotic tumor cells could recruit immature dendritic cells and memory T cells to the tumor site. Immune cell recruitment to the tumor site may also be enhanced by β-defensin-induced angiogenesis. Uptake and processing of tumor antigens by immature dendritic cells would result in a Th-1-type immune response with the subsequent activation and expansion of tumor-reactive cytotoxic T lymphocytes.

PRECLINICAL AND CLINICAL INVESTIGATIONS

Numerous preclinical studies on CAPs (native, modified, or synthetic) with *in vitro* anticancer activity have been conducted in different rodent models of human cancer to determine whether these peptides have the potential to be effective in the treatment of human malignancies. Although the results of preclinical testing in animals have been promising, these studies have not yet advanced to clinical trials in cancer patients. Nevertheless, results from Phase I clinical trials of other peptide-based cancer therapeutics suggest that CAPs with anticancer activity will be well tolerated and may well benefit cancer patients, providing that several technical obstacles can first be overcome.

Preclinical Studies on CAPs

LfcinB is among the CAPs with *in vitro* anticancer activity that has been investigated in animal models of cancer. LfcinB is a CAP that is cytotoxic for a variety of human and mouse cancer cell lines, although the mechanism of LfcinB-induced cytotoxicity varies depending on the type of cancer cell being investigated. LfcinB causes caspase-dependent apoptosis in human leukemia cells and breast cancer cells, whereas neuroblastoma and murine sarcoma cells are largely killed by membrane destabilization and necrosis (25, 46, 47). In mouse models of cancer, LfcinB inhibits the metastasis of lymphoma cells to the spleen and liver, as well as the metastasis of murine melanoma cells to the lungs (48). LfcinB also inhibits the growth of human neuroblastoma cells in immune-deficient rats (46).

Magainin is a CAP that was originally isolated from the skin of the African clawed frog *Xenopus laevis* (49). Magainin is a membranolytic α-helical CAP that is able to kill a variety of cancer cell lines (50). For example, tumors formed by human melanoma cells grown as xenografts in immune-deficient mice completely disappeared in six out of nine mice following intratumoral treatment with an all-D amino acid derivative of magainin, whereas a control peptide had no effect on tumor growth (51). Although the skin of magainin-treated mice initially showed an adverse response to the CAP, healing was complete within 2 weeks posttreatment with only minimal scarring.

Gomesin is an 18-amino acid CAP isolated from the tarantula spider and is similar in sequence to peptides of the tachyplesin family (52). Gomesin kills murine and human cancer cell lines *in vitro*, including human melanoma, breast, and colon carcinomas, by a direct mechanism with an IC_{50} of less than 5 μM (53). The cytotoxic activity of gomesin against cancer cells is dependent on the presence of at least one disulfide bond that forms a β-hairpin loop in the CAP; two disulfide bonds are required for enhanced stability of gomesin in the presence of serum. Rodrigues et al. recently showed that the application of gomesin as a topical ointment to tumors resulting from the subcutaneous injection of mouse melanoma cells into the hind flanks of mice significantly delayed subsequent tumor growth (53). Gomesin, and possibly other CAPs, might therefore be used as topical agents for the treatment of certain skin cancers. Anti-angiogenic activity might also contribute to the tumor inhibitory activity of gomesin since gomesin kills human endothelial cells *in vitro* with similar IC_{50} values (53); however, the importance of gomesin killing of endothelial cells in its *in vivo* anticancer activity has not been evaluated. Importantly, repeated topical applications of gomesin did not have any adverse affect on the skin of peptide-treated mice, suggesting that gomesin is not toxic to normal epithelial cells.

Clinical Studies on Peptide-Based Therapeutic Agents

Although CAPs with anticancer activities have not yet been evaluated in cancer patients, several other peptide-based therapies have been the subject of recent clinical trials. For example, Phase I clinical trials on the peptides cilengitide and ADH-1 are now complete (9, 54). Cilengitide is a synthetic anti-angiogenic peptide that contains the RGD integrin-binding motif and was designed to inhibit $\alpha_V\beta_3$ and $\alpha_V\beta_5$ integrin-mediated endothelial cell attachment and migration. Both of these processes are required for blood vessel formation. Cilengitide has a half-life of approximately 4 h following intravenous delivery and is well tolerated by patients with no dose-limiting toxicities. Interestingly, biopsies of tumors from several patients revealed an increase in tumor and endothelial cell apoptosis. However, the results did not reach statistical significance due to the small sample size and considerable tumor heterogeneity between and within patients. Although cilengitide is not a CAP (arginine and aspartic acid residues in cilengitide form a zwitterionic peptide), this study suggests that other peptide-based anticancer agents, including CAPs, may be well tolerated in humans.

ADH-1 (N-Ac-CHAVC-NH_2), also known as Exherin™ (Adherex Technologies, Chapel Hill, NC), is a synthetic peptide that contains two cysteine residues that form a disulfide bond, resulting in peptide cyclization (55).

ADH-1 is slightly cationic at neutral pH values and contains the sequence histidine–alanine–valine (HAV) that is highly conserved among cadherin proteins. HAV-containing peptides like ADH-1 or antibodies that recognize the HAV sequence are able to disrupt cadherin-mediated cellular adhesion (56). ADH-1 interferes with cadherin-mediated endothelial cell interactions, which results in cell death by apoptosis *in vitro* (56). ADH-1 recently underwent a Phase I clinical trial with 30 patients afflicted with N-cadherin-expressing tumors (57). Intravenous doses of ADH-1 ranging from 150 to 2400 mg/m^2 were well tolerated, although asthenia (weakness) was frequently reported as an adverse side effect. Interestingly, two patients with ovarian cancer exhibited prolonged disease stabilization. Unlike cilengitide, ADH-1 has a very short half-life, likely due to peptide digestion by proteases present in human serum. More frequent dosing, peptide modification, or combinatorial treatment regimens will be required if ADH-1 is to prove useful in a clinical setting.

To date, no clinical trials have been conducted on CAPs with anticancer activity; nevertheless, promising results from clinical trials that studied cilengitide and ADH-1 suggest that membranolytic and indirect-acting CAPs may be well tolerated by patients. However, the issue of serum stability remains a significant hurdle that must be overcome before CAPs can successfully be delivered by the intravenous route to humans.

ENHANCING ANTICANCER ACTIVITY AND CAP STABILITY THROUGH PEPTIDE MODIFICATION

Three major shortcomings that limit the potential therapeutic value of many CAPs in cancer treatment are (1) toxic effects on normal cells, (2) degradation of CAPs by proteases, and (3) cost of peptide production. Although most CAPs preferentially kill negatively charged cancer cells, certain healthy cells may also be targeted by CAPs at higher peptide concentrations if these normal cells also express some negatively charged cell surface molecules (19, 50). Modifying the positive charge and hydrophobicity of CAPs in order to enhance peptide binding to and anchoring in the membrane of the cancer cell, respectively, are strategies to improve CAP selectivity for cancer cells. Improved cancer cell selectivity may also be achieved by adding peptide sequences that recognize molecules expressed by cancer cells but not by normal cells. Proteolytic degradation by serum proteases occurs rapidly and can significantly reduce the anticancer activity of CAPs. Introducing enantiomeric amino acids (i.e., replacing L-amino acids with D-amino acids) is one technique that can improve CAP stability. Finally, it is important to note that the large-scale production of clinical grade CAPs will be expensive; hence, ongoing efforts are being made to reduce the cost of peptide production by engineering truncated CAPs with anticancer activity that is equivalent to their parent molecules. These modification strategies and others may enhance CAP-mediated cancer cell killing and peptide stability for therapeutic application.

CAP Modification Strategies to Enhance Cancer Cell Killing

Membranolytic cell death mediated by CAPs begins with peptide binding to the membrane of the cancer cell via electrostatic interactions between positively charged amino acids of the peptide and negatively charged molecules on the surface of the cancer cell (14). It therefore seems intuitive that replacement of neutral amino acids with basic amino acids should increase CAP attraction to negatively charged cancer cell membranes by increasing the overall positive charge of the peptide. However, care must be taken when selecting amino acids for replacement because hydrophobic amino acid residues are responsible for anchoring the CAP in the lipid bilayer, which leads to membrane destabilization and cytotoxicity (10, 58). Additionally, the presence of bulky hydrophobic amino acids may help to protect the peptide backbone from proteolytic digestion by masking protease cleavage sites. Although the effect of manipulating charge and hydrophobicity on the anticancer activity of CAPs has not been well studied, an investigation focused on the effect of manipulating the overall charge and hydrophobicity of $Lf_{(14-31)}$, the amino (N)-terminal α-helical region of lactoferrin corresponding to residues 14–31 (59). Helical wheel diagrams, which are models used to visualize the spatial organization of amino acids within α-helices, revealed that the cationic residues of $Lf_{(14-31)}$ are clustered in two spatially separated sections, the major and minor sectors, which consist of four and two cationic amino acids, respectively. The anticancer activity of $Lf_{(14-31)}$ is increased when the positively charged amino acids are moved from the minor sector to the major sector, thereby creating a cationic face and a predominantly hydrophobic face. However, the increase in cancer cell killing occurs at the expense of cancer cell selectivity because the modified peptide is toxic toward normal fibroblasts and red blood cells. These findings suggest that α-helices possessing a major sector in the absence of a minor sector exhibit reduced selectivity for cancer cells. Interestingly, increasing the charge of the CAP from +6 to +8 does not significantly increase toxicity toward normal cells, providing that two cationic amino acids remain in the minor sector. Unfortunately, this more cationic Lf-derived peptide exhibits significantly reduced cancer cell killing, presumably due to the loss of two bulky hydrophobic residues that are often vital for CAP-mediated membrane disruption (58).

Other strategies have been employed to increase the overall charge and/or hydrophobicity of CAPs. Endcapping is a technique whereby functional groups at either end of the peptide are chemically modified (60). Typically, endcapping refers to N-terminal acetylation and/or C-terminal amidation, which can be used to compensate for an inherent lack of hydrophobicity and positive charge, respectively. Both C-terminal amidation and N-terminal acetylation have been successfully used to increase the antibacterial and anticancer activities of certain peptides (61–64). Importantly, preliminary studies have also suggested that endcapping may decrease the susceptibility of CAPs to proteolytic degradation, thereby enhancing peptide stability in serum (60). This may be due to a protective effect by the bulky capping moiety that preserves the peptide backbone from digestion by endopeptidases, such as trypsin and chymotrypsin, as well as protecting the N-terminal and C-terminal of the CAP from degradation by exopeptidases, such as aminopeptidase and carboxypeptidase. Importantly, exopeptidases are involved in alimentary digestion and could severely limit the bioavailability of CAPs delivered by the oral route (65). Therefore, in the right context, endcapping may be a useful strategy for enhancing CAP stability; however, the influence of endcapping on the anticancer activity of CAPs has not yet been determined.

Addition of a Targeting Sequence to Enhance Cancer Cell Targeting

The selectivity of CAPs for cancer cells can be improved by the addition of targeting sequences identified through the use of phage display libraries in which virus particles are engineered to express short peptides on the surface of the mature virion. Virus particles that interact with specific cancer cells, but not normal cells, are identified, and the protein sequence is determined. This approach has resulted in the identification of peptide sequences such as leucine–threonine–valine–serine–proline–tryptophan-tyrosine (LTVSPWY), which interacts with breast cancer cells but not normal epithelial, endothelial, or hematopoetic cells (66). However, whether or not the addition of such targeting sequences improves the anticancer activity of CAPs has not yet been determined. Moreover, it is important to consider the secondary structure that the modified peptide will adopt both in aqueous solutions and in membrane-mimicking conditions. While the targeting sequence has to be accessible to its ligand in aqueous solutions in order to promote selective binding of the peptide to the cancer cell, it is essential that the secondary structure of the biologically active portion of the peptide remains intact and is not obstructed in any way by the targeting sequence; otherwise, the modified peptide will lose its cytotoxic activity. A detailed analysis of the secondary structure of a modified peptide will be essential to predict whether it is likely to be more or less cytotoxic than the parent molecule.

Peptide Modification Strategies to Alter Peptide Stability

CAPs contain an abundance of positively charged amino acids and aromatic ring-containing hydrophobic amino acids, which are substrates for trypsin and chymotrypsin digestion, respectively (67). Therefore, proteolysis can limit the therapeutic utility of CAPs because, in general, CAPs must be a certain length in order to directly kill cancer cells. Many strategies to decrease the susceptibility of CAPS to protease-mediated degradation have been evaluated, including the formation of peptidomimetics, which are protein-like molecules that are designed to mimic peptides and are thought to be resistant to proteases (68). Figure 14.3 shows one strategy that involves replacing L-amino acids with D-amino acids, which are the chiral opposites of naturally occurring L-amino

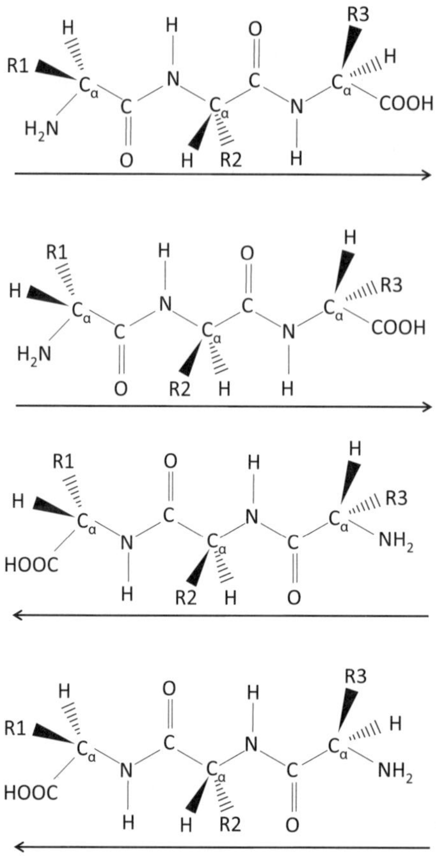

Figure 14.3 Chemical structures of peptide-mimicking molecules (peptidomimetics). Naturally occurring L-amino acids that comprise peptides can be replaced with D-amino acids to create all-D-amino acid peptides. The direction of the peptide can be inverted (note inversion of the N-terminal and C-terminal), which results in the formation of all-L-amino acid retro-peptides. All-D-amino acid retro–inverso-peptides are created by changing the direction of the peptide and the chirality at each α-carbon. The side chain topology is thought to be conserved when all-D-amino acid retro–inverso-peptides are produced. All of these peptidomimetics are thought to be resistant to protease-mediated digestion. R1–R3 denotes any amino acid side chain.

acids (15). Several groups have evaluated the effect of D-amino acid incorporation on CAP stability by using one of two strategies: forming the enantiomer of the peptide by replacing all L-amino acids with D-amino acids or forming a diastereomeric isomer by replacing those L-amino acids positioned at sites recognized by proteases (e.g., lysine and tryptophan recognized by trypsin and chymotrypsin, respectively) with D-amino acids. Several studies have evaluated the influence of D-amino acid incorporation on peptide-mediated cancer cell killing *in vitro* and *in vivo*. An *in vitro* comparison of the anticancer activity of two D-amino acid-containing magainin derivatives to that of unmodified magainin showed that the all-D-amino acid peptide is significantly more cytotoxic than the all-L-amino acid form of the peptide (69). Furthermore, after only a single injection, the all-D-amino acid magainin derivative significantly reduces the number and viability of P388D1 lymphoma cells in the ascitic fluid of tumor-bearing mice compared to control mice treated with the all-L-amino acid-containing form of magainin, suggesting that the all-D-amino acid CAP has improved stability *in vitro* and *in vivo*. Interestingly, a single injection of the all-D-amino acid magainin derivative is also able to significantly increase the life span of mice bearing spontaneous ovarian teratomas. Preservation of anticancer activity with all-D-amino acid CAPs is supported by a recent report that an all-D-amino acid form of gomesin has the same *in vitro* cytotoxic activity as naturally occurring gomesin (53). In addition, a synthetic diastereomeric CAP has recently been shown to possess *in vitro* and *in vivo* anticancer activity (15). Importantly, this diastereomeric peptide is resistant to degradation by trypsin, elastase, and proteinase-K. Moreover, this study suggests that selective replacement of L-amino acids with D-amino acids may have the same effect as replacing all L-amino acids with D-amino acids, which is important because D-amino acids are significantly more expensive than L-amino acids. However, if CAP-induced cancer cell killing is a consequence of receptor interactions with a chiral (α) center, then these isomers will be less potent than the naturally occurring all-L-amino acid form of the CAP unless a diastereomeric isomer is formed in which the receptor binding portion of the peptide is not changed. Thus, the mechanism by which a given CAP kills its target cell will largely determine the usefulness of this strategy to enhance CAP stability.

Peptidomimetics also include molecules containing amino acids that do not occur in nature and are thought to be less susceptible to proteolytic degradation. For example, Oh et al. synthesized a novel amino acid that is more positively charged and bulkier than naturally occurring amino acids (70). Incorporation of unnatural amino acids generally increases CAP stability in the presence of serum, perhaps by preventing proteases from accessing the peptide backbone and/or by removing the trypsin recognition sequence(s). However, this novel amino acid has a lower α-helical propensity than lysine, which may impact the biological activity of the CAP if the α-helical secondary structure is vital for cancer cell killing.

CAP stability may also be enhanced by alkylating or acylating lysine residues in an attempt to reduce digestion by trypsin (71, 72). Generating the peptoid equivalent of the peptide represents another possible strategy to enhance CAP stability. Peptoids are molecules that mimic the biological activity of peptides but possess enhanced resistance to proteases because the side chain is positioned on the nitrogen atom rather than the α carbon (73). Other examples of peptidomimetics include retro-isomers and retro–inverso-isomers (Figure 14.3). Retro-isomers are directional isomers of naturally occurring all-L-amino acid peptides, whereas retro–inverso-isomers are directional and chiral isomers of naturally occurring peptides (74). These pseudopeptides are believed to be resistant to proteolytic degradation because of chemical alterations in the peptide backbone. Additionally, retro–inverso-isomers are predicted to adopt the same side chain topology as the native peptide; therefore, the biological activity should be retained. These modification strategies are novel approaches to obtain enhanced CAP stability with a concomitant maintenance or enhancement of biological activity. However, studies on cancer cell killing by peptidomimetics are preliminary and often do not adequately address the issues of cancer cell selectivity, *in vivo* cytotoxicity, or the impact of these pseudopeptides on normal biological responses such as immune function. Further research is needed before the benefit of these modification strategies on the therapeutic application of CAPs in cancer treatment can be fully realized.

CONCLUSION

A wealth of data suggests that certain CAPs and their derivatives hold considerable promise as novel anticancer agents. Many CAPs kill a variety of cancer cell lines while sparing healthy cells. This characteristic, in addition to the predicted ability of CAPs to kill multidrug-resistant and indolent or slow-growing tumors makes CAPs ideal candidates for future development as a new class of therapeutic agents for cancer treatment. Preclinical studies demonstrate that certain CAPs and/or their derivatives are able to reduce or eliminate tumor burden in animal models of cancer, while Phase I and II clinical trials already indicate that peptide-based therapies are well tolerated by patients. Furthermore, Phase I–III clinical trials evaluating the antibacterial activity of numerous CAPs have been conducted and show promising results (14). There is every reason to expect that future clinical trials will establish CAPs as a new weapon in the oncologist's anticancer armamentarium.

REFERENCES

1. Donnelly J.G. 2004. Pharmacogenetics in cancer chemotherapy: balancing toxicity and response. Ther. Drug Monit. **26**: 231–235.
2. Naumov G.N., Townson J.L., MacDonald I.C., Wilson S.M., Bramwell V.H., Groom A.C., and Chambers A.F. 2003. Ineffectiveness of doxorubicin treatment on solitary dormant mammary carcinoma cells or late-developing metastases. Breast Cancer Res. Treat. **82**: 199–206.
3. Bush J.A., and Li G. 2002. Cancer chemoresistance: the relationship between p53 and multidrug transporters. Int. J. Cancer. **98**: 323–330.
4. Smith L.L., Brown K., Carthew P., Lim C.K., Martin E.A., Styles J., and White I.N. 2000. Chemoprevention of breast cancer by tamoxifen: risks and opportunities. Crit. Rev. Toxicol. **30**: 571–594.
5. Nagy P., Friedländer E., Tanner M., Kapanen A.I., Carraway K.L., Isola J., and Jovin T.M. 2005. Decreased accessibility and lack of activation of ErbB2 in JIMT-1, a herceptin-resistant, MUC4-expressing breast cancer cell line. Cancer Res. **65**: 473–482.

6. Hoskin D.W., and Ramamoorthy A. 2008. Studies on anticancer activities of antimicrobial peptides. Biochim. Biophys. Acta. **1778**: 357–375.
7. Mader J.S., and Hoskin D.W. 2006. Cationic antimicrobial peptides as novel cytotoxic agents for cancer treatment. Expert Opin. Investig. Drugs. **15**: 933–946.
8. Yeaman M.R., and Yount N.Y. 2003. Mechanisms of antimicrobial peptide action and resistance. Pharmacol Rev. **55**: 27–55.
9. Bhutia S.K., and Maiti T.K. 2008. Targeting tumors with peptides from natural sources. Trends Biotechnol. **26**: 210–217.
10. Powers J.P., and Hancock R.E. 2003. The relationship between peptide structure and antibacterial activity. Peptides. **24**: 1681–1691.
11. Patrzykat A., Friedrich C.L., Zhang L., Mendoza V., and Hancock R.E. 2002. Sublethal concentrations of pleurocidin-derived antimicrobial peptides inhibit macromolecular synthesis in *Escherichia coli*. Antimicrob. Agents Chemother. **46**: 605–614.
12. Papo N., and Shai Y. 2005. Host defense peptides as new weapons in cancer treatment. Cell Mol. Life Sci. **62**: 784–790.
13. Zwaal R.F., and Schroit A.J. 1997. Pathophysiologic implications of membrane phospholipid asymmetry in blood cells. Blood. **89**: 1121–1132.
14. Giuliani A., Pirri, G., and Nicoletto, S.F. 2007. Antimicrobial peptides: an overview of a promising class of therapeutics. Cent. Eur. J. Biol. **2**: 1–33.
15. Papo N., Shahar M., Eisenbach L., and Shai Y. 2003. A novel lytic peptide composed of DL-amino acids selectively kills cancer cells in culture and in mice. J. Biol. Chem. **278**: 21018–21023.
16. Gazit E., Boman A., Boman H.G., and Shai Y. 1995. Interaction of the mammalian antibacterial peptide cecropin P1 with phospholipid vesicles. Biochemistry. **34**: 11479–11488.
17. Gazit E., Lee W.J., Brey P.T., and Shai Y. 1994. Mode of action of the antibacterial cecropin B2: a spectrofluorometric study. Biochemistry. **33**: 10681–10692.
18. Matsuzaki K., Harada M., Handa T., Funakoshi S., Fujii N., Yajima H., and Miyajima K. 1989. Magainin 1-induced leakage of entrapped calcein out of negatively-charged lipid vesicles. Biochim. Biophys. Acta. **981**: 130–134.
19. Hancock R.E., and Sahl H.G. 2006. Antimicrobial and host-defense peptides as new anti-infective therapeutic strategies. Nat. Biotechnol. **24**: 1551–1557.
20. Lodish H., Berk A., Kaiser M., Scott M.P., Bretscher A., Ploegh H., and Matsudaira P. 2008. Molecular Cell Biology, 6th ed. New York: WH Freeman.
21. Yeagle P.L. 1985. Cholesterol and the cell membrane. Biochim. Biophys. Acta. **822**: 267–287.
22. Vaz Gomes A., de Waal A., Berden J.A., and Westerhoff H.V. 1993. Electric potentiation, cooperativity, and synergism of magainin peptides in protein-free liposomes. Biochemistry. **32**: 5365–5372.
23. Chan S.C., Hui L., and Chen H.M. 1998. Enhancement of the cytolytic effect of anti-bacterial cecropin by the microvilli of cancer cells. Anticancer Res. **18**: 4467–4474.
24. Leuschner C., and Hansel W. 2004. Membrane disrupting lytic peptides for cancer treatments. Curr. Pharm. Des. **10**: 2299–2310.
25. Mader J.S., Salsman J., Conrad D.M., and Hoskin D.W. 2005. Bovine lactoferricin selectively induces apoptosis in human leukemia and carcinoma cell lines. Mol. Cancer Ther. **4**: 612–624.

26. Mader J.S., Richardson A., Salsman J., Top D., de Antueno R., Duncan R., and Hoskin D.W. 2007. Bovine lactoferricin causes apoptosis in Jurkat T-leukemia cells by sequential permeabilization of the cell membrane and targeting of mitochondria. Exp. Cell Res. **313**: 2634–2650.
27. Mader J.S., Smyth D., Marshall J., and Hoskin D.W. 2006. Bovine lactoferricin inhibits basic fibroblast growth factor- and vascular endothelial growth factor$16_{5\text{-}}$ $_{i}$nduced angiogenesis by competing for heparin-like binding sites on endothelial cells. Am. J. Pathol. **169**: 1753–1766.
28. Takada Y., Ye X., and Simon S. 2007. The integrins. Genome Biol. **8**: 215.
29. Ruoslahti E., and Pierschbacher M.D. 1987. New perspectives in cell adhesion: RGD and integrins. Science. **238**: 491–497.
30. Zitzmann S., Ehemann V., and Schwab M. 2002. Arginine-glycine-aspartic acid (RGD)-peptide binds to both tumor and tumor-endothelial cells in vivo. Cancer Res. **62**: 5139–5143.
31. Chen Y., Xu X., Hong S., Chen J., Liu N., Underhill C.B., Creswell K., and Zhang L. 2001. RGD-tachyplesin inhibits tumor growth. Cancer Res. **61**: 2434–2438.
32. Boman H.G., Agerberth B., and Boman A. 1993. Mechanisms of action on *Escherichia coli* of cecropin P1 and PR-39, two antibacterial peptides from pig intestine. Infect. Immun. **61**: 2978–2984.
33. Cabiaux V., Agerberth B., Johansson J., Homble F., Goormaghtigh E., and Ruysschaert J.M. 1994. Secondary structure and membrane interaction of PR-39, a Pro+Arg-rich antibacterial peptide. Eur. J. Biochem. **224**: 1019–1027.
34. Tanaka K., Fujimoto Y., Suzuki M., Suzuki Y., Ohtake T., Saito H., and Kohgo Y. 2001. PI3-kinase p85α is a target molecule of proline-rich antimicrobial peptide to suppress proliferation of ras-transformed cells. Jpn. J. Cancer Res. **92**: 959–967.
35. Coffelt S.B., and Scandurro A.B. 2008. Tumors sound the alarmin(s). Cancer Res. **68**: 6482–6485.
36. Sorensen O.E., Cowland J.B., Theilgaard-Monch K., Liu L., Ganz T., and Borregaard N. 2003. Wound healing and expression of antimicrobial peptides/polypeptides in human keratinocytes, a consequence of common growth factors. J. Immunol. **170**: 5583–5589.
37. Ganz T. 2003. Defensins: antimicrobial peptides of innate immunity. Nat. Rev. Immunol. **3**: 710–720.
38. Liu L., Roberts A.A., and Ganz T. 2003. By IL-1 signaling, monocyte-derived cells dramatically enhance the epidermal antimicrobial response to lipopolysaccharide. J. Immunol. **170**: 575–580.
39. García J.R., Krause A., Schulz S., Rodríguez-Jiménez F.J., Klüver E., Adermann K., Forssmann U., Frimpong-Boateng A., Bals R., and Forssmann W.G. 2001. Human β-defensin 4: a novel inducible peptide with a specific salt-sensitive spectrum of antimicrobial activity. FASEB J. **15**: 1819–1821.
40. Yang D., Chertov O., Bykovskaia S.N., Chen Q., Buffo M.J., Shogan J., Anderson M., Schröder J.M., Wang J.M., Howard O.M., and Oppenheim J.J. 1999. β-defensins: linking innate and adaptive immunity through dendritic and T cell CCR6. Science. **286**: 525–528.
41. Biragyn A., Surenhu M., Yang D., Ruffini P.A., Haines B.A., Klyushnenkova E., Oppenheim J.J., and Kwak L.W. 2001. Mediators of innate immunity that target immature, but not mature, dendritic cells induce antitumor immunity when

genetically fused with nonimmunogenic tumor antigens. J. Immunol. **167**: 6644–6653.

42. Niyonsaba F., Iwabuchi K., Matsuda H., Ogawa H., and Nagaoka I. 2002. Epithelial cell-derived human β-defensin-2 acts as a chemotaxin for mast cells through a pertussis toxin-sensitive and phospholipase C-dependent pathway. Int. Immunol. **14**: 421–426.
43. Niyonsaba F., Ogawa H., and Nagaoka I. 2004. Human β-defensin-2 functions as a chemotactic agent for tumour necrosis factor-α-treated human neutrophils. Immunology. **111**: 273–281.
44. Biragyn A., Ruffini P.A., Leifer C.A., Klyushnenkova E., Shakhov A., Chertov O., Shirakawa A.K., Farber J.M., Segal D.M., Oppenheim J.J., and Kwak L.W. 2002. Toll-like receptor 4-dependent activation of dendritic cells by β-defensin 2. Science. **298**: 1025–1029.
45. Conejo-Garcia J.R., Benencia F., Courreges M.C., Kang E., Mohamed-Hadley A., Buckanovich R.J., Holtz D.O., Jenkins A., Na H., Zhang L., Wagner D.S., Katsaros D., Caroll R., and Coukos G. 2004. Tumor-infiltrating dendritic cell precursors recruited by a β-defensin contribute to vasculogenesis under the influence of Vegf-A. Nat. Med. **10**: 950–958.
46. Eliassen L.T., Berge G., Leknessund A., Wikman M., Lindin I., Løkke C., Ponthan F., Johnsen J.I., Sveinbjørnsson B., Kogner P., Flaegstad T., and Rekdal Ø. 2006. The antimicrobial peptide, lactoferricin B, is cytotoxic to neuroblastoma cells in vitro and inhibits xenograft growth in vivo. Int. J. Cancer. **119**: 493–500.
47. Eliassen L.T., Berge G., Sveinbjornsson B., Svendsen J.S., Vorland L.H., and Rekdal O. 2002. Evidence for a direct antitumor mechanism of action of bovine lactoferricin. Anticancer Res. **22**: 2703–2710.
48. Yoo Y.C., Watanabe S., Watanabe R., Hata K., Shimazaki K., and Azuma I. 1998. Bovine lactoferrin and lactoferricin inhibit tumor metastasis in mice. Adv. Exp. Med. Biol. **443**: 285–291.
49. Zasloff M. 1987. Magainins, a class of antimicrobial peptides from Xenopus skin: isolation, characterization of two active forms, and partial cDNA sequence of a precursor. Proc. Natl. Acad. Sci. U S A. **84**: 5449–5453.
50. Dennison S.R., Whittaker M., Harris F., and Phoenix D.A. 2006. Anticancer α-helical peptides and structure/function relationships underpinning their interactions with tumour cell membranes. Curr. Protein Pept. Sci. **7**: 487–499.
51. Soballe P.W., Maloy W.L., Myrga M.L., Jacob L.S., and Herlyn M. 1995. Experimental local therapy of human melanoma with lytic magainin peptides. Int. J. Cancer. **60**: 280–284.
52. Silva P.I., Jr., Daffre S., and Bulet P. 2000. Isolation and characterization of gomesin, an 18-residue cysteine-rich defense peptide from the spider *Acanthoscurria gomesiana* hemocytes with sequence similarities to horseshoe crab antimicrobial peptides of the tachyplesin family. J. Biol. Chem. **275**: 33464–33470.
53. Rodrigues E.G., Dobroff A.S., Cavarsan C.F., Paschoalin T., Nimrichter L., Mortara R.A., Santos E.L., Fázio M.A., Miranda A., Daffre S., and Travassos L.R. 2008. Effective topical treatment of subcutaneous murine B16F10-Nex2 melanoma by the antimicrobial peptide gomesin. Neoplasia. **10**: 61–68.
54. Hariharan S., Gustafson D., Holden S., McConkey D., Davis D., Morrow M., Basche M., Gore L., Zang C., O'Bryant C.L., Baron A., Gallemann D., Colevas D.,

and Eckhardt S.G. 2007. Assessment of the biological and pharmacological effects of the α nu β3 and α nu β5 integrin receptor antagonist, cilengitide (EMD 121974), in patients with advanced solid tumors. Ann. Oncol. **18**: 1400–1407.

55. Blaschuk O.W., Sullivan R., David S., and Pouliot Y. 1990. Identification of a cadherin cell adhesion recognition sequence. Dev. Biol. **139**: 227–229.
56. Erez N., Zamir E., Gour B.J., Blaschuk O.W., and Geiger B. 2004. Induction of apoptosis in cultured endothelial cells by a cadherin antagonist peptide: involvement of fibroblast growth factor receptor-mediated signalling. Exp. Cell Res. **294**: 366–378.
57. Perotti A., Sessa C., Mancuso A., Noberasco C., Cresta S., Locatelli A., Carcangiu M.L., Passera K., Braghetti A., Scaramuzza D., Zanaboni F., Fasolo A., Capri G., Miani M., Peters W.P., and Gianni L. 2009. Clinical and pharmacological phase I evaluation of Exherin (ADH-1), a selective anti-N-cadherin peptide in patients with N-cadherin-expressing solid tumours. Ann. Oncol. **20**: 741–745.
58. Schibli D.J., Epand R.F., Vogel H.J., and Epand R.M. 2002. Tryptophan-rich antimicrobial peptides: comparative properties and membrane interactions. Biochem. Cell Biol. **80**: 667–677.
59. Yang N., Stensen W., Svendsen J.S., and Rekdal O. 2002. Enhanced antitumor activity and selectivity of lactoferrin-derived peptides. J. Pept. Res. **60**: 187–197.
60. Svenson J., Stensen W., Brandsdal B.O., Haug B.E., Monrad J., and Svendsen J.S. 2008. Antimicrobial peptides with stability toward tryptic degradation. Biochemistry. **47**: 3777–3788.
61. Mayo K.H., Haseman J., Young H.C., and Mayo J.W. 2000. Structure-function relationships in novel peptide dodecamers with broad-spectrum bactericidal and endotoxin-neutralizing activities. Biochem J. **349**: 717–728.
62. Oren Z., and Shai Y. 1996. A class of highly potent antibacterial peptides derived from pardaxin, a pore-forming peptide isolated from Moses sole fish *Pardachirus marmoratus*. Eur. J. Biochem. **237**: 303–310.
63. Yang S.T., Shin S.Y., Hahm K.S., and Kim J.I. 2006. Design of perfectly symmetric Trp-rich peptides with potent and broad-spectrum antimicrobial activities. Int. J. Antimicrob. Agents. **27**: 325–330.
64. Avrahami D., and Shai Y. 2004. A new group of antifungal and antibacterials lipopeptides derived from non-membrane active peptides conjugated to palmitic acid. J. Biol. Chem. **279**: 12277–12285.
65. Bernkop-Schnurch A., and Schmitz T. 2007. Presystemic metabolism of orally administered peptide drugs and strategies to overcome it. Curr. Drug Metab. **8**: 509–517.
66. Shadidi M., and Sioud M. 2003. Identification of novel carrier peptides for the specific delivery of therapeutics into cancer cells. FASEB J. **17**: 256–258.
67. Vajda T., and Szabo T. 1976. Specificity of trypsin and α-chymotrypsin towards neutral substrates. Acta Biochim. Biophys. Acad. Sci. Hung. **11**: 287–294.
68. Fischer P.M. 2003. The design, synthesis and application of stereochemical and directional peptide isomers: a critical review. Curr. Protein Pept. Sci. **4**: 339–356.
69. Baker M.A., Maloy W.L., Zasloff M., and Jacob L.S. 1993. Anticancer efficacy of magainin2 and analogue peptides. Cancer Res. **53**: 3052–3057.

70. Oh J.E., and Lee K.H. 1999. Synthesis of novel unnatural amino acid as a building block and its incorporation into an antimicrobial peptide. Bioorg. Med. Chem. **7**: 2985–2990.

71. Pethe K., Bifani P., Drobecq H., Sergheraert C., Debrie A.S., Locht C., Menozzi F.D.2002. Mycobacterial heparin-binding hemagglutinin and laminin-binding protein share antigenic methyllysines that confer resistance to proteolysis. Proc. Natl. Acad. Sci. U S A. **99**: 10759–10764.

72. Radzishevsky I.S., Rotem S., Bourdetsky D., Navon-Venezia S., Carmeli Y., and Mor A. 2007. Improved antimicrobial peptides based on acyl-lysine oligomers. Nat. Biotechnol. **25**: 657–659.

73. Mas-Moruno C., Cruz L.J., Mora P., Francesch A., Messeguer A., Pérez-Paya E., and Albericio F. 2007. Smallest peptoids with antiproliferative activity on human neoplastic cells. J. Med. Chem. **50**: 2443–2449.

74. Chorev M., and Goodman M. 1995. Recent developments in retro peptides and proteins—an ongoing topochemical exploration. Trends Biotechnol. **13**: 438–445.

75. Risso A., Zanetti M., and Gennaro R. 1998. Cytotoxicity and apoptosis mediated by two peptides of innate immunity. Cell Immunol. **189**: 107–115.

76. Ghavami S., Asoodeh A., Klonisch T., Halayko A.J., Kadkhoda K., Kroczak T.J., Gibson S.B., Booy E.P., Naderi-Manesh H., and Los M. 2008. Brevinin-2R(1) semi-selectively kills cancer cells by a distinct mechanism, which involves the lysosomal-mitochondrial death pathway. J. Cell Mol. Med. **12**: 1005–1022.

77. Boman H.G., Faye I., von Hofsten P., Kockum K., Lee J.Y., Xanthopoulos K.G., Bennich H., Engström A., Merrifield R.B., and Andreu D. 1985. On the primary structures of lysozyme, cecropins and attacins from *Hyalophora cecropia*. Dev. Comp. Immunol. **9**: 551–558.

78. Lin W.J., Chien Y.L., Pan C.Y., Lin T.L., Chen J.Y., Chiu S.J., and Hui C.F. 2009. Epinecidin-1, an antimicrobial peptide from fish (*Epinephelus coioides*) which has an antitumor effect like lytic peptides in human fibrosarcoma cells. Peptides. **30**: 283–290.

79. Johnstone S.A., Gelmon K., Mayer L.D., Hancock R.E., and Bally M.B. 2000. In vitro characterization of the anticancer activity of membrane-active cationic peptides. I. Peptide-mediated cytotoxicity and peptide-enhanced cytotoxic activity of doxorubicin against wild-type and P-glycoprotein over-expressing tumor cell lines. Anticancer Drug Des. **15**: 151–160.

80. Silva M.T., do Vale A., and dos Santos N.M. 2008. Secondary necrosis in multicellular animals: an outcome of apoptosis with pathogenic implications. Apoptosis. **13**: 463–482.

81. Selsted M.E., Harwig S.S., Ganz T., Schilling J.W., and Lehrer R.I. 1985. Primary structures of three human neutrophil defensins. J. Clin. Invest. **76**: 1436–1439.

82. Chavakis T., Cines D.B., Rhee J.S., Liang O.D., Schubert U., Hammes H.P., Higazi A.A., Nawroth P.P., Preissner K.T., and Bdeir K. 2004. Regulation of neovascularization by human neutrophil peptides (α-defensins): a link between inflammation and angiogenesis. FASEB J. **18**: 1306–1308.

83. Wilde C.G., Griffith J.E., Marra M.N., Snable J.L., and Scott R.W. 1989. Purification and characterization of human neutrophil peptide 4, a novel member of the defensin family. J. Biol. Chem. **264**: 11200–11203.

84. Bensch K.W., Raida M., Magert H.J., Schulz-Knappe P., and Forssmann W.G. 1995. hBD-1: a novel β-defensin from human plasma. FEBS Lett. **368**: 331–335.

85. Bals R., Wang X., Wu Z., Freeman T., Bafna V., Zasloff M., and Wilson J.M. 1998. Human β-defensin 2 is a salt-sensitive peptide antibiotic expressed in human lung. J. Clin. Invest. **102**: 874–880.

86. Aarbiou J., Tjabringa G.S., Verhoosel R.M., Ninaber D.K., White S.R., Peltenburg L.T., Rabe K.F., and Hiemstra P.S. 2006. Mechanisms of cell death induced by the neutrophil antimicrobial peptides α-defensins and LL-37. Inflamm. Res. **55**: 119–127.

87. Barlow P.G., Li Y., Wilkinson T.S., Bowdish D.M., Lau Y.E., Cosseau C., Haslett C., Simpson A.J., Hancock R.E., and Davidson D.J. 2006. The human cationic host defense peptide LL-37 mediates contrasting effects on apoptotic pathways in different primary cells of the innate immune system. J. Leukoc. Biol. **80**: 509–520.

88. Zasloff M., Martin B., and Chen H.C. 1988. Antimicrobial activity of synthetic magainin peptides and several analogues. Proc. Natl. Acad. Sci. U S A. **85**: 910–913.

89. Lehmann J., Retz M., Sidhu S.S., Suttmann H., Sell M., Paulsen F., Harder J., Unteregger G., and Stöckle M. 2006. Antitumor activity of the antimicrobial peptide magainin II against bladder cancer cell lines. Eur. Urol. **50**: 141–147.

90. Kreil G. 1973. Biosynthesis of melittin, a toxic peptide from bee venom. Amino-acid sequence of the precursor. Eur. J. Biochem. **33**: 558–566.

91. Sui S.F., Wu H., Guo Y., and Chen KS. 1994. Conformational changes of melittin upon insertion into phospholipid monolayer and vesicle. J. Biochem. **116**: 482–487.

92. Saini S.S., Chopra A.K., and Peterson J.W. 1999. Melittin activates endogenous phospholipase D during cytolysis of human monocytic leukemia cells. Toxicon. **37**: 1605–1619.

93. Sharma S.V. 1992. Melittin resistance: a counterselection for ras transformation. Oncogene. **7**: 193–201.

94. Sharma S.V. 1993. Melittin-induced hyperactivation of phospholipase A2 activity and calcium influx in ras-transformed cells. Oncogene. **8**: 939–947.

95. Ohtake T., Fujimoto Y., Ikuta K., Saito H., Ohhira M., Ono M., and Kohgo Y. 1999. Proline-rich antimicrobial peptide, PR-39 gene transduction altered invasive activity and actin structure in human hepatocellular carcinoma cells. Br. J. Cancer. **81**: 393–403.

96. Chen J., Xu X.M., Underhill C.B., Yang S., Wang L., Chen Y., Hong S., Creswell K., and Zhang L. 2005. Tachyplesin activates the classic complement pathway to kill tumor cells. Cancer Res. **65**: 4614–4622.

97. Nakamura T., Furunaka H., Miyata T., Tokunaga F., Muta T., Iwanaga S., Niwa M., Takao T., and Shimonishi Y. 1988. Tachyplesin, a class of antimicrobial peptide from the hemocytes of the horseshoe crab (*Tachypleus tridentatus*). Isolation and chemical structure. J. Biol. Chem. **263**: 16709–16713.

15

PRODIGININES AND THEIR POTENTIAL UTILITY AS PROAPOPTOTIC ANTICANCER AGENTS

Neil R. Williamson,[1] Suresh Chawrai,[2] Finian J. Leeper,[2] and George P.C. Salmond[1]

[1]*Department of Biochemistry, University of Cambridge, Cambridge, UK*
[2]*Department of Chemistry, University of Cambridge, Cambridge, UK*

INTRODUCTION

There is a long history of exploitation of natural products produced by bacteria as sources of pharmaceutically important, bioactive compounds exhibiting antibiotic, antifungal, antiparasitic, immunosuppressive, and anticancer activities. Some of these natural products have been used directly, but they have also served as templates for the development of improved natural products generated either by semisynthesis, combinatorial biosynthesis, or chemical synthesis. This chapter will focus on a particularly interesting class of bacterial secondary metabolites with immunosuppressive and anticancer activity called the prodiginines which are characterized by a common pyrrolyldipyrromethene skeleton. Bacterial prodiginines can be divided into linear and cyclic derivatives (Table 15.1). Examples of the former include prodigiosin and undecylprodigiosin, and examples of the later include streptorubin B, cycloprodigiosin, and cyclononylprodigiosin (1–4). Prodiginines are hydrophobic, bioactive, tripyrrolic, red-pigmented antibiotics produced by a wide range of

Emerging Cancer Therapy: Microbial Approaches and Biotechnological Tools, Edited by Arsénio M. Fialho and Ananda M. Chakrabarty

TABLE 15.1. Natural Prodiginines and Related Compounds with Anticancer Activities

Prodigionine	Bacterial Species	Activitiy/Cancer Type
Prodigiosin 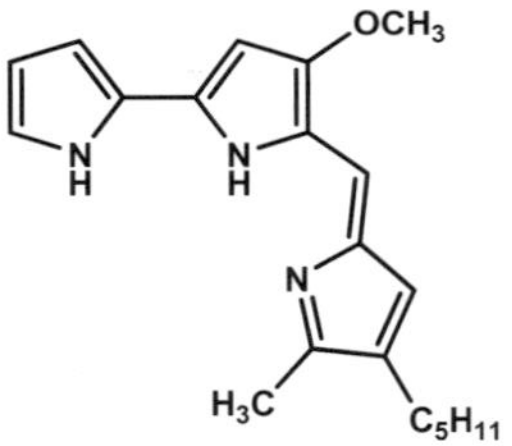	*Serratia* 39006 (13) *Serratia marcescens* (95) *Serratia plymuthica* (96, 97) *Hahella chejuensis* (98) *Pseudomonas magnesiorubra* (99) *Vibrio psychroerythreus* (99) *Vibrio gazogenes* ATCC 29988 (100) *Vibrio ruber* sp. nov. (101)	Madin-Darby canine kidney (MDCK), Chinese hamster ovary (CHO), Human epithelial carcinoma (HeLa), Green monkey kidney (Vero (-317), Human myeloid leukemia (U937), Human acute leukemia T cells (Jurkat-T), Human lung carcinoma (A549, NCI-H460), Human colon adenocarcinoma (DLD1, HT29, SW-620), Human breast carcinomas (MDA-MB-231), Human promyeloblast (HL60), Mouse fibroblast (MC-3T3-E1), Small cell lung cancer (GLC4), Human myeloma (NSO), Human Burkitt lymphoma (Ramos), Human gastric carcinoma (HGT-1), Human neuroblastomas (LAN-1, IMR-2, SH-SYSY, SK-N-AS) (11, 14, 15, 45, 47, 49, 60, 61, 102–107)
Undecylprodigiosin 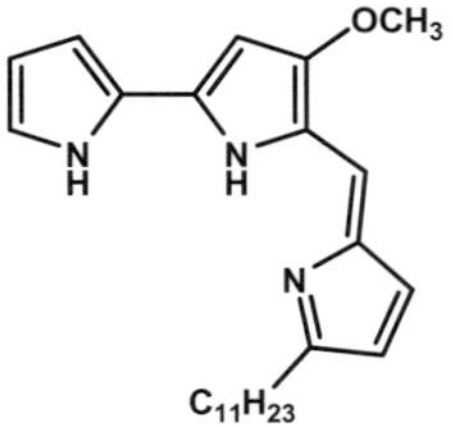	*Streptomyces coelicolor* A3(2) (108–110) *Streptomyces longisporus* ruber (111) *Saccharopolyspora* sp. nov. (112) *Actinomadura madurae* (113) *Streptomyces griseus* (114) *Streptomyces lividans* (114) *Streptomyces* sp. CP1130 (115)	Mouse lymphoma (P388), Human promyeloblast (HL60), Human lung carcinoma (A549, SPCA4), Human hepatic carcinoma (BEL-7402), Baby hamster kidney (BHK) (48, 116)
Streptorubin B 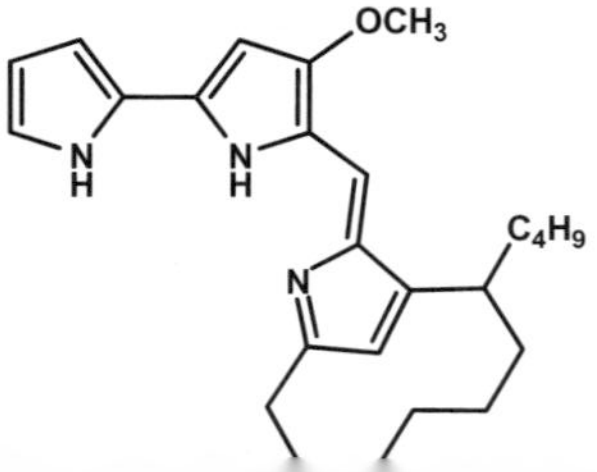	*Streptomyces coelicolor* A3(2) (108–110) *Saccharopolyspora* sp. nov. (112)	

Cycloprodigiosin 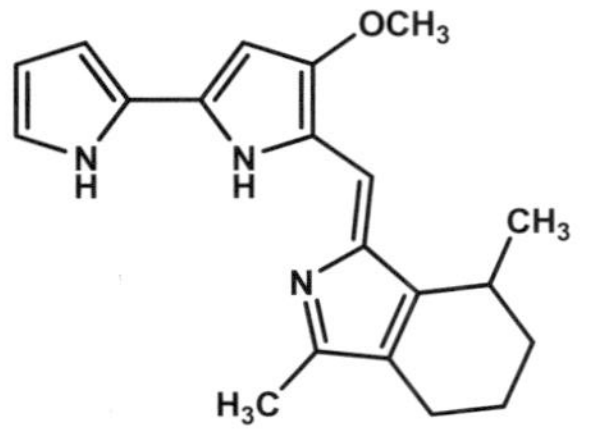	*Pseudoalteromonas denitrificans* (117) *Beneckea gazogenes* (118, 119) *Alteromonas rubra* (120)	Human acute leukemia T cells (Jurkat-T), Human promyeloblast (HL60), Human hepatic carcinoma (Huh-7, HCC-M, HCC-T), Human hepatoblastoma (HepG2), Rat hepatic carcinoma (dRLh-84, H-35), Human colon carcinoma (WiDr, SW480) (16, 66, 73, 121)
Cyclononylprodigiosin 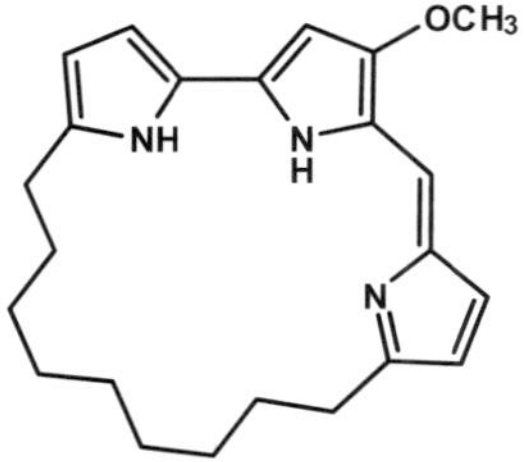	*Actinomadura pelletieri* (113) *Actinomadura madurae* (113)	
Metacycloprodigiosin 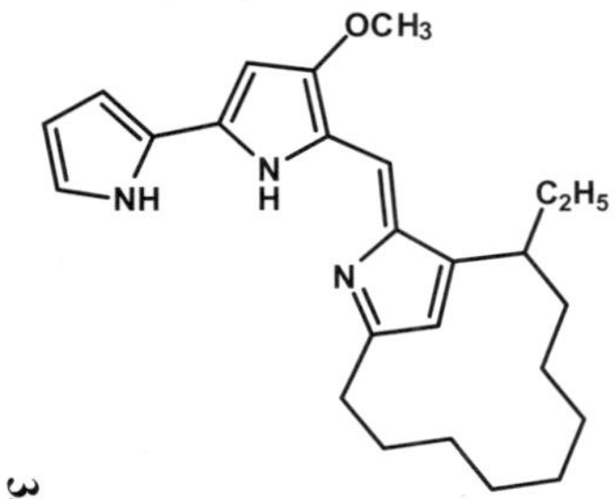	*Streptomyces longisporus* ruber (122) *Saccharopolyspora* sp. nov. (112) *Actinomadura madurae* (113) *Streptomyces* sp. CP1130 (115)	Mouse lymphoma (P388), Human promyeloblast (HL60), Human lung carcinoma (A549, SPCA4), Human hepatic carcinoma (BEL-7402), Mouse fibroblast (MC-3T3-E1), Green monkey kidney (Vero-317) (106, 116)

TABLE 15.1. *Continued*

Prodigionine	Bacterial Species	Activitiy/Cancer Type
Prodigiosin R1 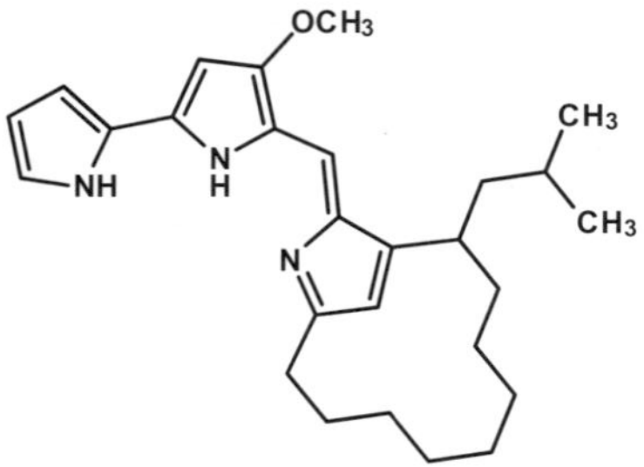	*Streptomyces griseoviridis* (123, 124)	
MAMPDM 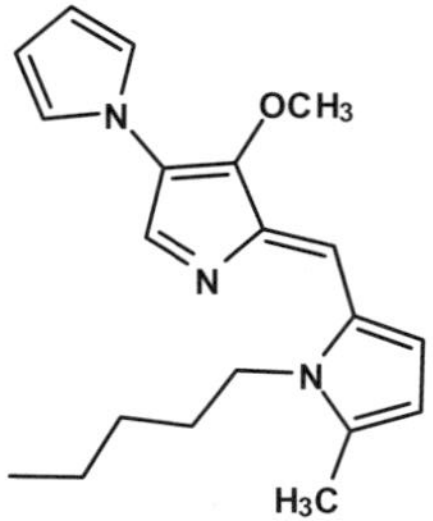	*Serratia marcescens* (125, 126)	Mouse lymphosarcoma ascites (LS-A), Human myeloid leukemia (U937), T lymphoma (EL-4) (127)

2-(*p*-Hydroxybenzyl) prodigiosin (HBPG) 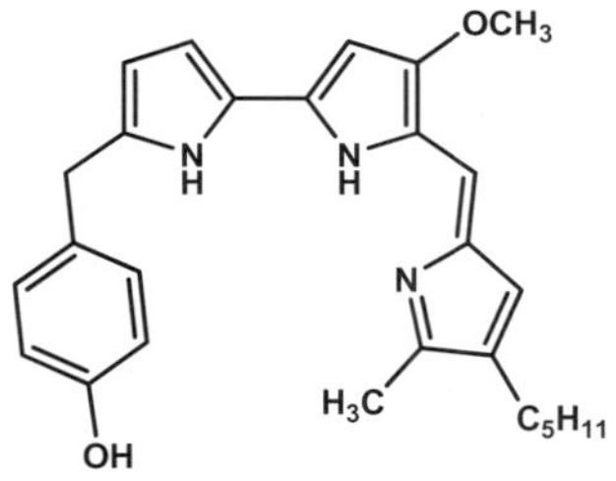 	*Pseudoalteromonas rubra* (128)	
Roseophilin 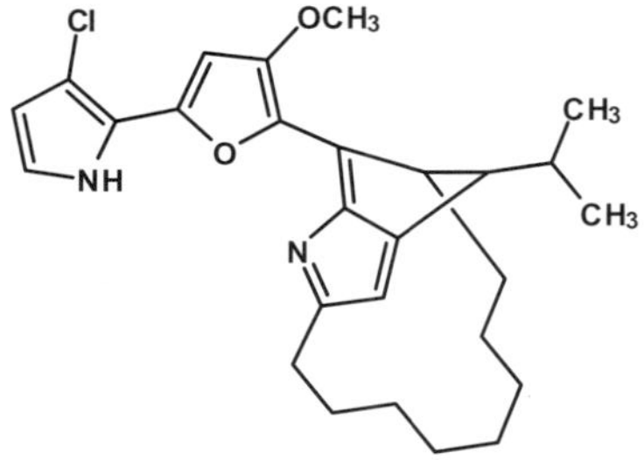 	*Streptomyces griseoviridis* (129)	Cytotoxicity (129, 130) Protein tyrosine phosphatase inhibitors (131) Mycobacterium tuberculosis protein tyrosine phosphatase A (MptpA) Inhibitors (132)

TABLE 15.1. *Continued*

Prodigionine	Bacterial Species	Activitiy/Cancer Type
Tambjamines 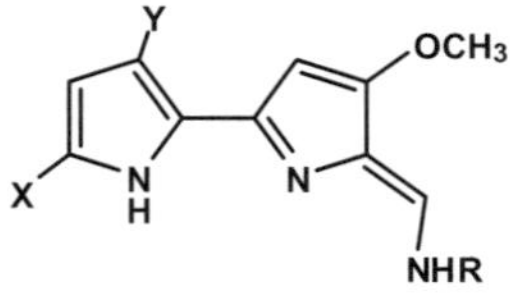 Tambjamine A, R = X = Y = H B, R = Y = H, X = Br C, R = CH_2CHMe_2, X = Y = H D, R = CH_2CHMe_2, X = H, Y = Br E, R = Et, X = Y = H F, R = CH_2CH_2Ph, X = Y = H G, R = Et, X = Br, Y = H H, R = Pr, X = Br, Y = H I, R = CH_2CHMe_2, X = Br, Y = H J, R = $CH_2CHMeCH_2Me$, X = Br, Y = H YP1, R = $(CH_2)_2CH{=}CH(CH_2)_7Me$, X = Y = H BE-18591, R = $(CH_2)_{11}Me$, X = Y = H	*Atapozoa* sp. (133) *Pseudoalteromonas tunicata* (YP1) (134) *Tambja eliora* (tambjamine D) (135)	(BE-18591)—Inhibits immunoproliferation and gastritis (136) Cytotoxic and genotoxic effects of tambjamine D (135) (YP1)-antifungal (134) (Tambjamine E)-cytotoxicity (137–139)
Blue tetrapyrrole 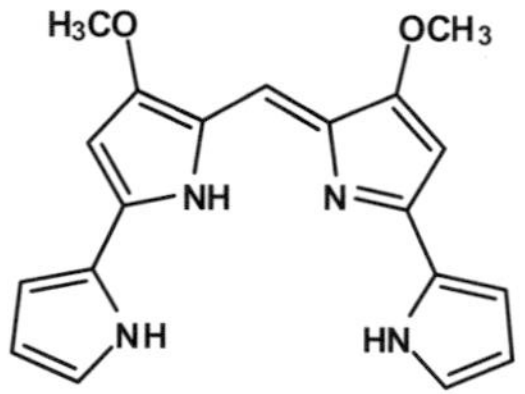	*Atapozoa* sp. (133) *Nembrotha* sp. (133)	Double-strand cleavage of DNA and cytotoxicity (61)

MAMPDM, 2,2′-[3-methoxy-1′amyl-5′-methyl-4-(1″-pyrryl)] dipyrrylmethene.

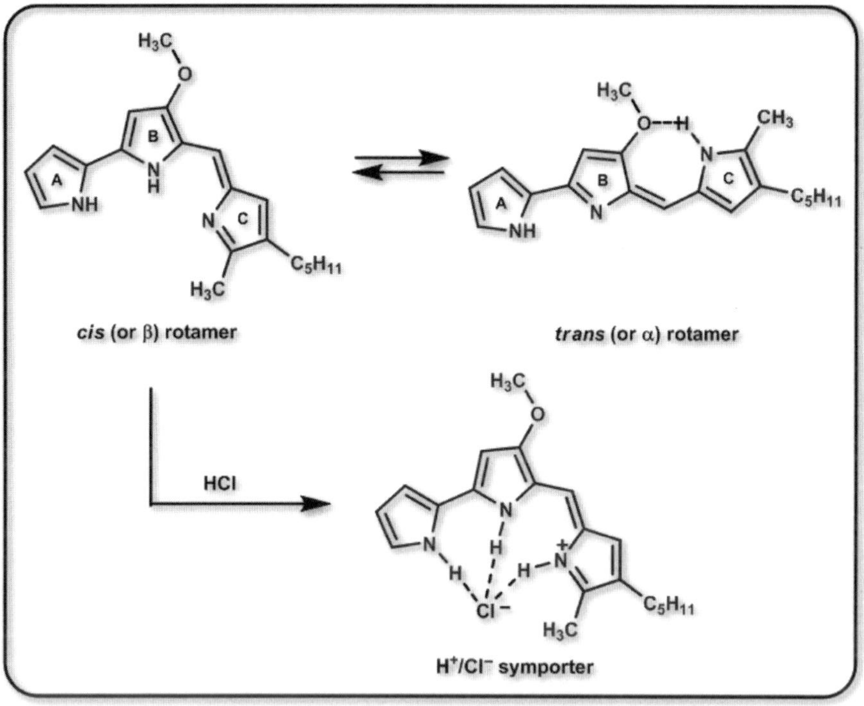

Figure 15.1 The structure of the archetypal prodiginine, prodigiosin. Prodigiosin exists in solution as a mixture of *cis* (or β) and *trans* (or α) rotamers in a ratio that is dependent on the pH of the solution (8). The three pyrrole rings of prodigiosin are labeled as rings A, B, and C. The lower structure shows the binding of all three pyrrolic nitrogens of prodigiosin with chloride ion when it is acting as a H^+/Cl^- symporter.

both gram-negative and gram-positive bacteria isolated from soil, water, and marine environments. Prodiginines have been isolated from terrestrial bacteria including actinomycetes, *Serratia* spp., *Pseudomonas* spp. and the marine bacteria *Vibrio* spp., *Pseudoalteromonas* spp., *Hahella* spp., and *Alteromonas* spp. (1–5).

The structure of prodigiosin, the archetypal prodiginine, was elucidated in the early 1960s by partial and total chemical synthesis revealing a pyrrolyldipyrromethene core skeleton (6, 7). The three pyrrolic rings of prodigiosin are conventionally labeled as pyrrolic ring A, ring B, and ring C (Figure 15.1). Prodigiosin exists in two interconverting rotamers, *cis* (or β) and *trans* (or α) (Figure 15.1). The balance between these forms is dependent on the pH of the solution as the *trans* form protonates more easily (8).

The biosynthesis and complex regulation of prodigiosin in *Serratia* sp. ATCC 39006 involves over 30 genes, which suggests that production of prodigiosin by the bacteria must confer some selective ecological advantage.

As with many secondary metabolites, the physiological function of prodiginines in the producing organism is unclear. Nonetheless, proposed physiological functions are numerous, and include functions associated with prodigiosin's antibacterial, antifungal, or antiprotozoal activity, a role as a metabolic sink, involvement in surface adherence, and enhancing bacterial dispersal (4). Recently, we proposed a further role for prodigiosin in niche colonization and defense through swarming in *Serratia* sp. ATCC 39006 (9). In support of this hypothesis, there is significant overlap in the regulatory networks controlling prodigiosin and surfactant production in this strain, consistent with a synergistic function (9).

Prodiginines have antialgal, antibacterial, antifungal, antiprotozoal, antimalarial (4, 10–12), and immunosuppressive activities (5, 13). However, the focus of this chapter is the impressive potential of prodiginines as anticancer agents. The prodiginines have been shown to be active against numerous cancer cell lines, including those of the liver, spleen, blood, colon, gastric, lung, and breast, in primary cultures of chronic lymphocytic leukemia, and in a liver cancer xenograft in mice (Table 12.1 (4, 5, 14–16). Little or no activity is observed against noncancerous cells (14–16). Because prodiginines have activity against numerous cancer cell lines *in vitro*, prodigiosin and a synthetic derivative (Obatoclax) are now in preclinical and Phase I and Phase II clinical trials, respectively. Prodigiosin is in preclinical trials for the treatment of pancreatic cancer and Obatoclax is in single- and dual-agent trials to treat multiple forms of leukemia, lymphoma, and solid tumor malignancies (17).

HISTORY OF PRODIGININES

Scientists and the general public have had a long-standing fascination with prodiginines and the producing organisms. Due to their pigmented nature, the prodiginines have been used as model secondary metabolites in the study of the biosynthesis and regulation of secondary metabolism in bacteria (4). One feature thought to be associated with the production of bright red prodiginines by the producing organisms is multiple reports of miraculous (prodigious) "bleeding bread." Instances of bleeding bread have been noted as far back as the time of Pythagoras, with the most famous example being the miracle of Bolsena, which is celebrated in the feast of Corpus Christi (18).

Because of the visual impact of the red prodigiosin, *Serratia marcescens* has been used as a tracer bacterium in infection transmission experiments in the U.K. House of Commons and by the U.S. army in the San Francisco Bay area and in New York subways (19). Prior to the development of synthetic dyes, prodigiosin was produced for dying silk and wool, and this has inspired the recent suggestion that prodigiosin might have utility as an antibacterial colorant (20).

The anticancer properties of prodigiosin may have been a contributory factor in the induction of tumor necrosis factor and the subsequent apoptosis of cancer cells when Coley's toxins were used to treat multiple forms of cancer between 1893 and 1960 (21). However, it has also been proposed that lipopolysaccharide released from the sterilized cultures of *Streptococcus* sp. and *S.marcescens* that were present in Coley's toxins caused induction of the tumor necrosis factor (21–23).

BIOSYNTHESIS OF PRODIGININES

The biosynthesis of prodiginines proceeds via a bifurcated pathway culminating in the enzyme-catalyzed condensation of 4-methoxy-2-2′-bipyrrole-5-carbaldehyde (MBC) with a monopyrrole. MBC is either condensed with 2-methyl-3-pentylpyrrole (MPP, known as MAP in older literature) to form prodigiosin, or, in the case of undecylprodigiosin, with a slightly different monopyrrole, 2-undecylpyrrole (Figure 15.2) (13). *Streptomyces coelicolor* produces the linear prodiginine undecylprodigiosin and a cyclic derivative, streptorubin B, in a 2:1 ratio (3, 24).

The prodiginine biosynthetic clusters of several producing organisms have been sequenced. These include two *Serratia* sp. (*pig* clusters) (25), *Str. coelicolor* (*red* cluster) (26), *Hahella chejuensis* (*hap* cluster) (27), the tambjamine cluster of *Pseudoalteromonas tunicata* (*tam* cluster) (28), and the prodigiosin cluster of the roseophilin producer, *Str. griseoviridis* (*rph* cluster) (29). Tambjamines, of which the anticancer compound BE18591 is a member, are bipyrrole compounds that are structurally similar to the prodiginines. All clusters contain a conserved set of genes which are homologous to each of the MBC biosynthetic enzymes of *Serratia*, suggesting a common route to the biosynthesis of MBC. An exception to this rule is the absence of a *pigN*/*redF* homologue in both the *rph* and *tam* clusters of *P. tunicata* and *S. griseoviridis*. *pigN* mutants of *Serratia* sp. ATCC 39006 produce a mixture of prodigiosin and norprodigiosin. This suggests that PigN is not essential, but may facilitate PigF-catalyzed methylation of HBC (13). Both *P. tunicata* and *S. griseoviridis* contain *pigF* homologues, *tamP* and *rphI*, suggesting that these enzymes alone are sufficient for methylation of MBC in these two pathways.

The biosynthesis of MBC has been studied in *Serratia* sp. and *Streptomyces* using genetic approaches, complementation, and *in vitro* analysis of the purified enzymes involved in the early steps of this pathway (3, 13, 30–34). The degree of conservation of the prodiginine biosynthetic clusters suggests that biosynthesis of MBC proceeds via a common route requiring proline, acetate/malonate, serine, and methionine (Figure 15.2) (13, 33, 35). Proline is incorporated to form pyrrole by a mechanism common to other pyrrole-containing compounds such as chlorobiocin, coumermycin A1, novobiocin, and pyoluteorin (30, 34, 36). In prodiginine biosynthesis, proline is incorporated to form pyrrolic ring A in a sequence of reactions catalyzed by PigA, PigG, and PigI and their homologues (30, 34). In the next steps, catalyzed by PigJ and PigH (and homologues), a C_2 unit from malonyl CoA and a C_2N unit from serine

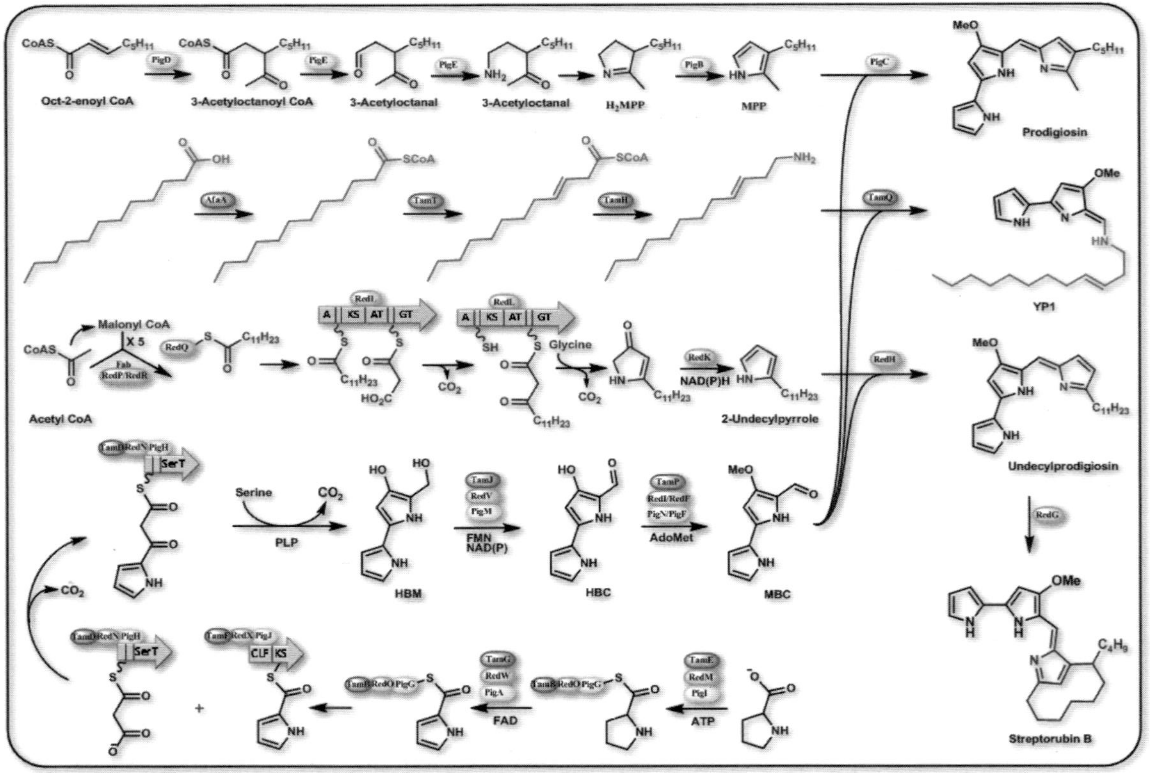

Figure 15.2 Proposed biosynthetic pathway of the prodiginines and the related 4-methoxybipyrrole, YP1. Biosynthesis of the prodiginines proceeds via a bifurcated pathway culminating in the enzymic condensation of the bipyrrole, 4-methoxy-2-2′-bipyrrole-5-carbaldehyde (MBC) with either 2-methyl-3-pentylpyrrole (MPP) to form prodigiosin, or, in the case of undecylprodigiosin, with a slightly different monopyrrole, 2-undecylpyrrole (3, 4, 13, 25, 26, 30–34). The proposed biosynthetic pathway of tambjamine YP1 is also shown (28).

are incorporated (both with concomitant decarboxylation) forming 4-hydroxy-2,2′-bipyrrole-5-methanol (HBM) (13, 33). The final steps in MBC biosynthesis are catalyzed by PigM and involve the oxidation of an alcohol group of HBM forming 4-hydroxy-2,2′-bipyrrole-5-carbaldehyde (Figure 15.2) (HBC). The hydroxyl group of HBC is then methylated to form MBC (13). This final methylation step is catalyzed by homologues of PigF, which, in some prodiginine producers, is facilitated by homologues of PigN.

Str. coelicolor and *Serratia* sp. condense MBC with different monopyrroles. These are synthesized using completely different enzymes, substrates, and pathways (Figure 15.2) (3, 13, 25, 26, 32). Currently, none of the enzymes involved in the biosynthesis of either monopyrrole have been purified, and the pathways and predicted functions of the enzymes are based entirely on homology with other enzymes, gene deletions, liquid chromatography–mass spectrometry (LC-MS) analysis of accumulating intermediates, and complementation experiments (3, 13, 32). The biosynthesis of MPP in *Serratia* is catalyzed by PigD, PigE, and PigB (Figure 15.2). However, the precise substrates required for MPP biosynthesis are uncertain. The biosynthetic intermediate H_2MPP was shown to accumulate in a Δ*pigB* mutant. Thus, PigB has been proposed to catalyze the oxidation of H_2MPP to form the terminal product of this pathway, MPP (13). In the biosynthesis of undecylprodiginine, the formation of the monopyrrole is catalyzed by RedP, RedQ, RedR, RedK, and RedL. RedP, RedQ, and RedR direct the biosynthesis of dodecanoic acid which is then transferred to the polyketide synthase RedL (Figure 15.2). This multifunctional enzyme catalyzes the incorporation of malonyl-CoA and glycine forming 4-keto-2-undecyl-4,5-dihydropyrrole. The reduction and dehydration of 4-keto-2-undecyl-4,5-dihydropyrrole is catalyzed by RedK giving the terminal product of this pathway, 2-undecylpyrrole (Figure 15.2) (3, 13, 32).

Biosynthesis of prodiginines and related compounds culminates in the condensation of MBC with either a monopyrrole, MPP (for prodigiosin) (13) or 2-undecylpyrrole (for undecylprodigiosin) (31), or, for the related bipyrrole tambjamine YP1, an amine (28). This condensation reaction is catalyzed by a novel family of pyrrole-condensing enzymes, characterized by PigC (Figure 15.2) (13). All of the biosynthetic clusters for prodiginines and related compounds sequenced to date contain a PigC homologue. PigC homologues include RedH (undecylprodigiosin), RphH (prodigiosin R1), HapC (prodigiosin), TamQ (tambjamine YP1), and a PigBC fusion enzyme from *Janthinobacterium lividum*. The exact nature of the compound produced by *J. lividum* is unknown, but the PigBC fusion enzyme is the only known example of a multifunctional enzyme that appears to catalyze both the final step of monopyrrole biosynthesis and the terminal condensation reaction (37). As more PigC homologues are identified, sequence alignments, site-directed mutagenesis, and crystallography may allow conserved residues to be identified that are critically involved in the condensation reaction. Knowledge of key residues of the active sites of the condensing enzymes, and enzymes involved in the earlier steps of the biosynthetic pathways, may allow directed evolution studies of the biosynthetic enzymes to generate novel prodiginines which might then be produced cheaply in large-scale fermentations. The prodiginine biosynthetic clusters of bacteria that produce cyclic prodiginine derivatives all contain at least one RedG homologue. RedG has been proposed to perform oxidative cyclization of undecylprodigiosin to form streptorubin B in *Str. coelicolor* (Figure 15.2) (26). The RedG homologue TamC is proposed to catalyze the cyclization of tambjamine YP1 to form a cyclic tambjamine (28). The roseophilin and prodigiosin R1 producer, *S. griseoviridis* contains four RedG homologues (29). This multiplicity of RedG homologues might be explained by the structural differences between prodigiosin R1 and roseophilin (29).

MUTASYNTHESIS TO GENERATE NOVEL PRODIGININES

As stated in the previous section, much progress has been made in understanding the biosynthesis of prodiginines, aided by the *in vitro* analysis of enzymes involved in the early steps in the biosynthesis of MBC (30, 34). A more detailed understanding of the enzymatic mechanisms of the prodiginine biosynthetic apparatus, including crystal structures of the enzymes, may enable the generation of novel prodiginines with improved pharmacological activity. Currently, only one prodiginine biosynthetic enzyme, HapK, has been crystallized (38). Analysis of the crystal structure identified a binding pocket capable of accommodating a bipyrrole, suggesting a role late in the biosynthesis of MBC for HapK and its Red and Pig homologues (13, 38).

It is known that some secondary metabolite biosynthetic clusters encode enzymes with relaxed substrate specificity, which, under selective pressure, may be capable of generating chemical diversity. Two related approaches, termed precursor-directed biosynthesis (PDB) and mutational biosynthesis (MBS), also known as mutasynthesis, can be used to generate large quantities of complex analogues of natural products (39–41). In PDB, the growth media of the secondary metabolite producer can be supplemented with analogues of early precursors. These analogues may outcompete the natural substrates to get incorporated into the secondary metabolite and a mixture of wild-type and novel analogues of the natural product will be produced. One disadvantage of this approach is that the mixture requires further purification, and this can be expensive and time-consuming (39). Mutasynthesis involves feeding biosynthetic intermediates to defined biosynthetic mutants. Where these intermediates are incorporated, novel analogues of the natural product can be produced, circumventing the need for separation of a mixture of products.

The prodiginine-condensing enzymes (PigC and homologues) are promiscuous, with the ability to condense the analogues of MBC and MPP to form prodiginine analogues (Figure 15.3) (4, 13). Two recent studies have expanded the known substrate specificity of PigC and RedH. MBC analogues with ring A altered were fed to defined MBC biosynthetic mutants yielding novel prodiginines (Figure 15.3) (31, 42). Both RedH and PigC were shown to be capable of accepting phenyl, 2-thienyl, 2-furyl, and 2-indolyl MBC analogues as substrates and condensing them with the monopyrrole to form prodiginine analogues (Figure 15.3). Chawrai et al. (2008) also demonstrated that, when an *Escherichia coli* strain expressing the PigC-condensing enzyme (along with the genes encoding the MPP biosynthetic enzymes) was fed with MBC

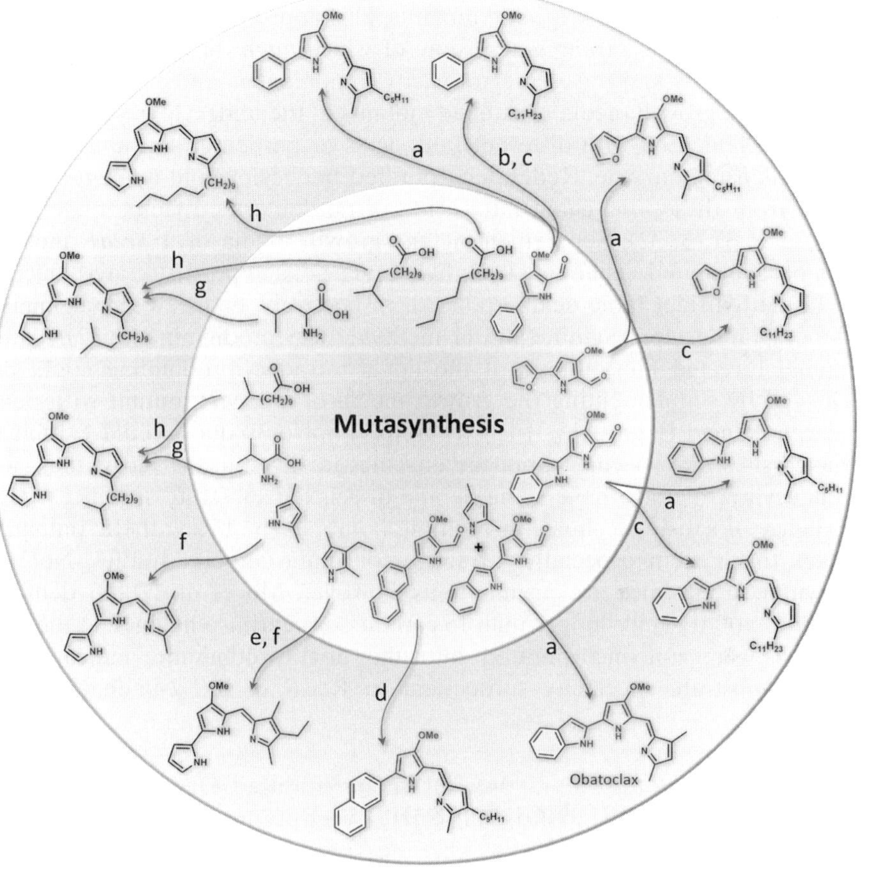

Figure 15.3 Examples of novel prodiginines produced in mutasynthesis experiments by feeding analogues of biosynthetic intermediates to defined biosynthetic mutants. Mutasynthesis experiments were performed by feeding biosynthetic intermediates to (a) *Serratia* 39006 Δ*pigM*, Δ*pigH57* (42), (b) *Str.coelicolor* A3(2) mutant classes A, B, C, and E; S*treptomyces collinus* Tu 105 and Tu 353; *Streptomyces griseus* ATCC 12648 (24), (c) *Str.coelicolor* W39 (31), (d) Chawrai, et al. (unpublished results), (e) *Serratia* 39006 Δ*pigB* (13), (f) *S.marcescens* strain 9-3-3 (92–94), (g) *Str.coelicolor* SJM1/pSW7 (32), and (h) *Str.coelicolor* SJM3 (3).

analogues, novel prodiginines could be produced in this heterologous host (42). Furthermore, by feeding the 2-indolyl derivative of MBC and 2,4-dimethylpyrrole (an analogue of MPP) to a *Serratia* MBC mutant, PigC could condense these two analogues to produce Obatoclax (42). These studies demonstrate the potential of mutasynthesis to generate novel prodiginines of pharmaceutical importance. These studies so far are largely of academic interest because bipyrrole and monopyrrole precursors can be readily condensed by chemical synthesis under acidic conditions. One goal of this mutasynthesis approach is to feed analogues of the early intermediates of the prodiginine biosynthetic pathways to defined biosynthetic mutants. This approach may be used in the generation of novel prodiginines, some of which might have improved anticancer specificity and reduced cytotoxicity to nonmalignant cells.

When the growth media of defined mutants in the undecylpyrrole pathway were supplemented with different fatty acids or perdeuterated amino acids, the condensing enzyme, RedH, incorporated undecylpyrrole analogues produced from these long-chain fatty acids, or perdeuterated valine or leucine (Figure 15.3) (3, 32). Supplementing the growth media of a Δ*redP* mutant with perdeuterated valine or 11-methyllauric acid, and perdeuterated leucine or 12-methyltridecanoic acid produced, respectively, two new prodiginines: 10-methylundecylprodiginine and 11-methyldodecylprodiginine (3, 32). Additionally, directed biosynthesis of the novel tetradecylprodiginine could be achieved by supplementing the growth media of a Δ*redR* mutant with pentadecanoic acid (Figure 15.3) (3). Therefore, not only does RedH exhibit a relaxed substrate specificity, but the enzymes involved in the biosynthesis of 2-undecylpyrrole also have some degree of relaxed substrate specificity. The flexibility of substrate choice at multiple points in the biosynthetic pathway makes this system especially attractive for mutasynthesis studies. Not all biosynthetic enzymes are promiscuous, however. Mo et al. (2008) demonstrated that fatty acids longer than 15 carbons, or bearing a halogen at the ω-2 position were not incorporated into the final prodiginine, indicating a stricter substrate specificity subsequent to RedR in the 2-undecylpyrrole pathway (3).

THE BIOACTIVE PRODIGININES DISPLAY ANTICANCER ACTIVITY

In total, prodigiosin has been tested against more than 60 cancer cell lines with an average inhibitory concentration of 2.1 μM (5, 43). The exact mode of action of prodiginines in inducing apoptosis is uncertain as they have multiple cellular targets (Figure 15.4). Prodiginines are also attractive options because they are not affected by several multidrug resistance pumps which can confer resistance to other anticancer agents (44, 45). Prodiginines have been described as proapoptotic anticancer compounds and have been shown to induce cellular stresses such as cell cycle arrest, DNA damage, and a change of intracellular

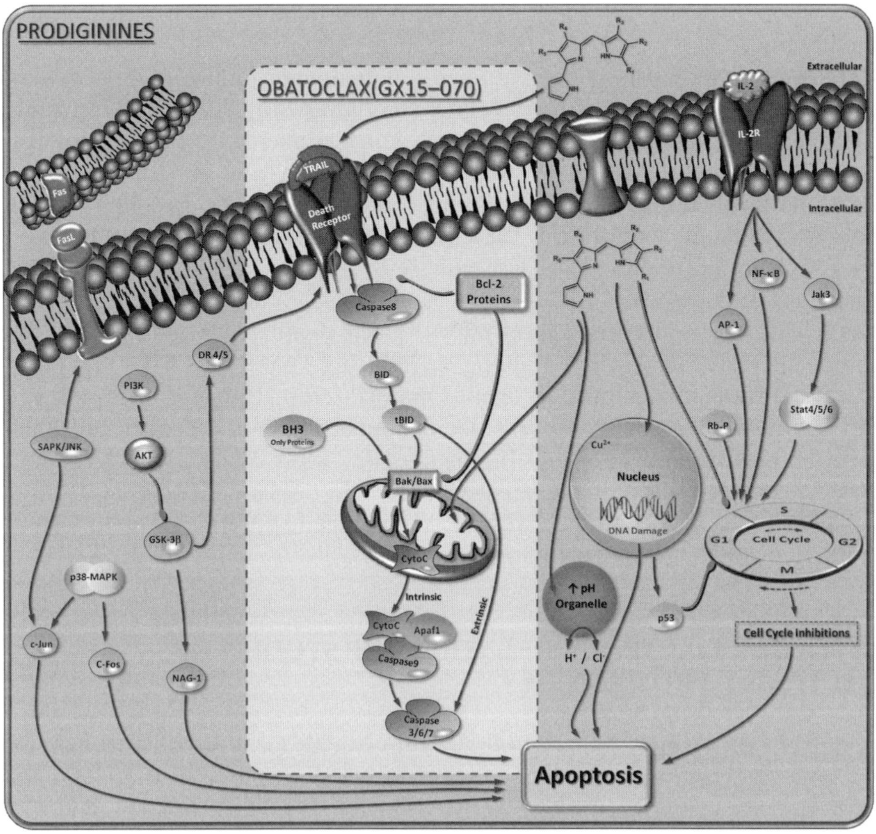

Figure 15.4 Proposed mechanisms by which prodiginines induce apoptosis in the cancer cell. The four different routes by which prodiginines have been proposed to induce apoptosis are shown, and these include interference with signal transduction, cell cycle arrest, intracellular acidification, and DNA cleavage. The mechanism of action of the lead prodiginine anticancer agent, Obatoclax is shown in the box.

pH (pHi), all of which can induce apoptosis (Figure 15.4). Prodiginines can also interfere with signaling pathways and inhibit the cell cycle. The proapoptotic mechanisms will be discussed in the following sections. The cellular location of prodigiosin, which gives clues to its mechanism of action, is a matter of debate with prodigiosin variously being detected in the nucleus (46), concentrated in the cytoplasm of cells (47), in granules near to the nucleus (48), and in the mitochondrial membrane (49). Presence of prodigiosin in the nucleus of the cell would support the hypothesis that DNA cleavage is important for its proapoptotic activity. However, the hydrophobic nature of the synthetic prodiginine, Obatoclax, is proposed to be important in the localization of the compound to the mitochondrial membrane which may promote interaction with the antiapoptotic Bcl-2 family members that reside in this membrane (17).

Defects in apoptosis or programmed cell death can contribute to oncogenesis and enhance resistance to current chemotherapeutic agents and radiation. There are two major apoptotic pathways which converge on the effector caspases. The intrinsic pathway, which is also known as the mitochondrial cell death pathway, can be activated by radiation, cytotoxic drugs, cellular stress, and growth factor withdrawal. The extrinsic cell death pathway can function by activating an initiator caspase, caspase-8, Bid cleavage and activation of caspase-3 and caspase-7 independently of mitochondria. The extrinsic cell death pathway is induced directly by the cell death receptors Fas and the tumor necrosis factor-related apoptosis-inducing ligand (TRAIL) receptors. The lead prodiginine anticancer agent, Obatoclax, targets the intrinsic pathway, but has also been shown to sensitize cholangiocarcinoma cancer cell lines, KMCH, KMBC, and TFK, and the pancreatic cancer cell lines, PANC-1 and BxPC-3 to TRAIL-mediated apoptosis (50). In cholangiocarcinoma cancer cell lines, Obatoclax combined with TRAIL induced apoptosis by up to 13-fold compared to Obatoclax treatment alone (50).

STRUCTURE–ACTIVITY RELATIONSHIP STUDIES OF BACTERIAL PRODIGININES

A number of structure–activity relationship studies have been performed on the core structure of the prodiginines to develop synthetic lead compounds with improved pharmacological activity (17, 51, 52). These studies have not only yielded two lead compounds of particular interest (Obatoclax and PNU156804, see below and Figure 15.5) but have also given insights into key structural features of the prodiginines which are important for their biological activity and cytotoxicity. For example, based on SAR studies, the alkyl side chain of pyrrole ring C of prodigiosin has been suggested to be important for prodigiosin to locate into biological membranes (53). Also, the N–H hydrogen bond donor groups of the three pyrrole-rings are important for the H^+/Cl^- symport activity of the prodiginines (8, 54–56) (Figure 15.1).

The methoxy moiety of pyrrolic ring B of undecylprodigiosin is important in cytotoxicity (51). This observation led D'Alessio and colleagues to focus SAR studies on replacing the methoxy group with larger moieties, eventually replacing this moiety with a benzyl group generating the lead immunosuppressant, PNU156804 (51). This compound had a therapeutic index threefold higher than the parent compound, undecylprodigiosin. Oral administration of PNU156804 in combination with cyclosporine A resulted in an increase in survival of heart-transplanted rats from a mean of 6.3 days without treatment to 85 days ± 27.4 days with treatment (51). While this study did not investigate the effects of PNU156804 on cancerous cell lines, it does display the dramatic outcomes that can be derived from informed SAR studies.

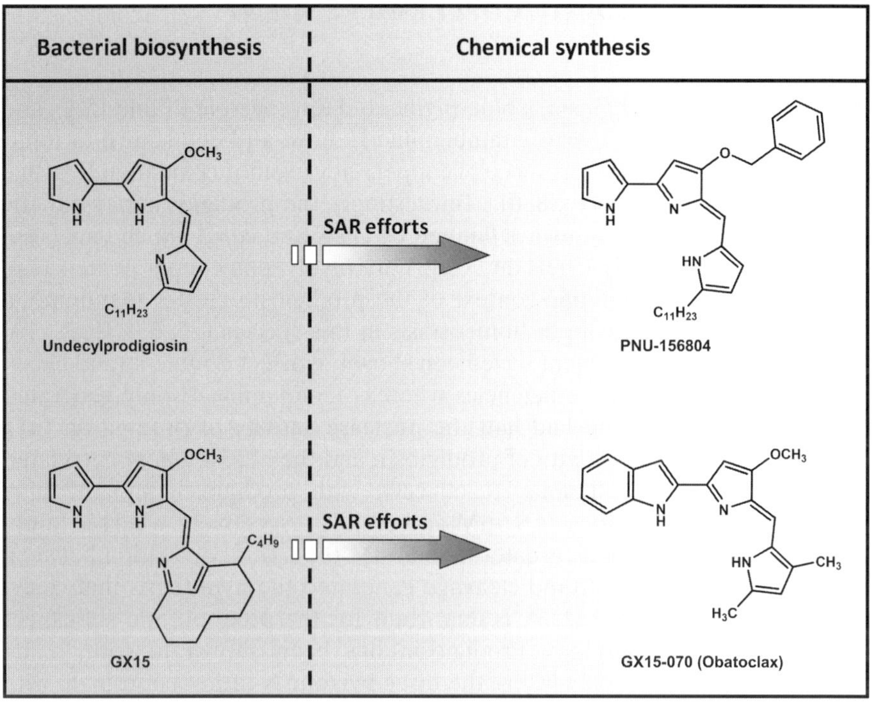

Figure 15.5 Examples of two lead therapeutically active synthetic prodiginines, Obatoclax and PNU156804. The lead immunosuppressive prodiginine, PNU156804 was developed by Pharmacia and Upjohn thorough SAR studies focused on the methoxy group of undecylprodigiosin (51). The lead anticancer prodiginine, Obatoclax developed by GeminX pharmaceuticals, was engineered through SAR studies of pyrrolic ring A of the bacterial prodiginine, GX15 (streptorubin B) (17, 52).

Various SAR studies have been performed with respect to the nuclease activity of prodiginines (discussed below). These studies have shown that the pyrrolyldipyrromethene chromophore is critical for double-strand DNA cleavage, whereas the bipyrrole is sufficient for single-strand DNA cleavage (57–61). Consistent with this hypothesis, prodigiosin and a blue tetrapyrrole (Table 15.1) cause double-strand cleavage whereas the bipyrrole tambjamine E and related bipyrroles can only nick single-stranded DNA (58). The ability of prodiginines to cause double-strand cleavage is dependent on all three nitrogen atoms of the three pyrrole rings forming a 1:1 complex with Cu(II) in which pyrrolic ring C is oxidized (62). The electron density of pyrrolic ring A is also important for nuclease activity, with a reduction in electron density resulting in a loss of activity and enhanced electron density correlated with increased potency (57). For example, attachment of an electron-withdrawing acetyl group reduces nuclease activity, and attachment of an electron donating dimethylamino (4-NMe_2) group resulted in increased nuclease activity (57).

PRODIGININES MEDIATE COPPER-DEPENDENT DNA CLEAVAGE

The prodiginines, prodigiosin, a blue-pigmented tetrapyrrole (Table 15.1), and the related 4-methoxybipyrrole tambjamine E, have all been shown to intercalate into dsDNA. In the presence of copper, they promote oxidative cleavage of the DNA (Figure 15.4) (58–61). Interestingly, the prodigiosin biosynthetic cluster of *S. marcescens* is always flanked by *cueR* and *copA* which encodes a copper efflux pump (25). CueR and CopA are involved in copper homeostasis in bacteria, and the genomic context of the prodigiosin cluster may imply a role for prodigiosin in copper homeostasis in this species (25, 63). Both prodigiosin and the blue pigment have been shown to affect double-strand cleavage of DNA with similar efficiencies whereas tambjamine E only generated single-stranded nicks and had half the nuclease activity of prodigiosin (61). The superior nuclease activity of prodigiosin and the blue pigment correlated with cytotoxicity in leukemia HL-60 cells, with prodigiosin and the blue pigment having IC_{50} values of 6.6 μM and 5.9 μM respectively, whereas tambjamine E had an IC_{50} value greater than 50 μM (61). This correlation between cytotoxicity and double-strand cleavage generated the hypothesis that cleavage of both strands of dsDNA is important for the proapoptotic anticancer activity of the prodiginines. Prodigiosin has been shown to form a 1:1 prodigiosin:Cu(II) complex where the three pyrrolic N-atoms coordinate with the Cu(II) (62). Pyrrolic ring A (57) and the alkyl side chain of the C-ring are important for nuclease activity (53, 64). Prodigiosin has also been shown to bind by intercalation, preferentially at AT sequences, from the minor groove (58). The mechanism of copper-dependent oxidative cleavage of dsDNA by prodigiosin is predicted to be due to cycling between reduced and oxidized copper, generating H_2O_2 and which results in dsDNA cleavage (60, 62).

If copper-mediated DNA cleavage were responsible for inducing apoptosis, the ability of prodiginines to target cancer cells with little, or no, activity against nonmalignant cells could be explained by the availability of copper. Cancer cells are reported to contain approximately 3.5-fold higher concentrations of Cu(II) than nonmalignant cells. Also, solid tumors are more acidic (pH 6.6–7.0) compared with normal cells where the pH is between pH 7.1 and pH 7.6. This could be important because prodigiosin cleaves optimally at pH 6.8 (46).

PRODIGININES AS H^+/CL^- SYMPORTERS AND A ROLE IN INTRACELLULAR ACIDIFICATION

The ability of prodiginines to disrupt the pH gradients across the membranes of cellular organelles, resulting in apoptosis, is discussed in this section. Mitochondrial alkalinization and cyotosolic acidification are thought to be early triggers in induction of the mitochondrial apoptotic pathway leading to the release of cytochrome *c*, the activation of caspases, and apoptotic cell death (Figure 15.4). Prodiginines are known for their ability to inhibit the establishment of a pH gradient by the plant and animal V-ATPase and the F_0F_1-ATPases of bacteria and mitochondria (48, 65–72). The ability of prodiginines to disrupt pH gradients across membranes is attributed to their functioning as H^+/Cl^- symporters which has the effect of uncoupling V-ATPases resulting in acidification of the cytoplasm and apoptosis (48, 56, 67, 70, 71). Cycloprodigiosin has been shown to uncouple the proton pump activity of V-ATPase from the ATP hydrolysis activity resulting in an increase in lysosomal pH (70). A recent study demonstrated that prodigiosin localized to the mitochondrial membrane of neuroblastoma N1 cells, and disrupted the intracellular pH gradient, uncoupling the mitochondrial ATP synthase (49). This action caused a 50% reduction in ATP levels. Other studies have demonstrated that the uncoupling effect on V-ATPases is not by inhibition of ATP hydrolysis or of ATP-dependent creation of a membrane potential (65, 70, 71). The IC_{50} of prodigiosin on neuroblastoma cell lines was lower than 1.5 μM, making it 30 times more potent than cisplatin—a current cancer therapy (49). Cycloprodigiosin has been shown to cause acidification of the cytoplasm and apoptosis through the activation of caspase-3, caspase-6, and caspase-8 (73). However, prodigiosin has been applied at low concentrations (resulting in cellular acidification) without induction of apoptosis (74).

SIGNAL TRANSDUCTION AND MAPKS

Mitogen activated protein kinase (MAPK) signaling cascades regulate cell proliferation, including cell differentiation, apoptosis, and cell migration. MAPKs can be divided into the extracellular (growth factor) regulated-signaling kinase (ERKs), the stress-induced c-jun N-terminal kinases (JNKs) and the p38 MAPK isoforms, which, like JNKs, can be activated by cellular stresses such as cytokines and UV irradiation. Treatment of the leukemia cell line HL-60 with cycloprodigiosin induced cellular acidification which resulted in increased expression of FasL and activation of SAPKs/JNKs and release of caspase-3-like, caspase-6, caspase-9, and induction of apoptosis (Figure 15.4) (73). Addition of the diffusible weak base imidazole could suppress the induction of caspases by cycloprodigiosin, indicating that cellular acidification was indeed responsible for apoptosis by cycloprodigiosin (73). Alternatively, prodigiosin purified from cultures of *S. marcescens* induced phosphorylation of p38-MAPK which led to activation of the p38 MAPK cascade (11, 75). The transcription factors c-jun and c-fos, which are in the p38 MAPK cascade, were upregulated in response to prodigiosin treatment, while an alternative signaling pathway, the SAPK/JNK cascade, was not affected. Prodigiosin-mediated apoptosis of Jurkat cells could be inhibited by the addition of phorbol myristate acetate, causing activation of protein kinase C and subsequently ERK1/2, conferring resistance to apoptosis. Prodigiosin has also been shown to upregulate death receptors 4 and 5, the nonsteroidal anti-inflamatory drug-activated gene 1 (NAG-1) and the cell cycle regulator p21 via activation of the growth factor β pathway (76, 77).

CELL CYCLE INHIBITION

The ability of prodiginines to inhibit the cell cycle has been exploited, at non-apoptotic doses, as an immunosuppressant. Prodigiosin has been shown in mouse models to suppresss graft versus host disease (GvHD) with no observable signs of toxicity (78). Prodigiosin also prevented GvHD, delayed autoimmune diabetes progression, and collagen-induced arthritis in mouse models (78). Undecylprodigiosin, metacycloprodigiosin, and cycloprodigiosin all selectively inhibit T cell proliferation. However, they all displayed varying levels of toxicity *in vivo*.

Prodiginines have been shown to arrest the cell cycle at multiple stages (Figure 15.4). The variety of effects has been attributed to their different structures and/or experimental systems (23). Prodigiosin inhibits proliferation of human Jurkat T cells in the G1/S phase transition, whereas undecylprodigiosin and PNU156804 do not inhibit proliferation of human Jurkat T cells. Undecylprodigiosin and PNU156804 have been shown to inhibit T and B lymphocyte proliferation in late G1 phase of the cell cycle. A role for prodigiosin in the late G1 phase was shown by decreased expression of cyclin E, cdk2, and cdk4, all of which are known to be expressed in mid-to-late G1 phase. In addition to their effect in late G1 phase, prodiginines have been shown to inhibit phosphorylation of retinoblastoma Rb protein (79, 80), an important mediator of the progression from G1 phase to S phase. Prodigiosin has been shown to inhibit cyclin E, cdk2, p27, p21, and Rb phosphorylation in leukemic Jurkat cells, resulting in apoptosis (Figure 15.4) (81). Prodigiosin has also been shown to affect the accumulation of p53, and induction of NAG-1, in the human Mcf-7 breast cancer cell line (76). At nonapoptotic doses, the indolyl-dipyrromethene, Obatoclax, caused cell cycle arrest at the S-G2 transition. This arrest occurred in cell lines lacking Bax and Bak/Bim, suggesting multiple targets for this pan-Bcl-2 inhibitor (82).

Prodiginines do not affect expression of IL-2, which forms a complex with its receptor (IL-2R). This IL-2R complex is another important checkpoint in T cell proliferation. Binding of IL-2 to IL-2R activates several protein kinases which are associated with the IL-2R complex, including jak1, jak3, p56k, and syk (Figure 15.4). Again, a range of effects specific to given prodiginines are observed at this stage of the cell cycle. Prodigiosin inhibits IL-2R expression whereas undecylprodigiosin and PNU156804 do not inhibit IL-2R expression (78, 79, 80). Cycloprodigiosin has been shown to inhibit nuclear factor (Nf)-κ activation in Hela cells, whereas PNU156804 triggers AP-1 and (Nf)-κB activation. Prodigiosin has no effect on (NF)-κB or AP-1 activation (78).

CASE STUDY

Obatoclax—A Small Molecule Pan Bcl-2 Inhibitor

Obatoclax (GX15-70) was developed by GeminX pharmaceuticals in a screen for inhibitors of the antiapoptotic B cell lymphoma type-2 proteins (Bcl-2) (17). Obatoclax is an indolyl-dipyrromethene (Figure 15.5) that was developed through structure activity relationship studies of pyrrolic ring A of GX15 (streptorubin B) (52). It is the leading prodiginine candidate for the treatment of multiple forms of cancer in the clinic. This compound also represents one of the first new anticancer drugs termed "BH3 mimetics," designed to antagonize the prosurvival Bcl-2 family proteins. Bcl-2 family members consist of six antiapoptotic members (Bcl-2, Bcl-X_L, Mcl-1, Bcl-W, Bfl-1/a-1, and Bcl-B). Bax and Bak are multidomain proapoptotic proteins and Bim, Bad, and Bid, also proapoptotic, represent the BH3-only domain proteins. Bcl-2 proteins are located in the mitochondrial outer membrane and have a central role in controlling the switch between life and death of the cell by regulating the intrinsic apoptotic pathway.

Currently, two models exist to explain how the propapoptotic, antiapoptotic, and BH3-only proteins interact to regulate apoptosis—the direct and indirect activation models (83). In the direct activation model, activator BH3-only proteins Bim and Bid interact directly with Bax and Bak activating the intrinsic apoptotic pathway causing the release of cytochrome *c*. Additional BH3-only proteins Bad, Bik, and Noxa act as sensitizers and bind to the hydrophobic groove of the prosurvival members, leading to the release of the proapoptotic proteins, Bax and Bak. In the indirect model, the BH3-only proteins directly inhibit binding of the proapoptotic Bax and Bak by binding to the 6 antiapoptotic, prosurvival proteins mentioned above. In this model, there is no direct binding of the BH3-only proteins to Bax and Bak (83). In both models, there is a role for anticancer agents that can bind to and sequester the prosurvival Bcl-2 proteins, releasing the proapoptotic Bax and Bak. This induces the mitochondrial apoptotic pathway and ultimately cancer cell death. The activation of the mitochondrial apoptotic pathway proceeds via the release of cytochrome *c*, which forms a multiprotein complex with apoptosis protease activating factor 1 and inactive procaspase-9, called the apoptosome. The activation of caspase-3, caspase-6, and caspase-9, the caspase cascade and, ultimately, apoptosis, is triggered by caspase-9.

The antiapoptotic Bcl-2 family proteins are also attracting much interest as potential anticancer targets for two reasons. First, these proteins are estimated to be upregulated in 50% of all cancers (84), and second upregulation of the prosurvival Bcl-2 proteins is associated with resistance to other proapoptotic anticancer agents.

Obatoclax had been predicted to occupy a hydrophobic pocket within the BH3 binding groove of the antiapoptotic Bcl-2 family members and inhibit binding of Bax and Bak (17, 85, 86). Inhibition by GX15 of the interaction between a BH3 peptide and Bcl-2, Bcl-X_L, Bcl-w, Bcl-b, and Mcl-1 has been demonstrated using fluorescence polarization, with K_i values of 1–7 μM (87). Obatoclax is similar to naturally occurring bacterial prodiginines in that it is a hydrophobic molecule. This hydrophobicity is believed to facilitate localization of Obatoclax to the mitochondrial outer membrane where the Bcl-2 family members reside (17). Obatoclax has recently been shown to directly, and potently, inhibit interactions between Mcl-1 and Bak in intact mitochondrial membranes (17). The hypothesis that Obatoclax acts by releasing Bak and Bax from complexes with Bcl-2 proteins was recently tested in yeast cells overexpressing Bax and Bak, which inhibits cell growth. This phenotype could be rescued by the expression of Mcl-1, Bcl-2, or Bcl-X_L. This effect could in turn be inhibited by the addition of 50 μM Obatoclax (17). The addition of 50 μM Obatoclax to control cells had no effect on cell growth, demonstrating that the mechanism of action of Obatoclax is due to antagonism of Bcl-2 proteins and is dependent on Bax and Bak (17). Treatment of a Bax and Bak double mutant kidney cell line transformed with E1A/p53DN with Obatoclax confirmed that no activation of caspase-3 was observed, further demonstrating the Bax and Bak dependence of Obatoclax action (17). Konopleva et al. (2008) demonstrated that Obatoclax induced apoptosis in acute myeloid leukemia (AML) cell lines and primary AML samples by the release of cytochrome *c* following release of Bak and Bim from Mcl-1 and release of Bim from Bcl-2. Most notably, this study demonstrated that AML cells transfected with siRNA directed against Bax and Bak or Bim (and therefore expressing low levels of these proteins) displayed reduced, but not abolished, apoptosis when treated with Obatoclax (82). This study suggests that, as with bacterial prodiginines, Obatoclax may have other apoptotic targets in the cell.

As mentioned earlier, resistance to several anticancer drugs is mediated by Mcl-1. For example, the BH3 mimetic ABT-737 does not antagonize Mcl-1, and hence Mcl-1 expression confers resistance to this drug. Mcl-1 is rapidly degraded by the 26S proteasome. Therefore, when the proteasome inhibitor Bortezomib is used to treat cancer cells, Mcl-1-mediated resistance to apoptosis can develop. A dual-agent study combining Obatoclax and Bortezomib was effective in inducing apoptosis in mantle cell lymphoma with Obatoclax acting synergistically by inhibiting the Bortezomib induced Mcl-1.

Obatoclax has been shown to be effective in single and dual-agent studies inducing apoptosis in multiple cancer cell lines including mantle cell lymphoma, human multiple myeloma, and chronic lymphocytic leukemia (CLL). In CLL cells, the Bcl-2 proteins are overexpressed, and Obatoclax induces apoptosis in the low micromolar range by disrupting Bcl-2/Bim and Mcl-1/Bak complexes (88). Importantly, it was effective in inducing apoptosis in CLL cells with altered p53, an important oncogene in cancer cells resistant to apoptosis (88). Therefore, its mechanism of action is independent of p53.

Obatoclax has been shown to disrupt the interaction of Bcl-2/Bim and Mcl-1/Bak in chronic lymphocyctic leukemia cells which resulted in induction of the mitochondrial apoptotic pathway (88). A lower sensitivity to Obatoclax was observed for CLL cells compared with primary MCL cells. This lower sensitivity correlated with higher levels of phosphorylation of serine 70 in Bcl-2 in CLL cells. Inhibition of ERK1/2 decreased phosphorylation of Bc1-2 and increased sensitivity of CLL cells to Obatoclax (88). If Bcl-2 dephosphorylation was inhibited with a PP2A antagonist, sensitivity of CLL cells to Obatoclax was reduced (88). Therefore, it has been proposed that phosphorylation of Bcl-2 is a mechanism of resistance to Obatoclax treatment which could be cirumvented by inhibiting Bcl-2 phosphorylation with ERK inhibitors (88).

Obatoclax is effective as a single agent and it synergistically enhances apoptosis, overcoming Mcl-1 resistance to ABT-737 in AML cell lines (82). Obatoclax has also been shown to potentiate the effect of the frontline chemotherapeutic agent, AraC, in decreasing leukemia cell proliferations in AML cell lines and primary samples (82). The effect of Obatoclax was shown to be time- and dose-dependent. Obatoclax can potentiate TRAIL-mediated apoptosis of the pancreatic cancer cell lines Panc-1 and bxPc-3 which is mediated by the release of cyotchrome *c*, Smac, and AIF (89). It is believed that Obatoclax and TRAIL enhance cross-talk between the mitochondrial and death receptor apoptotic pathways in pancreatic cancer cells. Obatoclax has been shown to act synergistically with Gefitinib in non-small cell lung cancer (NSCLC) cell lines that are dependent on EGFR for survival but not in cell lines that are not dependent on EFGR for survival. Obatoclax can act alone or synergistically in combination with cisplatin (used to treat advanced forms of lung cancer) to induce apoptosis in NSCLC cells, again by disrupting Mcl-1/Bak complexes (90). Finally, Obatoclax has been shown to have antiproliferative effects in breast cancer cell lines MCF-7, MCF/18, and MTR-13 when used in combination with the EGFR/HER-2 tyrosine kinase inhibitors Lapatinib and GW2974 (91).

In summary, Obatoclax is a pan-Bcl-2 inhibitor that is effective against several different cancer cell lines potently by disrupting the interaction of Bax and Bak with the prosurvival Bcl-2 proteins, releasing the proapoptotic Bax and Bak to induce apoptosis and cancer cell death (Figure 15.4). Moreover, Obatoclax can antagonize Mcl-1, a protein that is upregulated in 50% of all cancers and that has been shown to confer/affect resistance to other commonly used anticancer agents.

Due to the promising activity of Obatoclax against multiple cancer cell lines, this drug is currently in multiple, Phase I and Phase II, single and dual agent, clinical trials that are targeted against multiple different forms of cancer (5, 17, 85, 86). To date, Obatoclax has exhibited well-tolerated, dose-dependent clinical activity in patients with myelofibrosis, AML, CLL, and follicular lymphoma (FL). Following on from promising Phase I clinical studies, GeminX pharmaceuticals have launched, and plan to launch, further single-agent and dual-agent Phase II clinical studies. GeminX have recently started a Phase II dual-agent clinical study of Obatoclax with carboplatin and etoposide as a potential first-line treatment of small cell lung cancer (SCLC).

Obatoclax Phase II studies as a single agent in patients with mastocytosis, and dual-agent studies to treat patients with steroid-resistant acute lymphoblastic leukemia (ALL) are also planned. Finally, Phase III clinical trials of Obatoclax with Bortezomib are planned for the treatment of patients with relapsed mantle cell lymphoma.

SUMMARY

Bacterial prodiginines are a family of bioactive red-pigmented tripyrrolic compounds which have been shown to be effective proapoptotic anticancer agents with activity against multiple human cancer cell lines but with little activity against nonmalignant cells, and the precise mode of action of prodiginines is still debated. These mechanisms include interfering with signal transduction cascades, affecting the pH of cellular organelles, causing cell cycle arrest, and causing copper-dependent oxidative cleavage of DNA. All of these mechanisms can function to induce apoptosis of the cancerous cells.

Numerous biological activities have been attributed to the prodiginines, including antagonism against algae, bacteria, fungi, and protozoa, and immunosuppressive properties. Studies have been conducted on the potential of prodiginines as immunosuppressants, with a lead compound (PNU156804) developed through SAR studies focused on the methoxy group of pyrrolic ring B of undecylprodiosin. The direction of prodiginine research has now switched to the development of these compounds as novel anticancer agents. This is exemplified by the development of Obatoclax by GeminX pharmaceuticals. This leading anticancer prodiginine is an indolyl-dipyrromethene pan-Bcl-2 inhibitor and was developed through SAR studies of pyrrolic ring A of the bacterial prodiginine GX-15. It has demonstrated well-tolerated dose-dependent activity in Phase I clinical trials against multiple types of cancer.

In this chapter, recent progress in understanding the bifurcated biosynthetic pathway of the prodiginines and the substrate specificity of a number of the biosynthetic enzymes has been described. In particular, the terminal condensing enzyme has been shown to have a relaxed substrate specificity and may be of use in the synthesis of novel prodigiosin analogues. Interestingly, in one study, the lead anticancer prodiginine, Obatoclax, was produced using this enzyme. This demonstrates the potential of feeding analogues of prodiginine biosynthetic intermediates to defined biosynthetic mutants in a mutasynthesis approach to generate pharmaceutically important derivatives. As we gain a greater mechanistic understanding of the early prodiginine biosynthetic enzymes, directed evolution studies on these enzymes may also assist the generation of novel prodiginines. Thus, by a exploiting a combination of bacterial enzymology and synthetic chemistry, new chemotherapeutic agents may emerge for the treatment of diverse cancers.

REFERENCES

1. Bennett J.W., and Bentley R. 2000. Seeing red: the story of prodigiosin. Adv. Appl. Microbiol. **47**: 1–32.
2. Gerber N.N., and Stahly D.P. 1975. Prodiginine (prodigiosin-like) pigments from *Streptoverticillium rubrireticuli*, an organism that causes pink staining of polyvinyl chloride. Appl. Microbiol. **30**: 807–810.
3. Mo S., Sydor P.K., Corre C., Alhamadsheh M.M., Stanley A.E., Haynes S.W., Song L., Reynolds K.A., and Challis G.L. 2008. Elucidation of the *Streptomyces coelicolor* pathway to 2-undecylpyrrole, a key intermediate in undecylprodiginine and streptorubin B biosynthesis. Chem. Biol. **15**: 137–148.
4. Williamson N.R., Fineran P.C., Leeper F.J., and Salmond G.P. 2006. The biosynthesis and regulation of bacterial prodiginines. Nat. Rev. Microbiol. **4**: 887–899.
5. Williamson N.R., Fineran P.C., Gristwood T., Chawrai S.R., Leeper F.J., and Salmond G.P. 2007. Anticancer and immunosuppressive properties of bacterial prodiginines. Future Microbiol. **2**: 605–618.
6. Wasserman H.H., Mckeon J.E., and Smith L. 1960. Prodigiosin. Structure and partial synthesis. J. Am. Chem. Soc. **82**: 506–507.
7. Rapoport H., and Holden H.G. 1962. The synthesis of prodigiosin. J. Am. Chem. Soc. **84**: 635–642.
8. Furstner A., Grabowski J., Lehmann C.W., Kataoka T., and Nagai K. 2001. Synthesis and biological evaluation of nonylprodigiosin and macrocyclic prodigiosin analogues. Chembiochem. **2**: 60–68.
9. Williamson N.R., Fineran P.C., Ogawa W., Woodley L.R., and Salmond G.P. 2008. Integrated regulation involving quorum sensing, a two-component system, a GGDEF/EAL domain protein and a post-transcriptional regulator controls swarming and RhlA-dependent surfactant biosynthesis in *Serratia*. Environ. Microbiol. **10**: 1202–1217.
10. Demain A.L. 1995. Why do microorganisms produce antimicrobials? In: Hunter P.A., Darby G.K., Russell N.J., eds. Fifty Years of Antimicrobials: Past Perspectives and Future Trends. Cambridge: Society for General Microbiology, Symposium 53, pp. 205–228.
11. Nakashima T., Kurachi M., Kato Y., Yamaguchi K., and Oda T. 2005. Characterization of bacterium isolated from the sediment at coastal area of Omura Bay in Japan and several biological activities of pigment produced by this isolate. Microbiol. Immunol. **49**: 407–415.
12. Williams R.P., and Quadri S.M. 1980. The pigment of *Serratia*. In: Von Graevenitz A., Rubin S.J., eds. The Genus Serratia. Boca Raton, FL: CRC Press Inc, pp. 31–75.
13. Williamson N.R., Simonsen H.T., Ahmed R.A., Goldet G., Slater H., Woodley L., Leeper F.J., and Salmond G.P. 2005. Biosynthesis of the red antibiotic, prodigiosin, in *Serratia*: identification of a novel 2-methyl-3-n-amyl-pyrrole (MAP) assembly pathway, definition of the terminal condensing enzyme, and implications for undecylprodigiosin biosynthesis in *Streptomyces*. Mol. Microbiol. **56**: 971–989.
14. Montaner B., Navarro S., Piqué M., Vilaseca M., Martinell M., Giralt E., Gil J., and Pérez-Tomás R. 2000. Prodigiosin from the supernatant of *Serratia marcescens*

induces apoptosis in haematopoietic cancer cell lines. Br. J. Pharmacol. **131**: 585–593.

15. Montaner B., and Perez-Tomas R. 2001. Prodigiosin-induced apoptosis in human colon cancer cells. Life Sci. **68**: 2025–2036.
16. Yamamoto C., Takemoto H., Kuno K., Yamamoto D., Tsubura A., Kamata K., Hirata H., Yamamoto A., Kano H., Seki T., and Inoue K. 1999. Cycloprodigiosin hydrochloride, a new H(+)/Cl(−) symporter, induces apoptosis in human and rat hepatocellular cancer cell lines in vitro and inhibits the growth of hepatocellular carcinoma xenografts in nude mice. Hepatology. **30**: 894–902.
17. Nguyen M., Marcellus R.C., Roulston A., Watson M., Serfass L., Murthy Madiraju S.R., Goulet D., Viallet J., Bélec L., Billot X., Acoca S., Purisima E., Wiegmans A., Cluse L., Johnstone R.W., Beauparlant P., and Shore G.C. 2007. Small molecule obatoclax (GX15-070) antagonizes MCL-1 and overcomes MCL-1-mediated resistance to apoptosis. Proc. Natl. Acad. Sci. U S A. **104**: 19512–19517.
18. Cullen J.C. 1994. The miracle of Bolsena: growth of *Serratia* on sacramental bread and polenta may explain incidents in medieval Italy. ASM News. **60**: 187–191.
19. Yu V.L. 1979. *Serratia marcescens*: historical perspective and clinical review. N. Engl. J. Med. **300**: 887–893.
20. Alihosseini F., Ju K.S., Lango J., Hammock B.D., and Sun G. 2008. Antibacterial colorants: characterization of prodiginines and their applications on textile materials. Biotechnol Prog. **24**: 742–747.
21. Carswell E.A., Old L.J., Kassel R.L., Green S., Fiore N., and Williamson B. 1975. An endotoxin-induced serum factor that causes necrosis of tumors. Proc. Natl. Acad. Sci. U S A. **72**: 3666–3670.
22. Pennica D., Nedwin G.E., Hayflick J.S., Seeburg P.H., Derynck R., Palladino M.A., Kohr W.J., Aggarwal B.B., and Goeddel D.V. 1984. Human tumour necrosis factor: precursor structure, expression and homology to lymphotoxin. Nature. **312**: 724–729.
23. Perez-Tomas R., Montaner B., Llagostera E., and Soto-Cerrato V. The prodigiosins, proapoptotic drugs with anticancer properties. Biochem Pharmacol. **66**: 1447–1452.
24. Tsao S.W., Rudd B.A., He X.G., Chang C.J., and Floss HG. 1985. Identification of a red pigment from *Streptomyces coelicolor* A3(2) as a mixture of prodigiosin derivatives. J. Antibiot. (Tokyo). **38**: 128–131.
25. Harris A.K., Williamson N.R., Slater H., Cox A., Abbasi S., Foulds I., Simonsen H.T., Leeper F.J., and Salmond G.P. 2004. The *Serratia* gene cluster encoding biosynthesis of the red antibiotic, prodigiosin, shows species- and strain-dependent genome context variation. Microbiology **150**: 3547–3560.
26. Cerdeno A.M., Bibb M.J., and Challis G.L. 2001. Analysis of the prodiginine biosynthesis gene cluster of *Streptomyces coelicolor* A3(2): new mechanisms for chain initiation and termination in modular multienzymes. Chem. Biol. **8**: 817–829.
27. Kim D., Lee J.S., Park Y.K., Kim J.F., Jeong H., Oh T.K., Kim B.S., and Lee C.H. 2007. Biosynthesis of antibiotic prodiginines in the marine bacterium *Hahella chejuensis* KCTC 2396. J. Appl. Microbiol. **102**: 937–944.
28. Burke C., Thomas T., Egan S., and Kjelleberg S. 2007. The use of functional genomics for the identification of a gene cluster encoding for the biosynthesis of an

antifungal tambjamine in the marine bacterium *Pseudoalteromonas tunicata*. Environ. Microbiol. **9**: 814–818.

29. Kawasaki T., Sakurai F., Nagatsuka S.Y., and Hayakawa Y. 2009. Prodigiosin biosynthesis gene cluster in the roseophilin producer *Streptomyces griseoviridis*. J. Antibiot. (Tokyo). **62**: 271–276.
30. Garneau-Tsodikova S., Dorrestein P.C., Kelleher N.L., and Walsh C.T. 2006. Protein assembly line components in prodigiosin biosynthesis: characterization of PigA,G,H,I,J. J. Am. Chem. Soc. **128**: 12600–12601.
31. Haynes S.W., Sydor P.K., Stanley A.E., Song L., and Challis G.L. 2008. Role and substrate specificity of the *Streptomyces coelicolor* RedH enzyme in undecylprodiginine biosynthesis. Chem. Commun. (Camb). **28**: 1865–1867.
32. Mo S., Kim B.S., and Reynolds K.A. 2005. Production of branched-chain alkylprodiginines in *S.coelicolor* by replacement of the 3-ketoacyl ACP synthase III initiation enzyme, RedP. Chem. Biol. **12**: 191–200.
33. Stanley A.E., Walton L.J., Kourdi Zerikly M., Corre C., and Challis G.L. 2006. Elucidation of the *Streptomyces coelicolor* pathway to 4-methoxy-2,2′-bipyrrole-5-carboxaldehyde, an intermediate in prodiginine biosynthesis. Chem. Commun. (Camb). **14**: 3981–3983.
34. Thomas M.G., Burkart M.D., and Walsh C.T. 2002. Conversion of L-proline to pyrrolyl-2-carboxyl-S-PCP during undecylprodigiosin and pyoluteorin biosynthesis. Chem. Biol. **9**: 171–184.
35. Williams R.P. 1973. Biosynthesis of prodigiosin, a secondary metabolite of *Serratia marcescens*. Appl. Microbiol. **25**: 396–402.
36. Walsh C.T., Garneau-Tsodikova S., and Howard-Jones A.R. 2006. Biological formation of pyrroles: nature's logic and enzymatic machinery. Nat. Prod. Rep. **23**: 517–531.
37. Schloss P.D., and Handlesman, J. 2006. Psychrotrophic *Janthinobacterium lividum* strains from a microbial observatory on an island in the Tanana River (Fairbanks, AK). NCBI direct submission. EF063589–EF063592.
38. Cho H.J., Kim K.J., Kim M.H., and Kang B.S. 2008. Structural insight of the role of the *Hahella chejuensis* HapK protein in prodigiosin biosynthesis. Proteins. **70**: 257–262.
39. Kennedy J. 2008. Mutasynthesis, chemobiosynthesis, and back to semi-synthesis: combining synthetic chemistry and biosynthetic engineering for diversifying natural products. Nat. Prod. Rep. **25**: 25–34.
40. Weissman K.J. 2007. Mutasynthesis—uniting chemistry and genetics for drug discovery. Trends Biotechnol. **25**: 139–142.
41. Weist S., and Sussmuth R.D. 2005. Mutational biosynthesis—a tool for the generation of structural diversity in the biosynthesis of antibiotics. Appl. Microbiol. Biotechnol. **68**: 141–150.
42. Chawrai S.R., Williamson N.R., Salmond G.P., and Leeper F.J. 2008. Chemoenzymatic synthesis of prodigiosin analogues—exploring the substrate specificity of PigC. Chem. Commun. (Camb). **28**: 1862–1864.
43. Manderville R.A. 2001. Synthesis, proton-affinity and anti-cancer properties of the prodigiosin-group natural products. Curr. Med. Chem. Anti-Canc. Agents. **1**: 195–218.

44. Llagostera E., Soto-Cerrato V., Joshi R., Montaner B., Gimenez-Bonafe P., and Perez-Tomas R. 2005. High cytotoxic sensitivity of the human small cell lung doxorubicin-resistant carcinoma (GLC4/ADR) cell line to prodigiosin through apoptosis activation. Anticancer Drugs. **16**: 393–399.
45. Soto-Cerrato V., Llagostera E., Montaner B., Scheffer G.L., and Perez-Tomas R. 2004. Mitochondria-mediated apoptosis operating irrespective of multidrug resistance in breast cancer cells by the anticancer agent prodigiosin. Biochem. Pharmacol. **68**: 1345–1352.
46. Montaner B., Castillo-Avila W., Martinell M., Ollinger R., Aymami J., Giralt E., and Pérez-Tomás R. 2005. DNA interaction and dual topoisomerase I and II inhibition properties of the anti-tumor drug prodigiosin. Toxicol. Sci. **85**: 870–879.
47. Baldino C.M., Parr J., Wilson C.J., Ng S.C., Yohannes D., and Wasserman H.H. 2006. Indoloprodigiosins from the C-10 bipyrrolic precursor: new antiproliferative prodigiosin analogs. Bioorg. Med. Chem. Lett. **16**: 701–704.
48. Kataoka T., Muroi M., Ohkuma S., Waritani T., Magae J., Takatsuki A., Kondo S., Yamasaki M., and Nagai K. 1995. Prodigiosin 25-C uncouples vacuolar type H(+)-ATPase, inhibits vacuolar acidification and affects glycoprotein processing. FEBS Lett. **359**: 53–59.
49. Francisco R., Perez-Tomas R., Gimenez-Bonafe P., Soto-Cerrato V., Gimenez-Xavier P., and Ambrosio S. 2007. Mechanisms of prodigiosin cytotoxicity in human neuroblastoma cell lines. Eur. J. Pharmacol. **572**: 111–119.
50. Mott J.L., Bronk S.F., Mesa R.A., Kaufmann S.H., and Gores G.J. 2008. BH3-only protein mimetic obatoclax sensitizes cholangiocarcinoma cells to Apo2L/TRAIL-induced apoptosis. Mol. Cancer Ther. **7**: 2339–2347.
51. D'Alessio R., Bargiotti A., Carlini O., Colotta F., Ferrari M., Gnocchi P., Isetta A. Mongelli N., Motta P., Rossi A., Rossi M., Tibolla M., and Vanotti E. 2000. Synthesis and immunosuppressive activity of novel prodigiosin derivatives. J. Med. Chem. **43**: 2557–2565.
52. Rioux E., Billot X., and Dairi K. 2006. SAR study on aryl and heteroaryl bipyrrole inhibitors of BCl antiapoptotic proteins with potent antitumor acitivity in vivo. J. Mex. Chem. Soc. **50** Special Issue I, IUPAC ICOS-16:209.
53. Diaz R.I., Regourd J., Santacroce P.V., Davis J.T., Jakeman D.L., and Thompson A. 2007. Chloride anion transport and copper-mediated DNA cleavage by C-ring functionalized prodigiosenes. Chem. Commun. (Camb). **26**: 2701–2703.
54. Gale P.A. 2005. Amidopyrroles: from anion receptors to membrane transport agents. Chem. Commun. (Camb). **30**: 3761–3772.
55. Gale P.A., Light M.E., McNally B., Navakhun K., Sliwinski K.E., and Smith B.D. 2005. Co-transport of H+/Cl– by a synthetic prodigiosin mimic. Chem. Commun. (Camb). **30**: 3773–3775.
56. Davis J.T., Okunola O.A., Prados P., Iglesias-Sánchez J.C., Torroba T., and Quesada R. 2009. Using small molecules to facilitate exchange of bicarbonate and chloride anions across liposomal membranes. Nature Chem. **1**: 138–144.
57. Melvin M.S., Calcutt M.W., Noftle R.E., and Manderville R.A. 2002. Influence of the A-ring on the redox and nuclease properties of the prodigiosins: importance of the bipyrrole moiety in oxidative DNA cleavage. Chem. Res. Toxicol. **15**: 742–748.

58. Melvin M.S., Ferguson D.C., Lindquist N., and Manderville R.A. 1999. DNA binding by 4-methoxypyrrolic natural products. Preference for intercalation at AT sites by tambjamine E and prodigiosin. J. Org. Chem. **64**: 6861–6869.

59. Melvin M.S., Tomlinson J.T., Park G., Day C.S., Saluta G.R., Kucera G.L., and Manderville R.A. 2002. Influence of the A-ring on the proton affinity and anticancer properties of the prodigiosins. Chem. Res. Toxicol. **15**: 734–741.

60. Melvin M.S., Tomlinson J.T., Saluta G.R., Kucera G.L., Lindquist N., Manderville R.A. 2000. Double-strand DNA cleavage by copper-prodigiosin. J. Am. Chem. Soc. **122**: 6333–6334.

61. Melvin M.S., Wooton K.E., Rich C.C., Saluta G.R., Kucera G.L., Lindquist N., and Manderville R.A. 2001. Copper-nuclease efficiency correlates with cytotoxicity for the 4-methoxypyrrolic natural products. J. Inorg. Biochem. **87**: 129–135.

62. Park G., Tomlinson J.T., Melvin M.S., Wright M.W., Day C.S., and Manderville R.A. 2003. Zinc and copper complexes of prodigiosin: implications for copper-mediated double-strand DNA cleavage. Org. Lett. **5**: 113–116.

63. Williamson N.R., Simonsen H.T., Harris A.K., Leeper F.J., and Salmond G.P. Disruption of the copper efflux pump (CopA) of *Serratia marcescens* ATCC 274 pleiotropically affects copper sensitivity and production of the tripyrrole secondary metabolite, prodigiosin. J. Ind. Microbiol. Biotechnol. **33**: 151–158.

64. Furstner A. 2003. Chemistry and biology of roseophilin and the prodigiosin alkaloids: a survey of the last 2500 years. Angew. Chem. Int. Ed. Engl. **42**: 3582–3603.

65. Kataoka T., Magae J., Kasamo K., Yamanishi H., Endo A., Yamasaki M., and Nagai K. 1992. Effects of prodigiosin 25-C on cultured cell lines: its similarity to monovalent polyether ionophores and vacuolar type H(+)-ATPase inhibitors. J. Antibiot. (Tokyo). **45**: 1618–1625.

66. Kawauchi K., Shibutani K., Yagisawa H., Kamata H., Nakatsuji S., Anzai H., Yokoyama Y., Ikegami Y., Moriyama Y., and Hirata H. 1997. A possible immunosuppressant, cycloprodigiosin hydrochloride, obtained from *Pseudoalteromonas denitrificans*. Biochem. Biophys. Res. Commun. **237**: 543–547.

67. Konno H., Matsuya H., Okamoto M., Sato T., Tanaka Y., Yokoyama K., Kataoka T., Nagai K., Wasserman H.H., and Ohkuma S. 1998. Prodigiosins uncouple mitochondrial and bacterial F-ATPases: evidence for their H+/Cl– symport activity. J. Biochem. (Tokyo). **124**: 547–556.

68. Matsuya H., Okamoto M., Ochi T., Nishikawa A., Shimizu S., Kataoka T., Nagai K., Wasserman H.H., and Ohkuma S. 2000. Reversible and potent uncoupling of hog gastric (H(+)+K(+))-ATPase by prodigiosins. Biochem. Pharmacol. **60**: 1855–1863.

69. Nakayasu T., Kawauchi K., Hirata H., and Shimmen T. 2000. Demonstration of Cl- requirement for inhibition of vacuolar acidification by cycloprodigiosin in situ. Plant Cell Physiol. **41**: 857–863.

70. Ohkuma S., Sato T., Okamoto M., Matsuya H., Arai K., Kataoka T., Nagai K., Wasserman H.H. 1998. Prodigiosins uncouple lysosomal vacuolar-type ATPase through promotion of H+/Cl– symport. Biochem. J. **334**: 731–741.

71. Sato T., Konno H., Tanaka Y., Kataoka T., Nagai K., Wasserman H.H., and Ohkuma S. 1998. Prodigiosins as a new group of H+/Cl– symporters that uncouple proton translocators. J. Biol. Chem. **273**: 21455–21462.

72. Togashi K., Kataoka T., and Nagai K. 1997. Characterization of a series of vacuolar type H(+)-ATPase inhibitors on CTL-mediated cytotoxicity. Immunol. Lett. **55**: 139–144.
73. Yamamoto D., Uemura Y., Tanaka K., Nakai K., Yamamoto C., Takemoto H., Kamata K., Hirata H., and Hioki K. 2000. Cycloprodigiosin hydrochloride, H(+)/Cl(−) symporter, induces apoptosis and differentiation in HL-60 cells. Int. J. Cancer. **88**: 121–128.
74. Castillo-Avila W., Abal M., Robine S., and Perez-Tomas R. 2005. Non-apoptotic concentrations of prodigiosin (H+/Cl– symporter) inhibit the acidification of lysosomes and induce cell cycle blockage in colon cancer cells. Life Sci. **78**: 121–127.
75. Montaner B., and Perez-Tomas R. 2002. The cytotoxic prodigiosin induces phosphorylation of p38-MAPK but not of SAPK/JNK. Toxicol. Lett. **129**: 93–98.
76. Soto-Cerrato V., Vinals F., Lambert J.R., Kelly J.A., and Perez-Tomas R. 2007. Prodigiosin induces the proapoptotic gene NAG-1 via glycogen synthase kinase-3beta activity in human breast cancer cells. Mol. Cancer Ther. **6**: 362–369.
77. Soto-Cerrato V., Vinals F., Lambert J.R., and Perez-Tomas R. 2007. The anticancer agent prodigiosin induces p21WAF1/CIP1 expression via transforming growth factor-beta receptor pathway. Biochem. Pharmacol. **74**: 1340–1349.
78. Han S.B., Park S.H., Jeon Y.J., Kim Y.K., Kim H.M., and Yang K.H. 2001. Prodigiosin blocks T cell activation by inhibiting interleukin-2Ralpha expression and delays progression of autoimmune diabetes and collagen-induced arthritis. J. Pharmacol. Exp. Ther. **299**: 415–425.
79. Mortellaro A., Songia S., Gnocchi P., Ferrari M., Fornasiero C., D'Alessio R., Isetta A., Colotta F., and Golay J. 1999. New immunosuppressive drug PNU156804 blocks IL-2-dependent proliferation and NF-kappa B and AP-1 activation. J. Immunol. **162**: 7102–7109.
80. Songia S., Mortellaro A., Taverna S., Fornasiero C., Scheiber E.A., Erba E., Colotta F., Mantovani A., Isetta A.M., and Golay J. 1997. Characterization of the new immunosuppressive drug undecylprodigiosin in human lymphocytes: retinoblastoma protein, cyclin-dependent kinase-2, and cyclin-dependent kinase-4 as molecular targets. J. Immunol. **158**: 3987–3995.
81. Perez-Tomas R., and Montaner B. 2003. Effects of the proapoptotic drug prodigiosin on cell cycle-related proteins in Jurkat T cells. Histol. Histopathol. **18**: 379–385.
82. Konopleva M., Watt J., Contractor R., Tsao T., Harris D., Estrov Z., Bornmann W., Kantarjian H., Viallet J., Samudio I., and Andreeff M. 2008. Mechanisms of antileukemic activity of the novel Bcl-2 homology domain-3 mimetic GX15-070 (Obatoclax). Cancer Res. **68**: 3413–3420.
83. Kang M.H. 2009. Reynolds CP. Bcl-2 inhibitors: targeting mitochondrial apoptotic pathways in cancer therapy. Clin. Cancer Res. **15**: 1126–1132.
84. Reed J.C. 2003. Apoptosis-targeted therapies for cancer. Cancer Cell. **3**: 17–22.
85. Perez-Galan P., Roue G., Villamor N., Campo E., and Colomer D. 2007. The BH3-mimetic GX15-070 synergizes with bortezomib in mantle cell lymphoma by enhancing Noxa-mediated activation of Bak. Blood. **109**: 4441–4449.
86. Trudel S., Li Z.H., Rauw J., Tiedemann R.E., Wen X.Y., and Stewart A.K. 2007. Preclinical studies of the pan-Bcl inhibitor obatoclax (GX015-070) in multiple myeloma. Blood. **109**: 5430–5438.

87. Zhai D., Jin C., Satterthwait A.C., and Reed J.C. 2006. Comparison of chemical inhibitors of antiapoptotic Bcl-2-family proteins. Cell Death Differ. **13**: 1419–1421.

88. Pérez-Galán P., Roué G., López-Guerra M., Nguyen M., Villamor N., Montserrat E., Shore G.C., Campo E., and Colomer D. 2008. BCL-2 phosphorylation modulates sensitivity to the BH3 mimetic GX15-070 (Obatoclax) and reduces its synergistic interaction with bortezomib in chronic lymphocytic leukemia cells. Leukemia. **22**: 1712–1720.

89. Huang S., Okumura K., and Sinicrope F.A. BH3 mimetic obatoclax enhances TRAIL-mediated apoptosis in human pancreatic cancer cells. Clin. Cancer Res. **15**: 150–159.

90. Li J., Viallet J., and Haura E.B. 2008. A small molecule pan-Bcl-2 family inhibitor, GX15-070, induces apoptosis and enhances cisplatin-induced apoptosis in non-small cell lung cancer cells. Cancer Chemother. Pharmacol. **61**: 525–534.

91. Witters L.M., Witkoski A., Planas-Silva M.D., Berger M., Viallet J., and Lipton A. 2007. Synergistic inhibition of breast cancer cell lines with a dual inhibitor of EGFR-HER-2/neu and a Bcl-2 inhibitor. Oncol. Rep. **17**: 465–469.

92. Hearn W.R., Elson M.K., Williams R.H., and Medina-Castro J. 1970. Prodigiosene [5-(2-pyrryl)-2,2-'-dipyrrylmethene] and some substituted prodigiosenes. J. Org. Chem. **35**: 142–146.

93. Mukherjee P.P., Goldschmidt M.E., and Williams R.P. 1967. Enzymic formation of prodigiosin analog by a cell-free preparation from *Serratia marcescens*. Biochim. Biophys. Acta. **136**: 182–184.

94. Wasserman H.H., McKeon J.E., Smith L., and Forgione P. 1966. Studies of prodigiosin and the bipyrrole precursor. Tetrahedron Suppl. **22**(Suppl. 8): 647–662.

95. Williams R.P., and Quadri S.M. 1980. The pigments of *Serratia*. In: Von Graevenitz A., Rubin S.J., eds. The Genus Serratia. Boca Raton, FL: CRC Press, pp. 31–75.

96. Vivas J., González J.A., Barbeyto L., and Rodríguez L.A. 2000. Identification of environmental *Serratia plymuthica* strains with the new combo panels type 1S. Mem. Inst. Oswaldo Cruz. **95**: 227–229.

97. Grimont F. and Grimont P. 2006. The genus *Serratia*. In: M. Dwarkin, S. Falkow, E. Rosenberg, K.-H. Schleifer, and E. Stackebrandt, eds. The Prokaryotes, Vol. 6, pp. 219–244.

98. Jeong H., Yim J.H., Lee C., Choi S.H., Park Y.K., Yoon S.H., Hur C.G., Kang H.Y., Kim D., Lee H.H., Park K.H., Park S.H., Park H.S., Lee H.K., Oh T.K., Kim J.F. 2005. Genomic blueprint of *Hahella chejuensis*, a marine microbe producing an algicidal agent. Nucleic Acids Res. **33**: 7066–7073.

99. Gerber N.N. 1975. Prodigiosin-like pigments. CRC Crit. Rev. Microbiol. **3**: 469–485.

100. Allen G.R., Reichelt J.L., and Gray P.P. 1983. Influence of environmental factors and medium composition on *Vibrio gazogenes* growth and prodigiosin production. Appl. Environ. Microbiol. **45**: 1727–1732.

101. Shieh W.Y., Chen Y.-W., Chaw S.-M., Chiu H.-H. 2003. *Vibrio ruber* sp. nov., a red, facultatively anaerobic, marine bacterium isolated from sea water. Int. J. Syst. Evol. Microbiol. **53**: 479–484.

102. Diaz-Ruiz C., Montaner B., and Perez-Tomas R. 2001. Prodigiosin induces cell death and morphological changes indicative of apoptosis in gastric cancer cell line HGT-1. Histol. Histopathol. **16**: 415–421.

103. Kamata K., Okamoto S., Oka S., Kamata H., Yagisawa H., and Hirata H. 2001. Cycloprodigiosin hydrocloride suppresses tumor necrosis factor (TNF) alpha-induced transcriptional activation by NF-kappaB. FEBS Lett. **507**: 74–80.

104. Llagostera E., Soto-Cerrato V., Montaner B., and Perez-Tomas R. 2003. Prodigiosin induces apoptosis by acting on mitochondria in human lung cancer cells. Ann. N. Y. Acad. Sci. **1010**: 178–181.

105. Ramoneda B.M., and Perez-Tomas R. 2002. Activation of protein kinase C for protection of cells against apoptosis induced by the immunosuppressor prodigiosin. Biochem Pharmacol. **63**: 463–469.

106. Tanigaki K., Sasaki S., and Ohkuma S. 2003. In bafilomycin A1-resistant cells, bafilomycin A1 raised lysosomal pH and both prodigiosins and concanamycin A inhibited growth through apoptosis. FEBS Lett. **537**: 79–84.

107. Tomlinson J.T., Park G., Misenheimer J.A., Kucera G.L., Hesp K., and Manderville R.A. 2006. Photoinduced cytotoxicity and thioadduct formation by a prodigiosin analogue. Org. Lett. **8**: 4951–4954.

108. Cerdeno A.M., Bibb M.J., and Challis G.L. 2001. Analysis of the prodiginine biosynthesis gene cluster of *Streptomyces coelicolor* A3(2): new mechanisms for chain initiation and termination in modular multienzymes. Chem. Biol. **8**: 817–829.

109. Tsao S.W., Rudd B.A., He X.G., Chang C.J., and Floss H.G. 1985. Identification of a red pigment from *Streptomyces coelicolor* A3(2) as a mixture of prodigiosin derivatives. J. Antibiot. **38**: 128–131.

110. Umeyama T., Tanabe Y., Aigle B.D., and Horinouchi S. 1996. Expression of the *Streptomyces coelicolor* A3(2) ptpA gene encoding a phosphotyrosine protein phosphatase leads to overproduction of secondary metabolites in *S. lividans*. FEMS Microbiol. Lett. **144**: 177–184.

111. Wasserman H.H., Rodgers G.C., and Keith D.D. 1976. Undecyclprodigiosin. Tetrahedron. **32**: 1851–1854.

112. Liu R., Cui C.B., Duan L., Gu Q.Q., and Zhu W.M. 2005. Potent *in vitro* anticancer activity of metacycloprodigiosin and undecylprodigiosin from a sponge-derived actinomycete *Saccharopolyspora sp. nov*. Arch. Pharm. Res. **28**: 1341–1344.

113. Gerber N.N. 1969. Prodigiosin-Like pigments from *Actinomadura (Nocardia) pelletieri* and *Actinomadura madurae*. Appl. Microbiol. **18**: 1–3.

114. Xu D., Kim T.J., Park Z.Y., Lee S.K., Yang S.H., Kwon H.J., and Suh J.W. 2009. A DNA-binding factor, ArfA, interacts with the bldH promoter and affects undecylprodigiosin production in *Streptomyces lividans*. Biochem. Biophys. Res. Commun. **379**: 319–323.

115. Lewer P., Chapin E.L., Graupner P.R., Gilbert J.R., Peacock C., and Tartrolone C. 2003. A novel insecticidal macrodiolide produced by *Streptomyces* sp. CP1130. J. Nat. Prod. **66**: 143–145.

116. Liu R., Cui C.B., Duan L., Gu Q.Q., and Zhu W.M. 2005. Potent in vitro anticancer activity of metacycloprodigiosin and undecylprodigiosin from a sponge-derived actinomycete *Saccharopolyspora* sp. nov. Arch. Pharm. Res. **28**: 1341–1344.

117. Kawauchi K., Shibutani K., Yagisawa H., Kamata H., Nakatsuji S., Anzai H., Yokoyama Y., Ikegami Y., Moriyama Y., and Hirata H. 1997. A possible immunosuppressant, cycloprodigiosin hydrochloride, obtained from *Pseudoalteromonas denitrificans*. Biochem. Biophys. Res. Commun. **237**: 543–547.

118. Gerber N.N. 1983. Cycloprodigiosin from *Beneckea gazogenes*. Tetrahedron. Lett. **24**: 2797–2798.

119. Harwood C. 1978. *Beneckea gazogenes* sp. nov., a red, facultatively anaerobic, marine bacterium. Curr. Microbiol. **1**: 233–238.

120. Gerber N.N., and Gauthier M.J. 1979. New prodigiosin-like pigment from *Alteromonas rubra*. Appl. Environ. Microbiol. **37**: 1176–1179.

121. Yamamoto C., Takemoto H., Kuno K., Yamamoto D., Nakai K., Baden T., Kamata K., Hirata H., Watanabe T., and Inoue K. 2001. Cycloprodigiosin hydrochloride, a H+/Cl– symporter, induces apoptosis in human colon cancer cell lines in vitro. Oncol. Rep. **8**: 821–824.

122. Wasserman H.H, Rodgers G.C., and Keith D.D. 1969. Metacycloprodigiosin, a tripyrrole pigment from *Streptomyces longisporus ruber*. J. Am. Chem. Soc. **91**: 1263–1264.

123. Kawasaki T., Sakurai F., and Hayakawa Y. 2008. A prodigiosin from the roseophilin producer *Streptomyces griseoviridis*. J. Nat. Prod. **71**: 1265–1267.

124. Kawasaki T., Sakurai F., Nagatsuka S., and Hayakawa Y. 2009. Prodigiosin biosynthesis gene cluster in the roseophilin producer *Streptomyces griseoviridis*. J. Antibiot. **62**: 271–276.

125. Subramanian M., Chander R., and Chattopadhyay S. 2006. A novel naturally occurring tripyrrole with potential nuclease and anti-tumour properties. Bioorg. Med. Chem. **14**: 2480–2486.

126. Deorukhkar A.A., Chander R., Ghosh S.B., Sainis K.B. 2007. Identification of a red-pigmented bacterium producing a potent anti-tumor *N*-alkylated prodigiosin as *Serratia marcescens*. Res. Microbiol. **158**: 399–404.

127. Deorukhkar A.A., Chander R., Pandey R., Sainis K.B. 2008. A novel N-alkylated prodigiosin analogue induced death in tumour cell through apoptosis or necrosis depending upon the cell type. Cancer Chemother. Pharmacol. **61**: 355–363.

128. Feher D., Barlow R.S., Lorenzo P.S., Hemscheidt T.K. 2008. A 2-substituted prodiginine, 2-(p-Hydroxybenzyl)prodigiosin, from *Pseudoalteromonas rubra*. J. Nat. Prod. **71**: 1970–1972.

129. Hayakawa Y., Kawakami K., Seto H., Furihata K. 1992. Structure of a new antibiotic, roseophilin. Tetrahedron Lett. **33**: 2701–2704.

130. Furstner A., and Grabowski E.J. 2001. Studies on DNA cleavage by cytotoxic pyrrole alkaloids reveal the distinctly different behavior of roseophilin and prodigiosin derivatives. Chembiochem. **2**: 706–709.

131. Furstner A., Reinecke K., Prinz H., and Waldmann H. 2004. The core structures of roseophilin and the prodigiosin alkaloids define a new class of protein tyrosine phosphatase inhibitors. Chembiochem. **5**: 1575–1579.

132. Manger M., Scheck M., Prinz H., von Kries J.P., Langer T., Saxena K., Schwalbe H., Fürstner A., Rademann J., and Waldmann H. 2005. Discovery of *Mycobacterium tuberculosis* Protein Tyrosine Phosphatase A (MptpA) Inhibitors Based on Natural Products and a Fragment-Based Approach. Chembiochem. **6**: 1749–1753.

133. Lindquist N., and Fenical W. 1991. New tamjamine class alkaloids from the marine ascidian *Atapozoa* sp. and its nudibranch predators. Origin of the tambjamines in *Atapozoa*. Cell Molec. Life Sci. **47**: 504–506.

134. Burke C., Thomas T., Egan S., and Kjelleberg S. 2007. The use of functional genomics for the identification of a gene cluster encoding for the biosynthesis of an antifungal tambjamine in the marine bacterium *Pseudoalteromonas tunicata*. Environ. Microbiol. **9**: 814–818.

135. Cavalcanti B.C., Júnior H.V., Seleghim M.H., Berlinck R.G., Cunha G.M., Moraes M.O., and Pessoa C. 2008. Cytotoxic and genotoxic effects of tambjamine D, an alkaloid isolated from the nudibranch *Tambja eliora*, on Chinese hamster lung fibroblasts. Chem. Biol. Interact. **174**: 155–162.

136. Tanigaki K., Sato T., Tanaka Y., Ochi T., Nishikawa A., Nagai K., Kawashima H., and Ohkuma S. 2002. BE-18591 as a new H^+/Cl^- symport ionophore that inhibits immunoproliferation and gastritis. FEBS Lett. **524**: 37–42.

137. Borah S., Melvin M.S., Lindquist N., and Manderville RA. 1998. Copper-mediated nuclease activity of a tambjamine alkaloid. J. Am. Chem. Soc. **120**: 4557–4562.

138. Melvin M.S., Ferguson D.C., Lindquist N., and Manderville R.A. 1999. DNA binding by 4-methoxypyrrolic natural products. Preference for intercalation at AT sites by tambjamine E and prodigiosin. J. Org. Chem. **64**: 6861–6869.

139. Melvin M.S., Wooton K.E., Rich C.C., Saluta G.R., Kucera G.L., Lindquist N., and Manderville RA. 2001. Copper-nuclease efficiency correlates with cytotoxicity for the 4-methoxypyrrolic natural products. J. Inorg. Biochem. **87**: 129–135.

16

FARNESYLTRANSFERASE INHIBITORS OF MICROBIAL ORIGINS IN CANCER THERAPY

JINGXUAN PAN[1] AND SAI-CHING JIM YEUNG[2,3]

[1]*Department of Pathophysiology, Zhongshan School of Medicine, Sun Yat-Sen University, Guangzhou, People's Republic of China*
[2]*Department of General Internal Medicine, Ambulatory Treatment and Emergency Care, The University of Texas, M.D. Anderson Cancer Center, Houston, TX*
[3]*Department of Endocrine Neoplasia and Hormonal Disorders, The University of Texas, M.D. Anderson Cancer Center, Houston, TX*

RAS AND FARNESYLTRANSFERASE INHIBITORS (FTIS)

In 1964, Jennifer Harvey found that murine leukemia virus induced sarcomas in newborn rats. In the 1970s, the gene sequence responsible for inducing the sarcoma was identified and named H-ras (*ra*t *s*arcoma). Later, H-ras and additional isoforms (K-ras and N-ras) were found in mammalian genomes. This family of Ras proto-oncogenes encodes for a family of small GTP-binding proteins. In normal cells, Ras switches between an inactive GDP- and an active GTP-bound state. Oncogenic Ras mutations such as single amino acid substitutions at residues 12, 13, 61, inhibit GTP hydrolysis and consequent loss of GDP/GTP cycling. These constitutively active, GTP-bound mutants persistently stimulate downstream signaling pathways, resulting in loss of contact inhibition, anchorage-independent growth, ruffling of plasma membrane, and stimulation of DNA synthesis. Overall, Ras mutations are found in about 30% of malignancies.

Emerging Cancer Therapy: Microbial Approaches and Biotechnological Tools, Edited by Arsénio M. Fialho and Ananda M. Chakrabarty

To carry out their biological function, Ras must be anchored to the inner side of the plasma membrane. A series of posttranslational modifications is required for binding of Ras to the inner plasma membrane. Among these reactions, a critical rate-limiting step is farnesylation. The C-terminal CAAX motif (C = cysteine, A = aliphatic, X = serine, leucine, methionine, or glutamic acid) of Ras can be recognized by farnesyltransferase (FTase), and a 15-carbon farnesyl group from a farnesyl pyrophosphate (FPP) substrate is added to the cysteine of Ras. Subsequent steps in posttranslational modification are AAX proteolysis by Ras converting enzyme (Rce), alpha-carboxymethylation, and farnesylated cysteine residues by isoprenylcysteine carboxy methyltransferase. H-Ras and N-Ras are subsequently palmitoylated by palmitoyltransferase. Because farnesylation is the rate-limiting step, one can speculate that inhibiting FTase might perturb the Ras membrane localization and block the Ras signaling transduction in human malignancies.

Inhibition of hydroxymethylgutaryl-CoA reductase by statins (e.g., lovastatin, compactin [mevastatin]) decreases mevalonate which is the precursor to synthesis of the isoprenyl groups required for Ras function (1). Competition against the C-terminus of the H-ras protein using an octapeptide with similar sequence showed inhibitory effect on Ras function and pioneered the concept of peptidomimetic FTIs (1). Synthetic non-peptide small molecule analogs of CAAX peptidomimetic inhibitor are generally more potent than FTIs of microbial or other natural origins identified by screening. Tipifarnib (R115777, Zarnestra), Lonafarnib (SCH66336), and BMS-214662 have entered late phase clinical trials.

FTIS OF MICROBIAL ORIGINS

Microbes have been plentiful sources of pharmaceuticals. FTIs of microbial origins have recently been reviewed by Iwasaki and Omura (2). Over the last couple of decades, many FTIs of microbial origin have been identified, including pepticinnamins (3), andrastins (4, 5), kurasoins (6), gliotoxins (7), clavaric acids (8, 9), schizostatin (10), actinoplanic acids A and B (11), and TAN-1813 (12). *Pepticinnamins A to F* are produced by *Streptomyces* sp. OH-4652. Only the chemical structure of pepticinnamin E has been determined; it consists of five amino acids and an o-pentenylcinnamic acid, and the C-terminal glycylserine is in the cyclized diketopiperazine form (3). Pepticinnamin E inhibits FTase by competing against Ras (*Ki*, 1.76 mM). *Andrastins A to D* are produced by *Penicillium* sp. FO-3929. Andrastins are reversible FTIs with low to moderate inhibitory activities. Based on computational docking of Andrastins to FTase, they are likely to be peptidomimetic FTIs (13). *Kurasoins* are FTIs purified from *Paecilomyces* sp. FO-3684 (6) with no significant antimicrobial activity. Gliotoxins are mycotoxins from the culture of a fungus strain FO2047 with antimicrobial activity. *Gliotoxin and acetylgliotoxin* inhibit FTase with IC50 values of 1.1 mM and 4.4 mM, respectively (7). Although they are not very potent FTIs, they, together with 10′desmethoxystreptonigrin represented the early FTIs discovered from microbial metabolites. *Clavaric acid derivatives* have been isolated from *Clavariadelphus truncatus*. Clavaric acids are specific and reversible FTIs that are principally peptidomimetics, but they become FPP mimetics when a succinyl moiety is added at the 7α position. Clavaric acid A is peptidomimetic derivative while Clavaric acid B is FPP mimetic (13). *Schizostatins* are produced by *Schizophyllum commune* and are good inhibitors of squalene synthase but are weak FTIs (10). Only the *Z*-isomer is active against FTase. This stereoselectivity is corroborated by computational docking to FTase, and they may compete with FPP as the potential mechanism of inhibition (13). *Actinoplanic acids* A and B are produced by *Actinoplanes* sp. MA 7066 while actinoplanic acid B is isolated from both MA7066 and *Streptomyces* sp. MA 7099 (11). The FTI *TAN-1813* is isolated from the culture broth of a plant endophyte fungus strain, *Phoma sp.* FL-41510. TAN-1813 inhibited rat brain FTase and geranylgeranyltransferase (GGPTase) I with IC50s of 23 μg/mL and 47 μg/mL, respectively. TAN-1813 showed mixed-type inhibition with respect to FPP and noncompetitive inhibition with respect to a Ras C-terminal peptide (12).

Hara et al. used a yeast strain with conditional deficiency in the GPA1 gene to search FTIs and found that *Streptomyces pavulus* produced three active inhibitors (UCF1-A, B, and C) (14). UCF1-C was identified as *manumycin* A (Table 16.1), a known antibiotic with a molecule weight of 550.7 Da. UCF1-A

TABLE 16.1. Farnesyltransferase Inhibitors of Microbial Origin with Antineoplastic Effects in Animal Models

Compound Name	Formula	FTase IC50	GGPTase	Microorganism
Manumycin A	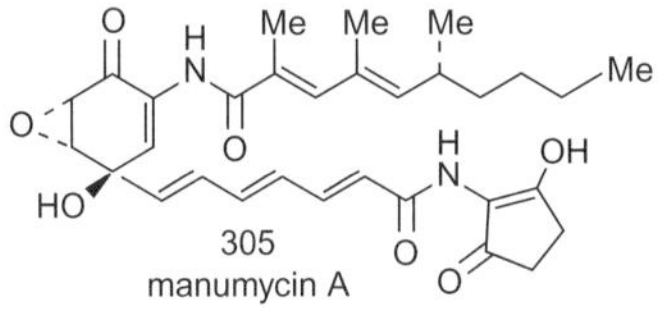 305 manumycin A	5 μM	180 μM	*Streptomyces parvulus*
Gliotoxin	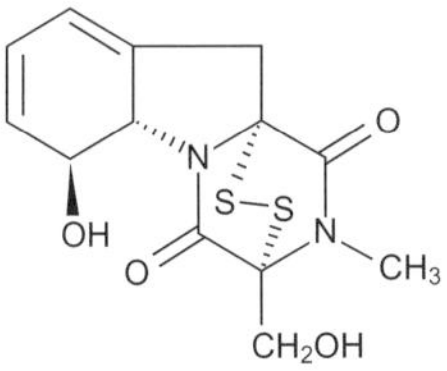	1.1 mM	—	Fungus strain FO2047 or *Gliocladium fimbriatum*
TAN-1813	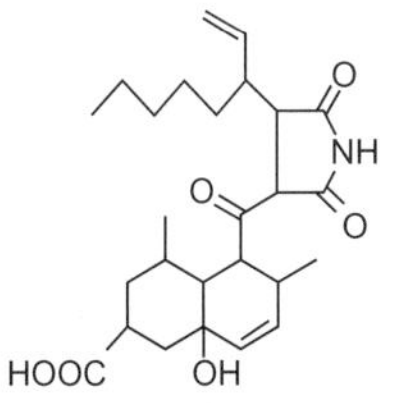 TAN-1813(2000)	23 μg/mL	47 μg/mL	*Phoma sp.* FL-41510

and UCF1-B are related to manumycin in chemical structures, but manumycin showed the strongest inhibition (IC50: 5 μM) against FTase among the three compounds. Manumycin is selective against FTase as it would inhibit GGPTase only at high concentrations (IC50: 180 μM). Manumycin is a competitive inhibitor of FTase competing against FPP with a Ki of 1.2 μM. Manumycins are structurally closely related to asukamycin which is also isolated from the culture of *Streptomyces* sp. Other antibiotics isolated from *Streptomyces* strains such as frenolicin B and nanaomycins A and D also inhibit FTase. Manumycin A has also been discovered to inhibit neutral sphingomyelinase (15).

Although some of these FTIs of microbial origin were found to be effective in suppressing tumor growth both in cells and in animals (nude mice transplanted with *ras*-transformed cells and cancer mouse models), no FTI of microbial origin has been selected for clinical development for human use. Most of these FTIs inhibit FTase competitively with respect to FPP; it is disappointing that the potency of these microbial FTIs are relatively low compared with that of the synthetic CAAX mimetic FTIs. As long as FTase remains a therapeutic target for cancer, the search for a more potent and effective FTI from microbial metabolites would go on.

ANTITUMOR ACTIVITY OF FTIS OF MICROBIAL ORIGINS

Only a few FTIs of microbial origin have been demonstrated to have antitumor activity in animal models. Among these, manumycin A is the compound that has been most studied. Manumycin A demonstrates antitumor activity against a variety of human cancer cell lines (e.g., human pancreatic tumor [16, 17], anaplastic thyroid carcinoma [18], leukemias [19], myeloma [20], hepatocellular carcinoma cells [21, 22]). In combination, manumycin enhanced the apoptotic effect of paclitaxel (18). Manumycin-plus-paclitaxel-induced DNA fragmentation was blocked by the inhibitors of caspase-9, caspase-8, and caspase-3 (23). The drug combination enhanced the activation of caspase-9, caspase-8, and caspase-3, and cytochrome *c* release into the cytosol. Cytochrome *c* release was not affected by the inhibitors of caspase-9, caspase-8, and caspase-3 (23). Thus, the cytochrome *c* release is upstream of the activation of caspase-9, caspase-8, and caspase-3 in the enhanced apoptosis of cancer cells treated with manumycin plus paclitaxel, and that the interaction between manumycin and paclitaxel occurred at or upstream of cytochrome *c* in the apoptosis regulatory pathway in cancer cells. The drug-induced apoptosis may be enhanced by p21Waf-1/Cip1 (24) and may involve Bax translocation to the mitochondria (25). Overall, manumycin-induced apoptosis followed a typical "xenobiotic apoptosis pathway," the hallmarks of which are induction of oxidative stress, mitogen-activated protein kinase (MAPK) signaling, and cytochrome *c* release, which activates the intrinsic apoptosis pathway (26). Manumycin-induced reactive oxygen species (ROS) can damage DNA, leading to DNA damage response (27), and inhibition of base-excision repair by methoxyamine may enhance manumycin-induced apoptosis (19). In some cell lines, manumycin induces apoptosis (22, 23) while in other cell lines, it induces autophagy (28).

Manumycin has *in vivo* activity against various human cancer cell lines in xenograft models: pancreatic cancer (16) and thyroid cancer (18, 29). We examined *in vivo* effect of combined manumycin and paclitaxel treatments in a nude mouse xenograft model using anaplastic thyroid cancer cells (29). Both manumycin and paclitaxel had significant inhibitory effects on tumor growth (18, 29). Combined manumycin and paclitaxel treatments exhibited more tumor inhibition than either manumycin or paclitaxel alone. Inhibition of tumor angiogenesis appears to be an important *in vivo* antineoplastic mechanism involved (29). FTIs have inhibitory effect on vascular endothelial growth factor (VEGF) (29), and may add to the antineoplastic effect of combretastatin A4 phosphate (CA4P), a vascular disrupting agent, in nude mice (30).

TAN-1813 inhibited farnesylation of Ras in a K-ras transformant of mouse embryonic fibroblast cell line NIH3T3 and the proliferation of various human cancer cells, regardless of the Ras mutational status. TAN-1813 arrested NIH3T3/K-ras cells at both G1 and G2/M phases of the cell cycle. It induced morphological reversion of NIH3T3/K-ras cells from the transformed phenotype. Moreover, TAN-1813 inhibited human fibrosarcoma HT-1080 xenografts and NIH3T3/K-ras tumors in nude mice (12).

Gliotoxin inhibits both FTase (IC50: 80 μM) and geranylgeranyltransferase I (IC50: 17 μM). Gliotoxin inhibited proliferation of six breast cancer cell lines in culture with IC50s ranging from 38 to 985 μM. Using a carcinogen-induced rat mammary carcinoma model, gliotoxin at 10 mg/kg by subcutaneous injection weekly for 4 weeks showed antitumor activity with little systemic toxicity compared with control mice. The mice can tolerate up to 25 mg/kg as a single administration (31).

TARGETS OF FTIS

In the proteome, there are close to 300 proteins with a CAAX motif that are potentially farnesylated and over 20 proteins have been proven to be farnesylated (32). Among these, FTIs have been shown to inhibit the farnesylation of H-Ras, Rheb, RhoB, and centromere proteins (CENP-E and CENP-F), and the contributions of these proteins to the antineoplastic action of FTIs have been evident. These proteins may be relevant in intracellular signaling regulating cell proliferation, cell cycling, apoptosis, and autophagy (Figure 16.1). In a cell culture study, 31 of 42 cancer cell lines of various tumor types and oncogene makeups (including wild-type Ras) were sensitive to L-744832. This supported the notion that FTIs may exert anticancer activity via mechanisms in addition to blocking the function of Ras. In addition, inhibition of PI-3 kinase/Akt-mediated cell survival and adhesion pathway by FTIs has been shown to play a role in the induction of apoptosis. The mechanisms of action of FTIs have recently been reviewed (32).

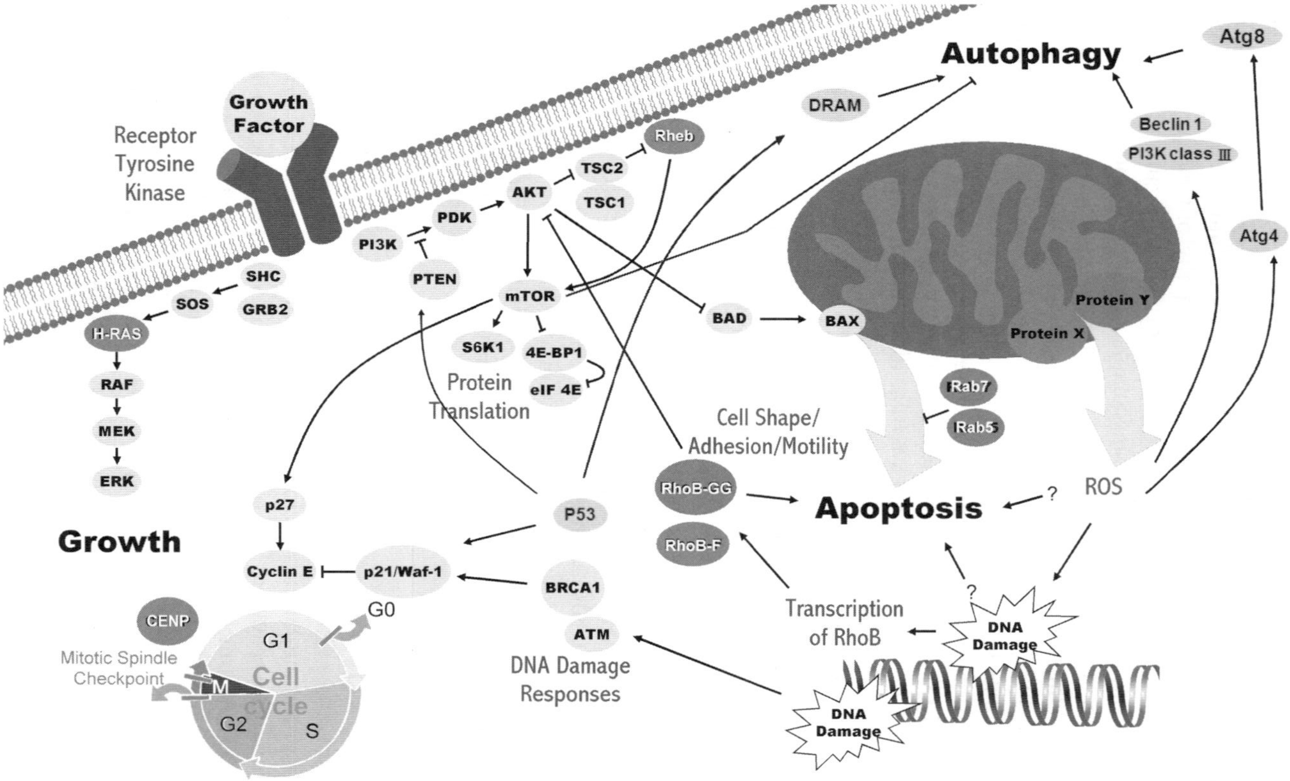
Growth Factor
Receptor Tyrosine Kinase
SHC
GRB2
SOS
H-RAS
RAF
MEK
ERK
Growth
PI3K
PDK
PTEN
AKT
TSC2
TSC1
Rheb
mTOR
S6K1
4E-BP1
eIF 4E
Protein Translation
BAD
BAX
DRAM
Autophagy
Atg8
Beclin 1
PI3K class III
Atg4
Protein Y
Protein X
Rab7
Rab5
Cell Shape/ Adhesion/Motility
RhoB-GG
RhoB-F
Apoptosis
ROS
?
?
DNA Damage
DNA Damage
Transcription of RhoB
P53
BRCA1
ATM
DNA Damage Responses
p27
Cyclin E
p21/Waf-1
G0
G1
Cell cycle
S
G2
M
CENP
Mitotic Spindle Checkpoint

FTIs may inhibit cancer growth by interfering with bipolar spindle formation during transition from prophase to metaphase in mitosis. Centromere proteins CENP-E and CENP-F are substrates for FTase but not GGTase I. FTIs do not affect the localization of these proteins to the kinetochores (33). However, FTI-treated cells have ring-shaped chromosomes that are not aligned at the metaphase plate, and the association between CENP-E and the microtubules is altered.

RhoB, which is localized primarily in endosomes which regulate cytokine trafficking and cell survival, may represent an important target of FTIs. Knockout mouse studies have implicated RhoB as an inhibitory regulator of integrin and growth factor signaling in oncogenesis (34). RhoB can be prenylated in cells by FTase or GGTase I to form farnesylated RhoB (RhoB-F) or geranylgeranylated RhoB (RhoB-GG), respectively. RhoB-GG is distributed in multivesicular late endosomes and RhoB-F at the plasma membrane (35). When farnesylation is blocked by FTIs, a shift of RhoB to RhoB-GG occurs. The antineoplastic action of FTIs may be mediated by a gain of RhoB-GG.

Our group has shown that FTIs induce the production of reactive oxygen species (27). Induction of ROS by FTIs may be a common phenomenon in malignant cells but not in normal cells. Among the consequences of ROS induction by FTIs was DNA damage (caspase-independent double-strand breaks, abasic sites, and generation of oxo-8-deoxyguanidine). Quenching ROS with *N*-acetyl-L-cysteine abrogated the FTI-induced DNA damage and antagonized the inhibitory effect of the FTI manumycin on accumulation of protein biomass. DNA damage initiates various DNA damage response pathways. Several proteins are known to be involved in double-strand break repair (e.g., Brca1 functions in homologous recombination; DNA-dependent protein kinase and Ku70 function in nonhomologous end-joining). In FTI-treated cells, phosphorylation and nuclear foci of Brca1 were observed, suggesting that FTI may affect key components of the homologous recombination repair pathway. We also found that ataxia telangiectasia-mutated and DNA-dependent protein kinase, but not ataxia telangiectasia-mutated and Rad3-related, were involved in sensing the FTI-induced DNA damage. These DNA damage response proteins, particularly

Figure 16.1 The mechanisms of antineoplastic actions of FTIs. The diagram depicts molecular targets of FTIs in cancer cells. The farnesylated proteins (white fonts) occupy key positions in pathways regulating proliferation, cell survival, cell cycling, and cytoskeletal regulation. Increase in RhoB-GG is an important event in FTI-induced apoptosis. GGTaseII, which prenylates the Rab proteins, may be a molecular target of certain FTIs in addition to FTase. Induction of ROS with ROS-mediated DNA damage subsequently links to induction of RhoB. The mechanism of ROS induction is unclear and may involve an unidentified farnesylated protein X or geranylgeranylated protein Y. Disruption of various farnesylated proteins lead to induction of apoptosis or autophagy.

ataxia telangiectasia-mutated and Brca1, also provide links of FTI-induced DNA damage to cell cycle arrest and apoptosis.

FTI-induced ROS damage DNA and contribute to the antineoplastic effect of FTIs. DNA damage responses include activation of DNA repair proteins and induction of RhoB. RhoB may sensitize cancer cells to DNA damage-induced apoptosis following genotoxic stress. Quenching ROS did prevent induction of RhoB by the FTIs, suggesting that RhoB was induced by FTI-induced ROS-mediated DNA damage. Therefore, FTI-induced ROS, DNA damage, and DNA damage responses are linked to RhoB, linking these events in the complex mechanism of action of FTIs.

In autophagy regulation, Rheb (*R*as *h*omologue *e*nriched in *b*rain) situates downstream of tuberous sclerosis complex 1/2 (TSC1/2) and upstream of mTOR to regulate cell growth. Protein kinase B (PKB or Akt) phosphorylates the TSC1/TSC2 complex to suppress its activity which is to turn off Rheb, an activator of mTOR. Like other members of the Ras family, farnesylation of Rheb is limiting inner membrane localization and the biological activity of Rheb. Rheb mutated at the farnesylation site stimulates S6 kinase-1 less efficiently than wild-type (36). FTIs can completely inhibit Rheb farnesylation (36) and, since Rheb is not prenylated by alternate pathways (such as GGTs), there is complete downstream abrogation of S6 phosphorylation and activation. FTIs induce autophagy in several solid tumor cell lines (37). siRNA suppression of Rheb expression also induces the appearance of autophagosomes in U2OS cells transfected with GFP-LC3, visualized by fluorescent microscopy. mTOR is an established negative regulator of autophagy. Inhibition of mTOR by FTIs would contribute to autophagy. Consistent with the involvement of Rheb in FTI-induced autophagy, FTIs decrease phosphorylation (i.e., activation) of mTOR in a dose-dependent manner, and phosphorylation of S6 kinase downstream of mTOR (37). Although it is not clear yet whether FTIs have alternative pathways or mechanisms to regulate the activity of mTOR other than going through Rheb, FTIs induce autophagy by interfering with the function of Rheb. Alternative pathways that link FTIs to autophagy are possible but have not yet been explored experimentally: (1) Ras-PI3K class I-Akt-mTOR pathway, (2) FTI-induced ROS and DNA damage response pathways (38).

In summary, FTIs can induce apoptosis and autophagy in cancer cells. The mechanisms may involve disruption of farnesylation of many farnesylated proteins. Further insights into the molecular mechanism for FTIs to induce cancer cell death will probably help to better define their optimal use in combination with other chemotherapeutic agents in the treatment of cancer patients.

FUTURE PROSPECTS

Tipifarnib (R115777, Zarnestra), Lonafarnib (SCH66336), and BMS-214662 have been under clinical trials in diverse advanced malignancies (e.g., breast, lung, esophageal cancer, and leukemia). Tipifarnib appears to have demonstrable clinical activity in AML, CML, MDS, and breast cancer. Karp and Lancet had reviewed the clinical trials involving these agents (39). The results of early Phase II/III studies suggest that the activity of FTIs as single agents is modest and is generally lower than that obtained by standard cytotoxic drugs (40). Ongoing clinical studies are assessing the role of FTIs for early-stage disease or in combination with cytotoxic agents or with other molecular targeted therapies for advanced stage tumors. Although the clinical utility of FTIs in cancer remains to be clearly demonstrated, FTIs may also be useful in the treatment of infectious protozoan diseases such as Chaga's disease and malaria (41). Many FTIs of microbial origins inhibit other targets. A dirty agent may be more toxic, but it may also be more effective. However, such a conclusion should not be reached a priori (42). Evaluation of clinical safety and efficacy in clinical trials is justified when safety and efficacy are supported by animal data.

ACKNOWLEDGMENTS

This study was supported by grants from the National Natural Science Fund of China (90713036 to J.P.), the National High Technology Research and Development Program of China (863 Program 2008 AA02Z420 to J.P.), National Basic Research Program of China (973 Program 2009 CB825506 to J.P.), Department of Defense, Breast Cancer Research Program (BCRP) of the Office of the Congressionally Directed Medical Research Programs (CDMRP) Synergistic Idea Development Award (BC062166 to S.C.Y.) and Susan G. Komen Foundation Promise Grant (2008 KG081048 to S.C.Y.).

REFERENCES

1. Kim R., Rine J., and Kim S.H. 1990. Prenylation of mammalian Ras protein in Xenopus oocytes. Mol. Cell Biol. **10**: 5945–5949.
2. Iwasaki S., and Omura S. 2007. Search for protein farnesyltransferase inhibitors of microbial origin: our strategy and results as well as the results obtained by other groups. J. Antibiot. (Tokyo). **60**: 1–12.
3. Shiomi K., Yang H., Inokoshi J., Van der Pyl D., Nakagawa A., Takeshima H., and Omura S. 1993. Pepticinnamins, new farnesyl-protein transferase inhibitors produced by an actinomycete. II. Structural elucidation of pepticinnamin E. J. Antibiot. (Tokyo). **46**: 229–234.
4. Omura S., Inokoshi J., Uchida R., Shiomi K., Masuma R., Kawakubo T., Tanaka H., Iwai Y., Kosemura S., and Yamamura S. 1996. Andrastins A-C, new protein farnesyltransferase inhibitors produced by *Penicillium* sp. FO-3929. I. Producing strain, fermentation, isolation, and biological activities. J. Antibiot. (Tokyo). **49**: 414–417.

5. Uchida R., Shiomi K., Inokoshi J., Tanaka H., Iwai Y., and Omura S. 1996. Andrastin D, novel protein farnesyltransferase inhibitor produced by *Penicillium* sp. FO-3929. J. Antibiot. (Tokyo). **49**: 1278–1280.
6. Sunazuka T., Hirose T., Tian Z.M., Uchida R., Shiomi K., Harigaya Y., and Omura S. 1997. Syntheses and absolute structures of novel protein farnesyltransferase inhibitors, kurasoins A and B. J. Antibiot. (Tokyo). **50**: 453–455.
7. Van der Pyl D., Inokoshi J., Shiomi K., Yang H., Takeshima H., and Omura S. 1992. Inhibition of farnesyl-protein transferase by gliotoxin and acetylgliotoxin. J. Antibiot. (Tokyo). **45**: 1802–1805.
8. Lingham R.B., Silverman K.C., Jayasuriya H., Kim B.M., Amo S.E., Wilson F.R., Rew D.J., Schaber M.D., Bergstrom J.D., Koblan K.S., Graham S.L., Kohl N.E., Gibbs J.B., and Singh S.B. 1998. Clavaric acid and steroidal analogues as Ras- and FPP-directed inhibitors of human farnesyl-protein transferase. J. Med. Chem. **41**: 4492–4501.
9. Jayasuriya H., Silverman K.C., Zink D.L., Jenkins R.G., Sanchez M., Pelaez F., Vilella D., Lingham R.B., and Singh S.B. 1998. Clavaric acid: a triterpenoid inhibitor of farnesyl-protein transferase from *Clavariadelphus truncatus*. J. Nat. Prod. **61** 1568–1570.
10. Tanimoto T., Onodera K., Hosoya T., Takamatsu Y., Kinoshita T., Tago K., Kogen H., Fujioka T., Hamano K., and Tsujita Y. 1996. Schizostatin, a novel squalene synthase inhibitor produced by the mushroom, *Schizophyllum commune*. I. Taxonomy, fermentation, isolation, physico-chemical properties and biological activities. J. Antibiot. (Tokyo). **49**: 617–623.
11. Silverman K.C., Cascales C., Genilloud O., Sigmund J.M., Gartner S.E., Koch G.E., Gagliardi M.M., Heimbuch B.K., Nallin-Omstead M., Sanchez M., Diez M.T., Martin I., Garrity G.M., Hirsch C.F., Gibbs J.B., Singh S.B., and Lingham R.B. 1995. Actinoplanic acids A and B as novel inhibitors of farnesyl-protein transferase. Appl. Microbiol. Biotechnol. **43**: 610–616.
12. Ishii T., Hayashi K., Hida T., Yamamoto Y., and Nozaki Y. 2000. TAN-1813, a novel Ras-farnesyltransferase inhibitor produced by Phoma sp. taxonomy, fermentation, isolation and biological activities in vitro and in vivo. J. Antibiot. (Tokyo). **53**: 765–778.
13. Pedretti A., Villa L., and Vistoli G. 2002. Modeling of binding modes and inhibition mechanism of some natural ligands of farnesyl transferase using molecular docking. J. Med. Chem. **45**: 1460–1465.
14. Hara M., Akasaka K., Akinaga S., Okabe M., Nakano H., Gomez R., Wood D., Uh M., and Tamanoi F. 1993. Identification of Ras farnesyltransferase inhibitors by microbial screening. Proc. Natl. Acad. Sci. U S A. **90**: 281–285.
15. Arenz C., Thutewohl M., Block O., Waldmann H., Altenbach H.J., and Giannis A. 2001. Manumycin A and its analogues are irreversible inhibitors of neutral sphingomyelinase. Chembiochem. **2**: 141–143.
16. Ito T., Kawata S., Tamura S., Igura T., Nagase T., Miyagawa J.I., Yamazaki E., Ishiguro H., and Matasuzawa Y. 1996. Suppression of human pancreatic cancer growth in BALB/c nude mice by manumycin, a farnesyl:protein transferase inhibitor. Jpn. J. Cancer Res. **87**: 113–116.
17. Kainuma O., Asano T., Hasegawa M., Kenmochi T., Nakagohri T., Tokoro Y., and Isono K. 1997. Inhibition of growth and invasive activity of human pancreatic cancer cells by a farnesyltransferase inhibitor, manumycin. Pancreas. **15**: 379–83.

18. Yeung S.C., Xu G., Pan J., Christgen M., and Bamiagis A. 2000. Manumycin enhances the cytotoxic effect of paclitaxel on anaplastic thyroid carcinoma cells. Cancer Res. **60**: 650–656.
19. She M., Pan I., Sun L., and Yeung S.C. 2005. Enhancement of manumycin A-induced apoptosis by methoxyamine in myeloid leukemia cells. Leukemia. **19**: 595–602.
20. Frassanito M.A., Cusmai A., Piccoli C., and Dammacco F. 2002. Manumycin inhibits farnesyltransferase and induces apoptosis of drug-resistant interleukin 6-producing myeloma cells. Br. J. Haematol. **118**: 157–165.
21. Zhou J.M., Zhu X.F., Pan Q.C., Liao D.F., Li Z.M., and Liu Z.C. 2003. Manumycin inhibits cell proliferation and the Ras signal transduction pathway in human hepatocellular carcinoma cells. Int. J. Mol. Med. **11**: 767–771.
22. Zhou J.M., Zhu X.F., Pan Q.C., Liao D.F., Li Z.M., and Liu Z.C. 2003. Manumycin induces apoptosis in human hepatocellular carcinoma HepG2 cells. Int. J. Mol. Med. **12**: 955–959.
23. Pan J., Xu G., and Yeung S.C. 2001. Cytochrome c release is upstream to activation of caspase-9, caspase-8, and caspase-3 in the enhanced apoptosis of anaplastic thyroid cancer cells induced by manumycin and paclitaxel. J. Clin. Endocrinol. Metab. **86**: 4731–4740.
24. Yang H.L., Pan J.X., Sun L., and Yeung S.C. 2003. p21 Waf-1 (Cip-1) enhances apoptosis induced by manumycin and paclitaxel in anaplastic thyroid cancer cells. J. Clin. Endocrinol. Metab. **88**: 763–772.
25. Pan J., Huang H., Sun L., Fang B., and Yeung S.C. 2005. Bcl-2-associated X protein is the main mediator of manumycin a-induced apoptosis in anaplastic thyroid cancer cells. J. Clin. Endocrinol. Metab. **90**: 3583–3591.
26. She M., Yang H., Sun L., and Yeung S.C. 2006. Redox control of manumycin A-induced apoptosis in anaplastic thyroid cancer cells: involvement of the xenobiotic apoptotic pathway. Cancer Biol. Ther. **5**: 275–280.
27. Pan J., She M., Xu Z.X., Sun L., and Yeung S.C. 2005. Farnesyltransferase inhibitors induce DNA damage via reactive oxygen species in human cancer cells. Cancer Res. **65**: 3671–3681.
28. Pan J., Chen B., Su C.H., Zhao R., Xu Z.X., Sun L., Lee M.H., and Yeung S.C. 2008. Autophagy induced by farnesyltransferase inhibitors in cancer cells. Cancer Biol. Ther. **7**: 1679–1684.
29. Xu G., Pan J., Martin C., and Yeung S.C. 2001. Angiogenesis inhibition in the in vivo antineoplastic effect of manumycin and paclitaxel against anaplastic thyroid carcinoma. J. Clinical Endocrinol. Metab. **86**: 1769–1777.
30. Yeung S.C., She M., Yang H., Pan J., Sun L., and Chaplin D. 2007. Combination chemotherapy including combretastatin A4 phosphate and paclitaxel is effective against anaplastic thyroid cancer in a nude mouse xenograft model. J. Clinical Endocrinol. Metab. **92**: 2902–2909.
31. Vigushin D.M., Mirsaidi N., Brooke G., Sun C., Pace P., Inman L., Moody C.J., and Coombes R.C. 2004. Gliotoxin is a dual inhibitor of farnesyltransferase and geranylgeranyltransferase I with antitumor activity against breast cancer in vivo. Med. Oncol. **21**: 21–30.
32. Pan J., and Yeung S.C. 2005. Recent advances in understanding the antineoplastic mechanisms of farnesyltransferase inhibitors. Cancer Res. **65**: 9109–9112.
33. Crespo N.C., Ohkanda J., Yen T.J., Hamilton A.D., and Sebti S.M. 2001. The farnesyltransferase inhibitor, FTI-2153, blocks bipolar spindle formation and

chromosome alignment and causes prometaphase accumulation during mitosis of human lung cancer cells. J. Biol. Chem. **276**: 16161–16167.

34. Liu Ax., Cerniglia G.J., Bernhard E.J., and Prendergast G.C. 2001. RhoB is required to mediate apoptosis in neoplastically transformed cells after DNA damage. Proc. Natl. Acad. Sci. U S A. **98**: 6192–6197.
35. Wherlock M., Gampel A., Futter C., and Mellor H. 2004. Farnesyltransferase inhibitors disrupt EGF receptor traffic through modulation of the RhoB GTPase. J. Cell Sci. **117**: 3221–3231.
36. Basso A.D., Mirza A., Liu G., Long B.J., Bishop W.R., and Kirschmeier P. 2005. The farnesyl transferase inhibitor (FTI) SCH66336 (lonafarnib) inhibits Rheb farnesylation and mTOR signaling. Role in FTI enhancement of taxane and tamoxifen anti-tumor activity. J. Biol. Chem. **280**: 31101–31108.
37. Pan J., Chen B., Su C.H., Zhao R., Xu Z.X., Sun L., Lee M.H., and Yeung S.C. 2008. Autophagy induced by farnesyltransferase inhibitors in cancer cells. Cancer Biol. Ther. **7**: 1679–1684.
38. Pan J., Song E., Cheng C., Lee M.H., and Yeung S.C. 2009. Farnesyltransferase inhibitors-induced autophagy: alternative mechanisms? Autophagy **5**: 129–131.
39. Karp J.E., and Lancet J.E. 2007. Development of farnesyltransferase inhibitors for clinical cancer therapy: focus on hematologic malignancies. Cancer Invest. **25**: 484–494.
40. Braun T., and Fenaux P. 2008. Farnesyltransferase inhibitors and their potential role in therapy for myelodysplastic syndromes and acute myeloid leukaemia. Br. J. Haematol. **141**: 576–586.
41. Adjei A.A. 2006. Farnesyltransferase inhibitors. Update Cancer Ther. **1**: 17–23.
42. Fojo T. 2008. Commentary: Novel therapies for cancer: why dirty might be better. Oncologist. **13**: 277–283.

17

THE USE OF RNA AND CpG DNA AS NUCLEIC ACID-BASED THERAPEUTICS

JÖRG VOLLMER
Coley Pharmaceutical GmbH—A Pfizer Company, Düsseldorf, Germany

TOLL-LIKE RECEPTOR 9 AND CpG DNA SEQUENCE RECOGNITION

Toll-like receptors (TLRs) recognize bacterial and viral pathogenic ligands, and stimulate innate immune responses to fight against invading microorganisms. Certain pathogenic ligands are shared by specific classes of bacteria and viruses, making it possible to respond to specific pathogen classes. Three of the TLRs are stimulated by natural single-stranded nucleic acids and are localized in intracellular endolysosomal compartments: Single-stranded viral RNA is recognized by TLR7 and TLR8 (1, 2), and TLR9 senses bacterial and viral single-stranded DNA (3–6). In contrast, other TLRs are expressed on the cell surface and recognize bacterial flagellin (TLR5) or interact with lipopolysaccharide (LPS) derived from gram-negative bacteria (TLR4) (7). A distinct compartmentalization of these receptors is an important factor for the ability to mount specific innate immune responses to the tremendous variety of potentially harmful intracellular and extracellular pathogens.

Innate immune stimulation by bacterial DNA can be mimicked, at least in part, by short single-stranded synthetic oligodeoxynucleotides (ODNs) (5, 8–10). Phosphorothioate (PS)-modified ODNs are usually used as synthetic TLR9 ligands because ODNs with a "natural" phosphodiester (PO) backbone

Emerging Cancer Therapy: Microbial Approaches and Biotechnological Tools, Edited by Arsénio M. Fialho and Ananda M. Chakrabarty

have low stability and are not readily taken up by target cells (11). Detailed studies on the influence of sequence, sugar, base, or backbone modifications on immune modulation identified the hexameric CpG DNA motif 5′-GTCGTT-3′ preferentially located near the 5′ end being responsible for strongest TLR9 stimulation of human immune cells (9, 10, 12). The first identified ODNs with such a nucleotide composition were termed B-Class due to their capacity to activate B cells (10). TLR9 is preferentially expressed in B cells, as well as in a subclass of dendritic cells, the so-called plasmacytoid dendritic cells (pDCs) (13, 14). pDCs are unique in their ability to respond to infection with pathogens by secreting large amounts of type I interferons. B-Class CpG ODN immune stimulation results in a number of effects characterized by strong B cell stimulation and production of Th-1 and Th-1-like cytokines, as well as upregulation of costimulatory and major histocompatibility complex (MHC) molecules on professional antigen-presenting cells (APCs) (15). However, compared to other CpG ODN classes, B-Class CpG ODNs stimulate moderate amounts of type I interferons from pDCs. Very high levels of interferons (e.g., IFN-α) are stimulated by A-Class CpG ODNs, which contain G-rich PS 5′ and 3′ sequences that surround a palindromic PO CpG motif (16–18). Due to the G-rich ends, A-Class ODNs form higher ordered structures (G-tetrads) that contribute to enhanced stability, increased endosomal uptake, and are the major contributors to the strong pDC interferon production (19–21). In contrast, the CpG C-Class combines the characteristics of the A- and B-Classes (17, 21, 22), consists of 5′ CpG motifs linked to palindromic GC-rich 3′ regions, and stimulates strong IFN-α production and B cell stimulation. A fourth CpG ODN class is the P-Class: P-Class CpG ODNs contain two palindromic sequences at their 5′ and 3′ ends, which enable them to form concatamers, multimeric units, where each molecule is bound via Watson–Crick base pairing to a second and a third palindrome (23). Double-palindromic P-Class ODNs induce stronger type I interferon secretion compared to the C-Class by retaining a B cell stimulatory capacity similar to the B-Class and, upon *in vivo* application, lead to strongest type I interferon production. In addition, P-Class ODNs can induce remarkable IFN-α production in peripheral blood mononuclear cells (PBMCs) from HIV-1-infected subjects, and the impaired phenotype and function of pDCs in HIV-1 patients may be improved, in particular, upon treatment with P-Class CpG ODNs (24).

CpG motif-mediated binding of immune stimulatory ODNs to TLR9 results in intracellular signaling (25, 26). The TLRs are type I integral membrane glycoproteins, which contain leucine-rich repeats (LRR) in their extracellular domain and a cytoplasmic Toll/interleukin-1 receptor (TIR) signaling domain (27). Binding of the ligand induces the formation of TLR9 homodimers or heterodimers, as was demonstrated for TLR4 or TLR2/6 (28–30). However, the formation of TLR9 dimers appears not to be sufficient for efficient cellular activation. Only conformational changes in the TLR9 cytoplasmic signaling domain upon binding of CpG DNA to TLR9 dimers result in efficient signaling (31). The conformational changes probably result in the recruitment of signaling molecules, for example, the adaptor molecule MyD88, which initiates

CpG-dependent effects via signal-transducing proteins such as members of the interleukin (IL)-1 receptor-associated kinase (IRAK) family, mitogen-activated kinases (MAPK), or IFN regulatory factors (IRFs) (32). The resulting activation of nuclear factor-κB (NF-κB) transcription factors converts the signaling cascade into cytokine production or expression of costimulatory molecules in human B cells and pDCs (15, 33). The adaptor molecule MyD88 is central to TLR9 signaling and is as well involved in TLR9-mediated type I IFN production in pDCs (34). Type I interferon gene expression involves the formation of complexes with members of the IRAK family, tumor necrosis factor (TNF) receptor-associated factor 6 (TRAF6), and interferon regulatory factor 7 (IRF-7) (35, 36).

TLR7/8 AND RNA SEQUENCE RECOGNITION

Single-stranded RNA from pathogens such as RNA viruses is recognized by human TLR7 and TLR8, and murine TLR7. Single-stranded RNA sequences function as physiological pathogen-derived ligands for TLR7 and TLR8 (1, 37, 38). Certain synthetic antiviral guanosine derivatives and imidazoquinolines were also demonstrated to stimulate TLR7- and TLR8-dependent signaling (39–41). Signaling induced via TLR7 and TLR8 results in innate immune events with some similarities to TLR9 (42). Receptor stimulation causes the production of type I interferons (TLR7 only), Th-1, and proinflammatory cytokines, but lacks strong induction of Th-2 cytokines (38, 43, 44). Similar effects are observed upon intravenous injection of lipid-encapsulated oligoribonucleotides (ORNs) in mice, resulting in the production of Th-1 and proinflammatory cytokines (38, 45, 46). In contrast to TLR9, the cellular and tissue expression of human TLR7 and TLR8 appears to be broader (reviewed in Reference 47). TLR7 and TLR9 are both expressed in pDCs and B cells; TLR8 is observed in myeloid cells with the strongest TLR8 RNA expression in human monocytes. Sequence-specific recognition of single-stranded RNA via human and murine TLR7 was demonstrated for sequences derived from single-stranded RNA viruses, such as HIV, Sendai virus, or influenza virus (1, 2, 48–50), and synthetic RNA sequences derived from these viruses or alphaviruses were demonstrated to induce TLR7 and/or TLR8 activation (1, 2, 51). In addition, self-RNA derived from U1 small nuclear RNA (snRNA) containing U and G nucleotides (52, 53), and certain siRNA sequences are able to stimulate TLR7/8 signaling (46, 54). A search for potential short single-stranded RNA sequence motifs identified specific GU-rich 4mer sequences like UUGU, GUUU, or UGUU to activate human TLR7/8 by inducing IFN-α and proinflammatory cytokines (38, 51). In contrast, AU-rich 4mer sequences like UAUA, AUAU, or UUAU were found to induce strongest TNF-α production by lacking substantial IFN-α secretion, and revealed target cell and receptor selectivity by stimulating monocytes or myeloid dendritic cells (mDCs) via TLR8, but not pDCs via TLR7 (51). The observed selectivity for very similar AU- versus GU-rich RNA sequences may have evolutionary reasons: Both TLRs lie in close vicinity to each other on the X chromosome, suggesting that they did arise from a tandem duplication of an ancestral gene (55–57).

HOMEOSTASIS OF TLR RESPONSES: HOW TO DISTINGUISH FOREIGN FROM SELF

Immune responses stimulated upon TLR engagement must be tightly controlled to avoid excessive stimulation by self nucleic acids. One factor enabling TLR9 to discriminate between self and foreign DNA is the higher frequency and presence of unmethylated CpG dinucleotides in specific sequence contexts in pathogen compared to mammalian DNA (15, 58). However, self-DNA containing C5-methylated cytosines in principle still has the capacity to stimulate TLR9 signaling (59). Therefore, other mechanisms must have evolved to limit or avoid immune stimulation by self nucleic acids to protect from the potential induction of autoimmune diseases (60). One of these mechanisms is the expression of TLRs 7, 8, and 9 at locations at which self-RNA and self-DNA are not expected under physiological conditions, or at which they induce only specific immune effects. TLR9 is only detected in intracellular compartments where self nucleic acids rarely can be found. Furthermore, the CpG A- and C-Classes localize to different intracellular compartments than the B-Class CpG ODN (61): They trigger IRF-7-mediated intracellular signaling pathways from early endosomes, resulting in strong IFN-α induction, whereas the B-Classes mainly stimulate NF-κB-mediated signaling from late endosomes, leading to strong B cell activation. Another factor contributing to the modulation of TLR9 signaling pathways appears to be compartmental retention of CpG ODNs (36, 61). CpG A-Class ODNs are retained longer in endosomal compartments together with the MyD88–IRF-7 signaling complex compared to B-Class ODNs, probably due to the higher ordered structure of A-Class ODNs. In addition, different mechanisms appear to account for the specific intracellular endosomal localization of and signaling via TLR9. For example, TLR9 trafficks directly to endosomes due to specific structural features (62, 63). Even more importantly, it was found that the ectodomain of TLR9 is cleaved only upon lysosomal proteolysis to result in a truncated receptor form that appears to be functional to recruit MyD88 upon ligand binding to result in intracellular signaling (64, 65).

Similar observations were made for self and foreign RNAs (66). Specific natural RNA modifications abrogate RNA innate immune effects via TLRs (67–70) and appear more prevalent in vertebrate RNAs than in virus RNAs (68). In addition, the expression of TLR7 and TLR8 is confined to intracellular endolysosomal compartments (30, 54, 66, 71) to allow for discrimination between self and foreign RNAs. Unprotected self-RNA that can be found, for example, in the interstitial fluid or on the skin surface is rapidly degraded by RNAses (72), making it most difficult to reach intracellular compartments. However, pathogen single-stranded RNA is able to reach TLR-containing compartments, for example, after the transport of viral replication intermediates by autophagy or during virus budding (60).

TLRs 7, 8, AND 9 SINGLE NUCLEOTIDE POLYMORPHISMS AND SUSCEPTIBILITY TO ATOPIES, INFECTIOUS DISEASES, OR SYSTEMIC LUPUS ERYTHEMATOSUS

Single nucleotide polymorphisms (SNPs) are gene variants that can be found at a frequency of about 1% or higher. SNPs may have different effects depending on their position in a gene: They may influence protein expression, may even completely silence gene expression, or affect promoter characteristics. For example, polymorphisms were described in the TLR4 or TLR5 gene (73, 74), and association of these polymorphisms with different diseases was suggested (reviewed in References 74–76). Several recent studies also identified SNPs in the human TLR9 gene: 20 SNPs were discovered, some of which are relatively common, that can influence, for example, the expression levels of the TLR9 protein or result in higher promotor activity (77, 78). Common SNPs are at positions C-1237T in the TLR9 promoter region and A2848G in the coding region (79). In a study involving 71 subjects of different ethnicity, the C allele at –1237 was suggested to be possibly associated with asthma in Europeans (79). In contrast, a study with 32 asthmatic Japanese children and a European study with 527 volunteers could not find common TLR9 gene polymorphisms likely to be associated with atopy (80, 81). However, a recent study investigated the potential involvement of SNPs found in TLR7 and TLR8 in the etiology of asthma and related disorders (82). The authors found strong evidence for each one SNP in TLR7 and TLR8 to be associated with susceptibility to asthma or atopic dermatitis. It may be interesting to note that an association of a SNP (A1237E) to atopic dermatitis was also uncovered for TLR9 (78). The skin of patients with atopic dermatitis reacts easily to irritants and other allergens. Contact hypersensitivity is a skin allergy often observed upon skin contact with certain natural or synthetic chemicals, metals, or irritants (83). The inflammatory reaction observed during the course of this disease resembles that of the local inflammatory response to TLR agonists (84), and indeed, mice lacking TLR4 and IL-12 are completely resistant to the induction of contact hypersensitivity. Moreover, TLR9 or other TLRs appear as well to have a function during the sensitization phase to contact allergens (84).

An association of SNPs in TLR9 was also suggested for certain autoimmune diseases. In a study involving patients with Crohn's disease or ulcerative colitis, the frequency of the –1237C allele was reported to be significantly increased in patients with Crohn's disease when compared to controls (85). In addition, patients with ulcerative colitis developing chronic relapsing pouchitis after surgery showed a significant increase of the combined carriership of the CD14 –260T and TLR9 –1237C alleles (86). TLRs are involved in the control of intestinal epithelial homeostasis and protection from injury (87), so that polymorphism in the TLR9 gene could be associated with the pathogenesis of intestinal diseases like Crohn's disease. Systemic lupus erythematosus (SLE) is a complex and chronic disease. Its genetic analysis is hindered by environmental factors that can influence disease development (88). Three different studies could not correlate a certain set of polymorphisms in TLR9 with SLE (89–91). However, two groups investigating the involvement of TLR9 SNPs in SLE in Chinese and Japanese populations suggested two different SNPs to potentially increase the risk of SLE (77, 92), although the same SNPs were found not to be associated with SLE in the other studies described above (89–91). In mouse lupus models and in human SLE patients, disease has been associated with reduced clearance of apoptotic cells and increased circulating levels of nucleosomes (93–95), indicating that exposure to elevated levels of nucleosomes or snRNPs can induce autoimmunity to the nucleic acid and/or to the associated proteins. It has long been known that SLE patients have increased serum levels of IFN-α, which correlate with disease activity (60). An etiopathogenic role for the interferon overexpression is suggested by the observation that recombinant IFN-α therapy can induce SLE (reviewed in Reference 96), with pDCs as a potential source (97–99). pDCs express both TLR7 and TLR9, ligands to either of which can induce IFN-α secretion. Therefore, DNA-containing immune complexes may activate pDCs through TLR9, and RNA-containing immune complexes may activate pDCs through TLR7, inducing IFN-α secretion and disease development. Studies of TLR9-deficient mice and SLE mouse models indeed suggest that the production of autoantibodies to DNA and DNA-associated antigens is driven through TLR9 activation by endogenous DNA in immune complexes. Surprisingly, despite the lower levels of anti-DNA antibodies, TLR9-deficient MRL/Mplpr/lpr mice had increased IFN-α production and clinical autoimmune disease (100, 101). This suggests that the net effects of TLR9 activation by endogenous ligands are anti-inflammatory, preventing or controlling the development of SLE by a possible protective mechanism for constitutive TLR9 expression in immune regulation. In contrast, TLR7-deficient lupus-prone mice have reduced severity of autoimmune disease, and antagonism of TLR7 prevented autoimmune lung and kidney injury (100, 102). Taken together, these data point to a more prominent role for TLR7 in the pathogenesis of SLE and to an immune regulatory effect of TLR9.

Fewer data are available for SNPs in TLR7 or TLR8, and their potential involvement in disease development or susceptibility. Interestingly, two studies described the involvement of TLR7 or TLR8 SNPs in the clinical course of HIV disease. SNP G11L in TLR7 was suggested to result in a more severe disease course, and this SNP resulted in significantly less IFN-α production following TLR7 activation (103). IFN-α possesses antiviral and immune modulatory activity, and is an important innate immune response to viral infections. Therefore, it is possible that a defective TLR7 signaling results in a weaker immune response to pathogen infections. Moreover, two SNPs in TLR9

(A1635G and G1174A) were suggested to correlate with higher viral load and more rapid HIV progression (104, 105). In contrast, a SNP in TLR8 (A1G) appears to have a protective effect for HIV disease progression. TLR8 activation does not result in IFN-α production due to its cellular expression (51), but in strong secretion of proinflammatory cytokines, so that it may be likely that a decrease in such cytokines would result in a net positive effect on disease progression.

A limitation to most of these studies is the relatively small number of blood donors and the complex experimental setup. For example, functional analysis of the CpG ODN-mediated IFN-α production from peripheral blood cells of 220 healthy blood donors did not indicate any influence of TLR9 SNPs on *in vitro* CpG-dependent type I IFN production, although interindividual differences within the strength of the IFN-α response could be observed (81). In summary, TLRs and TLR SNPs can be involved in disease susceptibility or the course of diseases, and caution may be taken when TLR agonists or antagonists are used as therapies.

CpG DNA MOTIFS AND BEYOND

Optimal stimulation of TLR9 with synthetic PS ODNs does require an intact CpG motif. For example, C5 modifications at cytosines in the CpG dinucleotides are not well tolerated in the interaction with TLR9. Although several other chemically modified CpG ODNs can stimulate relatively strong stimulation of B cell responses, most of them lack efficient type I IFN production from human pDCs (106). Even certain ODNs lacking CpG dinucleotides can induce B cell activation (107). Although such non-CpG ODNs do not contain stimulatory CpG dinucleotides or CpG motifs, they appear to still require specific sequence compositions such as a thymidine-rich sequence. A non-CpG motif, PyNTTTTGT, was previously reported that is a strong stimulator of B cell responses (108). An ODN containing this non-CpG motif was also demonstrated to increase the Th-1-associated immunoglobulin IgG2b subclass more than the other subclasses in a primate hepatitis B vaccine study (109) and was shown to induce Ig secretion by B cells and T cell proliferation *in vivo* in a murine foot-and-mouth disease DNA vaccine adjuvant study (110). However, although the PyNTTTTGT motif was suggested to stimulate Th-1 antibody production, very similar non-CpG PS ODNs containing a 5′-TC motif in a thymidine-rich background did not stimulate Th-1 cytokines, but induced a Th-2-biased antibody response in a tetanus toxoid vaccine adjuvant mouse model (107). Additional studies also demonstrated that non-CpG ODNs bias Ag-specific antibody responses to Th-2-type responses in contrast to CpG ODNs that stimulate Th-1-type Ag-specific responses (109, 111, 112). Common to single-stranded non-CpG and CpG ODNs is their localization to intracellular endosomal compartments where TLR9 resides. However, such single-stranded ODNs, but also long single-stranded DNA, are not able to stimulate nucleic acid receptors found in the cytoplasm (113). In contrast, cytoplasmic double-stranded DNA is sensed by the innate immune system. A candidate receptor (DNA-dependent activator of IRFs, DAI) has been implicated in the recognition of long double-stranded DNA (114). Most recently, the PYHIN (pyrin and HIN domain-containing protein) family member absent in melanoma 2 (AIM2) was identified as a receptor for cytosolic double-stranded DNA, which regulates caspase-1 (115).

Two more recent reports suggest non-DNA molecules to modulate TLR(9)-mediated activation of innate immune responses. Thymosin alpha 1, a naturally occurring peptide from the thymus, primarily augments T cell function, which was recently attributed, at least in part, to the stimulation of human TLR9 signaling (116). However, at least, purified human B cells expressing TLR9 do not respond by cytokine secretion to stimulation with thymosin alpha 1 (117). However, innate immune modulation in the presence of thymosin alpha 1 may occur under certain conditions, for example, during fungal infection, contributing to the eradication of the invading pathogen. Another molecule that was described to be a non-DNA ligand is hemozoin (HZ), a detoxification product of heme molecules persisting in the food vacuoles of *Plasmodium falciparum* parasites (reviewed in Reference 118). Blood-stage schizonts and schizont extracts activate human pDCs as well as murine DCs through MyD88- and TLR9-dependent pathways (119). HZ activation of murine immune cells was suggested to depend on the presence of MyD88 and to be mediated by TLR9 (120). However, natural HZ is coated with malarial DNA, and only malarial DNA but not purified HZ activated TLR9 when the DNA was targeted directly to the endosome (121). Therefore, the malarial DNA is highly proinflammatory, with the potential to induce cytokinemia and fever during disease.

In contrast to potential non-DNA TLR9 ligands, a recent report demonstrated that human bacterial pathogens can target TLR signaling in order to survive and spread (122). A protein found in an uropathogenic *Escherichia coli* strain was shown to bind MyD88 (also involved in TLR9 signaling) directly and limit innate immune activation induced by this bacterial strain. Similar TIR domain-containing proteins are also present in certain other *E. coli* and *Brucella* strains. Such proteins may give rise for developing new treatments for bacterial infections or inflammatory diseases involving TLR signaling (123).

Not only proteins were suggested to probably be suitable for certain anti-inflammatory therapies. Inhibitory or suppressive DNA oligonucleotides containing defined G-rich sequence motifs or “antimalarial” small molecules such as chloroquine can also block TLR9 immune stimulatory effects with potential therapeutic implications (60). In addition, suppressive ODNs can inhibit TLR7- and TLR8-mediated responses stimulated by self-RNA or viral RNA. Therefore, suppressive ODNs or small molecules as TLR9 antagonists appear to be promising candidates for therapeutic development in certain inflammatory autoimmune diseases.

CELL SURFACE AND CYTOPLASMIC RNA RECEPTORS: TLR3 AND RETINOIC ACID-INDUCIBLE PROTEIN-I-LIKE RECEPTORS

TLR7, TLR8, and TLR9 are expressed in endolysosomal compartments and respond to stimulation by pathogen or synthetic single-stranded RNA or DNA. However, viral RNA can be as well found in the cytoplasm, and therefore, the innate immune system was developed to recognize such cytoplasmic pathogen molecules in this cellular compartment. Receptors in the cytoplasm responding to viral double-stranded RNA (dsRNA) include dsRNA-dependent protein kinase (PKR) (124), 2′-5′-oligoadenylate synthetase (OAS) (125), or receptors containing the DEXD/H box helicase motifs, the retinoic acid-inducible protein I (RIG-I)-like receptors (RLR), RIG-I, melanoma differentiation-associated protein 5 (MDA5), and LGP2 (126). The natural RIG-I-mediated antiviral response targets the family of negative strand viruses, including vesicular stomatitis virus (VSV) influenza, and rabies virus, whereas MDA5 responds to infection with picornaviruses, such as polio virus or rhino virus. Signaling induced upon stimulation of these receptors via IFN-β promotor stimulator-1 (IPS-1) activates transcription factors, including IRF3, IRF7, and NF-κB. The resulting effects are the induction of proinflammatory cytokines and type I interferon, the latter mainly from monocytes, which usually do not respond by interferon production to stimulation with nucleic acids (38, 113). Several different natural and synthetic nucleic acids were suggested to constitute the ligand for RIG-I and MDA5, although specific sequence motifs as identified for TLR7, TLR8, and TLR9 were not yet described. RIG-I appears to be promiscuous in its nucleic acid recognition, and different ligands stimulating RIG-I signaling were described: single-stranded RNA with a 5′-triphosphate that may contain A- and U-rich 3′ sequences, double-stranded RNA with a 5′-triphosphate, short blunt-end double-stranded RNA (23–30 bp), and double-stranded RNA of intermediate length, both without a 5′-triphosphate (126, 127). However, the presence of a triphosphate group at the 5′ end of the RNA is a feature that is employed by the innate immune system to distinguish between foreign and self-RNA (113, 128). Indeed, processing of the 5′-triphosphate can be used by viruses to escape immune recognition (129). In contrast, MDA5 was reported to be activated by long double-stranded RNA (>2000 bp) (127, 130). As for the TLRs, RIG-I and MDA5 also can be stimulated by synthetic nucleic acids, for example, short synthetic double-stranded RNA with a 5′-triphosphate (131), which appears to be a much stronger ligand for RIG-I as compared to such double-strands without a 5′-triphosphate (126), and long synthetic double-stranded RNA such as polyI:C (130).

From all the TLRs recognizing nucleic acids, only one is responding to double-stranded RNA. TLR3 is not only expressed in endolysosomal compartments where the other nucleic acid-sensing TLRs can be found, but TLR3 can also be expressed on the cell surface (32, 132). Stimulation of signaling via TLR3 results in a variety of innate immune effects, most prominently, the strong production of type I interferons (32, 42). TLR3 can respond to natural long viral double-stranded RNA as shown for dsRNA viruses such as reovirus (133) but as well to synthetic long (polyI:C and polyA:U) and short double-stranded RNA (134, 135). Interestingly, siRNAs can bind to TLR3 in a sequence-independent manner and can stimulate cytokine production such as IL-12 and IFN-γ, which may, in some instances, account for the antiviral and antitumor effects observed with some siRNAs in animal models (136).

FROM PRECLINICAL STUDIES TO CLINICAL APPLICATION

CpG ODN stimulation results in orchestrated immune responses via TLR9-dependent signaling, and CpG ODNs are attractive drugs to trigger Th-1 effects and stimulate efficient antigen-specific B cell and T cell responses. Animal models in infectious diseases, cancer, and asthma and allergy have demonstrated that CpG ODNs are capable to enhance the immunogenicity of vaccines, elicit efficient antiviral responses, or eradicate tumors (reviewed in References 137–140). In human clinical trials, similar effects were observed (141–145). For example, the combination of B-Class CpG ODNs with commercial hepatitis B vaccine resulted in increased rates of seroconversion and higher antigen-specific antibody titers in healthy human volunteers or immunocompromised HIV-infected patients. In addition, a B-Class CpG ODN severely accelerated seroconversion in a randomized controlled trial in healthy volunteers when used as an adjuvant to the approved anthrax vaccine. The peak titer of toxin-neutralizing antibody was statistically significantly increased in the subjects receiving the B-Class ODN and was achieved already at day 22, more than 20 days earlier than control subjects (47). Furthermore, the proportion of subjects achieving a strong IgG response to the anthrax protective antigen was increased from 61% to 100%. Additionally, CpG ODNs can promote a strong tumor antigen-specific CD8 T cell response in cancer patients (146). In a Phase I tumor vaccine trial, a 1 mg dose of a B-Class CpG ODN was used as adjuvant to recombinant MAGE-3 tumor antigen for triweekly vaccination in six patients with metastatic melanoma. Two stable diseases and two partial responses beginning after 7–10 vaccinations were observed and did last at least 1 year (147). This CpG ODN is momentarily incorporated into a MAGE-A3 melanoma vaccine that has progressed into Phase III clinical trials for the treatment of patients with early-stage completely resected non-small cell lung cancer whose tumors express the antigen (47). Furthermore, additional clinical studies with CpG ODNs as monotherapy or in combination with antibody therapy or chemotherapy in basal cell carcinoma, melanoma, cutaneous T cell lymphoma, non-Hodgkin's lymphoma, renal cell cancer, breast cancer, or non-small cell lung cancer strongly suggested their ability to induce antitumor effects in cancer patients (148).

Synthetic TLR7/8 ORN agonists complexed to liposomal or similar formulations were demonstrated to stimulate strong innate immune responses *in vitro* and *in vivo* (1, 2, 44, 45). Immune stimulatory ORNs are efficient vaccine adjuvants in mouse models and induce potent innate and adaptive immune responses (45, 149, 150). Three mouse vaccine studies demonstrated that the addition of the model antigen ovalbumin (OVA) to immune stimulatory single-stranded ORNs encapsulated in DOTAP or polylactide microparticles induces enhanced levels of antigen-specific antibodies. However, whereas in two of these studies, humoral responses characterized by OVA-specific antibodies of the IgG1 type, suggesting a Th-2 immune response, were observed (149, 151), another study demonstrated that an immune stimulatory ORN with a stabilized PS backbone triggers the stimulation of an IgG2a-biased antibody response similar to TLR9 CpG ODN agonists (45). Indeed, in this latter study, PS ORNs were shown to be more efficient to induce Th-1 cytokines and to more efficiently activate DCs compared to the same unmodified PO RNA sequence, which may explain the differences in the induction of Th-1 or Th-2 antigen responses observed in these vaccine studies. However, all these studies demonstrated increased numbers of antigen-specific IFN-γ-producing T cells or stronger antigen-specific cytotoxic T cell (CTL) responses in the presence of the ORN adjuvant compared to mice immunized with lipid-encapsulated antigen alone. In other studies, vaccination against the hepatitis B virus in the presence of immune stimulatory RNA or a GU-rich ORN as well resulted in humoral and cellular adaptive immune responses (152, 153). Single-stranded immune stimulatory ORNs also can have antitumor effects when applied alone or together with chemotherapy in mouse tumor models (150; A.P. Vicari and J. Vollmer, unpublished observations). For example, in a mouse tumor model using a glioma tumor cell line, intratumoral or even distant injections of protamine-protected RNA resulted in anti-tumor immunity (150). All together, these studies suggest that RNA immune modulators can be used as potent and safe TLR agonists for tumor immune therapy or as adjuvants for efficient vaccination.

Other RNA-based immune modulators are also used in preclinical or clinical studies in tumor and other indications. TLR3 or MDA5 is stimulated by the synthetic double-stranded polyI:C mimicking viral RNA, and several studies showed that polyI:C or analogues are effective as vaccine adjuvant (135, 154), for example, in antitumor vaccination. In addition, in a recent report, polyI:C was demonstrated to be an effective vaccine adjuvant in rhesus macaques when applied with human papillomavirus (HPV)16 capsomeres supporting the induction of Th-1 and antibody responses (155). Short siRNAs with a 5′-triphosphate were also demonstrated to induce immune effects via RIG-I. The combination of gene silencing and immune stimulatory properties in one double-stranded RNA molecule was shown to be beneficial *in vivo*. In a mouse lung tumor model, a siRNA directed to the tumor target Bcl-2 and containing a 5′-triphosphate stimulating RIG-I resulted not only in innate immune activation, but also in apoptosis of tumor cells and antitumor effects (131).

ACKNOWLEDGMENT

I want to thank Silke Fähndrich for excellent assistance in preparing the manuscript.

REFERENCES

1. Heil F., Hemmi H., Hochrein H., Ampenberger F., Kirschning C., Akira S., Lipford G., Wagner H., and Bauer S. 2004. Species-specific recognition of single-stranded RNA via toll-like receptor 7 and 8. Science. **303**: 1526–1529.
2. Diebold S.S., Kaisho T., Hemmi H., Akira S., and Reis e Sousa C. 2004. Innate antiviral responses by means of TLR7-mediated recognition of single-stranded RNA. Science. **303**: 1529–1531.
3. Hemmi H., Takeuchi O., Kawai T., Kaisho T., Sato S., Sanjo H., Matsumoto M., Hoshino K., Wagner H., Takeda K., and Akira S. 2000. A Toll-like receptor recognizes bacterial DNA. Nature. **408**: 740–745.
4. Sun S., Beard C., Jaenisch R., Jones P., and Sprent J. 1997. Mitogenicity of DNA from different organisms for murine B cells. J. Immunol. **159**: 3119–3125.
5. Krieg A.M., Yi A.K., Matson S., Waldschmidt T.J., Bishop G.A., Teasdale R., Koretzky G.A., and Klinman D.M. 1995. CpG motifs in bacterial DNA trigger direct B-cell activation. Nature. **374**: 546–549.
6. Krug A., Luker G.D., Barchet W., Leib D.A., Akira S., and Colonna M. 2004. Herpes simplex virus type 1 activates murine natural interferon-producing cells through toll-like receptor 9. Blood. **103**: 1433–1437.
7. Krieg A.M. 2003. CpG motifs: the active ingredient in bacterial extracts? Nat. Med. **9**: 831–835.
8. Pisetsky D.S., and Reich C.F., 3rd. 1998. The influence of base sequence on the immunological properties of defined oligonucleotides. Immunopharmacology. **40**:n 199–208.
9. Hartmann G., and Krieg A.M. 2000. Mechanism and function of a newly identified CpG DNA motif in human primary B cells. J. Immunol. **164**: 944–953.
10. Hartmann G., Weeratna R.D., Ballas Z.K., Payette P., Blackwell S., Suparto I., Rasmussen W.L., Waldschmidt M., Sajuthi D., Purcell R.H., Davis H.L., and Krieg A.M. 2000. Delineation of a CpG phosphorothioate oligodeoxynucleotide for activating primate immune responses in vitro and in vivo. J. Immunol. **164**: 1617–1624.
11. Sester D.P., Naik S., Beasley S.J., Hume D.A., and Stacey K.J. 2000. Phosphorothioate backbone modification modulates macrophage activation by CpG DNA. J. Immunol. **165**: 4165–4173.
12. Bauer S., Kirschning C.J., Häcker H., Redecke V., Hausmann S., Akira S., Wagner H., and Lipford G.B. 2001. Human TLR9 confers responsiveness to bacterial DNA via species-specific CpG motif recognition. Proc. Natl. Acad. Sci. U S A. **98**: 9237–9242.
13. Hornung V., Rothenfusser S., Britsch S., Krug A., Jahrsdörfer B., Giese T., Endres S., and Hartmann G. 2002. Quantitative expression of toll-like receptor 1-10 mRNA in cellular subsets of human peripheral blood mononuclear cells and sensitivity to CpG oligodeoxynucleotides. J. Immunol. **168**: 4531–4537.

14. Kadowaki N., Ho S., Antonenko S., Malefyt R.W., Kastelein R.A., Bazan F., and Liu Y.J. 2001. Subsets of human dendritic cell precursors express different toll-like receptors and respond to different microbial antigens. J. Exp. Med. **194**: 863–869.
15. Krieg A.M. 2002. CpG motifs in bacterial DNA and their immune effects. Annu. Rev. Immunol. **20**: 709–760.
16. Krug A., Rothenfusser S., Hornung V., Jahrsdörfer B., Blackwell S., Ballas Z.K., Endres S., Krieg A.M., and Hartmann G. 2001. Identification of CpG oligonucleotide sequences with high induction of IFN-alpha/beta in plasmacytoid dendritic cells. Eur. J. Immunol. **31**: 2154–2163.
17. Vollmer J., Weeratna R., Payette P., Jurk M., Schetter C., Laucht M., Wader T., Tluk S., Liu M., Davis H.L., and Krieg AM. 2004. Characterization of three CpG oligodeoxynucleotide classes with distinct immunostimulatory activities. Eur. J. Immunol. **34**: 251–262.
18. Ballas Z.K., Rasmussen W.L., and Krieg A.M. 1996. Induction of NK activity in murine and human cells by CpG motifs in oligodeoxynucleotides and bacterial DNA. J. Immunol. **157**: 1840–1845.
19. Kerkmann M., Costa L.T., Richter C., Rothenfusser S., Battiany J., Hornung V., Johnson J., Englert S., Ketterer T., Heckl W., Thalhammer S., Endres S., and Hartmann G. 2005. Spontaneous formation of nucleic acid-based nanoparticles is responsible for high interferon-alpha induction by CpG-A in plasmacytoid dendritic cells. J. Biol. Chem. **280**: 8086–8093.
20. Verthelyi D., Ishii K.J., Gursel M., Takeshita F., and Klinman D.M. 2001. Human peripheral blood cells differentially recognize and respond to two distinct CPG motifs. J. Immunol. **166**: 2372–2377.
21. Marshall J.D., Fearon K., Abbate C., Subramanian S., Yee P., Gregorio J., Coffman R.L., and Van Nest G. 2003. Identification of a novel CpG DNA class and motif that optimally stimulate B cell and plasmacytoid dendritic cell functions. J. Leukoc. Biol. **73**: 781–792.
22. Hartmann G., Battiany J., Poeck H., Wagner M., Kerkmann M., Lubenow N., Rothenfusser S., and Endres S. 2003. Rational design of new CpG oligonucleotides that combine B cell activation with high IFN-A induction in plasmacytoid dendritic cells. Eur. J. Immunol. **33**: 1633–1641.
23. Samulowitz U., Weber M., Weeratna R.D., Uhlmann E., Noll B., Krieg A.M., and Vollmer J. 2010. A novel class of immune stimulatory CpG oligodeoxynucleotides unifies high potency in type I interferon induction with preferred structural properties. In press.
24. Donhauser N., Helm M., Pritschet K., Schuster P., Ries M., Korn K., Vollmer J., and Schmidt B. 2010. Differential effects of P-class versus other CpG oligodeoxynucleotide classes on the impaired innate immunity of plasmacytoid dendritic cells in HIV type 1 infection. AIDS Res. Hum. Retroviruses **26**: 161–171.
25. Rutz M., Metzger J., Gellert T., Luppa P., Lipford G.B., Wagner H., and Bauer S. 2004. Toll-like receptor 9 binds single-stranded CpG-DNA in a sequence- and pH-dependent manner. Eur. J. Immunol. **34**: 2541–2550.
26. Latz E., Schoenemeyer A., Visintin A., Fitzgerald K.A., Monks B.G., Knetter C.F., Lien E., Nilsen N.J., Espevik T., and Golenbock D.T. 2004. TLR9 signals after translocating from the ER to CpG DNA in the lysosome. Nat. Immunol. **5**: 190–198.

27. Bell J.K., Mullen G.E., Leifer C.A., Mazzoni A., Davies D.R., and Segal D.M. 2003. Leucine-rich repeats and pathogen recognition in Toll-like receptors. Trends Immunol. **24**: 528–533.
28. Zhang H., Tay P.N., Cao W., Li W., and Lu J. 2002. Integrin-nucleated Toll-like receptor (TLR) dimerization reveals subcellular targeting of TLRs and distinct mechanisms of TLR4 activation and signaling. FEBS Lett. **532**: 171–176.
29. Kindrachuk J., Potter J.E., Brownlie R., Ficzycz A.D., Griebel P.J., Mookherjee N., Mutwiri G.K., Babiuk L.A., and Napper S. 2007. Nucleic acids exert a sequence-independent cooperative effect on sequence-dependent activation of Toll-like receptor 9. J. Biol. Chem. **282**: 13944–13953.
30. Nishiya T., and DeFranco A.L. 2004. Ligand-regulated chimeric receptor approach reveals distinctive subcellular localization and signaling properties of the Toll-like receptors. J. Biol. Chem. **279**: 19008–19017.
31. Latz E., Verma A., Visintin A., Gong M., Sirois C.M., Klein D.C., Monks B.G., McKnight C.J., Lamphier M.S., Duprex W.P., Espevik T., and Golenbock D.T. 2007. Ligand-induced conformational changes allosterically activate Toll-like receptor 9. Nat. Immunol. **8**: 772–779.
32. Kawai T., and Akira S. 2008. Toll-like receptor and RIG-I-like receptor signaling. Ann. N. Y. Acad. Sci. **1143**: 1–20.
33. Vollmer J. 2006. TLR9 in health and disease. Int. Rev. Immunol. **25**: 155–181.
34. Schnare M., Holt A.C., Takeda K., Akira S., and Medzhitov R. 2000. Recognition of CpG DNA is mediated by signaling pathways dependent on the adaptor protein MyD88. Curr. Biol. **10**: 1139–1142.
35. Kawai T., Sato S., Ishii K.J., Coban C., Hemmi H., Yamamoto M., Terai K., Matsuda M., Inoue J., Uematsu S., Takeuchi O., and Akira S. 2004. Interferon-alpha induction through Toll-like receptors involves a direct interaction of IRF7 with MyD88 and TRAF6. Nat. Immunol. **5**: 1061–1068.
36. Honda K., Yanai H., Mizutani T., Negishi H., Shimada N., Suzuki N., Ohba Y., Takaoka A., Yeh W.C., and Taniguchi T. 2004. Role of a transductional-transcriptional processor complex involving MyD88 and IRF-7 in Toll-like receptor signaling. Proc. Natl. Acad. Sci. U S A. **101**: 15416–15421.
37. Diebold S.S., Massacrier C., Akira S., Paturel C., Morel Y., and Reis e Sousa C. 2006. Nucleic acid agonists for Toll-like receptor 7 are defined by the presence of uridine ribonucleotides. Eur. J. Immunol. **36**: 3256–3267.
38. Forsbach A., Nemorin J.G., Völp K., Samulowitz U., Montino C., Müller C., Tluk S., Hamm S., Bauer S., Lipford G.B., and Vollmer J. 2007. Characterization of conserved viral leader RNA sequences that stimulate innate immunity through TLRs. Oligonucleotides. **17**: 405–417.
39. Gorden K.B., Gorski K.S., Gibson S.J., Kedl R.M., Kieper W.C., Qiu X., Tomai M.A., Alkan S.S., and Vasilakos J.P. 2005. Synthetic TLR agonists reveal functional differences between human TLR7 and TLR8. J. Immunol. **174**: 1259–1268.
40. Jurk M., Kritzler A., Schulte B., Tluk S., Schetter C., Krieg A.M., and Vollmer J. 2006. Modulating responsiveness of human TLR7 and 8 to small molecule ligands with T-rich phosphorothiate oligodeoxynucleotides. Eur. J. Immunol. **36**: 1815–1826.
41. Smits E.L., Ponsaerts P., Berneman Z.N., and Van Tendeloo V.F. 2008. The use of TLR7 and TLR8 ligands for the enhancement of cancer immunotherapy. Oncologist. **13**: 859–875.

42. Kawai T., and Akira S. 2007. Antiviral signaling through pattern recognition receptors. J. Biochem. **141**: 137–145.
43. Heil F., Ahmad-Nejad P., Hemmi H., Hochrein H., Ampenberger F., Gellert T., Dietrich H., Lipford G., Takeda K., Akira S., Wagner H., and Bauer S. 2003. The Toll-like receptor 7 (TLR7)-specific stimulus loxoribine uncovers a strong relationship within the TLR7, 8 and 9 subfamily. Eur. J. Immunol. **33**: 2987–2997.
44. Scheel B., Braedel S., Probst J., Carralot J.P., Wagner H., Schild H., Jung G., Rammensee H.G., and Pascolo S. 2004. Immunostimulating capacities of stabilized RNA molecules. Eur. J. Immunol.. **34**: 537–547.
45. Bourquin C., Schmidt L., Hornung V., Wurzenberger C., Anz D., Sandholzer N., Schreiber S., Voelkl A., Hartmann G., and Endres S. 2007. Immunostimulatory RNA oligonucleotides trigger an antigen-specific cytotoxic T-cell and IgG2a response. Blood. **109**: 2953–2960.
46. Hornung V., Guenthner-Biller M., Bourquin C., Ablasser A., Schlee M., Uematsu S., Noronha A., Manoharan M., Akira S., de Fougerolles A., Endres S., and Hartmann G. 2005. Sequence-specific potent induction of IFN-alpha by short interfering RNA in plasmacytoid dendritic cells through TLR7. Nat. Med. **11**: 263–270.
47. Vollmer J., and Krieg A.M. 2009. Immunotherapeutic applications of CpG oligodeoxynucleotide TLR9 agonists. Adv. Drug Deliv. Rev. **61**: 195–204.
48. Wang J.P., Liu P., Latz E., Golenbock D.T., Finberg R.W., and Libraty D.H.2006. Flavivirus activation of plasmacytoid dendritic cells delineates key elements of TLR7 signaling beyond endosomal recognition. J. Immunol. **177**: 7114–7121.
49. Barchet W., Krug A., Cella M., Newby C., Fischer J.A., Dzionek A., Pekosz A., and Colonna M. 2004. Dendritic cells respond to influenza virus through TLR7- and PKR-independent pathways. Eur. J. Immunol. **35**: 236–242.
50. Melchjorsen J., Jensen S.B., Malmgaard L., Rasmussen S.B., Weber F., Bowie A.G., Matikainen S., and Paludan S.R. 2005. Activation of innate defense against a paramyxovirus is mediated by RIG-I and TLR7 and TLR8 in a cell-type-specific manner. J. Virol. **79**: 12944–12951.
51. Forsbach A., Nemorin J.G., Montino C., Müller C., Samulowitz U., Vicari A.P., Jurk M., Mutwiri G.K., Krieg A.M., Lipford G.B., and Vollmer J. 2008. Identification of RNA sequence motifs stimulating sequence-specific TLR8-dependent immune responses. J. Immunol. **180**: 3729–3738.
52. Vollmer J., Tluk S., Schmitz C., Hamm S., Jurk M., Forsbach A., Akira S., Kelly K.M., Reeves W.H., Bauer S., and Krieg A.M. 2005. Immune stimulation mediated by autoantigen binding sites within small nuclear RNAs involves Toll-like receptors 7 and 8. J. Exp. Med. **202**: 1575–1585.
53. Savarese E., Chae O.W., Trowitzsch S., Weber G., Kastner B., Akira S., Wagner H., Schmid R.M., Bauer S., and Krug A. 2005. U1 small nuclear ribonucleoprotein immune complexes induce type I interferon in plasmacytoid dendritic cells via TLR7. Blood. **107**: 3229–3234.
54. Sioud M. 2005. Induction of inflammatory cytokines and interferon responses by double-stranded and single-stranded siRNAs is sequence-dependent and requires endosomal localization. J. Mol. Biol. **348**: 1079–1090.
55. Du X., Poltorak A., Wei Y., and Beutler B. 2000. Three novel mammalian toll-like receptors: gene structure, expression, and evolution. Eur. Cytokine Netw. **11**: 362–371.

56. Matsushima N., Tanaka T., Enkhbayar P., Mikami T., Taga M., Yamada K., and Kuroki Y. 2007. Comparative sequence analysis of leucine-rich repeats (LRRs) within vertebrate toll-like receptors. BMC Genomics. **8**: 124.
57. Chuang T.H., and Ulevitch R.J. 2000. Cloning and characterization of a sub-family of human toll-like receptors: hTLR7, hTLR8 and hTLR9. Eur. Cytokine Netw. **11**: 372–378.
58. Stacey K.J., Young G.R., Clark F., Sester D.P., Roberts T.L., Naik S., Sweet M.J., and Hume D.A. 2003. The molecular basis for the lack of immunostimulatory activity of vertebrate DNA. J. Immunol. **170**: 3614–3620.
59. Vollmer J., Janosch A., Laucht M., Ballas Z.K., Schetter C., and Krieg A.M. 2002. Highly immunostimulatory CpG-free oligodeoxynucleotides for activation of human leukocytes. Antisense Nucleic Acid Drug Dev. **12**: 165–175.
60. Krieg A.M., and Vollmer J. 2007. Toll-like receptors 7, 8, and 9: linking innate immunity to autoimmunity. Immunol. Rev. **220**: 251–269.
61. Honda K., Ohba Y., Yanai H., Negishi H., Mizutani T., Takaoka A., Taya C., and Taniguchi T. 2005. Spatiotemporal regulation of MyD88-IRF-7 signalling for robust type-I interferon induction. Nature. **434**: 1035–1040.
62. Leifer C.A., Kennedy M.N., Mazzoni A., Lee C., Kruhlak M.J., and Segal D.M. 2004. TLR9 is localized in the endoplasmic reticulum prior to stimulation. J. Immunol. **173**: 1179–1183.
63. Kajita E., Nishiya T., and Miwa S. 2006. The transmembrane domain directs TLR9 to intracellular compartments that contain TLR3. Biochem. Biophys. Res. Commun. **343**: 578–584.
64. Ewald S.E., Lee B.L., Lau L., Wickliffe K.E., Shi G.P., Chapman H.A., and Barton G.M. 2008. The ectodomain of Toll-like receptor 9 is cleaved to generate a functional receptor. Nature. **456**: 658–662.
65. Park B., Brinkmann M.M., Spooner E., Lee C.C., Kim Y.M., and Ploegh H.L. 2008. Proteolytic cleavage in an endolysosomal compartment is required for activation of Toll-like receptor 9. Nat. Immunol. **9**: 1407–1414.
66. Nishiya T., Kajita E., Miwa S., and DeFranco A.L. 2005. TLR3 and TLR7 are targeted to the same intracellular compartments by distinct regulatory elements. J. Biol. Chem. **280**: 37107–37117.
67. Sioud M., Furset G., and Cekaite L. 2007. Suppression of immunostimulatory siRNA-driven innate immune activation by 2'-modified RNAs. Biochem. Biophys. Res. Commun. **361**: 122–126.
68. Kariko K., Buckstein M., Ni H., and Weissman D. 2005. Suppression of RNA recognition by Toll-like receptors: the impact of nucleoside modification and the evolutionary origin of RNA. Immunity. **23**: 165–175.
69. Judge A.D., Bola G., Lee A.C., and Maclachlan I. 2006. Design of noninflammatory synthetic siRNA mediating potent gene silencing in vivo. Mol. Ther. **13**: 494–505.
70. Tluk S., Jurk M., Forsbach A., Weeratna R., Samulowitz U., Krieg A.M., Bauer S., and Vollmer J. 2009. Sequences derived from self-RNA containing certain natural modifications act as suppressors of RNA-mediated inflammatory immune responses. Int. Immunol. **21**: 607–619.
71. Barton G.M., Kagan J.C., and Medzhitov R. 2006. Intracellular localization of Toll-like receptor 9 prevents recognition of self DNA but facilitates access to viral DNA. Nat. Immunol. **7**: 49–56.

72. Probst J., Brechtel S., Scheel B., Hoerr I., Jung G., Rammensee H.G., and Pascolo S. 2006. Characterization of the ribonuclease activity on the skin surface. Genet. Vaccines Ther. **4**: 4.

73. Arbour N.C., Lorenz E., Schutte B.C., Zabner J., Kline J.N., Jones M., Frees K., Watt J.L., and Schwartz D.A. 2000. TLR4 mutations are associated with endotoxin hyporesponsiveness in humans. Nat. Genet. **25**: 187–191.

74. Hawn T.R., Wu H., Grossman J.M., Hahn B.H., Tsao B.P., and Aderem A. 2005. A stop codon polymorphism of Toll-like receptor 5 is associated with resistance to systemic lupus erythematosus. Proc. Natl. Acad. Sci. U S A. **102**: 10593–10597.

75. Schroder N.W., and Schumann R.R. 2005. Single nucleotide polymorphisms of Toll-like receptors and susceptibility to infectious disease. Lancet Infect. Dis. **5**: 156–164.

76. Youinou P. 2007. Do Toll-like receptors contribute to the pathogenesis of lupus? J Rheumatol. **34**: 1641–1642.

77. Tao K., Fujii M., Tsukumo S., Maekawa Y., Kishihara K., Kimoto Y., Horiuchi T., Hisaeda H., Akira S., Kagami S., and Yasutomo K. 2007. Genetic variations of Toll-like receptor 9 predispose to systemic lupus erythematosus in Japanese population. Ann. Rheum. Dis. **66**: 905–909.

78. Novak N., Yu C.F., Bussmann C., Maintz L., Peng W.M., Hart J., Hagemann T., Diaz-Lacava A., Baurecht H.J., Klopp N., Wagenpfeil S., Behrendt H., Bieber T., Ring J., Illig T., and Weidinger S. 2007. Putative association of a TLR9 promoter polymorphism with atopic eczema. Allergy. **62**: 766–772.

79. Lazarus R., Klimecki W.T., Raby B.A., Vercelli D., Palmer L.J., Kwiatkowski D.J., Silverman E.K., Martinez F., and Weiss S.T. 2003. Single-nucleotide polymorphisms in the Toll-like receptor 9 gene (TLR9): frequencies, pairwise linkage disequilibrium, and haplotypes in three U.S. ethnic groups and exploratory case-control disease association studies (small star, filled). Genomics. **81**: 85–91.

80. Noguchi E., Nishimura F., Fukai H., Kim J., Ichikawa K., Shibasaki M., and Arinami T. 2004. An association study of asthma and total serum immunoglobin E levels for Toll-like receptor polymorphisms in a Japanese population. Clin. Exp. Allergy. **34**: 177–183.

81. Berghöfer B., Frommer T., König I.R., Ziegler A., Chakraborty T., Bein G., and Hackstein H. 2005. Common human Toll-like receptor 9 polymorphisms and haplotypes: association with atopy and functional relevance. Clin. Exp. Allergy. **35**: 1147–1154.

82. Moller-Larsen S., Nyegaard M., Haagerup A., Vestbo J., Kruse T.A., and Borglum A.D. 2008. Association analysis identifies TLR7 and TLR8 as novel risk genes in asthma and related disorders. Thorax. **63**: 1064–1069.

83. Vollmer J. 2010. Immunotoxicity: technologies to predicting immune stimulation, a focus on nucleic acids and haptens. In: Xu J., and Urban L., eds. Predictive Toxicology in Drug Safety Assessment. Cambridge: Cambridge University Press, in press.

84. Martin S.F., Dudda J.C., Bachtanian E., Lembo A., Liller S., Dürr C., Heimesaat M.M., Bereswill S., Fejer G., Vassileva R., Jakob T., Freudenberg N., Termeer C.C., Johner C., Galanos C., and Freudenberg M.A. 2008. Toll-like receptor and IL-12 signaling control susceptibility to contact hypersensitivity. J. Exp. Med. **205**: 2151–2162.

85. Torok H.P., Glas J., Tonenchi L., Bruennler G., Folwaczny M., and Folwaczny C. 2004. Crohn's disease is associated with a toll-like receptor-9 polymorphism. Gastroenterology. **127**: 365–366.

86. Lammers K.M., Ouburg S., Morré S.A., Crusius J.B., Gionchett P., Rizzello F., Morselli C., Caramelli E., Conte R., Poggioli G., Campieri M., and Peña A.S. 2005. Combined carriership of TLR9-1237C and CD14-260T alleles enhances the risk of developing chronic relapsing pouchitis. World J. Gastroenterol. **11**: 7323–7329.

87. Rakoff-Nahoum S., Paglino J., Eslami-Varzaneh F., Edberg S., and Medzhitov R. 2004. Recognition of commensal microflora by toll-like receptors is required for intestinal homeostasis. Cell. **118**: 229–241.

88. Yu P., Musette P., and Peng S.L. 2008. Toll-like receptor 9 in murine lupus: more friend than foe! Immunobiology. **213**: 151–157.

89. De Jager P.L., Richardson A., Vyse T.J., and Rioux J.D. 2006. Genetic variation in toll-like receptor 9 and susceptibility to systemic lupus erythematosus. Arthritis Rheum. **54**: 1279–1282.

90. Demirci F.Y., Manzi S., Ramsey-Goldman R., Kenney M., Shaw P.S., Dunlop-Thomas C.M., Kao A.H., Rhew E.Y., Bontempo F., Kammerer C., and Kamboh M.I. 2007. Association study of Toll-like receptor 5 (TLR5) and Toll-like receptor 9 (TLR9) polymorphisms in systemic lupus erythematosus. J. Rheumatol. **34**: 1708–1711.

91. Hur J.W., Shin H.D., Park B.L., Kim L.H., Kim S.Y., and Bae S.C. 2005. Association study of Toll-like receptor 9 gene polymorphism in Korean patients with systemic lupus erythematosus. Tissue Antigens. **65**: 266–270.

92. Xu C.J., Zhang W.H., Pan H.F., Li X.P., Xu J.H., and Ye D.Q. 2009. Association study of a single nucleotide polymorphism in the exon 2 region of toll-like receptor 9 (TLR9) gene with susceptibility to systemic lupus erythematosus among Chinese. Mol. Biol. Rep. **36**: 2245–2248.

93. Rumore P.M., and Steinman C.R. 1990. Endogenous circulating DNA in systemic lupus erythematosus. Occurrence as multimeric complexes bound to histone. J. Clin. Invest. **86**: 69–74.

94. Herrmann M., Voll R.E., Zoller O.M., Hagenhofer M., Ponner B.B., and Kalden J.R. 1998. Impaired phagocytosis of apoptotic cell material by monocyte-derived macrophages from patients with systemic lupus erythematosus. Arthritis Rheum. **41**: 1241–1250.

95. Amoura Z., Piette J.C., Chabre H., Cacoub P., Papo T., Wechsler B., Bach J.F., and Koutouzov S. 1997. Circulating plasma levels of nucleosomes in patients with systemic lupus erythematosus: correlation with serum antinucleosome antibody titers and absence of clear association with disease activity. Arthritis Rheum. **40**: 2217–2225.

96. Ioannou Y., and Isenberg D.A. 2000. Current evidence for the induction of autoimmune rheumatic manifestations by cytokine therapy. Arthritis Rheum. **43**: 1431–1442.

97. Baechler E.C., Gregersen P.K., and Behrens T.W. 2004. The emerging role of interferon in human systemic lupus erythematosus. Curr. Opin. Immunol. **16**: 801–807.

98. Blanco P., Palucka A.K., Gill M., Pascual V., and Banchereau J. 2001. Induction of dendritic cell differentiation by IFN-alpha in systemic lupus erythematosus. Science. **294**: 1540–1543.
99. Ronnblom L., and Alm G.V. 2001. A pivotal role for the natural interferon alpha-producing cells (plasmacytoid dendritic cells) in the pathogenesis of lupus. J. Exp. Med. **194**: F59–F63.
100. Christensen S.R., Shupe J., Nickerson K., Kashgarian M., Flavell R.A., and Shlomchik M.J. 2006. Toll-like receptor 7 and TLR9 dictate autoantibody specificity and have opposing inflammatory and regulatory roles in a murine model of lupus. Immunity. **25**: 417–428.
101. Wu X., and Peng S.L. 2006. Toll-like receptor 9 signaling protects against murine lupus. Arthritis Rheum. **54**: 336–342.
102. Pawar R.D., Ramanjaneyulu A., Kulkarni O.P., Lech M., Segerer S., and Anders H.J. 2007. Inhibition of Toll-like receptor-7 (TLR-7) or TLR-7 plus TLR-9 attenuates glomerulonephritis and lung injury in experimental lupus. J. Am. Soc. Nephrol. **18**: 1721–1731.
103. Oh D.Y., Taube S., Hamouda O., Kücherer C., Poggensee G., Jessen H., Eckert J.K., Neumann K., Storek A., Pouliot M., Borgeat P., Oh N., Schreier E., Pruss A., Hattermann K., and Schumann R.R. 2008. A functional toll-like receptor 8 variant is associated with HIV disease restriction. J. Infect. Dis. **198**: 701–709.
104. Soriano-Sarabia N., Vallejo A., Ramírez-Lorca R., Rodríguez Mdel M., Salinas A., Pulido I., Sáez M.E., and Leal M. 2008. Influence of the Toll-like receptor 9 1635A/G polymorphism on the CD4 count, HIV viral load, and clinical progression. J. Acquir. Immune Defic. Syndr. **49**: 128–135.
105. Bochud P.Y., Hersberger M., Taffé P., Bochud M., Stein C.M., Rodrigues S.D., Calandra T., Francioli P., Telenti A., Speck R.F., Aderem A., and Swiss HIV Cohort Study. 2007. Polymorphisms in Toll-like receptor 9 influence the clinical course of HIV-1 infection. AIDS. **21**: 441–446.
106. Vollmer J., Weeratna R.D., Jurk M., Davis H.L., Schetter C., Wüllner M., Wader T., Liu M., Kritzler A., and Krieg A.M. 2004. Impact of modifications of heterocyclic bases in CpG dinucleotides on their immune-modulatory activity. J. Leukoc. Biol. **76**: 1–9.
107. Vollmer J., Weeratna R.D., Jurk M., Samulowitz U., McCluskie M.J., Payette P., Davis H.L., Schetter C., and Krieg A.M. 2004. Oligodeoxynucleotides lacking CpG dinucleotides mediate Toll-like receptor 9 dependent T helper type 2 biased immune stimulation. Immunology. **113**: 212–223.
108. Elias F., Flo J., Lopez R.A., Zorzopulos J., Montaner A., and Rodriguez J.M. 2003. Strong cytosine-guanosine-independent immunostimulation in humans and other primates by synthetic oligodeoxynucleotides with PyNTTTTGT motifs. J. Immunol. **171**: 3697–3704.
109. Elias F., Flo J., Rodriguez J.M., De Nichilo A., Lopez R.A., Zorzopulos J., Nagle C., Lahoz M., and Montaner A. 2005. PyNTTTTGT prototype oligonucleotide IMT504 is a potent adjuvant for the recombinant hepatitis B vaccine that enhances the Th1 response. Vaccine. **23**: 3597–3603.
110. Zhao G., Jin H., Li J., Su B., Du X., Kang Y., Wang X., and Wang B. 2009. PyNTTTTGT prototype oligonucleotide IMT504, a novel effective adjuvant of the FMDV DNA vaccine. Viral Immunol. **22**: 131–138.

111. McCluskie M.J., and Davis H.L. 2000. Oral, intrarectal and intranasal immunizations using CpG and non-CpG oligodeoxynucleotides as adjuvants. Vaccine. **19**: 413–422.

112. Sano K., Shirota H., Terui T., Hattori T., and Tamura G. 2003. Oligodeoxynucleotides without CpG motifs work as adjuvant for the induction of Th2 differentiation in a sequence-independent manner. J. Immunol. **170**: 2367–2373.

113. Hornung V., Ellegast J., Kim S., Brzózka K., Jung A., Kato H., Poeck H., Akira S., Conzelmann K.K., Schlee M., Endres S., and Hartmann G. 2006. 5'-Triphosphate RNA is the ligand for RIG-I. Science. **314**: 994–997.

114. Takaoka A., Wang Z., Choi M.K., Yanai H., Negishi H., Ban T., Lu Y., Miyagishi M., Kodama T., Honda K., Ohba Y., and Taniguchi T. 2007. DAI (DLM-1/ZBP1) is a cytosolic DNA sensor and an activator of innate immune response. Nature. **448**: 501–505.

115. Hornung V., Ablasser A., Charrel-Dennis M., Bauernfeind F., Horvath G., Caffrey D.R., Latz E., and Fitzgerald K.A. 2009. AIM2 recognizes cytosolic dsDNA and forms a caspase-1-activating inflammasome with ASC. Nature. **458**: 514–518.

116. Bozza S., Gaziano R., Bonifazi P., Zelante T., Pitzurra L., Montagnoli C., Moretti S., Castronari R., Sinibaldi P., Rasi G., Garaci E., Bistoni F., and Romani L. 2007. Thymosin alpha1: an endogenous regulator of inflammation, immunity, and tolerance. Ann. N. Y. Acad. Sci. **1112**: 326–338.

117. Vollmer J., and Bellert H. 2009. In vitro effects of adjuvants on B cells. In: Davies G., ed. Vaccine Adjuvants. Totowa, NJ: Humana Press Inc., pp. 131–148.

118. Arese P., and Schwarzer E. 1997. Malarial pigment (haemozoin): a very active "inert" substance. Ann. Trop. Med. Parasitol. **91**: 501–516.

119. Pichyangkul S., Yongvanitchit K., Kum-arb U., Hemmi H., Akira S., Krieg A.M., Heppner D.G., Stewart V.A., Hasegawa H., Looareesuwan S., Shanks G.D., and Miller R.S. 2004. Malaria blood stage parasites activate human plasmacytoid dendritic cells and murine dendritic cells through a Toll-like receptor 9-dependent pathway. J. Immunol. **172**: 4926–4933.

120. Coban C., Ishii K.J., Kawai T., Hemmi H., Sato S., Uematsu S., Yamamoto M., Takeuchi O., Itagaki S., Kumar N., Horii T., and Akira S. 2005. Toll-like receptor 9 mediates innate immune activation by the malaria pigment hemozoin. J. Exp. Med. **201**: 19–25.

121. Parroche P., Lauw F.N., Goutagny N., Latz E., Monks B.G., Visintin A., Halmen K.A., Lamphier M., Olivier M., Bartholomeu D.C., Gazzinelli R.T., and Golenbock D.T. 2007. Malaria hemozoin is immunologically inert but radically enhances innate responses by presenting malaria DNA to Toll-like receptor 9. Proc. Natl. Acad. Sci. U S A. **104**: 1919–1924.

122. Cirl C., Wieser A., Yadav M., Duerr S., Schubert S., Fischer H., Stappert D., Wantia N., Rodriguez N., Wagner H., Svanborg C., and Miethke T. 2008. Subversion of Toll-like receptor signaling by a unique family of bacterial Toll/interleukin-1 receptor domain-containing proteins. Nat. Med. **14**: 399–406.

123. O'Neill L.A. 2008. Bacteria fight back against Toll-like receptors. Nat. Med. **14**: 370–372.

124. Manche L., Green S.R., Schmedt C., and Mathews M.B. 1992. Interactions between double-stranded RNA regulators and the protein kinase DAI. Mol. Cell Biol. **12**: 5238–5248.

125. Hovanessian A.G. 2007. On the discovery of interferon-inducible, double-stranded RNA activated enzymes: the 2′-5′oligoadenylate synthetases and the protein kinase PKR. Cytokine Growth Factor Rev. **18**: 351–361.

126. Schlee M., Hartmann E., Coch C., Wimmenauer V., Janke M., Barchet W., and Hartmann G. 2009. Approaching the RNA ligand for RIG-I? Immunol. Rev. **227**: 66–74.

127. Ranjith-Kumar C.T., Murali A., Dong W., Srisathiyanarayanan D., Vaughan R., Ortiz-Alacantara J., Bhardwaj K., Li X., Li P., and Kao C.C. 2009. Agonist and antagonist recognition by RIG-I, a cytoplasmic innate immunity receptor. J. Biol. Chem. **284**: 1155–1165.

128. Pichlmair A., Schulz O., Tan C.P., Näslund T.I., Liljeström P., Weber F., Reis e Sousa C. 2006. RIG-I-mediated antiviral responses to single-stranded RNA bearing 5′-phosphates. Science. **314**: 997–1001.

129. Habjan M., Andersson I., Klingström J., Schümann M., Martin A., Zimmermann P., Wagner V., Pichlmair A., Schneider U., Mühlberger E., Mirazimi A., and Weber F. 2008. Processing of genome 5′ termini as a strategy of negative-strand RNA viruses to avoid RIG-I-dependent interferon induction. PLoS One. **3**: e2032.

130. Kato H., Takeuchi O., Mikamo-Satoh E., Hirai R., Kawai T., Matsushita K., Hiiragi A., Dermody T.S., Fujita T., and Akira S. 2008. Length-dependent recognition of double-stranded ribonucleic acids by retinoic acid-inducible gene-I and melanoma differentiation-associated gene 5. J. Exp. Med. **205**: 1601–1610.

131. Poeck H., Besch R., Maihoefer C., Renn M., Tormo D., Morskaya S.S., Kirschnek S., Gaffal E., Landsberg J., Hellmuth J., Schmidt A., Anz D., Bscheider M., Schwerd T., Berking C., Bourquin C., Kalinke U., Kremmer E., Kato H., Akira S., Meyers R., Häcker G., Neuenhahn M., Busch D., Ruland J., Rothenfusser S., Prinz M., Hornung V., Endres S., Tüting T., and Hartmann G. 2008. 5’-triphosphate-siRNA: turning gene silencing and Rig-I activation against melanoma. Nat. Med. **14**: 1256–1263.

132. Vollmer J., and Krieg A. 2007. Mechanisms and therapeutic applications of immune modulatory oligodeoxynucleotide and oligoribonucleotide ligands for toll-like receptors. In: S.T. Crooke (ed.), Antisense Drug Technology Book (Second Edition). Boca Raton, FL: CRC Press, pp. 747–772.

133. Alexopoulou L., Holt A.C., Medzhitov R., and Flavell R.A. 2001. Recognition of double-stranded RNA and activation of NF-kappaB by Toll-like receptor 3. Nature. **413**: 732–738.

134. Pirher N., Ivicak K., Pohar J., Bencina M., and Jerala R. 2008. A second binding site for double-stranded RNA in TLR3 and consequences for interferon activation. Nat. Struct. Mol. Biol. **15**: 761–763.

135. Jasani B., Navabi H., and Adams M. 2009. Ampligen: a potential toll-like 3 receptor adjuvant for immunotherapy of cancer. Vaccine. **27**: 3401–3404.

136. Kleinman M.E., Yamada K., Takeda A., Chandrasekaran V., Nozaki M., Baffi J.Z., Albuquerque R.J., Yamasaki S., Itaya M., Pan Y., Appukuttan B., Gibbs D., Yang Z., Karikó K., Ambati B.K., Wilgus T.A., DiPietro L.A., Sakurai E., Zhang K., Smith J.R., Taylor E.W., and Ambati J. 2008. Sequence- and target-independent angiogenesis suppression by siRNA via TLR3. Nature. **452**: 591–597.

137. Klinman D.M. 2004. Use of CpG oligodeoxynucleotides as immunoprotective agents. Expert Opin. Biol. Ther. **4**: 937–946.

138. Hussain I., and Kline J.N. 2003. CpG oligodeoxynucleotides: a novel therapeutic approach for atopic disorders. Curr. Drug Targets Inflamm. Allergy. **2**: 199–205.

139. Krieg A.M. 2004. Antitumor applications of stimulating toll-like receptor 9 with CpG oligodeoxynucleotides. Curr. Oncol. Rep. **6**: 88–95.

140. Jurk M., and Vollmer J. 2007. Therapeutic applications of synthetic CpG oligodeoxynucleotides as TLR9 agonists for immune modulation. BioDrugs. **21**: 387–401.

141. Siegrist C.A., Pihlgren M., Tougne C., Efler S.M., Morris M.L., AlAdhami M.J., Cameron D.W., Cooper C.L., Heathcote J., Davis H.L., and Lambert P.H. 2004. Co-administration of CpG oligonucleotides enhances the late affinity maturation process of human anti-hepatitis B vaccine response. Vaccine. **23**: 615–622.

142. Cooper C.L., Davis H.L., Angel J.B., Morris M.L., Elfer S.M., Seguin I., Krieg A.M., and Cameron D.W. 2005. CPG 7909 adjuvant improves hepatitis B virus vaccine seroprotection in antiretroviral-treated HIV-infected adults. AIDS. **19**: 1473–1479.

143. Cooper C.L., Davis H.L., Morris M.L., Efler S.M., Krieg A.M., Li Y., Laframboise C., Al Adhami M.J., Khaliq Y., Seguin I., and Cameron D.W. 2004. Safety and immunogenicity of CPG 7909 injection as an adjuvant to Fluarix influenza vaccine. Vaccine. **22**: 3136–3143.

144. Halperin S.A., Van Nest G., Smith B., Abtahi S., Whiley H., and Eiden J.J. 2003. A phase I study of the safety and immunogenicity of recombinant hepatitis B surface antigen co-administered with an immunostimulatory phosphorothioate oligonucleotide adjuvant. Vaccine. **21**: 2461–2467.

145. Cooper C.L., Davis H.L., Morris M.L., Efler S.M., Adhami M.A., Krieg A.M., Cameron D.W., and Heathcote J. 2004. CPG 7909, an immunostimulatory TLR9 agonist oligodeoxynucleotide, as adjuvant to Engerix-B((R)) HBV vaccine in healthy adults: a double-blind phase I/II study. J. Clin. Immunol. **24**: 693–701.

146. Speiser D.E., Liénard D., Rufer N., Rubio-Godoy V., Rimoldi D., Lejeune F., Krieg A.M., Cerottini J.C., and Romero P. 2005. Rapid and strong human CD8(+) T cell responses to vaccination with peptide, IFA, and CpG oligodeoxynucleotide 7909. J. Clin. Invest. **115**: 739–746.

147. van Ojik H., Kruit W., Portielje J., Brichard V., Verloes R., and Delire M. 2002. Phase I/II study with CpG 7909 as adjuvant to vaccination with MAGA-3 protein in patients with MAGE-3 positive tumors. Ann.Oncol. **13**: 157.

148. Vollmer J. 2005. Progress in drug development of immunostimulatory CpG oligodeoxynucleotide ligands for TLR9. Expert Opin. Biol. Ther. **5**: 673–682.

149. Hamm S., Heit A., Koffler M., Huster K.M., Akira S., Busch D.H., Wagner H., and Bauer S. 2007. Immunostimulatory RNA is a potent inducer of antigen-specific cytotoxic and humoral immune response in vivo. Int. Immunol. **19**: 297–304.

150. Scheel B., Aulwurm S., Probst J., Stitz L., Hoerr I., Rammensee H.G., Weller M., and Pascolo S. 2006. Therapeutic anti-tumor immunity triggered by injections of immunostimulating single-stranded RNA. Eur. J. Immunol. **36**: 2807–2816.

151. Westwood A., Elvin S.J., Healey G.D., Williamson E.D., and Eyles J.E. 2006. Immunological responses after immunisation of mice with microparticles containing antigen and single stranded RNA (polyuridylic acid). Vaccine. **24**: 1736–1743.

152. Nemorin J.G., Lampron C., McCluskie M., Weeratna R., Forsbach A., and Vollmer J. 2005. Adjuvant activity of ssRNA in hepatitis B vaccine. First Meeting of the Oligonucleotide Therapeutics Society.

153. Riedl P., Stober D., Oehninger C., Melber K., Reimann J., and Schirmbeck R. 2002. Priming Th1 immunity to viral core particles is facilitated by trace amounts of RNA bound to its arginine-rich domain. J. Immunol. **168**: 4951–4959.

154. Seya T., and Matsumoto M. 2009. The extrinsic RNA-sensing pathway for adjuvant immunotherapy of cancer. Cancer Immunol. Immunother. **58**: 1175–1184.

155. Stahl-Hennig C., Eisenblätter M., Jasny E., Rzehak T., Tenner-Racz K., Trumpfheller C., Salazar A.M., Uberla K., Nieto K., Kleinschmidt J., Schulte R., Gissmann L., Müller M., Sacher A., Racz P., Steinman R.M., Uguccioni M., and Ignatius R. 2009. Synthetic double-stranded RNAs are adjuvants for the induction of T helper 1 and humoral immune responses to human papillomavirus in rhesus macaques. PLoS Pathog. **5**: e1000373.

PART III

PATENTS ON BACTERIA/ BACTERIAL PRODUCTS AS ANTICANCER AGENTS

18

THE ROLE AND IMPORTANCE OF INTELLECTUAL PROPERTY GENERATION AND PROTECTION IN DRUG DEVELOPMENT

ARSÉNIO M. FIALHO[1] AND ANANDA M. CHAKRABARTY[2]

[1]*Institute for Biotechnology and Bioengineering (IBB), Center for Biological and Chemical Engineering, Instituto Superior Tecnico, Lisbon, Portugal*
[2]*Department of Microbiology and Immunology, University of Illinois College of Medicine, Chicago, IL*

INTRODUCTION

Academic research encompasses many disciplines including biomedical science, agricultural science, engineering, literature, economics, business, arts, architecture, computer science, and music. Innovations in any of these areas may have important economic consequences and therefore need to be protected from copying. The modern-day patent laws are thought to go back to a Greek colony known as Sybarns in the fifth century BC where a confectioner with a new dish preparation was given 1 year of exclusive right to sell the dish so long as the recipe was clearly disclosed. There are now national laws in various countries to allow such protections, commonly known as intellectual property rights (IPR) laws. Such laws may encompass patents, which cover scientific or technological innovations involving machines, processes, composition of matter, and improvements thereof, while innovations in arts, literature, music, computer software, and so on are usually covered under copyright laws.

Emerging Cancer Therapy: Microbial Approaches and Biotechnological Tools, Edited by Arsénio M. Fialho and Ananda M. Chakrabarty

Certain innovations may involve designs or even secret formulas and are covered under trademark or trade secret laws. In this article, we will primarily discuss drug development and pharmaceutical business, which are guided by patent laws. These laws allow inventors exclusive rights for a limited period of time, usually 20 years, in exchange for a complete disclosure of the invention. Such disclosures then benefit the society by making the knowledge public. All patents are issued by national governments and there are no worldwide patent laws or patent-granting body, although attempts are being made by the World Intellectual Property Organization (WIPO) to harmonize various diverse national patent laws. The patent laws in the United States are a part of the U.S. Constitution and are designed, in the words of Thomas Jefferson written in 1793, "to promote the progress of science and useful arts, by securing for limited times to authors and inventors the exclusive right to their respective writings and discoveries" (Article I, Section 8).

There are statutory requirements for modern-day patents to be granted. An invention must be novel and not previously known, although in the United States, there is a 1-year grace period to allow a patent to be issued after publication. Most other countries have no such allowance. Depending on whether it is issued in Europe or in the United States, a patentable invention must be novel and not be obvious, must have an inventive step, must have utility and industrial application, and must be fully disclosed. Omission of any important step or condition in the written description may invalidate the patent. Certain subject matters are not patentable and there are important differences in the patent laws of the United States and Europe in this regard. Abstract ideas, methods of treatment, laws of nature, algorithms, and physical phenomena are not usually patentable, including in Europe any inventions that are considered contrary to public order or morality, including human cloning and uses of human embryos for industrial or commercial purposes (rule 23[d] [e], European Patent Convention [EPC]). In the United States, "anything under the sun that is made by man" is patentable (447 U.S. 303, at 309 n.6, 1980), while in Europe and in many other countries, higher forms of life (excluding microorganisms, rule 23[c]; EPC) or inventions contrary to public order (Article 53, EPC) are not patentable. Special exemptions or laws sometimes allow patentability of some new processes. The U.S. Congress in 1970 promulgated the "Plant Variety Protection and Breeders' Rights" law to protect new varieties of plants. In 1983, the "Orphan Drug" law was enacted to allow a non-patented drug that demonstrates efficacy in the treatment of life-threatening rare diseases to be marketed. This exclusivity provides the incentives to develop and manufacture drugs that are important in the treatment of rare diseases for which the market is small and incentives are needed. In 1984, the U.S. Congress passed the Drug Price Competition and Patent Term Restoration Act, the so-called Hatch–Waxman legislation, to authorize the U.S. Food and Drug Administration (FDA) to approve generic drugs without providing efficacy and safety data for known drugs where the patents expired. The legislation allows the patent holder certain extension periods based on the time lost during the FDA review and approval process. The legislation also limited the FDA's ability to approve a generic version of a patented drug until the time the patent expired. Previously, FDA could approve generic drugs irrespective of patent protection. However, patent infringement cases are plentiful, as will be discussed later, and the U.S. Supreme Court has affirmed the gatekeeping function of the judges to evaluate the quality of science (*Daubert v. Merrell Dow Pharmaceuticals*, 509 U.S. 579, 1993) as well as the scope of the patent claims and their limitations before the liability of patent infringement can be decided (*Markman v. Westview Instruments*, 116 S. Ct. 1384, 1996). The U.S. Supreme Court in *Markman v. Westview Instruments*, 517 U.S. 370 (1996) has delegated to the federal district court judges the interpretation of the claims and their validity, while the Court of Appeals for the Federal Circuit (CAFC) in *Phillips v. AWH Corp.*, 415 F. 3d 1303 (Fed. Cir. 2005), has provided useful guidance to the trial judges and patent applicants in constructing and interpreting appropriate claims, which are the heart of the patent. Since generics are normally low-molecular-weight synthetic compounds, a growing congressional concern now relates to the Patent Reform Act to address the problems of generic versions of protein drugs such as monoclonal antibodies (mAbs). The Patent Reform Act is intended to address issues of patent reform, including perhaps a post-grant opposition that will allow a patent to be challenged even after it has been granted. The act is also intended to devise a route to grant approval of generic versions of biologics, normally protein pharmaceutical products derived from living cells. Since the isolation procedures for biologics are complex, purity lacks uniformity with cellular contaminants, and potential immunogenicity as a possible problem, the generic versions, often called biosimilars, present complex issues with regard to their standards for safety, raising questions if biosimilars should undergo clinical trials. This will, of course, allow the patented biologics to enjoy longer data exclusivity and delayed copying. No wonder, then, that the pharmaceutical and biotechnology industries insist that data exclusivity periods, where the clinical data are submitted by an inventor to the FDA for marketing approval and which remains proprietary, be extended significantly beyond 5 years before the generic makers can copy such drugs. Unlike small-molecule synthetic generics that are regulated by the FDA through an abbreviated new drug application (ANDA) under the Hatch–Waxman Act, biologics are regulated through Biologic License Application (BLA) by the FDA under the Public Health Service Act, thus denying their generic versions using 505 (b)(2) applications for regular generics. In contrast, many generic versions of biologics have been approved in Europe and a similar trend is seen in Canada. Since the isolation and purification of biologics from mammalian sources are expensive, even without clinical trials, the generic versions of biologics are not expected to be as cheap as regular generics, and given the use of different cell lines or slightly different procedures for the isolation of biologics, the safety issues are not as clear-cut as those of synthetic generics. In October, 2008, the PRO-IP Act of 2008 was signed into law in the United States. In addition to combating counterfeiting and piracy, the law will provide resources at the state and federal levels to battle intellectual property (IP) theft and is intended to prevent an infringer from avoiding liability based on a harmless error in a registration by a rights holder. The courts in the United States also try to limit the number of lawsuits that are considered frivolous. In a recent case, the District Court of Colorado awarded US$4.5 million in attorney fees against Medtronic and its law firm, because of perceived baseless litigation by the patentee against an accused infringer. Often, courts try to provide some guidelines for sham litigation. In *re Ciprofloxacin Antitrust Litigation*, Bayer, the patent holder of ciprofloxacin, sued several generic makers for patent infringement. Even after resolution, some of the plaintiff-side lawyers sued Bayer, alleging anticompetitive sham litigation by Bayer. These lawsuits were rejected by CAFC, and in similar cases by other courts, to uphold valid patents.

U.S. AND EUROPEAN PATENT LAWS HAVE SIMILARITIES AND DIFFERENCES

Most of the statutory requirements for patentability, namely, nonobviousness, utility or industrial application, complete written description, novelty, and so on, are common in both the United States and Europe including countries that are signatory to the Patent Cooperation Treaty (PCT). The U.S. laws are somewhat more stringent pertaining to the best mode to practice and complete the written description of the invention. The U.S. law also requires that only inventors contributing materially to the invention be included in the patent application. Any deviation in this regard can invalidate the patent. As opposed to the PCT, where the prior art search is conducted by the examiner (but patents issued by individual national countries), the U.S. law requires the applicants to bring to the attention of the patent examiner any prior art material relevant to the invention even after the application has been filed. There are some major differences, however. In the United States, the first to invent something gets a patent on it. In Europe and essentially in all other countries, the person who files a patent application first is eligible for the patent. It is much harder, and more contentious, to determine who invented something first, than just from the date of filing. The U.S. system is based on rewarding individual inventors, who are normally slow to file a patent application, than major corporations that have the resources to file quickly. As mentioned earlier, the U.S. system offers a 1-year grace period to file an application after public disclosure of the invention, again to encourage individual inventors who may not be routinely filing patent applications and who may not be aware of the filing rules. Europe and other countries do not allow such grace period. In Europe, once a patent application is published (previously in the United States, patent applications were not published, giving rise to submarine patents), it can be opposed by a third party during a 9-month period on issues of novelty or industrial application. Also, each patent must be validated in each European country. This system of each individual European country granting its own patent often creates interesting litigation cases where the validity of a patent issued in United Kingdom may be revoked while the validity of basically the same patent is upheld in the Netherlands. In the case where a U.S. company, Document Security Systems (DSS), accused the European Central Bank (ECB) of infringing its European patent on anticounterfeiting technology to produce euro bank notes, the U.K. Court of Appeals revoked the patent in 2008, followed by a court in France, but the patent was upheld in trial courts in Germany and in the Netherlands, exemplifying a divided system of justice in patent litigation in the European Union (EU). In the United States, a patent is considered valid and can only be challenged on prior art. European patent offices (EPOs) are also much more flexible with regard to the number of times an inventor can amend the claims or can provide rebuttal than in the United States. Since patent prosecution is usually faster and less cumbersome in Europe than in the United States, a dialogue in this area, called the Patent Prosecution Highway (PPH), is underway between the patent offices of Europe, Japan, and the United States. A major difference in patent laws between the United States and Europe, as mentioned earlier, is that anything against public order or morality cannot be patented in Europe. Thus, patents on human embryonic stem cells were issued in the United States in 1998, 2001, and 2006, covering both the methods for isolating stem cells from fertilized embryos as well as the embryonic stem cells themselves. While these patents were revoked in 2007 for some prior art considerations and were later reinstated with amendments, no such patents have been issued in Europe on the basis of moral grounds, even though the British Parliament recently approved the use of human–nonhuman hybrid embryos for medical research. Indeed, in a recent ruling in November, 2008, the Enlarged Board of Appeal (EBoA) of the EPO rejected the so-called Wisconsin Alumni Research Fund (WARF)/Thomson stem cell application because the invention involves the use and destruction of human embryos. This EBoA ruling is based on the provisions of the EPC and on the EU Biotechnology Directive (98/44/EC) implemented in 1999. Another difference between the U.S. and European patents involves the interpretation of the broadness of the claims. Myriad Genetics, based in Utah, has obtained U.S. patents on diagnostic methods for detecting predisposition to breast cancers because of mutations in the BRCA1 type of genes. An European patent, EP 699754, was initially granted to Myriad Genetics in 2001 by the EPO. The patented diagnostic methods for BRCA1 mutations do not detect some genetic deletions or other rearrangements in the gene that could be detected by other techniques developed in France. Because the Myriad patent covered the BRCA1 gene itself or its mutated forms, several French institutes and national centers for human genetics filed objections to the broad patent claims, citing EPC rules. The patent was therefore revoked in May 2004 by the EPO. On appeal from Myriad, the Technical Board of Appeal of the EPO subsequently decided that a patent, EP 705902, involving the BRCA1 gene can be maintained in a severely limited form. In November, 2008, the EPO Technical Board of Appeal allowed an amended version of the two patents, restricting them to diagnostic methods for the detection of predisposition to breast and ovarian cancers caused by frameshift mutations and eliminated claims involving the BRCA1 gene itself or other mutated forms. The amended patent cannot be further contested at the EPO level.

EXTENDING THE REACH OF GRANTED PATENTS

Patents have a limited life of 20 years and by the time drugs or other regulated innovations obtain marketing approval from the FDA or other regulatory agencies, several years have passed, making a patent's de facto life period much shorter. Patent holders, therefore, try to extend the life period of their patents by making incremental improvements or by finding new uses of their patented invention, a process often known as evergreening. The 1984 Hatch–Waxman legislation tried to compensate for this lost time by authorizing the FDA to extend patent validity up to several months if the regulatory process lasted for a long time before marketing approval. The generic makers often contest such marketing extensions of patented drugs to bring cheap copies to the market on patent expiration. A recent court case illustrates this point. AstraZeneca was granted U.S. Patents 4,786,505 and 4,853,230 that cover an inert subcoating protecting the antiulcer drug Prilosec (omeprazole). In a continuing effort to improve the stability and efficacy of omeprazole in the acidic environment of the stomach, alkaline-reacting compounds were used as subcoating in the Prilosec tablet along with the enteric coating. Just before the two patents expired, Astra commissioned some clinical studies for which the FDA granted Astra a 6-month period of market exclusivity after the expiry of the patents. Two generic makers, Apotex Corp. and Impax Laboratories, submitted an application (ANDA) to the FDA to sell generic versions of Prilosec. Upon notification from the FDA, Astra sued for infringement of the two patents. After the bench trials, but before the District Court ruled, the two patents expired, whereupon Impax asked the court to dismiss Astra's lawsuit since the relevant patents were no longer in force. The District Court, however, denied the motion and ruled that Astra's patents were valid and infringed by Impax because of the FDA's extension of market exclusivity. On appeal, the CAFC in 2008 upheld the District Court's decision, thus demonstrating that patent extension by the FDA is a reasonable and lawful way to prolong market accessibility of patented drugs. A similar supplementary protection certificate (SPC) system was introduced within the EU by regulation 1768/92/EEC, where the lifetime of a patent on drugs can be extended for a maximum of 5 years, sometimes with interesting consequences. For example, the basic patent for Prozac expired in 1995 and the European SPC expired in 2000. In the United Kingdom, 80% of total sales revenue was generated to the rights holder during the SPC term, whereas in Germany, where no SPC was available for Prozac, sales declined steadily from 1995 because of generic versions of Prozac.

In spite of patent validity and extension, patents are not always enforceable. In many countries, including the United States, there are compulsory licensure laws that allow licensing of a patented invention to a third party without the consent or a valid licensing agreement from the patent holder. Such laws are intended to allow licensing of a patent if necessary for public health or for national emergency (1,2). Although rarely enforced, such national compulsory licensing laws have allowed many developing or even developed countries to entice multinational corporations to reduce the price of their anti-infective drugs for use in the treatment of patients infected with anthrax spores or with the HIV/AIDS virus. The Government of Thailand (state-owned Government Pharmaceutical Organization) has imposed compulsory licensure on five life-saving drugs, producing and selling them at cheap prices, and thereby making a profit.

PATENTING HUMAN GENES AND PARTS THEREOF

Patenting genes, particularly human genes, has always evoked mixed reactions (3). While arguments can be made about the usefulness and importance of patenting DNA sequences, just as any other inventions, to promote commercial ventures and predict predispositions to genetic diseases (4), many critics consider patenting of expressed sequence tags (ESTs) as lacking utility and therefore not meeting one of the statutory requirements of patentability. Of greater concern has been the impact of gene patenting, including mutations leading to disease causations, in raising the cost of genetic testing (5, 6). Our society values the gene as "belonging to everyone and not the proprietary material of a single human being or a company," and such patents often do not appear to meet another statutory requirement of enablement as in *Wallach* 378 F. 3d 1330 (Fed. Cir. 2004). In this case, the inventors sought a patent on isolated DNA molecules encoding tumor necrosis factor-binding proteins (TNF-BP) where they had only a partial amino acid sequence of the N-terminal portion of the protein (TNF-BPII) with known size and function. Both the patent office and the CAFC rejected the patentability because of a lack of description of the complete amino acid sequence, and therefore a complete written description. The granting of patents to a whole spectrum of ESTs in the early days of gene patenting, where the ESTs had no known function or utility, but on the presumption that such utility will eventually be found, greatly undermined the statutory requirements of patenting (*Brenner v. Manson*, 383 U.S. 519, 1966). Another case, *Laboratory Corporation v. Metabolite Laboratories*, has raised similar questions. In this case, the question raised involves if a patent can be issued over a basic and obvious scientific fact, namely, a correlation between higher levels of homocysteine and a deficiency of vitamin B12 or folic acid in the blood. In gene patenting, the CAFC, in *re Deuel*, 34 USPQ2d 1210, reversed the objection of the PTO in granting a patent on a cloning method that the PTO based on the principle of obviousness. Since the Supreme Court decision in *KSR v. Teleflex*, however, the situation has changed and fewer gene patents were granted (7) because of the more stringent standard of obviousness. As opposed to *Deuel*, in case of *Kubin* 83 USPQ2d 1410, where the PTO rejected a patent application on a gene encoding a protein NAIL on the ground of obviousness, the PTO's Board of Appeals ignored the CAFC decision on *Deuel* and rejected the patent application because of the Supreme Court's *KSR* decision (see below). Indeed, an analysis of human gene patent litigations from 1987 to 2008, as compared to litigations over drug-related therapeutic proteins, has shown very few enduring litigations over genetic diagnostic testing, suggesting that gene or gene mutation patents have not had the kind of commercial impact as was earlier presumed in creating new viable businesses (8).

In the United States, patent applications are examined and patents issued or rejected by the United States Patent and Trademark Office (USPTO). The Technology Center 1600, as it is known, has the responsibility of accepting and issuing or rejecting patents in biotechnology, organic chemistry, and pharmaceutical products with about 400 patent examiners. Fifty percent of these examiners have a PhD degree. The USPTO tries to respond to an applicant within 14 months after receiving an application and, once the questions raised by the patent examiner are answered to the USPTO's satisfaction, a patent is issued within 36 months.

OBVIOUSNESS IN THE ISSUANCE OF PATENTS

A statutory requirement for the issuance of a patent is that the invention must not be obvious. As we will see later in the case of inherent anticipation, it is not easy to determine how a previous invention, called a prior art, may make another invention with incremental or insignificant changes or improvements quite obvious. Many patent infringement cases in the courts involve the issue of obviousness. In general, the CAFC has indicated its guidelines that an invention is obvious only if the prior art has a significant role in the teaching, suggesting or motivating an inventor to come up with a similar invention. Many inventors, as well as the U.S. Supreme Court, found such guidelines to be too lax and inflexible, leading to issuance of patents that perhaps should not have been issued. Two court cases will illustrate this point. In *KSR International Co. v. Teleflex Inc.*, gas pedal manufacturer Teleflex accused another manufacturer, KSR, of infringing its patents on gas pedals that combined electronic sensor pedals with adjustable accelerator pedals. Such gas pedals individually were known, but, since there was no obvious teaching, suggestion, or motivation, the CAFC concluded that the invention was nonobvious. The Supreme Court rejected these guidelines used by the CAFC, emphasizing that a common sense approach, where a combination of known inventions produces predictable results, makes such inventions obvious.

While the KSR decision involved gas pedals of automobiles, the Supreme Court's common sense approach of determining obviousness has interesting implications in biotechnological or pharmaceutical patents. A case in point is the challenge to the patent on an antiulcer drug of Eisai Company. Eisai came up with a pyridine derivative rabeprazole with a methoxypropoxy group having strong activity as an inhibitor of gastric acid secretion. A widely known prior art compound, lansoprazole, also a pyridine derivative with a different side chain trifluoroethoxy group, was known to have similar activity. Both the District Court and the CAFC affirmed the finding of nonobviousness, using the KSR decision that there must be enough reasons to make improvements for patentability. The CAFC decision was based on the fact that in spite of the structural similarity, the changes in the side chains to confer improved inhibition of proton pump and stomach acid secretion could not be predicted and, therefore, were not obvious.

There are other instances where a challenge to obviousness of an invention can be countered by showing that an invention has beneficial properties that were unforeseen at the time of patent issuance and hence were nonobvious. This is exemplified by *Knoll Pharmaceutical Co. v. Teva Pharmaceuticals USA*, where Knoll marketed a 1986 patented combination drug, hydrocodone, a narcotic, and ibuprofen, a nonsteroidal anti-inflammatory drug called Vicoprofen, receiving FDA approval in 1997. In 2000, the generic maker filed an ANDA to the FDA for marketing a generic version, arguing that the Knoll invention was obvious because other such combination drugs existed and the patent was therefore invalid. However, in several clinical trials conducted independently, Knoll's combination drug was shown to be much better than the individual drugs, and there was a synergistic analgesic effect when the two drugs were used in combination. The trial court ruled in favor of the generic company, asserting that the patent was invalid on grounds of obviousness. This decision was reversed on appeal, when the CAFC held that new and unexpected benefits, seen even after the patent was issued, can be taken into account to determine the obviousness of an invention. Such incidences of unexpected benefits of combination of drugs, where individual component drugs are manufactured by generic makers and are cheap, are often used by pharmaceutical industries to extend patent protection and to introduce a new, more profitable version of two or more old drugs. An example is Pfizer's prescription drug Caduet, a combination of blood pressure-reducing pill Norvasc and cholesterol-reducing drug Lipitor. Abbott Laboratories have introduced a new combination drug, Simcor, which combines HDL-raising Niaspan and a statin Zocor that lowers LDL, the bad form of cholesterol.

INHERENT ANTICIPATION

A criterion of patentability, besides nonobviousness, is novelty. Thus, a patentable subject matter should not be inherently anticipated. Many patent infringement cases involve issues where a patent should not have been issued because it should have been anticipated from a prior art. A court case decided by the CAFC in 2003, *Schering Corporation v. Geneva Pharmaceutical Inc.*, and others illustrate the problem of inherent anticipation involving a drug. The patent for loratadine, a drug for the treatment of allergies, expired in 2002. In the human body, loratadine is known to be converted to a metabolite desloratadine, which is the active ingredient. A U.S. patent for desloratadine, number 4,659,716, was listed in the FDA Orange Book along with loratadine. Thus, generic makers who wanted to manufacture loratadine upon the patent's expiry had to declare that the desloratadine patent was invalid and therefore was not infringed because it was inherently anticipated as a metabolite of loratadine. When the FDA issued a notice to the patent holder, Schering Corporation, the company filed a lawsuit claiming patent infringement.

After determining that desloratadine was known to be a metabolite of loratadine and was inherently anticipated to have loratadine drug property, the District Court, where the lawsuit was filed, found the desloratadine patent lacking novelty and therefore invalid. On appeal, the CAFC affirmed the District Court decision finding that inherent anticipation does not necessarily be explicitly recognized in the prior art, in this case the loratadine patent, and even though desloratadine's chemical structure was different from that of loratadine, the mode of administration of loratadine as described in prior art was predictive of the formation of desloratadine. The CAFC emphasized that although the Schering infringement case seemingly made metabolites inherently anticipated and therefore lacking novelty, patent protection could be afforded to other metabolites that are not specifically described in the claims of the prior art.

ARE ALL PATENTS NECESSARILY ENFORCED?

A patent gives the patent holder the right, for a limited period of time, to exclude others from making commercial gains using the patented invention. This raises two important questions. What about universities that conduct basic research to further knowledge in all areas of science and technology? Can they use a patented process or product in their research without paying royalty? The situation is complicated by the fact that not all university research is dedicated to fundamental knowledge generation. Increasingly, universities and members of the faculty or students file for patent protection of their inventions and either license such patents to other companies for gaining royalty income or initiate start-up companies for commercial exploitation of their inventions, thereby acting more like business entities. Should such researchers or the university be able to freely use patented processes or products without paying any royalty? A second problem is often the high cost of patented products, particularly life-saving drugs. While patents certainly promote innovation and therefore industrial and economic well-being, high costs often tend to negate the purpose for which the patent system stands for the benefit of common people. To address such problems, as previously mentioned, the U.S. Congress passed the Drug Price Competition and Patent Term Restoration Act, known as the Hatch–Waxman Act, to extend certain patent terms due to regulatory delays for marketing but to allow generic makers to file ANDAs for developing cheaper versions of the drug after the expiry of the patent. The Act also allows competitors to make, use, or sell a patented product/process that is needed to obtain the data for regulatory submission and approval. Such exemptions from infringement are often known as FDA exemption, FDA safe harbor, or experimental use exemption. Congress has often acted, not necessarily with a great deal of clarity, to legislate research exemptions for patented materials to make it easier for researchers or business entities to perform genetic diagnostic or other tests for public good by exempting them from infringement actions. Two interesting cases involving university research and patented products/processes involve court cases. One case relates to the desire of Hoffman-La Roche to enforce the patent on Taq polymerase that is almost universally used by university or other researchers as a research tool without any royalty payment to Roche. The question of how to enforce such patents points out to some other similar problems. An example is the *Madey v. Duke University* case, where a professor of Duke University developed and patented lasers. On termination from the university, the professor sued Duke University for the infringement of his patents. In this instance, the CAFC ruled that Duke behaved like a business entity and should therefore pay royalty to the patent holder on the use of the lasers.

The FDA safe harbor of noninfringement of 35 U.S.C Section 271(e)(1) exempts drug screening and related preclinical research from patent enforcement so long as such activities are directed to data submission to the FDA for seeking regulatory approval. Can early research phases, not yet ready for enough data generation for submission to the FDA, fall under the protection of the FDA safe harbor? There is a case where the U.S. Supreme Court stepped in to clarify some of the uncertainties in such provisions. In the mid-1980s, a research institute in California developed specific arginine-glycine-aspartic acid (RGD) motif-containing peptides that promote adhesion to certain integrin receptors on endothelial cells. These receptors are involved in angiogenesis and therefore binding of the RGD-containing peptides could inhibit angiogenesis, which is important for cancer cells to grow. These RGD peptides were patented by the institute and in the mid-1990s, the patents were assigned to the company Integra Life Sciences. In the late 1980s, the pharmaceutical company Merck provided funding to another California institute which, in the mid-1990s, demonstrated that a cyclic RGD-containing peptide could inhibit angiogenesis, which in turn allowed Merck and the institute to develop more substantial data for submission to the FDA as an investigational new drug (IND) application. This use of the patented RGD motif-containing peptides in preclinical research, which ultimately led to the data for IND application submission to the FDA, became the subject matter of a lawsuit known as *Merck KGaA v. Integra Life Sciences Ltd.* The CAFC narrowly interpreted the Hatch–Waxman Act, ruling that early drug design and development is not protected under the FDA safe harbor. This ruling was reversed in 2005 by the

U.S. Supreme Court, which took a much broader view of the law, emphasizing that all preclinical research and research tools that are reasonably related to the IND submission are protected from patent infringement. However, the Supreme Court was vague on the use of other patented research tools that are often used by universities and research institutions but that may not be related to ultimate FDA submission for marketing approval. Thus, unlicensed research on pharmaceutical products, reasonably related to FDA submission, falls under the protection of the FDA safe harbor.

What about a patented invention or medical device that itself is not subject to FDA review but is used by a company as part of a broader invention or device that can be used for drug testing for generating data for FDA submission? In a case decided by the CAFC in August 2005, *Proveris Scientific Corp. v. Innovasystems Inc.*, the Appeals Court ruled that the safe harbor immunity is provided only to products that are involved in FDA regulation for market approval. In this instance, Proveris had a patented device (U.S. Patent 6,785,400) that was used in an aerosol inhalation system often used in aerosolized drug delivery in asthma or similar patients. Innovasystems used this patented device as part of its own inhalation drug therapy device. When Proveris sued Innovasystems for patent infringement, Innovasystems invoked the safe harbor exemption of Section 271(e)(1), as its inhalation drug delivery device was important for performing various drug efficacy tests that were subject to FDA approval, even though the device itself was not. The Federal Circuit ruled that such immunity did not exist for alleged patent infringement liability as the device itself was not subject to FDA approval. Since many research tools that are not subject to regulatory review but are used by universities and business entities are patented but are not necessarily involved in drug efficacy testing for securing FDA marketing approval, the CAFC decision imposes certain limits to the broad scope of immunity recommended by the Supreme Court *re Merck v. Integra*.

COPYING OR MAKING INCONSEQUENTIAL CHANGES IN A PATENT—THE DOCTRINE OF EQUIVALENTS

Often, when first written and submitted, a patent application is broad and tries to cover many anticipated developments. During prosecution, the USPTO rejects many of the claims so that the allowed claims are often fewer and are more narrowly focused, even though the disclosed text is still very broad. Any subject matter mentioned in the text but not specifically claimed is then considered to be dedicated to the public. What happens when or if somebody practices the disclosed but unclaimed subject matter? An even more difficult case is when an accused infringer makes insubstantial or inconsequential changes in the patented invention to circumvent infringement. To address such issues, the courts have developed what is commonly known as "doctrine of equivalents." The doctrine of equivalents mandates that a patentee has an obligation to draft claims that cover future unforeseeable developments so as to deter copiers from making inconsequential changes to avoid patent infringement because of the amended narrowly focused claims. The celebrated case in this regard is known as *Festo Corporation v. Shoketsu Kinzoku Kogyo Kabushiki Co. Ltd.* In this case, the CAFC in 2000 ruled that narrowing a claim by a patentee creates estoppel, a legal term meaning barring an action, such that the patentee can no longer claim the surrendered subject matter as an unforeseen equivalent. In May 2002, the Supreme Court reversed this rigid guideline of the CAFC, basically ruling that the surrendered claims are not necessarily without protection from infringement. However, the subject matter is not a complete equivalent and the burden is on the patentee to demonstrate that the claim amendment is unrelated to patentability. The idea is to protect patent holders from copiers making inconsequential changes in the patented inventions while at the same time defining the parameters of patent monopoly and broader prosecution history estoppel.

WRITTEN DESCRIPTION AND ENABLEMENT

The primary purpose of granting patents by a government is not only to encourage and reward innovation, but also to help disseminate the knowledge so that further improvements can be made and practiced during or after the expiry of patent. This is only possible if all the critical parameters of the invention, particularly for drugs, are fully described, including methods of isolation or design, physical/chemical properties, and functional characteristics, to meet the statutory requirement of 35 U.S.C. art. 112. After the CAFC decision in 1997 of the case *Regents of University of California v. Eli Lilly*, relevant to complete or partial sequence description of nucleic acids, particularly cDNAs encoding proteins such as insulin, the USPTO issued specific guidelines in 1999 to address the requirements for written descriptions. Since the evolution of genes often occurs through gene duplication and subsequent sequence divergence, even genes with completely different functions may demonstrate significant sequence identity, making it difficult to assign functional characteristics of a gene, unless extensive mutational alterations of specific bases in the nucleic acid have been implicated in the functionality of the gene products. With the emergence of many new cDNAs, partial genes, and protein products, the problem of full and acceptable written description was revised by the USPTO in 2008, leading to the issuance of a set of new guidelines to address the emergence of new and complex technologies. The purpose of such continuing guidelines is to ensure that written descriptions are enabling; that is, a person skilled in the art should be able to reproduce the entire invention, and that no important or essential steps have been omitted in the written description. For example, many mAbs, such as trastuzumab (Herceptin), bevacizumab (Avastin), and cetuximab (Erbitux), are now used in cancer therapy to bind and neutralize components such as HER2, vascular endothelial growth factor

(VEGF), and epidermal growth factor receptor (EGFR), which are often hyperexpressed in cancer cells to promote cancer growth (see Chapter 8 for details). More recently, bavituximab, a mAb that targets cellular membrane-located phosphatidylserine that lines the blood vessels of tumors but that are normally intracellular in normal cells, is undergoing clinical trials in patients with advanced breast cancer. Many of such mAbs have been patented and FDA approved; for example, cetuximab was approved by FDA in 2004 for metastatic colorectal cancer and rituximab in 1997 for B cell non-Hodgkin's lymphoma. For patenting, an antibody must be defined clearly with regard to the antigen it binds. Thus, the written description should disclose the method of purification of the antigen, its physiological properties, and its clinical relevance as well as its structure, including complete amino acid sequence if it is a protein. Similar information regarding the isolation and functionality of the mAb should also be incorporated. The newly issued guidelines, which encompass 17 examples, 14 of which relate to biotechnological inventions, can be accessed at the PTO website (http://www.uspto.gov/web/menu/written.pdf). The question of obviousness is also an issue. For example, in *re Saul Tzipori*, a patent application claiming a mAb against Shiga-like toxin II, known for causing hemolytic uremic syndrome (HUS), was rejected by the Board of Patent Appeals as an obvious variant of similar antibodies directed against the same target. On appeal, the CAFC upheld the rejection in 2008 on grounds of obviousness.

CONCLUDING REMARKS

Patents are an essential part in the business of drug design and development as well as their marketing. Thus, today's pharmaceutical industry not only depends on hundreds of researchers conducting cutting-edge research on drug design, formulations, clinical toxicity, and efficacy testing but also on a battery of lawyers, both patent attorneys and patent litigators, to generate patents and to protect the patents from being copied by patent infringers. This article provides only a glimpse of the confounding and complex patent laws, court cases, and legal maneuvers. It is quite clear, however, that patents have played, and will continue to play, very important roles in drug design and discovery, as well as global marketing, in the treatment of cancer and many other diseases.

REFERENCES

1. Resnick D.B. and De Ville K.A. 2002. Bioterrorism and patent rights: compulsory licensure and the case of Cipro. Am. J. Bioeth. **2**: 29–39.
2. Chakrabarty A.M. 2002. Compulsory licensure: the case of Cipro and beyond. Am. J. Bioeth. **2**: 40.

3. Eisenberg R.S. 2002. How can you patent genes? Am. J. Bioeth. **2**: 3–11.
4. Doll J.J. 1998. The patenting of DNA. Science. **280**: 689–690.
5. Chakrabarty A.M. 2003. Patenting life forms: yesterday, today and tomorrow. In: Kieff F.S., ed. Perspectives on Properties of the Human Genome Project. San Diego, CA: Elsevier Academic Press, pp. 3–11.
6. Caulfield T., Cook-Deegan R.M., Kieff F.S., and Walsh J.P. 2006. Evidence and anecdotes: an analysis of human gene patenting controversies. Nat. Biotechnol. **24**: 1091–1094.
7. Hopkins M.M., Mahdi S., Patel P., and Thomas S.M. 2007. DNA patenting: the end of an era? Nat. Biotechnol. **25**: 185–187.
8. Holman C.M. 2008. Trends in human gene patent litigation. Science **322**: 198–199.

INDEX

Emerging Cancer Therapy: Microbial Approaches and Biotechnological Tools, Edited by Arsénio M. Fialho and Ananda M. Chakrabarty